Lecture Notes in Artificial Intelligence   5243

Edited by R. Goebel, J. Siekmann, and W. Wahlster

Subseries of Lecture Notes in Computer Science

Andreas R. Dengel   Karsten Berns
Thomas M. Breuel   Frank Bomarius
Thomas R. Roth-Berghofer (Eds.)

KI 2008: Advances in Artificial Intelligence

31st Annual German Conference on AI, KI 2008
Kaiserslautern, Germany, September 23-26, 2008
Proceedings

 Springer

Series Editors

Randy Goebel, University of Alberta, Edmonton, Canada
Jörg Siekmann, University of Saarland, Saarbrücken, Germany
Wolfgang Wahlster, DFKI and University of Saarland, Saarbrücken, Germany

Volume Editors

Andreas R. Dengel
Thomas R. Roth-Berghofer
German Research Center for Artificial Intelligence DFKI GmbH
67663, Kaiserslautern, Germany
E-mail: {andreas.dengel, thomas.roth-berghofer}@dfki.de

Karsten Berns
University of Kaiserslautern
Robotics Research Lab.
67663 Kaiserslautern, Germany
E-mail: berns@informatik.uni-kl.de

Thomas M. Breuel
German Research Center for Artificial Intelligence DFKI GmbH
Image Understanding and Pattern Recognition Group
67663 Kaiserslautern, Germany
E-mail: thomas.breuel@dfki.de

Frank Bomarius
Fraunhofer Institute for Experimental Software Engineering (IESE)
67663 Kaiserslautern , Germany
E-mail: frank.bomarius@iese.fraunhofer.de

Library of Congress Control Number: 2008934869

CR Subject Classification (1998): I.2, I.2.6, F.1.1, I.5.1, H.5.2

LNCS Sublibrary: SL 7 – Artificial Intelligence

ISSN   0302-9743
ISBN-10   3-540-85844-X Springer Berlin Heidelberg New York
ISBN-13   978-3-540-85844-7 Springer Berlin Heidelberg New York

Springer is a part of Springer Science+Business Media

springer.com

© Springer-Verlag Berlin Heidelberg 2008
Printed in Germany

Typesetting: Camera-ready by author, data conversion by Scientific Publishing Services, Chennai, India
Printed on acid-free paper   SPIN: 12521739   06/3180   5 4 3 2 1 0

Preface

KI 2008 was the 31st Annual German Conference on Artificial Intelligence held September 23–26 at the University of Kaiserslautern and the German Research Center for Artificial Intelligence DFKI GmbH in Kaiserslautern, Germany. The conference series started in 1975 with the German Workshop on AI (GWAI), which took place in Bonn, and represents the first forum of its type for the German AI Community. Over the years AI has become a major field in computer science in Germany involving a number of successful projects that received much international attention. Today KI conferences are international forums where participants from academia and industry from all over the world meet to exchange their recent research results and to discuss trends in the field. Since 1993 the meeting has been called the "Annual German Conference on Artificial Intelligence," designated by the German acronym KI.

This volume contains the papers selected out of 77 submissions, including a number of submissions from outside German-speaking countries. In total, 15 submissions (19%) were accepted for oral and 30 (39%) for poster presentation. Oral presentations at the conference were single track. Because of this, the choice of presentation form (oral, poster) was based on how well reviews indicated that the paper would fit into one or the other format. The proceedings allocate the same space to both types of papers.

In addition, we selected six papers that show high application potential describing systems or prototypical implementations of innovative AI technologies. They are also included in this volume as two-page extended abstracts.

All papers cover important areas such as pattern recognition, multi-agent systems, machine learning, natural language processing, constraint reasoning, knowledge representation and management, planning, and temporal reasoning. In the afternoon of September 25, we organized a session in which the best papers selected by the Program Committee were presented. Congratulations to the authors for their excellent contribution. The best paper selected for the Springer Best Paper Award was honored by a special ceremony during the conference dinner.

In addition to the technical papers, this volume contains an abstract and a paper of the two invited presentations of the conference:

- Randy Goebel (University of Alberta, Edmonton, Canada): Folk Reducibility and AI-Complete Problems
- Yuzuru Tanaka (Hokkaido University, Sapporo, Japan): Meme Media and Knowledge Federation

A large number of people were involved in making this conference a success. As for the technical program, we thank all those authors who submitted papers to the conference and provided the basis from which we could select a high-quality technical program. The members of the Program Committee and the additional reviewers helped us in this process by providing timely, qualified reviews and participating in the discussion during the paper selection process. Many thanks to all of you! We also wish to acknowledge the generosity of our sponsors, namely, Microsoft Deutschland, Deutsche Telekom Laboratories, IDS Scheer, SAP, Living-e, Springer, and DFKI, which invited us to its 20th anniversary dinner in the historical Fruchthalle in downtown Kaiserslautern.

Last but not least, we are especially very grateful to Stefan Zinsmeister who did a tremendous job as the Local Arrangements Chair and the communication hub between all of us. We would also like to thank Jane Bensch and Brigitte Selzer as members of the team taking care of the many important but time-consuming details that made up KI 2008.

This volume has been produced using the EasyChair system[1]. We would like to express our gratitude to its author Andrei Voronkov. Finally, we thank Springer for its continuing support in publishing this series of conference proceedings.

July 2008

Andreas Dengel
Karsten Berns
Thomas Breuel
Frank Bomarius
Thomas R. Roth-Berghofer

---

[1] http://www.easychair.org

Organization

Conference Chair

Andreas Dengel   DFKI GmbH, Kaiserslautern, Germany

Program Chairs

Karsten Berns   University of Kaiserslautern, Germany
Thomas Breuel   DFKI GmbH, Kaiserslautern, Germany

Workshop Chair

Thomas R. Roth-Berghofer   DFKI GmbH, Kaiserslautern, Germany

Tutorial Chair

Frank Bomarius   Fraunhofer Institute for Experimental Software Engineering / University of Applied Sciences Kaiserslautern

Conference Secretariat

Christine Harms   ccHa, Sankt Augustin

Local Organization Chair

Stefan Zinsmeister   DFKI GmbH, Kaiserslautern, Germany

Local Arrangements Support

Jane Bensch   Brigitte Selzer
(all with DFKI GmbH, Kaiserslautern)

Program Committee

Elisabeth Andre	University of Augsburg
Michael Beetz	TU Munich
Stephan Busemann	DFKI GmbH, Saarbrücken
Rüdiger Dillmann	University of Karlsruhe
Christian Freksa	University of Bremen
Ulrich Furbach	University of Koblenz
Nicola Henze	University of Hannover
Joachim Hertzberg	University of Osnabrück
Gabriele Kern-Isberner	University of Dortmund
Frank Kirchner	University of Bremen
Michael Kohlhase	IUB, Bremen
Bernhard Nebel	University of Freiburg
Bernd Neumann	University of Hamburg
Martin Riedmiller	University of Osnabrück
Raul Rojas	University of Berlin
Gerhard Sagerer	University of Bielefeld
Bernt Schiele	TU Darmstadt
Ute Schmid	University of Bamberg
Stefan Wrobel	Fraunhofer IAS, St. Augustin
Sandra Zilles	University of Alberta, Canada

Additional Reviewers

John Bateman	Heni Ben Amor	Andreas Birk
Stephan Busemann	Thomas Gabel	Ingo Glöckner
Markus Goldstein	Helmar Gust	Roland Hafner
Tetyana Ivanovska	Alexander Kleiner	Ivana Kruijff-Korbayova
Franz Kummert	Sascha Lange	Kai Lingemann
Martin Memmel	Reinhard Moratz	Bernhard Nebel
Andreas Nüchter	Johannes Pellenz	Florian Rabe
Achim Rettinger	Christoph Ringlstetter	Paul Rosenthal
Joachim Schmidt	Klaus Stein	Gerhard Strube
Agnes Swadzba	Marko Tscherepanow	Sven Wachsmuth
Katja Windt	Oliver Wirjadi	Stefan Woelfl
Johannes Wolter	Britta Wrede	Christoph Zetzsche
Dapeng Zhang		

Sponsors

Microsoft Deutschland GmbH
Deutsche Telekom AG Laboratories
IDS Scheer AG
SAP AG
Living-e AG
Springer-Verlag GmbH
DFKI GmbH

Table of Contents

Posters

Demos

Folk Reducibility
and AI-Complete Problems

Randy Goebel

Department of Computer Science,
University of Alberta, Edmonton, Canada
goebel@cs.ualberta.ca
http://web.cs.ualberta.ca/~goebel/

Abstract. The idea of an "AI-complete" problem has been around since at least the late 1970s, and refers to the more formal idea of the technique used to confirm the computational complexity of NP-complete problems. In the more formal context, the technique of reducibility was used to transform one problem into another that had already been proved to be NP-complete.

Our presentation takes a closer look at what we call "Folk Reducibility", as an approximation to reducibility, in order to try and improve coherence regarding what constitutes tough AI problems. We argue that the traditional AI-complete problems like "the vision problem" and "the natural language problem" are too vague. We provide examples of more precisely specified problems, and argue that relationships amongst them provide a little more insight regarding where and how valuable problem relationships might emerge.

A. Dengel et al. (Eds.): KI 2008, LNAI 5243, p. 1, 2008.

Meme Media and Knowledge Federation

Yuzuru Tanaka, Jun Fujima, and Micke Kuwahara

Meme Media Laboratory, Hokkaido University
N13 W8, Sapporo 0608628, Japan
{tanaka,fujima,mkwahara}@meme.hokudai.ac.jp

Abstract. This paper first clarifies two major difficulties that prevents our maximum utilization of web resources; the difficulty to find appropriate ones, and the difficulty to make them work together. It focuses on the latter problem, and proposes *ad hoc* knowledge federation technologies as key technologies for its solution. Then it reviews our group's 15 year research on meme media, as well as his 6 year research on their application to *ad hoc* knowledge federation of web resources. Then it proposes an extension of the Web to the memetic Web. This extension is brought by the recent significant revision of the 2D meme media architecture. Finally it shows some new applications of these technologies.

1 Introduction

1.1 What Prevents Our Maximum Utilization of Web Resources?

The Internet has been and still is continuously changing our lives. It is pervading the economy and the private sphere, thus boosting the pace at which modern information and communication technologies are penetrating everyone's life, every science and engineering discipline, and almost every business. Despite the fact that billions of Web resources are available world-wide and no slowdown of the Web's growth is in sight, users typically make little use of the richness of the Internet.

One well-known problem is the difficulty of finding what a user is interested in. A second problem is the difficulty of making mutually related Web resources work together, due to a lack of suitable interfaces or conversely a plethora of interfaces, formats, and standards. While some big projects in the US, Europe, and Japan, such as Digital Library Initiative (DLI) Projects in US [1], and Information Grand Voyage Project [2] in Japan, have been focusing on the former problem, no such major projects have focused on the latter except those on GRID computing, which basically focuses on the interoperability of distributed resources to perform routine or frequently requested jobs, and requires no *ad hoc* immediate definition of interoperation among resources. For those web resources provided as web applications, users can combine their functions and contents only by manually transferring the outputs from one resource to the inputs of another. A tiny proportion of web applications have been augmented with web service interfaces, but even then only a programmer can make such resources work

A. Dengel et al. (Eds.): KI 2008, LNAI 5243, pp. 2–21, 2008.

together. Web mashup tools also aim at the same goal, but are still premature as end-user tools, and lack generic architecture.

There is a similar prominent problem in practically every large enterprise, in the larger healthcare institutions and in literally every governmental agency: enterprise application integration. The key problem is that legacy systems have been introduced over many decades. Most of these systems have never been designed to work together, and the potential for synergetic effects dwindles away.

The diversity, heterogeneity, autonomy, and openness of both resources and their usages are the inherent characteristics of current information system environments. These characteristics are inherently out-of-control, and incompatible with conventional top-down system integration methods.

It seems that human creativity will always lead to newly designed systems and solutions that go beyond earlier standards, and are thus not fully compatible with the established and widely used systems. Creating and updating standards in information and communication technologies is of course needed, but there is little hope that this alone could ever solve the enterprise application integration problem, or alleviate the above-mentioned problems that face individuals, enterprises and communities when dealing with the diversity of the Web.

1.2 Federation as a Key Technology to Develop

One of the research and development efforts undertaken to make systems communicate with each other without explicitly having been designed for such communication is called 'federation.' The term was probably first introduced to IT areas by Dennis Heimbigner in the context of a federated database architecture [3], and then secondarily in late 90s, by Bill Joy in a different context, namely, federation of services [4]. In a federated database architecture, a collection of independent database systems are united into a loosely coupled federation in order to share and exchange information. In the context of service federation, a federation is a set of services that can work together to perform a task. A service, in turn, is an entity that sits on the network ready to perform some kind of useful function. A service can be anything: a hardware device, a piece of software, a communications channel, or even a human user.

Federation in general is the process of combining multiple technology elements into a single virtual entity; it is being driven by complexity (i.e., size, heterogeneity, and distributedness), and by IT's need to make sense of technology systems that have sprawled to an unprecedented scale and have become extremely expensive to maintain and manage. Federation is different from integration. Federation assumes an open networked environment of diverse, heterogeneous, autonomous, and distributed resources, and deals with open scenarios of information processing. Integration basically targets local and centralized management, and interoperation of resources in a closed environment, and deals with closed scenarios of information processing. While it is not difficult to design, from the outset, an integration scheme for elements of a given closed environment, no such integration scheme can remain compatible with future evolution in the case of an open environment.

1.3   Federation of Shared Knowledge Resources

In our daily intellectual activities, whenever we conceive an idea, we try it out, observe the result to evaluate it, and eventually proceed to conceive a new idea. We repeat such a process of 'think', 'try', and 'see.' In the context of problem-solving, researchers call such a repetitive process a plan-do-see loop. Many feel that computers can be applied to support this process, speeding it up and driving progress around this cycle. This requires seamless support of the three phases: think, try, and see. The first 'think' phase is mainly performed by humans. The second 'try' phase requires the representation of each idea and its execution. The third 'see' phase requires various evaluation and analysis tools. The first and the third phases are the application areas of artificial-intelligence systems and computation-intensive systems, but we do not have appropriate systems that effectively support the second phase.

On the other hand, the current Web is becoming an open repository of a huge variety of knowledge resources including databases, and synthesis/analysis tools of almost all areas of intellectual activities. The problem we face today is how to make it easy to select some set of resources including web resources and local ones, to customize them, and to make them work together. We want to try our new idea for its evaluation immediately whenever it comes into our mind. A trial of a new idea requires a composition of a new tool from available resources, and its application to a data set extracted from available resources. It may require a repetition of such processes. We also want to republish some of these composed tools and extracted data sets as useful web resources for further reuse by other people.

Commercially available drug design software systems are good examples of the complex integrated application systems available today. Each system provides an integrated environment comprising modules such as a protein database, a homology search system, a ligand structure prediction system, and a docking simulator. While such integrated systems are very expensive, for each module you can find a free Web service, or a downloadable freeware program. Similar situations are frequently observed in a wide range of academic areas. Nonetheless, finding appropriate data sources and applications from the Web, customizing them, and making them work together are as difficult and expensive as developing an integrated commercial product with similar complexity.

1.4   Science Infrastructure vs. Conventional e-Science

Recently a new research methodology has been developed for the advanced automation of scientific data acquisition tools, based on collecting data for all cases without any specific purposes, then later retrieving and analyzing appropriate data sets to support a specific hypothesis or conjecture. This new methodology is referred to as 'data-based science', or 'data-centric science' [5]. Data-based science is becoming more and more popular in scientific areas such as genomics, proteomics, brain science, clinical trials, nuclear physics, astrophysics, material science, meteorology, and seismology. Data-based science makes data acquisition independent from data analysis; it leads to large numbers of huge, independent accumulations of data, and a great variety of data analysis tools. Many such data accumulations and analysis tools have been made available over the Web to encourage their reuse. Advanced utilization of such resources requires flexible extraction of appropriate data sets from these diverse sources, flexible application of appropriate analysis tools to extracted data sets, and flexible coordination of such analyses to evaluate a hypothesis. Such an approach, although currently performed with lots of manual operations and programming, has enabled the comparative analysis of seemingly unrelated large data sets, leading to the discovery of interesting correlations such as that between the El Niño-Southern Oscillation and the weather in Europe, volcanic eruptions, and diseases. These studies required extraction of appropriate data sets from multiple seemingly unrelated sources, their compilation or customization, and the application of data analysis and/or data visualization tools to these data sets. Such operations are inherently performed in an *ad hoc* manner, and should be executable rapidly enough to follow up on fleeting ideas.

The creation of information systems that may work as a science infrastructure to accelerate further development of science and technology requires a new generic technology for 'knowledge federation' as well as the development and extension of databases, simulators, and analysis tools in each area. Knowledge federation denotes flexible, *ad hoc*, and instantaneous selection, customization, and combination of knowledge resources by users to meet the dynamically changing demands of creative activities. For routine or frequently requested jobs that require interoperation of multiple knowledge resources, we may use work-flow and/or resource-orchestration technologies based on GRID computing. These technologies were extensively studied and developed during the last decade in the US, Europe and Japan, and are now becoming off-the-shelf technologies. But in R&D activities, where repetitive trial-and-error operations are a vital part of creative thinking, we need support for *ad hoc*, instantaneous implementation of passing ideas. By contrast, most of the preceding e-science projects were based on GRID computing technologies that are suited only to routine work.

1.5 Ad Hoc Knowledge Federation and Meme Media

If there were an easy, generic way for users to edit open web resources instantaneously without any coding for extracting appropriate data sets and application tools, for composing the desired complex applications, for applying composite tools to extracted data sets, and for coordinating these extractions, compositions and applications, we would be able to accelerate substantially the evolution of the knowledge resources shared by our societies, and therefore, the knowledge evolution of our societies.

The existing proposals for federation, such as Jini [4], however, provide no support for instantaneous extraction, customization, or composition of knowledge resources. We call such definition of knowledge federation by end users without any coding '*ad hoc* knowledge federation.'

In 2004 we proposed concepts and architecture of *ad hoc* knowledge federation [6] based on our R&D on 2D and 3D meme media since 1987 [7]. Meme media systems are component-based media systems. The 2D meme media system, proposed in 1989 [8], represents every object as a pad - a card-like visual component that has a pin plug and connection jacks to support plug-and-play capabilities. Users can paste pads together to compose them, and peel apart such compositions to reuse the components. The 3D meme media system, proposed in 1995 [9], represents every object as a box - a 3D visual component that again has a pin plug and connection jacks. A meme pool, proposed in 1996 [10], works as a world-wide repository of meme media objects, i.e., pads and boxes. Users can drag and drop meme media objects into and out of a meme pool to publish and to reuse them. In 1999 and 2001 we proposed application frameworks for database visualization based on meme media technologies [11,12], and in 2004 we extended this framework to cope with knowledge resources over the Web [13].

Recently, our group has significantly revised our 2D meme media architecture. The new 2D meme media architecture is a web-top architecture requiring no additional kernel installation at each client. It fully exploits the functionality of *de facto* standard advanced browsers and Silverlight plug-in. It allows us to extract fragments of web resources as pads, to paste them together with local pads for composing a new composite pad, and to republish any composite pad into the Web. It enables us to perform all these operations through direct operation on web top environments. This new 2D meme media architecture makes the Web work as a meme pool. We call this extension 'the memetic Web.'

This paper reviews our group's 15 year research on meme media and their application frameworks as well as his 6 year research on their application to knowledge federation of resources over the Web, and then shows the recent significant extension of 2D meme media architecture and the concept of memetic Web. Finally, it shows some new applications of these technologies.

2   Knowledge Media, Meme Media, and Knowledge Federation

Before discussing technological details, here we clarify our terminology in this paper.

2.1   Knowledge, Knowledge Resources, and Knowledge Media

The word 'knowledge' in computer science was initially used in a very limited meaning in 70s and 80s to denote what can be described in some "knowledge representation language" for the problem solving by machines. The same word in dairy conversation denotes facts, information and skills acquired through experience or education, theoretical or practical understanding of a subject, language, etc., or awareness or familiarity gained by experience of a fact or situation. Today, we use computers to deal with all these different types of knowledge by representing them as some kinds of objects such as documents, tools and services. IO relations and functions of applications and services can be also considered as knowledge fragments. The Web is expanding its capability to deal with all these things for their sharing in a society and/or in an enterprise, and to promote their reuse in different contexts for different purposes. Recently, some philosophers such as Davis Baird are also trying to expand the definition of knowledge to denote not only linguistically expressible knowledge, but also 'thing knowledge', i.e., knowledge embodied as some artifact object [14]. Computerization of the design and manufacturing of artifacts is also accumulating their thing knowledge as reusable and sharable knowledge. Our definition of 'knowledge' in this paper includes both linguistically expressible knowledge and thing knowledge.

'Knowledge resources' denote sharable and reusable different types of such knowledge represented on computers. The web represents knowledge resources as compound documents with embedded nonfunctional multimedia contents and/or functional contents such as application tools and services. The compound document architecture was developed in 1980s as an extension of multimedia documents. A compound document looks like a multimedia document and can embed in itself not only multimedia contents but also any interactive visual object with some functionality. Web documents including web applications are compound documents. Generally, knowledge resources require some media to externalize them as interactive visual objects, and to publish them for their reuse by other people. 'Knowledge media' denote such media since they externalize knowledge resources. Knowledge resources treated in the same system exploit the same knowledge media architecture. Compound document architectures are examples of knowledge media architectures. While compound documents are two dimensional visual representations of knowledge resources, some knowledge media architectures represent knowledge resources as three dimensional visual objects or spaces.

2.2   From Knowledge Media to Meme Media

The Web provides both a standard knowledge media architecture to publish knowledge resources as web documents, and an open repository of web documents for publishing, sharing, and reusing them among different people distributed over the world.

Human users access what appears to be easily accessible and what fits their expectation. Large amounts of data and potentially useful knowledge in the Web remain literally invisible. Web resources can be reused only as they are, whereas the reusing of a resource in general means its reuse in a different context, and hence requires its modification or interoperation with other resources. Web resources represented as web applications cannot interoperate with each other since they have no IO interfaces to any other programs. Only human users can input parameters to them and read their output results.

'Meme media' denote new knowledge media that enable users not only to publish knowledge resources, but also to reedit some of them into a new composite knowledge resource for the reuse in a different context [7,10,15]. Figure 1 shows the difference between the current Web and its extension in which knowledge resources are published as meme media objects. The repository in the latter case is called a meme pool [7,10]. A meme pool allows a large number of people to get some published meme media objects, to reedit them to compose a

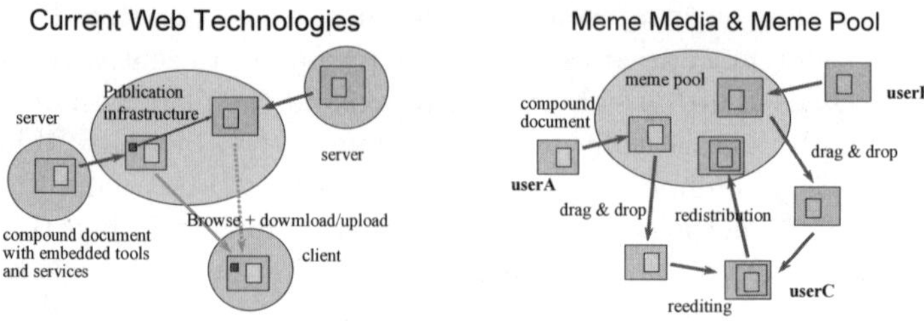

Fig. 1. Difference between the current Web and a meme pool of meme media objects

new meme media object, and to publish the result as a new sharable knowledge resource. Such a meme pool will rapidly evolve the variety of its knowledge resources through people's frequent reediting of them. Meme media and a meme pool accelerate the evolution of knowledge resources. The reediting of knowledge resources denotes both the rearrangement of objects in two-dimensional layout or in three-dimensional configuration, and the new definition of interoperation among objects. Knowledge resources carried by meme media work as what Richard Dowkins called memes [16]. He observed some similarity between genetic evolution and cultural evolution. New ideas are replicated and transferred from people to people. More than one idea can be recombined to compose a new idea. An idea may not be correctly transferred from one person to another, but the received idea may turn out to be a useful new idea. Finally, ideas are evaluated by our society and useful ones may stay longer. These correspond to replication, recombination, mutation, and natural selection of genes. Based on this observation, he coined the word 'meme' by combining two words, 'gene' and 'mimesis' to denote a virtual or imaginary entity that carries a fragment of human's knowledge.

2.3   Knowledge Federation

The Web is becoming an open repository of a huge variety of knowledge resources ranging from entertainment contents to advanced science and/or technology databases and analysis tools. Their reuse in different contexts and for different purposes requires their customization for partially using their functions, making them interoperate with other knowledge resources, and designing new composed resources. Many organizations are facing the similar problem; enterprise application integration. Such a problem can be commonly observed even in those enterprises which, in 70s and 80s, made large investments to fully integrate their data and applications by introducing database management systems. These paradoxical situations were brought by rapid and *ad hoc* introduction of many different types of client applications, and also by the needs to make such applications interoperate with web resources. The diversity, heterogeneity, and openness of both resources and their usages are the inherent characteristics of current information system environments. These characteristics are inherently out-of-control, and incompatible with conventional top-down system integration methods.

It is an important way to aim at standards in information and communication technologies. But there is not much hope that standardization might ever solve the enterprise application integration problem or alleviate the mentioned problems of individuals, enterprises and communities when facing the diversity of the Web.

As mentioned in Chapter 1, 'federation' technologies are expected to give such problems one possible solution. Federation is becoming fundamental to IT. 'Knowledge federation' denotes a way to select and to combine multiple knowledge resources from an open environment such as the Web and pervasive computing environments into a single virtual entity representing a new knowledge resource [6]. 'Knowledge federation' in this paper is different from knowledge reorganization and knowledge mediation. The latter two respectively denote structural/semantic reorganization of knowledge resources, and semantic bridging across more than one coherent organization of knowledge. Knowledge federation in broader sense may include these latter two.

3   Meme Media as Basic Enabling Technologies

3.1   Meme Media Architectures as Component-Based Knowledge Media Architectures

We have been conducting the research and development of 'meme media' and 'meme pool' architectures since 1987. We developed 2D and 3D meme media architectures 'IntelligentPad' [6,8,10] and 'IntelligentBox' [9] respectively in 1989 and in 1995, and have been working on their meme-pool architectures, as well as on their applications and revisions. 'IntelligentPad' represents each component as a pad, a sheet of paper on the screen. A pad can be pasted on another pad to define both a physical containment relationship and a functional linkage between them. When a pad P2 is pasted on another pad P1, the pad P2 becomes a child of P1, and P1 becomes the parent of P2. No pad may have more than one parent pad. Pads can be pasted together to define various multimedia documents and application tools. Unless otherwise specified, composite pads are always decomposable and reeditable.

In object-oriented component architectures, all types of knowledge fragments are defined as objects. IntelligentPad exploits both object-oriented component architecture and wrapper architecture. Instead of directly dealing with component objects, IntelligentPad wraps each object with a standard pad wrapper and treats it as a pad. Each pad has both a standard user interface and a standard functional connection interface. The user interface of a pad has a card like view on the screen and a standard set of operations like 'move', 'resize', 'copy', 'paste', and 'peel'. Users can easily replicate any pad, paste a pad onto another, and peel a pad off a composite pad. Pads are decomposable persistent objects. You can easily decompose any composite pad by simply peeling off the primitive or composite pad from its parent pad. As its functional connection interface, each pad provides a list of slots that work in a similar way as connection jacks of an AV-system component, and a single pin-plug to be connected to a slot of its parent pad. Each pad uses two standard messages 'set' and 'gimme' to access a single slot of its parent pad, and another standard message 'update' to propagate its state change to its child pads. In their default definitions, a 'set' message sends its parameter value to its recipient slot, while a 'gimme' message requests a value from its recipient slot.

3.2   Wrapping Web Resources as Meme Media Objects

From 2003 to 2004, we had developed new technologies for users to easily wrap web applications and web services into meme media objects [17,18]. C3W [19,20,21,22] is a wrapper framework that enables users to open a browser showing a web application page, to clip out some input portions and some output portions as pads, and to make such pads clipped out from more than one web page interoperate with each other. C3W enables users to make these definitions only through direct manipulation without writing any codes. For a given web service with a WSDL description of its interface, our web service wrapper guides users to define its proxy object as a pad with a specified subset of IO signals. Our web service wrapper first pops up the IO signal list of the web service for users to arbitrary select some of them as the slots of the pad to define, and to specify default values for some other input signals. Users can make these specifications only through direct manipulation. Then the wrapper automatically creates a pad that wraps the original web service and works as its proxy object.

Figure 2 shows, at its bottom right corner, a web application by US Naval Observatory showing day and night over the earth. The left object is a composite pad showing the difference of arbitrarily chosen two seasons; the same time on summer solstice and winter solstice for example. This was constructed by just clipping out the date input forms and the simulated result as pads from the Naval Observatory web application, drag-and-dropping them on a special pad, and applying the multiplexing to the input to obtain multiple outputs.

All these operations were performed through direct manipulation of pads. For browsing the original web page, we use a web browser pad, which dynamically frames different extractable document portions for different mouse locations so that its user may move the mouse cursor around to see every extractable document portion. When it frames a desired object, you can just drag the mouse to clip out this object as a pad. All the pads thus clipped out from web pages in a single navigation process keep their original IO functional relationship even after their arrangement on the same special pad C3WSheet. The C3WSheet gives a cell name to each pad pasted on it. The pads clipped out from date input forms are named as E, F, G, H, I cells. They respectively correspond to year, month, day, hour, and minute input forms. The simulation output pad is named as J cell. Whenever you input a new date to the extracted date input forms, the corresponding extracted output pad showing a simulated result will change its display. This simulation is performed by the server corresponding to the source web application from which all these pads are extracted.

The multiplexer pad, when inserted between the base C3WSheet and the extracted date input form pad, automatically inserts another multiplexer pad

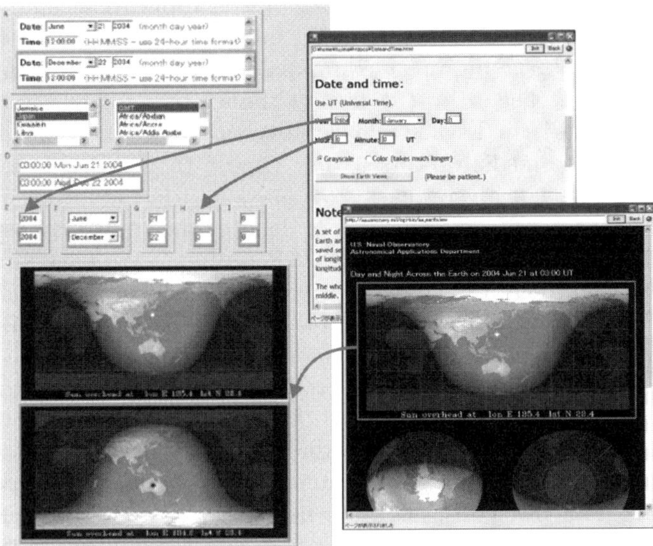

Fig. 2. Construction of a composite pad, using clips extracted from web pages as pads

between the base C3WSheet and every extracted pad whose value may depend on the input to this date input form. If you make a copy of the extracted date input form pad on the multiplexer pad, each copy of this multiplexer pad automatically makes a copy of its child pad that was also extracted from the same web application and is dependent on the update of the date value. Mutually related multiplexer pads maintain the relationship among the copies of their child pads. In Figure 2, two simulation results are obtained for two different days.

The above mentioned composite pad allows us to specify any time in GST. You can easily modify this content to accept any local time by combining this with another public web application for time conversion. The C3WSheet in Figure 2 has four more multiplexed pads named as A, B. C. and D cells. They are extracted from the time conversion service. The pads named as A, B, C cells are the extracted input forms, respectively specifying the input local time, the source time zone, and the destination time zone. The pad named as D shows the local time in the destination time zone. The destination time zone is fixed to GST in the cell C. You can just specify equations in the cells E, F, G, H, I to make their values always equal to the year, month, day, hour, and minute of the cell value of D. Such a direct manipulation process customizes existing web resources, and makes some of them interoperate with each other to define a new knowledge resource as a composite pad.

3.3   Meme Media and Meme Pool Architectures for Web Resources

Figure 3 shows our basic idea of wrapping different types of knowledge resources into meme media objects for achieving interoperability among them. The arrows

(1) and (2) respectively show the wrapping of web applications and web services into meme media objects. The arrow (6) denotes the combination of more than one meme media object by direct manipulation to compose a new composite meme media object. The arrow (5) denotes the wrapping of legacy applications into meme media objects. Legacy applications without GUI can be always easily wrapped into meme media objects if they expose their APIs. Legacy applications with GUI are not always easily wrapped into meme media objects even if they expose their APIs. The arrow (3) denotes that any composite pad can be accessed as a web service through SOAP protocol. We have already developed a tool that automatically creates a corresponding web service for any given composite meme media object. The arrow (4) denotes the conversion of an arbitrary composite meme media object into a web application. This conversion is called 'flattening' since it converts a composite media object into a flat web page viewable with any web browser [7]. Since composite meme media objects are not HTML documents, they are not directly viewable with web browsers. We have already developed a tool for the automatic flattening of composite meme media objects. This mechanism however makes it easy to develop phishing web sites just by pasting a password stealing pad over a wrapped trustworthy web application to define a composite pad. Because of the inherent fragility of the current web technology against phishing trials, we have made a decision that we should not provide this flattening conversion mechanism to the public.

While the flattening mechanism enables us to use the Web itself as a meme pool, we developed another world-wide repository Piazza that works as a meme pool for pads. Piazza was developed on top of Wiki, and uses Wiki servers [6]. Piazza provides its own browser to access different pages. Each page that can be identified by a URL may contain arbitrary number of pads. Users can open an arbitrary Piazza page, and drag and drop pads between this page and a local

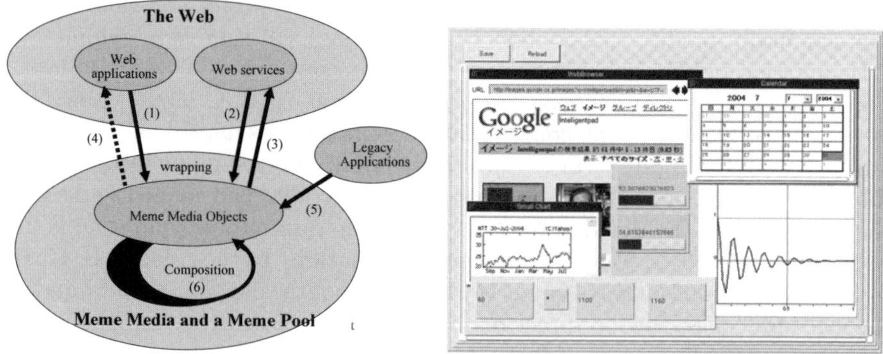

Fig. 3. Basic architecture to wrap different types of knowledge resources into meme media objects for their mutual interoperability

Fig. 4. A worldwide repository of pads developed by applying C3W framework to Wiki

pad environment. By clicking the page registration button, users can reflect the page update to the servers. By clicking the page refresh button, users can reflect the current state of the server to the displayed Piazza page. We have already shown that Piazza can be easily developed by applying C3W framework to Wiki system [6]. Figure 4 shows the Piazza browser displaying a Piazza page with several registered pads.

Piazza allows people not only to publish and share knowledge resources represented as pads, but also to compose new knowledge resources by combining components of those pads already published in it. People can publish such newly composed knowledge resources into the same Piazza system. The collaborative reediting and redistribution of knowledge resources in a shared publishing repository by a community or a society of people will accelerate the memetic evolution of knowledge resources in this repository, and make it work as a meme pool.

4   Extending the Web with Meme Media Technologies

The most recent 2008 version of IntelligentPad fully exploits Microsoft Silverlight technology, and runs on Internet Explorer 6 and 7, Mozilla Firefox, or Safari browser empowered by Silverlight plug-in. It is a web-top system that allows us to directly manipulate pads on a web page.

It is one important feature of this version that its pad may not have its rectangular canvas. Any graphical object including a line segment can be treated as a pad. Such a pad is called a canvas-free pad. Our preceding versions could not deal with canvas-free pads. This new version exploits SVG (Scalable Vector Graphics) to describe graphical objects. When a pad is pasted on a canvas-free pad to work as its child pad, the former is bound by the local coordinate system spanned by the latter pad, instead of being clipped by the canvas area of the parent pad.

It is another important feature that we need no IntelligentPad kernel running on clients. Any browser empowered by Silverlight plug-in can provide a web-top environment of pads, and render any composite pad as an extended web page. Users accessing through such browsers can directly manipulate pads on a web page, and do not recognize any difference between web resources and composite pads. Composite pads can be also registered in HTTP servers for other people to access them as extended web resources.

Figure 5 shows a single web page shown by a browser. This page shows an application environment composed as composite pads. This is a display snapshot of Trial Outline Builder; a subsystem of ObTiMa (Oncology-based Trial Management System) we are developing in collaboration with the Department for Pediatric Hematology and Oncology at the University Hospital of the Saarland and Fraunhofer IAIS in EU's FP6 Integrated Project ACGT (Advancing Clinico-Genomic Trials on Cancer). It shows a trial plan diagram, which is actually a composite pad constructed by pasting various primitive pads representing different treatment events such as radio therapies, chemotherapies, and surgeries. Each treatment process proceeds from left to right in this diagram. Each

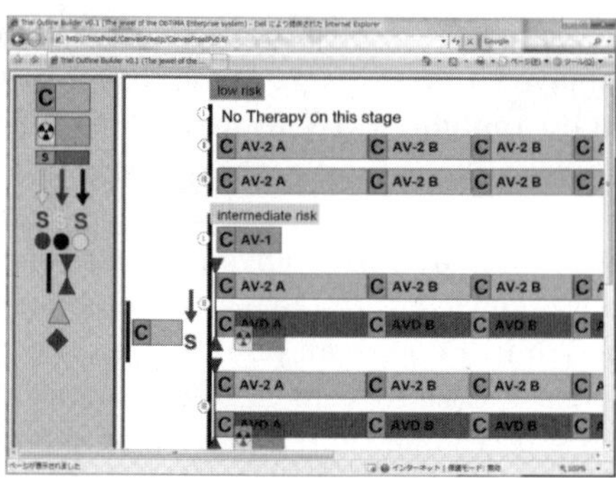

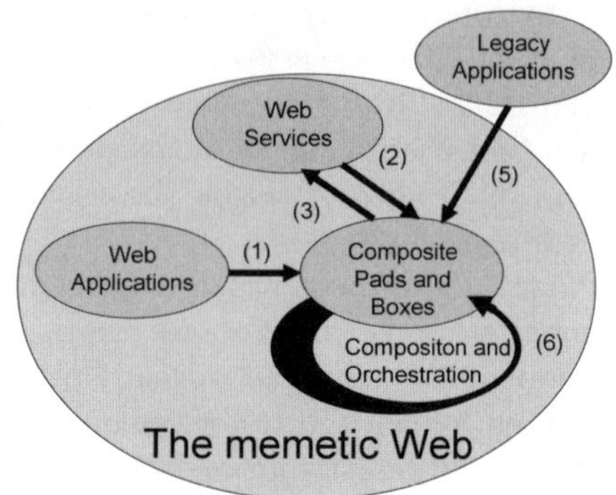

Fig. 5. The Trial Outline Builder running on a web top environment

Fig. 6. The memetic Web and different types of knowledge resources

event is associated with a CRF (Case Report Form). By clicking each treatment event, users can open the corresponding CRF. Each CRF is a web application enabling users to specify some item values for retrieving the corresponding CRF record, and to input new values to some other entries for updating the record. CRF records are managed by database servers. CRF servers and their browser are being developed by the Fraunhofer team. The trial plan diagram uses two different types of branching pads. A stratification pad, represented as a vertical line segment, classifies each case depending on patient's status, and selects one of the different treatment branches. A randomization pad, represented as a vertical line segment with two triangles at both ends, randomly assigns each input case to one of the different treatment branches.

The Trial Outline Builder enables users to graphically design a new trial just by pasting copies of primitive pads representing different treatment events. It also enables users to input treatment data at each treatment event through its corresponding CRF. After sufficient accumulation of data, it allows users to specify special cases to retrieve the related data for the evaluation of different treatments. All these activities are supported by the system in an integrated way using the same trial plan diagram. The exploitation of the IntelligentPad architecture makes it easy to extend the system function by associating related web resources to each treatment event, or by making new modules extract some treatment data for their analysis.

The recent 2008 version of IntelligentPad system has made it unnecessary to use such a pad repository as Piazza. It runs pads on web pages browsed through *de facto* standard browsers empowered by Silverlight plug-in. As shown in Figure 6, the extended Web enables us not only to use web pages as Piazza pages, but also to publish any composite pad as a web page. The extended Web works as a world-wide repository of both web resources and composite pads (Figure 6). Users need not distinguish composite pads from web documents. We call this extension 'the memetic Web.' The memetic Web uses the C3W framework for users to extract fragments of web resources, to combine them, and to compose a composite pad.

Such pads clipped out from web applications use the rendering function of the browser. The handling of more than one pad of this type on the web top requires cross-frame scripting, which is not allowed by any browser. The memetic Web uses a proxy server called a C3W server through which each pad of this type accesses the original web application. The access of different web applications through the same proxy server solves the cross-frame scripting problem. 3D meme media objects represented as composite boxes can run on a special pad that works as an IntelligentBox run-time environment. Their browsing and reuse however require client-side installation of the IntelligentBox kernel.

5   Meme Media and Knowledge Federation vs. Portals and Web Mashups

Meme media, when applied to the Web, can aggregate data, texts, multimedia contents, and services that are provided by more than one different web applications and web services to compose a single compound document as a composite meme media object, namely in a reeditable form. Since the recent versions of IntelligentPad can run on *de facto* standard web browsers empowered by Silverlight plug-in, composite pads can be published into the Web, and be browsed by *de facto* standard web browsers.

Web portal and mashup technologies are also developed to aggregate data and contents from more than one web resource into a single web application. A portal graphically arranges portlets into a single web page without overlaying their graphics. Each portlet represents a visual portion of a source web page or a visual IO interface of a web service. Basically, portal technologies do not provide any standard method for users to define interoperation among these portlets.

A web mashup is a web application that combines data from more than one source into a single integrated tool. A famous example is the use of cartographic data from Google Maps to add location information to real-estate data, thereby creating a new and distinct web application that was not originally provided by either source. Each web mashup performs the following three functions, i.e., the data extraction from some web resources, the processing of extracted data for integration, and finally the visual presentation as a web application. Each mashup tool provides users with an easy way to define, at least, one of these three different functions. Definition of the remaining functions requires programming.

Web mashup tools in the first generation used some specific web application such as Google Maps as the base application, and used its API to add, on its page, some graphical objects that represent the data retrieved from other web services. The server program of this base web application needs to provide such a function through its API. The addition of graphical objects corresponding to the external data is performed by the server program of this web application. The first generation mashup technologies focused on the visual presentation function, and provided no tools for data extraction and data processing functions. They use the API of such specified web applications as Google Maps and Google Chart to render a map or a chart with a data set extracted from a different web resource. Google My Map [23], and MapCruncher [24] are examples of such mash up tools focusing on presentation functions.

Some mashup tools such as Yahoo!Pipes [25] focus on the GUI support of implementing the data processing function to filter and to merge more than one data source on the Web. Yahoo!Pipes provides users with various components for reusing RSS feeds, filtering their data, merging them, and defining new RSS feed. The visual presentation by rendering the result RSS data needs to use other tools or script programming.

The data extraction from web resources may use standard API in the case of web services, or some web wrapper technologies for web applications and web documents. There are lots of proposals on web wrapper technologies ranging from simple practical ones with limited capabilities to theoretical sophisticated ones. Dapper [26] is used to easily extract chunks of data called Dapps from any web resource, convert it to various formats including XML, and RSS, and create applications out of them.

Web mashup tools in the second generation use the web page rewriting capabilities at the client-side. The recent advances of browser technologies have enabled such capabilities. Microsoft Popfly [27] provides a library consisting of visual data-editing components for filtering and merging source data, and visual data-presentation components for defining and rendering web pages. It allows users to define a work flow among them. In addition to these system-providing components, users are allowed to define their own components by programming. Google Mashup Editor [28] provides a developers' visual environment and toolkit for programming the data-editing and presentation functions of mashups in HTML and JavaScript. Plagger [29] is a pluggable RSS feed aggregator written in Perl. Everything is implemented as a small plug-in and users can mash them up together to build a new application to handle RSS feeds. Plagger supports both the data editing and the presentation in defining mashups.

DataMashups [30] is an Ajax-based enterprise-web-application development service, and offers an integrated development environment to rapidly create mashups and web applications with limited or no need for server side code. DataMashups lets users use a number of predefined applications called widgets to easily create personalized portals for clients. DataMashups provides no tools for users to make these widgets communicate with each other, OpenKapow [31] works with the concept of robots. A desktop application called Robomaker is used to gather data from websites. Robomaker allows us to automate complex processes and simulate a real person's behavior in a web browser to retrieve the data we need. It can create three different types of robots - RSS, REST or Web Clip robots, which enable us to either create RSS feeds, create an API out of a web site or simply collect one piece of functionality from a site and use it somewhere else. Users can share and reuse tool component robots for data extraction, data editing, and presentation. They can be registered into and retrieved from a shared repository.

Intel Mash Maker [32] provides users with a tool to construct extractors that extract structures of both source web site and target web site. The target web site is the one into which you will embed data and portals. The extracted structures are visualized as tree views. Each node represents some extractable data or portlet. You can select a node of the tree view of a source web site, and drag-and-drop its copy at any location of the tree view of the target web site. You can define a mashup by repeating such operations. The query to access the web page from which a portlet is extracted can be treated as the input of this portlet, and any extracted portion of this portlet can be treated as its output. By combining such IO functions of portlets and web services, users can define interoperability among them in the composed web application.

While mashup technologies focus on composing a new web resource from available web resources, our meme media architectures focuses on the repetitive reediting, reuse, and republication of knowledge resources by end users in instantaneous and *ad hoc* ways, and aim to extend the Web to work as a meme pool for accelerating the co-evolution of both knowledge resources and their usage. The first generation mashup tools are applicable to specific target web applications. Although the second generation mashup tools are applicable to any target web applications, they use client-side composition of widgets extracted from other web resources. Their mashups are no longer web applications described in HTML. Therefore, we cannot apply the same mashup tool to further reedit mashups to compose another mashup. Mashup technologies cannot extend the Web to the memetic Web because they cannot enable us to reedit mashups.

6   Knowledge Federation beyond the Web

Recently, the Web works not only a world-wide repository of documents, but also an infrastructure to make mobile phones, sensor/actuator networks, home electronic appliances, and expensive scientific equipments accessible to and from web resources. Users can access and/or control these objects through the Web from anywhere, and these objects can get useful information from anywhere through the Web, and/or update some web resources. The Web provides a basic network of knowledge resources. Mobile phones are connected to some web servers. A mobile phone can access any web resources via such a server. It can be also accessed from some web resources via such a server. A sensor/actuator network with ZigBee protocol can dynamically span a network of a coordinator and sensor/actuator nodes. The coordinator can read the value of any sensor, and/or write the value of any actuator, in any node. By defining a web application or a web service for the computer with this coordinator node, we can make the value of each sensor of every node accessible through this web application or web service. In each house, home electronic appliances will be connected to a home network, which is connected to the Internet through a network node. Therefore, these appliances will be accessible to and from the Web for their monitoring and control.

All different types of functions that are provided by mobile devices, sensor/actuator nodes, and electronic appliances are accessible through the Web as web resources, namely as web applications or web services. Therefore, we may

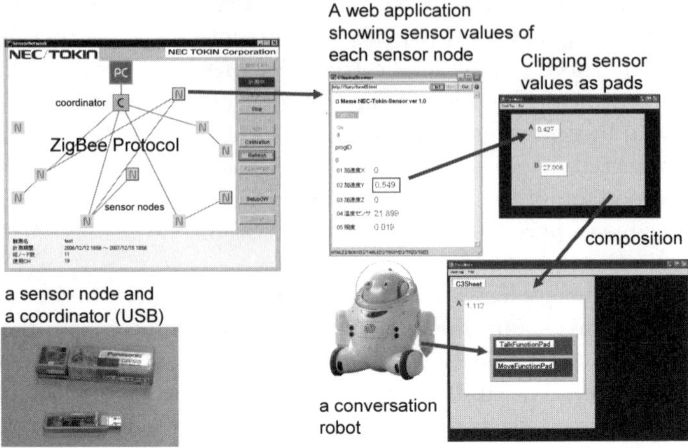

Fig. 7. *Ad hoc* federation of sensor nodes with a conversation robot

consider them as knowledge resources. We can apply our knowledge federation technology to extract some of their sub functions, to make them interoperate with each other, and to compose a new knowledge resource as a web application or a web service.

Figure 7 shows a sensor node module with WiFi communication capability. This node module assembled in a cube of 1.5 cm in each dimension consists of an 8 bit processor and several sensors including a temperature sensor, a light sensor, and three acceleration sensors. This prototype version uses an external battery, the final version will use a solar battery installed in the main cube. Two copies of this node module can communicate with each other within the range of 100 m from each other. A coordinator node uses ZigBee protocol to coordinates multiple copies of this node module to span a network among them dynamically. This sensor node module and the coordinator module were provided by NEC TOKIN Corporation for research purposes. Figure 7 also shows a conversation robot. We developed primitive pads for the flexible and immediate composition of the robot speech and gesture. The speech pad, for example, has a slot for inputting a sentence, and controls the robot to make the speech of this sentence.

Figure 7 also shows a web application that shows the current values of all the sensors in the sensor node specified by a user. You can use the C3W technology to clip out both the input form for a sensor-node id, and the output value of the Y-axis acceleration sensor. Using standard primitive pads such as the text-number conversion pad and the threshold overrun detection pad together with these clipped out pads and robot speech and gesture pads, you can easily compose a composite pad that detects the overrun of the Y-axis acceleration value beyond a specified threshold value, and makes the robot say, for example, 'The door of the living room has just opened'. This composite pad accesses the sensor node in an *ad hoc* sensor network through the web application, and controls a robot through its primitive proxy pads of its various functions. Suppose that the web application of sensor nodes is already defined using a tool provided by the sensor node provider, and that the primitive proxy pads of robot functions are provided by the company. Then, you can immediately define such a composition just by clipping out pads from the web application, and by pasting them together with robot proxy pads and standard primitive pads.

7   Concluding Remarks

The current Web is becoming an open repository of a huge variety of knowledge resources including databases, and synthesis/analysis tools of almost all areas of intellectual activities. The problem we face today is how to make it easy to select some set of resources including web resources and local ones, to customize them, and to make them work together. We want to try our new idea for its evaluation immediately whenever it comes into our mind. We also want to republish some of these composed tools and extracted data sets as useful web resources for further reuse by other people.

This paper has pointed out the importance of extending the current Web so that each people may freely and immediately reedit some of the knowledge resources that others have already published, compose a new one for his or her own reuse, and publish it again into the Web for further reuse by others. To achieve this goal, our group has been trying to apply meme media technologies to the Web to make it work as a meme pool. Meme media and meme pool architectures have been studied since 1993 based on our preceding research on the synthetic media architecture since 1987. Their R&D started even before the proposal of the Web, and focused on media architectures that enable users to reedit and to redistribute knowledge resources.

The application of meme media technologies to the Web initially assumed both the installation of a 2D meme media kernel system at each client, and the additional use of a world-wide repository of composite pads. Our preceding R&D efforts proposed the C3W framework as *ad hoc* knowledge federation framework, and Piazza as a meme pool architecture. C3W enabled us to construct the Piazza system from the Wiki system instantaneously through direct manipulation without any coding.

It is with the 2008 version of IntelligentPad that we have successfully removed all such assumptions of having a 2D meme media kernel system at each client, and of using a world-wide repository of composite pads. The 2008 version of IntelligentPad extends the Web to the memetic Web. Users can freely publish both web resources and composite pads into the memetic Web, extract resource fragments from some resources in the memetic Web, make them work together to compose a composite pad, and publish it into the memetic Web for its further reuse by others.

Meme media technologies enable us to define *ad hoc* knowledge federation instantaneously among knowledge resources including both web resources and composite pads. While mashup technologies focus on composing a new web resource from available web resources, our meme media architecture focuses on the repetitive reediting, reuse, and republication of knowledge resources by end users in instantaneous and *ad hoc* ways, and aims to extend the Web to work as a meme pool for accelerating the co-evolution of both knowledge resources and their usage.

References

1. Fox, E.A.: Digital Libraries Initiative (DLI) Projects 1994-1999. Bulletin of the Ametican Society for Information Science, 7–11 (October/November 1999)
2. Information Grand Voyage Project (2007), http://www.igvpj.jp/e/index.html
3. Heimbigner, D., McLeod, D.: A federated architecture for information management. ACM Trans. Inf. Syst. 3(3), 253–278 (1985)
4. Edwards, W.K., Joy, B., Murphy, B.: Core JINI. Prentice Hall Professional Technical Reference, Englewood Cliffs (2000)
5. Erbach, G.: Data-centric view in e-science information systems. Data Science Journal 5, 219–222 (2006)
6. Tanaka, Y., Fujima, J., Ohigashi, M.: Meme media for the knowledge federation over the web and pervasive computing environments. In: Maher, M.J. (ed.) ASIAN 2004. LNCS, vol. 3321, pp. 33–47. Springer, Heidelberg (2004)
7. Tanaka, Y.: Meme Media and Meme Market Architectures: Knowledge Media for Editing, Distributing, and Managing Intellectual Resources. IEEE Press & Wiley-Interscience, NJ (2003)
8. Tanaka, Y., Imataki, T.: Intelligent Pad: A hypermedia system allowing functional compositions of active media objects through direct manipulations. In: Proceedings of the IFIP 11th World Computer Congress, pp. 541–546 (1989)
9. Okada, Y., Tanaka, Y.: Intelligentbox: a constructive visual software development system for interactive 3d graphic applications. In: Proc. of the Computer Animation 1995 Conference, pp. 114–125. IEEE Computer Society, Los Alamitos (1995)
10. Tanaka, Y.: Meme media and a world-wide meme pool. In: MULTIMEDIA 1996: Proceedings of the fourth ACM international conference on Multimedia, pp. 175–186. ACM, New York (1996)
11. Ohigashi, M., Tanaka, Y.: From visualization to interactive animation of database records. In: Arikawa, S., Furukawa, K. (eds.) DS 1999. LNCS (LNAI), vol. 1721, pp. 349–350. Springer, Heidelberg (1999)
12. Tanaka, Y., Sugibuchi, T.: Component-based framework for virtual information materialization. In: Jantke, K.P., Shinohara, A. (eds.) DS 2001. LNCS (LNAI), vol. 2226, pp. 458–463. Springer, Heidelberg (2001)
13. Sugibuchi, T., Tanaka, Y.: Integrated visualization framework for relational databases and web resources. In: Grieser, G., Tanaka, Y. (eds.) Dagstuhl Seminar 2004. LNCS (LNAI), vol. 3359, pp. 159–174. Springer, Heidelberg (2005)
14. Baird, D.: Thing Knowledge: A philosophy of scientific instruments. University of California Press (2004)
15. Tanaka, Y.: Intelligentpad as meme media and its application to multimedia databases. Information & Software Technology 38(3), 201–211 (1996)
16. Dawkins, R.: The Selfish Gene. Oxford Univ. Press, Oxford (1976)
17. Ito, K., Tanaka, Y.: A visual environment for dynamic web application composition. In: HYPERTEXT 2003: Proceedings of the fourteenth ACM conference on Hypertext and hypermedia, pp. 184–193. ACM, New York (2003)

18. Tanaka, Y., Ito, K., Kurosaki, D.: Meme media architectures for re-editing and redistributing intellectual assets over the web. International Journal of Human Computer Studies 60(4), 489–526 (2004)
19. Fujima, J., Lunzer, A., Hornbæk, K., Tanaka, Y.: Clip, connect, clone: combining application elements to build custom interfaces for information access. In: UIST 2004: Proceedings of the 17th annual ACM symposium on User interface software and technology, pp. 175–184. ACM, New York (2004)
20. Tanaka, Y., Ito, K., Fujima, J.: Meme media for clipping and combining web resources. World Wide Web 9(2), 175–184 (2004)
21. Tanaka, Y.: Knowledge media and meme media architectures from the viewpoint of the phenotype-genotype mapping. In: SIGDOC 2006: Proceedings of the 24th annual ACM international conference on Design of communication, pp. 3–10. ACM, New York (2006)
22. Tanaka, Y., Fujima, J., Sugibuchi, T.: Meme media and meme pools for re-editing and redistributing intellectual assets. In: Reich, S., Tzagarakis, M.M., De Bra, P.M.E. (eds.) AH-WS 2001, SC 2001, and OHS 2001. LNCS, vol. 2266, pp. 28–46. Springer, Heidelberg (2002)
23. My Maps – Google Maps User Guide,
    `http://maps.google.com/support/bin/answer.py?hl=en&answer=68480`
24. MSR MapCruncher for Virtual Earth,
    `http://research.microsoft.com/MapCruncher/`
25. Yahoo! Pipes, `http://pipes.yahoo.com/pipes/`
26. Dapper, `http://www.dapper.net/`
27. Microsoft Popfly, `http://www.popfly.com/`
28. Google Mashup Editor, `http://code.google.com/gme/`
29. Plagger: the UNIX pipe programming for Web 2.0,
    `http://plagger.org/trac`
30. DataMashups: Online Mashups for Enterprises,
    `http://datamashups.com/entmashups.html`
31. OpenKapow, `http://openkapow.com/`
32. Intel Mash Maker, `http://mashmaker.intel.com/web/`

Navidgator - Similarity Based Browsing for Image and Video Databases

Damian Borth, Christian Schulze, Adrian Ulges, and Thomas M. Breuel

University of Kaiserslautern,
German Research Center for Artificial Intelligence (DFKI),
67663 Kaiserslautern, Germany
{d_borth,a_ulges,tmb}@informatik.uni-kl.de,
christian.schulze@dfki.de
http://www.iupr.org

Abstract. A main problem with the handling of multimedia databases is the navigation through and the search within the content of a database. The problem arises from the difference between the possible textual description (annotation) of the database content and its visual appearance. Overcoming the so called - semantic gap - has been in the focus of research for some time. This paper presents a new system for similarity-based browsing of multimedia databases. The system aims at decreasing the semantic gap by using a tree structure, built up on balanced hierarchical clustering. Using this approach, operators are provided with an intuitive and easy-to-use browsing tool. An important objective of this paper is not only on the description of the database organization and retrieval structure, but also how the illustrated techniques might be integrated into a single system.

Our main contribution is the direct use of a balanced tree structure for navigating through the database of keyframes, paired with an easy-to-use interface, offering a coarse to fine similarity-based view of the grouped database content.

Keywords: hierarchical clustering, image databases, video databases, browsing, multimedia retrieval.

1   Introduction

Nowadays, content-based image and video retrieval (CBIR/CBVR) are getting more and more into focus with the rapidly growing amount of image and video information being stored and published. Online portals providing image and video content like *flickr.com, youtube.com, revver.com, etc.* offer data in large amounts which creates the need for an efficient way of searching through the content. But the need for an efficient search and browsing method is not bound to those online portals. Archives of TV stations storing increasingly more digital content also need appropriate tools to find the material they are looking for. Additionally, with the already widely spread availability of digital acquisition

A. Dengel et al. (Eds.): KI 2008, LNAI 5243, pp. 22–29, 2008.

devices (still image and video cameras) it is easy for everybody to acquire large amounts of digital data in short amounts of time.

Currently the search for specific content in such collections is mostly done through a *query-by-text* approach, exploiting manual annotation of the stored data. This approach suffers from a few drawbacks which arise from the nature of this method. *First*: manual annotation is a very time consuming process which *second*: might lead to a rather subjective result, depending on the person doing the annotation. *Third*: the result of the query depends highly on the quality of the annotations. Visual content that has not been transcribed into the meta-data can therefor not be retrieved afterwards. *Fourth*: The result of a query can be manipulated by the type and number of tags associated with the visual data. This might mostly apply to online portals where currently the owners of uploaded content provide the meta-data. With such tag manipulation it is possible to assure that a specific content appears in most of the query results, which then in turn reduces the quality of the search result.

To prevent such drawbacks it is necessary to use different approaches for search in large multimedia databases. One of these approaches is the description of the database content in a specific feature space. Unfortunately, a major problem of this approach is the formulation of a query. This is bypassed by using the *query-by-example* approach [1], where a sample image or video is representing a query forming a visual concept for search.

Our approach is to build up a similarity based structure of the multimedia database (here we focus on video as content) to overcome the need for an appropriate search example. Doing this, the user is enabled to browse through the database by picking an entity as starting point which represents the query most. During browsing the user is able to zoom in and out of the database content with variable step size representing the similarity of appearance. Entities showing up during the navigation, giving a better representation of the users query, provide the opportunity to narrow down the selection of possible matches.

1.1 Related Work

With the development of multimedia retrieval/browsing systems quickly the problem of formulating a proper query arose. This lead to the definition of the so called *semantic gap* - 'a lack of coincidence between the information that one can extract from the visual data and the interpretation that the same data have for a user in a given situation.'[2]. Different approaches have been examined to bridge this semantic gap.[3,4,5,6] and lead to multiple approaches towards user interface design i.e. the RotorBrowser [7] or VideoSOM [8].

When we talk about similarity based browsing, we focus on the visual content of the video. This means we want to allow the user to find temporally independent shots of video with similar content, which is different to cluster-temporal browsing where causal relations in storytelling are used for browsing [9]. We purely utilize content based similarity between different videos in the video database for browsing the database. A similar approach can be found in [10,11], where the concept of a similarity pyramid is introduced for browsing large image databases. The idea of similarity pyramids was also applied to video databases in a system called ViBE [12].

Our main contribution is the direct use of a balanced tree structure for navigating through the database of keyframes, paired with an easy to use interface, offering a coarse to fine view on the grouped database content based on similarity.

2   Balanced Hierarchical Clustering

To find structures of strong visual similarity we use unsupervised learning methods like clustering. In particular we are not only interested in the pure cluster partitioning, even more we want to capture the relationship between different clustering levels i.e. the clustering structure in the video database including their cluster and their subcluster partitions. To achieve this we use a standard agglomerative hierarchical clustering [13], which runs through a series of partitions starting from singelton clusters containing a single image and terminating with a final cluster, containing all images of the database. This structure, usually represented as a dendogram, is postprocessed into a binary tree and used by our system for continuous browsing between different coarseness levels of similarity.

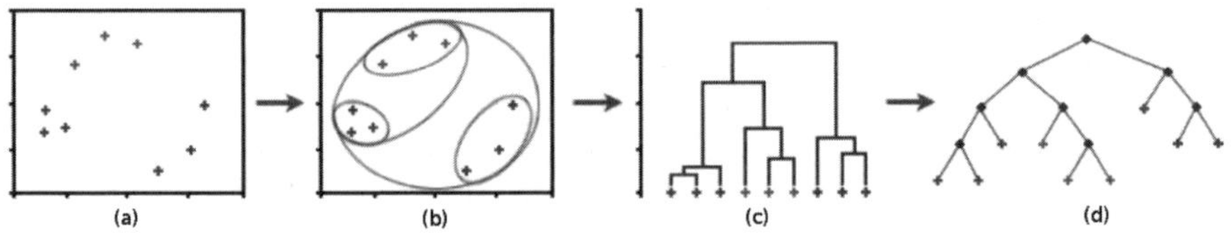

Fig. 1. A feature space representation of keyframes (a) leads to a hierarchical clustering (b). This clustering can be viewed as a dendogram with similarity measurement at y-axis (c) and will be postprocessed to a binary tree for our GUI (d).

2.1   Feature Extraction

Let the videos in the database be denoted as $X_1, ..., X_n$. Every video is represented by a set of keyframes $\{x_{i1}, ..., x_{im}\}$, resulting in a total keyframe set $\{x_1, ..., x_k\}$ for the entire video database. For every keyframe $x_i \in \{x_1, ..., x_k\}$ a feature vector $z_i \in \mathbb{R}^D$ is extracted.

A common way to extract keyframes for each video is to segment the data into shots and analyze the shots individually for representative keyframes. We use a divide-and-conquer approach that delivers multiple keyframes per shot in respect to its visual complexity. To achieve this, we compute MPEG-7 color layout descriptors [14] for each frame of the shot and fit a Gaussian mixture model to the feature set using k-means [15] in combination with the Bayesian Information Criterion (BIC) for determining the number of clusters. [16]. The entire procedure is illustrated in more detail in [17].

The computation of the feature vector z_i for every keyframe $x_i \in \{x_1...x_k\}$ is based on our Baseline System definition [18]. There we are using Color and

Texture features with a equally weighted early fusion (i.e concatenated). In particular, we use color histograms which are quantized to $8 \times 8 \times 8$ bins and also Tamura texture features [19] to form the feature vector.

2.2 Agglomerative Hierarchical Clustering

The first step in using conventional agglomerative hierarchical clustering algorithm is the computation of a distance matrix $D = [d(z_i, z_j)]$ where $i, j = 1...k$, with k the number of keyframes. The distance matrix D represents similarities between all pairs of keyframes and is used for successive fusion of clusters [13]. As distance function $d(z_i, z_j)$ serves the Euclidian distance.

The agglomerative hierarchical clustering creates a cluster $c_i = \{x_i\}$ for each keyframe $x_i \in \{x_1...x_k\}$, resulting in $C_0 = \{c_1...c_k\}$ disjoint singelton clusters, where $C_0 \subset C$. In this first step the distance matrix D is equal to the distances between the feature vectors of the keyframes. The cluster c_i c_j with the smallest distance $d(z_i, z_j)$ are fused together to form a new cluster $c_k = \{c_i, c_j\}$. After creating a new cluster, the distance matrix D has to be updated to represent the distance between the new cluster c_k to each other cluster in $C_0 \backslash \{c_k\}$. This recalculation of distances is usually done with one of the known linkage methods [11]. Considering the used linkage method the entire clustering structure will be more or less *dilating* i.e. individual elements not yet grouped are more likely to form new groups instead of being fused to existing groups. According to [11], the *complete linkage* method tends to be dilating and therefore resulting in more balanced dendograms compared to the *single linkage* method, which chains clusters together and therefore results in deep unbalanced dendograms. However, our goal is to use the resulting clustering structure for continuous similarity browsing of video databases. In order to achieve this, two points are important: **First**, the clustering must produce clusters with visual similar content and **Second**, the dendrogram produced by the clustering has to be as balanced as possible.According to our experience, *average linkage* produces the visually most similar clusters. Unfortunately, the clustering structure is not as balanced as desired for usable browsing. Therefore a modification of the average linkage method was needed to achieve the desired properties.

Viewing the resulting dendogram of the clustering as a tree structure, this structure will hierarchically organize keyframes into groups of visual similar content, thereby retaining the relationship between different coarseness levels of similarity and tree depth. Let S denote the set of all tree nodes. Each node of the tree $s \in S$ is associated with a set of keyframes $c_s \subset C \wedge c_s \notin C_0$ representing the cluster of keyframes. The number of elements in the cluster c_s is denoted by $|c_s|$. The children of a node $s \in S$ denoted by $ch(s) \subset S$ will partition the keyframes of the parent node so that

$$c_s = \bigcup_{r \in ch(s)} c_r$$

The leaf nodes of the tree correspond to the extracted keyframes and are indexed by the set S_0. Each leaf node contains a single keyframe, so for all $s_i \in S_0$ we have $c_i = \{x_i\}$ with $|c_i| = 1$ implying that $|S_0| = |C_0|$. This notation is derived from the notations of [10,11,12].

Furthermore we define $D = [d(c_i, c_j)]$ as the updated distance matrix of distances between the pairwise different clusters c_i and c_j. The linkage method for calculating distances between clusters with $|c_i| > 1 \wedge |c_j| > 1$ is in our case the *average linkage* method enhanced by a weighted penalty, which depends on the amount of elements in both clusters

$$d(c_i, c_j) = \frac{1}{|c_i| * |c_j|} * \sum_{z_i \in c_i} \sum_{z_j \in c_j} d(z_i, z_j) + \alpha * (|c_i| + |c_j|)$$

We are naming this method: *balanced linkage* due to its tendency to form balanced trees with clusters of consistent visual properties. The weight can be set to $0 \leq \alpha \leq 1$, which either results in having no effect to the average linkage or totally forcing the algorithm to produce balanced trees without any visual similarity. The chosen α value was empirically evaluated and set to $\alpha = 0.01$.

2.3   Binary Tree Construction

The usually chosen representation of hierarchical clustering is a dendogram, which illustrates the fusions made at each successive stage of analysis. In a dendogram, the elements being clustered are placed usually at the bottom of the diagram and show fusions of clusters through connecting lines. Another representation for hierarchical clustering is using sets showing the elements being clustered in their feature space. Such sets represent one particular step in the clustering process and may contain subsets illustrating previous clustering steps.

Because we want to use the clustering structure for navigation, we postprocess the dendogram into a binary tree. The binary tree structure enables us to efficiently follow a keyframe from the root of the tree, which contains the entire database, to the leaf of the tree, which only contains the selected keyframe. With every *zoom in* step, the system is presenting a more similar subtree considering the selected keyframe and leaving out the frames, which are less similar to the selected keyframe. The *zoom out* step, lets the system present a tree containing more dissimilar keyframes, which might be useful for refinement of the query.

For the notation of the constructed binary tree, we refer to section 2.2. Let $T = S$ be the binary tree, then the nodes $n_i \in T$ are the fusion points where clusters c_i, c_j are being fused together to form a new cluster $c_k = \{c_i, c_j\}$.

An interesting outcome of the binary tree postprocessing is the creation of so called *content stripes*. These structures represent clusters within the binary tree, in such a way that the keyframes of subclusters are ordered in stripes according to their similarity and therefore providing a more intuitive way of visualizing clusters at a particular level (Fig. 2). Content stripes replace the need to additionally compute spatial arrangemets for keyframes within a cluster like done with similarity pyramids [10,11]. Therefore our method is not only able to construct a hierarchical database structure but also to build up a

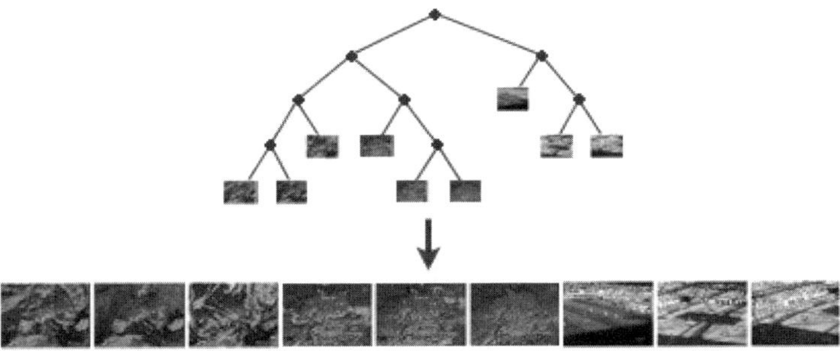

Fig. 2. A binary tree representation of a sample clustering, transformed to a content stripe displaying similarity clusters in an one-dimensional order

similarity based order within a cluster in one single step instead of separating these tasks.

3   Graphical User Interface

Retrieval systems based on keyframes and best-match similarity tend to present a localized view of the database to the user, rather then providing an overview of the entire database. For uses who do not clearly know what exactly they are searching for, it would be more efficient to let them browse through the database and allow them to dynamically redefine their search query.

In this section we would like to present the Navidgator graphical user interface[1], which allows a user to easily and efficiently browse a video database in respect to his selected query. In our system a browsing process is initialized with starting at the root of the hierarchical clustered binary tree. First, the user has to select his first keyframe out of a randomly sampled set from the entire database to formulate his query [1]. This keyframe will represent the visual concept, which will guide the user while browsing. The user is also able to dynamically refine his visual concept in every point during browsing by selecting another keyframe.

Browsing itself is performed by the given zooming tools. The user can either *zoom-in* or *zoom-out* in the database. A *zoom-in* action will narrow down the available keyframe according to his visual concept and a *zoom-out* action will display a coarser level of the database to the user. For better usability the interface provides a *multi-level-zoom-in* action and a *multi-level-zoom-out* action. Furthermore the user can perform a *max-zoom-in* action, which brings him straight to the most similar keyframes in the database or a *max-zoom-out* action, which brings him back to the root of the binary tree i.e. the top of the database. The depth of the database and the users current position are visualized by a vertical bar next to the zooming tools enabling an intuitive orientation. Additionally the user can utilize a click history to jump back to particular points of his browsing process. The Navidgator browsing interface is displayed in Fig. 3.

---

[1] http://demo.iupr.org:8180/navidgator-tv (10.000 keyframes database).

Fig. 3. The Navidgator browsing interface. The selected visual concept is displayed at the center. The lower area displays the cluster preview box, where the visual concept might be refined. Next to the selected keyframe the navigation tools are arranged.

4 Conclusion

In this paper we have presented a new system for similarity-based browsing of multimedia databases. By using a balanced tree based on hierarchical clustering of the database content it is possible to supply users with an intuitive and easy-to-use browsing tool. We were able to improve search results by providing a set of navigation tools which support the decision tree like structure of the clustering. Because of our concept of an offline clustering and online retrieval we are able to efficiently perform a search on the entire database. The system offers coarse and detailed views on the database content with the opportunity to change the focus of search at any time. This enables the user to start navigation with a fuzzy visual concept and improve relevance incrementally while browsing.

Our future work will focus on advanced tree building and on dealing with growing databases, which will basically cover merging of new data into the binary tree. Additionally, to bridge the semantic gap and include high-level semantics we want to enhance the browser with an automatic tagging system like in [18].

References

1. Flickner, M., Sawhney, H., Niblack, W., Ashley, J., Huang, Q., Dom, B., Gorkani, M., Hafner, J., Lee, D., Petkovic, D., Steele, D., Yanker, P.: Query by image and video content: the qbic system. Computer 28(9), 23–32 (1995)
2. Smeulders, A., Worring, M., Santini, S., Gupta, A., Jain, R.: Contentbased image retrieval at the end of the early years. IEEE transactions on pattern analysis and machine intelligence 22(12), 1349–1379 (2000)

3. Broecker, L., Bogen, M., Cremers, A.B.: Bridging the semantic gap in content-based image retrieval systems. In: Internet Multimedia Management Systems II. Volume 4519 of the Society of Photo-Optical Instrumentation Engineers (SPIE) Conference, 54–62 (2001)
4. Zhao, R., Grosky, W.: Narrowing the semantic gap-improved text-based web document retrieval using visual features. Multimedia, IEEE Transactions on 4(2), 189–200 (2002)
5. Zhao, R., Grosky, W.: Bridging the Semantic Gap in Image Retrieval. Distributed Multimedia Databases: Techniques and Applications (2003)
6. Dorai, C., Venkatesh, S.: Bridging the semantic gap with computational media aesthetics. Multimedia, IEEE 10(2), 15–17 (2003)
7. de Rooij, O., Snoek, C.G.M., Worring, M.: Mediamill: semantic video search using the rotorbrowser. In: [7], p. 649 (2007)
8. Barecke, T., Kijak, E., Nurnberger, A., Detyniecki, M.: Videosom: A som-based interface for video browsing. Image And Video Retrieval, Proceedings 4071, 506–509 (2006)
9. Rautiainen, M., Ojala, T., Seppanen, T.: Cluster-temporal browsing of large news video databases. IEEE Int. Conference on Multimedia and Expo. 2, 751–754 (2004)
10. Chen, J., Bouman, C., Dalton, J.: Similarity pyramids for browsing and organization of large image databases. SPIE Human Vision and Electronic Imaging III 3299 (1998)
11. Chen, J.Y., Bouman, C., Dalton, J.: Hierarchical browsing and search of large image databases. Image Processing, IEEE Transactions on 9(3), 442–455 (2000)
12. Taskiran, C., Chen, J., Albiol, A., Torres, L., Bouman, C., Delp, E.: Vibe: A compressed video database structured for active browsing and search. IEEE Transactions on Multimedia 6(1), 103–118 (2004)
13. Johnson, S.C.: Hierarchical clustering schemes. Psychometrika 32(3), 241–241 (1967)
14. Manjunath, B., Ohm, J., Vasudevan, V., Yamada, A.: Color and texture descriptors. IEEE Trans. on Circuits Syst. for Video Techn. 11(6) (2001)
15. McQueen, J.: Some methods for classification and analysis of multivariate observations. In: Proceedings of the Fifth Berkeley Symposium on Mathematical Statistics and Probability, pp. 281–297 (1967)
16. Schwarz, G.: Estimating the dimension of a model. Annals of Statistics 6(2), 461–464 (1978)
17. Borth, D., Ulges, A., Schulze, C., Breuel, T.M.: Keyframe extraction for video taggging and summarization. In: Informatiktage 2008, pp. 45–48 (2008)
18. Ulges, A., Schulze, C., Keysers, D., Breuel, T.M.: Content-based video tagging for online video portals. In: MUSCLE/Image-CLEF Workshop (2007)
19. Tamura, H., Mori, S., Yamawaki, T.: Textual features corresponding to visual perception. IEEE Transactions on Systems, Man, and Cybernetics SMC-8(6) (1978)

Automating Interactive Protocol Verification

Lassaad Cheikhrouhou[1], Andreas Nonnengart[1], Werner Stephan[1],
Frank Koob[2], and Georg Rock[3]

[1] German Research Center for Artificial Intelligence, DFKI GmbH
[2] Federal Office for Information Security, BSI
[3] PROSTEP IMP GmbH

Abstract. Showing the *absence* of security hazards in cryptographic protocols is of major interest in the area of protocol security analysis. Standard model checking techniques - despite their advantages of being both fast and automatic - serve as mere debuggers that allow the user at best to detect security risks if they exist at all. In general they are not able to guarantee that *all* such potential hazards can be found, though. A full verification usually involves induction and therefore can hardly be fully automatic. Therefore the definition and application of suitable heuristics has turned out to become a central necessity. This paper describes how we attack this problem with the help of the Verification Support Environment (VSE) and how we nevertheless arrive at a high degree of automation.

1 Introduction

Protocols that (try to) provide certain security properties in open network environments by using cryptographic primitives like encryption, signing, and hashing, play a crucial role in many emerging application scenarios. Machine readable travel documents are a typical example of a technology whose common acceptance heavily depends on how far security properties like confidentiality and authenticity can really be *guaranteed* by their designers [1]. However, if carried out on an informal basis, the analysis of cryptographic protocols has turned out to be fairly error prone. Justifiably so, the *formal* specification and analysis of cryptographic protocols has become one of the fastest growing research areas, although with a hardly manageable variety of different approaches.

Formal protocol models usually incorporate the exchanged messages between protocol participants. These messages are sent via an open network which is accessible by an attacker who can intercept, and even change or forge (some of) these messages. The capabilities of the attacker as well as of honest protocol participants are restricted by the assumption that encryption and decryption is impossible, unless the correct cryptographic key is at hand. In this setting we speak of the so-called Dolev-Yao model as introduced (rather implicitly) in [2].

Formal analysis approaches to protocol analysis can roughly be divided into two main categories: *model checking* and *(interactive) verification*. Whereas model checking has the big advantage that it is both fast and automatic, it

A. Dengel et al. (Eds.): KI 2008, LNAI 5243, pp. 30–37, 2008.

in general relies on fixed scenarios with a given number of protocol participants and a given number of protocol runs. They thus serve well as a *systematic* way of debugging security protocols. The general validity of security properties, however, can only be guaranteed under certain circumstances.

(Interactive) verification methods, on the other hand, not only try to find existing security hazards; they are also able to show the *non-existence* of such risks. A typical example for such an approach can be found in Paulson's inductive verification approach [3]. There neither the number of participants nor the length of protocol traces is restricted in any way. Obviously, reasoning about such (inductively defined) *event traces* and the knowledge gained by an attacker (e. g. while eavesdropping on exchanged messages) heavily relies on *induction* proofs. This seriously complicates matters and one can hardly expect to get a fully automatic verification system to solve such problems.

In this paper we focus on the techniques we developed to adopt Paulson's approach for the Verification Support Environment (VSE). VSE is a kind of case tool for formal software development that closely combines a front end for specification (including refinement) and the management of structured developments with an interactive theorem prover [4]. We emphasize on our attempt to lower the burden of interactive proof generation to an extent that makes the VSE framework for protocol verification applicable within the limited time frames of commercial developments. In order to do this, we exploited the specific struture of the inductive proofs in our domain and implemented several proof heuristics to carry out most of the proof tasks automatically. This allowed us to reach an automation degree of about 90%.

2   BAC - A Cryptographic Protocol Example

One of the commercial protocols that were verified in VSE was the Extended Access Control (EAC) protocol. It includes (in 12 steps) three related phases in the inspection procedure of an electronic passport (ePass A) by a terminal (B): Basic Access Control (BAC), Chip Authentication and Terminal Authentication. In this paper we emphasize on the steps of the first phase to examplify our work.

2.1   The Protocol and Its Properties

The inspection procedure starts with the BAC protocol to protect the data on the ePass chip from unauthorized access and to guarantee (for the terminal) that this chip corresponds to the machine readable zone (MRZ) of the inspected ePass. The protocol comprises the following steps:

1. $B \longleftarrow A :$ $K_{BAC}(A)$
2. $B \longrightarrow A :$ B, Get_Random_Number
3. $A \longrightarrow B :$ r_A
4. $B \longrightarrow A :$ $\{r_B, r_A, K_B\}_{K_{BAC}(A)}$
5. $A \longrightarrow B :$ $\{r_A, r_B, K_A\}_{K_{BAC}(A)}$

32   L. Cheikhrouhou et al.

This protocol realizes a mutual authentication between A and B, if both participants have access to the basic access key $K_{BAC}(A)$. The key is stored on the chip of A and can be computed by a terminal B from the MRZ. Such a BAC run can only be successful if B has physical (optical) access on the ePass for it has to read the MRZ in order to be able to compute the correct key. This is represented in the first protocol step by the right-to-left arrow. It indicates that B is both the active part of the communication and the receiver of the message. After the second step which signals that the terminal determined the key $K_{BAC}(A)$, the participants A and B authenticate each other in two interleaved challenge-responses using the random numbers (nonces) r_A and r_B. Here, $\{M\}_K$ denotes the cipher of a message M by a key K. The ciphers in the fourth and the fifth message are obtained by symmetric encryption. Encrypting as well as decrypting these messages obviously requires to possess the key $K_{BAC}(A)$.

In addition to the used nonces, the fresh key materials (session keys) K_B and K_A are exchanged. These will be used for the computation of the session keys that are needed for secure messaging in the subsequent phase of the inspection procedure.

The main properties of the BAC protocol are:

- BAC1: The ePass A authenticates the terminal B and agrees with B on the session key K_B.[1]
- BAC2: The terminal B authenticates the ePass A and agrees with A on the session keys K_B and K_A.
- BAC3: The session keys K_A and K_B are confidential.

2.2   Specification of Protocol Runs and Properties

The above properties are defined and verified for all possible traces of the BAC protocol. The set of these traces is associated with a corresponding predicate (here BAC) which holds for the empty trace ϵ and is inductively defined for every extension of a BAC-trace tr with an admissible event ev:

$$BAC(ev\#tr) \Leftrightarrow (BAC(tr) \wedge (Reads1(ev,tr) \vee Says2(ev,tr) \vee Says3(ev,tr)$$
$$\vee Says4(ev,tr) \vee Says5(ev,tr) \vee Gets\_event(ev,tr) \vee Fake\_event(ev,tr)))$$

This inductive definition covers all the alternatives to extend an arbitrary BAC-trace by a new event: five protocol steps ($Reads1, Says2, \ldots, Says5$), message reception ($Gets\_event$), and sending a faked message by the attacker ($Fake\_event$). The definitions of these predicates determine the structure of the corresponding events and express the corresponding conditions depending on the extended trace tr and the event parameters.

The definitions of the gets-case and the fake-case are *generic*, since they do not depend on the considered protocol. For instance, a send-event $Says(spy, ag, m)$

---

[1] Note that these first five steps of the protocol do not allow A to deduce that B receives the session key K_A, since this key is sent to B in the last protocol step. This is proven in the complete version in subsequent steps where B confirms the reception of the session key K_A.

from the attacker (spy) to any agent ag may occur in the fake-case, when the message m can be constructed from the attacker's extended knowledge. This condition is expressed by $isSynth(m, analz(spies(tr)))$, where $spies(tr)$ contains the messages that are collected by the attacker and $analz$ extends this set by message decomposition and decryption.

The definitions of the remaining cases are protocol specific, since they encode the conditions of the corresponding protocol steps according to the protocol rules. These conditions express the occurrence of previous events, the existence of certain (initial) information and the freshness of newly generated information. For instance, the definition of the second step requires that the extended trace tr contains an event $Reads(B, A, K_{BAC}(A))$ which represents the access by B of the initial information $K_{BAC}(A)$ of A via a secure channel.

The protocol properties are formalized for arbitrary traces tr for which the predicate BAC holds. For instance, confidentiality properties are expressed with the help of the attacker's extended knowledge. Basically, a message m is confidential if it is not contained in and can not be constructed from this knowledge. Consider for instance the formalization of the confidentiality property BAC3:

$$(BAC(tr) \land \{r_1, r_2, aK\}_{K_{BAC}(A)} \in parts(spies(tr)) \land$$
$$\neg bad(A) \land \forall ag : (Reads(ag, A, K_{BAC}(A)) \in tr \Rightarrow \neg bad(ag)))$$
$$\Rightarrow aK \notin analz(spies(tr))$$

This secrecy property is defined for every session key aK which occurs in a message of the form $\{r_1, r_2, aK\}_{K_{BAC}(A)}$. It holds for the session key K_B of step 4, as well as for the session key K_A of step 5, assuming that A is not compromised, i.e. $bad(A)^2$ does not hold, and its basic access key is not accessible to the attacker. In contrast to $analz$, the function $parts$ includes the encrypted sub-messages even if the keys needed for decryption can not be obtained from the given message set. It is used, especially in authenticity properties, to express the occurrence of sub-messages including the encrypted ones.

2.3   Specification and Verification Support in VSE

The formal specification of cryptographic protocols in VSE is supported by

- A library of VSE theories that define the basic data types for messages, events etc. and the analysis operators, e.g., $analz$, which are needed to formalize protocols (from a given class) in the VSE specification language (VSE-SL),
- And a more user friendly specification language called VSE-CAPSL (as an extension of CAPSL [5]) with an automatic translation to VSE-SL.

The translation of VSE-CAPSL specifications leads to definitions of the protocol traces in the above style and formalizes some of the stated protocol properties. Additionally, certain lemmata like possibility, regularity, forwarding and unicity

---

[2] The attacker is bad ($bad(spy)$) and possesses the knowledge of any other bad agent.

lemmata, that give rise to the generic proof structure as discussed in [3] are generated.

The main protocol properties and the lemmata generated by the system serve as proof obligations for the VSE prover. For each proof obligation a complete proof has to be constructed in the (VSE) sequent calculus. Besides the (basic and derived) inference rules, including the application of lemmata and axioms, powerful simplification routines can be selected by the user for this purpose. The number of the (protocol specific) proof obligations and the size of their proofs definitely shows that the inductive approach will only be successful in real world environments if the burden of user interaction is lowered drastically. Therefore we have concentrated on a heuristic proof search that applies higher-level knowledge available from a systematic analysis of the given class of proof obligations. This approach will be described in a detailed way in the next section.

3 Proof Automation by Heuristics

Heuristics are intended as a means to incrementally provide support for the user of an interactive system. Our approach is bottom up in the sense that starting with routine tasks whose automatization just saves some clicks we proceed by building heuristics that cover more complex proof decisions using the lower level ones as primitives. In particular the heuristics do not constitute a fixed, closed proof procedure: They can freely be called on certain subtasks to obtain goals whose further reduction requires a decision from the user before the next heuristics can continue to generate partial proofs for the new subgoals. Since full automation can hardly be reached, this kind of *mixed initiative* is our ultimate goal, although some of our most recent high-level heuristics were even able to prove certain lemmata completely on their own.

3.1 Protocol Properties and Their Proofs

The main properties of cryptographic protocols that we are interested in are *confidentiality* and *authenticity*. *Sensitive* data, like session keys or nonces that are often used for the generation of new data items that were not used in the current protocol run, like for example new session keys, have to be protected against a malicious attacker.

Authentication properties are often formulated from the perspective of a particular participant. In the protocol presented in section 2.1 the authentication property BAC2 is formulated from the perspective of the terminal. Authentication proofs as the one for the BAC2 property are often based on *authenticity* properties of certain messages. The authenticity of such a message is used to identify the sender of that message.

In addition to the top-level properties we have to prove so called *structuring lemmata*. These are generated automatically by the system and used by the heuristics. Their proof uses the same basic scheme that is used for confidentiality and authenticity.

This general scheme for structural induction on the traces of a particular protocol is more or less straightforward. The heuristics discussed in the following implement an application specific refinement of this scheme. This refinement takes into account both the type of the proof goals and the available lemmata and axioms.

The organization of the independent building blocks in the general scheme described below was left to the user in the beginning of our work. Meanwhile more complex tasks can be achieved automatically by high-level heuristics that implement more global proof plans.

3.2 The Top-Level Proof Scheme

All inductive proofs of protocol properties are structured into the following (proof-) tasks. For each task there is a collection of heuristics that are potentially applicable in these situations.

1. Set up a proof by structural induction on traces.
2. Handle the base case.
3. Handle the step cases:
   (a) Reduce certain formulas to *negative* assumptions in the induction hypothesis.
   (b) Add information about individual protocol steps.
   (c) Reduce the remaining differences and apply the induction hypothesis.

An inductive proof about traces of a given protocol is initialized by a heuristic (`traceInd`) which selects the induction variable representing the protocol trace and reduces the proof goals representing the base case and the step case to a simplified normal form.

Base Case: The proof goals of the base case will contain assumptions that contradict properties of the empty trace ϵ. These assumptions are searched for by a heuristic that applies appropriately instantiated lemmata (axioms) to close the proof goals by contradiction. For instance, an assumption of the form $ev \in \epsilon$, stating that the event ev is part of the empty trace, obviously contradicts an axiom about traces. This is detected by the corresponding heuristic (`nullEvent`). Similar, but more complex heuristics for this task are for example based on the fact that nonces and session keys do not exist before starting a protocol run.

Step Case: Like in all other mechanized induction systems in the step case(s) we try to reduce the given goal(s) to a situation where the inductive hypothesis can be applied.

For this purpose the VSE strategy provides heuristics from three groups.

Negative assumptions are used to impose certain restrictions on the set of traces under consideration. For example, we might be interested only in traces without certain events. In the BAC protocol most of the properties do not include the optical reading of the considered basic access key by a compromised agent. This is formulated by the negative assumption $\forall ag : (Reads(ag, A, K_{BAC}(A)) \in$

$tr \Rightarrow \neg bad(ag))$. The difference between this formula and the corresponding assumption in the step case, i.e. in $\forall ag : ((Reads(ag, A, K_{BAC}(A)) \in (ev\#tr)) \Rightarrow \neg bad(ag))$, can be eliminated without knowing any details about the event ev added to the trace (induction variable) tr. This kind of difference reduction is performed by a heuristic (`eventNotInTrace`) which treats negative assumptions about the membership of events in traces.

In the next step(s) we exploit the assumptions under which a certain event ev can be added to a trace tr. This is achieved by a heuristic called `protocolSteps`. First a *case split* is carried out according to the protocol rules as formalized in section 2.2. Next the conditions for each particular extension is added to the goals which are simplified to a normal form.

The remaining differences stem from formulas of the form (i) $ev' \in (ev\#tr)$, (ii) $m \in analz(spies(ev\#tr))$, and (iii) $m \in parts(spies(ev\#tr))$ in the goals. For the three cases there are heuristics that start the difference reduction. For *analz* and *parts* they rely on *symbolic evaluation*.

Sometimes the induction hypothesis can be applied directly in the resulting goals. In other cases it is necessary to apply appropriate *structuring lemmata* and to make use of specific *goal assumptions*. These goal assumptions can originate from the definition of the corresponding protocol step (by `protocolSteps` in (b)) or during the symbolic evaluation itself.

While the proof tasks (a) and (b) are carried out in a canonical way the last proof task is much more complex and, for the time being, typically requires user interaction. The treatment of certain subgoals resulting from symbolic evaluation of *analz* or *parts* often involves proof decisions, like:

- Which goal assumptions should be considered?
- Which structuring lemmata have to be applied?
- Do we need a *new* structuring lemma?

However, also in those cases where *some* user interaction is necessary, heuristics are available to continue (and complete) the proof afterwards.

First steps towards an automatization of complex proof decisions in subtasks, tasks, and even complete proofs were made by designing high-level heuristics that combine the ones discussed so far.

3.3   Definition of Heuristics

Heuristics achieve their proof tasks by expanding the proof tree starting with a given goal. The basic steps that are carried out are the same as those the user might select (see section 2.3). In addition, heuristics might backtrack and choose to continue with another (open) subgoal.

Heuristics in VSE, like tactics in other systems, determine algorithmically the execution of the mentioned steps. In addition to the purely logical information, heuristics in VSE have access to local (goals) and global *control information* that is read to decide about the next step to be performed and updated to influence the further execution. The additional information slots are used to model the internal control flow of a single heuristic as well as to organize the composition of heuristics.

The existing VSE module for writing and executing heuristics was extended to meet the requirements of semi-automatic protocol verification. This concerns in particular those parts of the heuristics that depend on certain theories.

4 Results and Future Work

We developed about 25 heuristics which we used in the formal verification of several real world protocols, e.g., different versions of the EAC protocol (12 protocol steps, 41 properties) [6] and of a Chip-card-based Biometric Identification (CBI) protocol (15 protocol steps, 28 properties) [7]. Typical proofs of these properties consist of 1500 to 2000 proof nodes (steps) represented in a VSE proof tree. A proof of this magnitude would require at least 400 user interactions if the heuristics were not utilized. The heuristics thus allowed us to save up to 90% of user interactions in average.

So far the developed higher-level heuristics are especially tailored for the proof of confidentiality and possibility properties. For this kind of properties an even better automatization degree (up to full automatic proof generation) is reached. In order to increase the automation degree in the verification of the other protocol properties, i.e. of authenticity properties and other structuring lemmata, we plan to define more heuristics: higher-level heuristics tailored for these proofs and probably lower level heuristics when needed.

Although several proof structuring lemmata are formulated automatically, certain proof attempts force the user to define additional lemmata that depend on the remaining open proof goals. We therefore also plan to develop suitable lemma speculation heuristics.

References

1. Advanced Security Mechanisms for Machine Readable Travel Documents – Extended Access Control (EAC) – Version 1.11 Technical Guideline TR-03110, Federal Office for Information Security (BSI)
2. Dolev, D., Yao, A.: On the security of public-key protocols. IEEE Transactions on Information Theory 2(29) (1983)
3. Paulson, L.C.: The inductive approach to verifying cryptographic protocols. Journal of Computer Security 6, 85–128 (1998)
4. Hutter, D., Rock, G., Siekmann, J.H., Stephan, W., Vogt, R.: Formal Software Development in the Verification Support Environment (VSE). In: Manaris, B., Etheredge, J. (eds.) Proceedings of the FLAIRS 2000. AAAI Press, Menlo Park (2000)
5. Denker, G., Millen, J., Rueß, H.: The CAPSL Integrated Protocol Environment. SRI Technical Report SRI-CSL-2000-02 (October 2000)
6. Formal Verification of the Cryptographic Protocols for Extended Access Control on Machine Readable Travel Documents. Technical Report, German Research Center for Artificial Intelligence and Federal Office for Information Security
7. Cheikhrouhou, L., Rock, G., Stephan, W., Schwan, M., Lassmann, G.: Verifying a chip-card-based biometric identification protocol in VSE. In: Górski, J. (ed.) SAFECOMP 2006. LNCS, vol. 4166, pp. 42–56. Springer, Heidelberg (2006)

Iterative Search for Similar Documents on Mobile Devices

Kristóf Csorba and István Vajk

Budapest University of Technology and Economics
Department of Automation and Applied Informatics
{kristof,vajk}@aut.bme.hu

Abstract. This paper presents a new method for searching documents which have similar topics to an already present document set. It is designed to help mobile device users to search for documents in a peer-to-peer environment which have similar topic to the ones on the users own device. The algorithms are designed for slower processors, smaller memory and small data traffic between the devices. These features allow the application in an environment of mobile devices like phones or PDA-s. The keyword list based topic comparison is enhanced with cascading, leading to a series of document searching elements specialized on documents not selected by previous stages. The architecture, the employed algorithms, and the experimental results are presented in this paper.

1 Introduction

Searching for documents is a frequent task today. The most common approach is based on keywords given by the user as search criteria. Another possibility is the "topic by example" [1] approach, where the user presents documents for which similar ones should be retrieved. In this case, document topics have to be compared. The most common solutions are using the bag-of-words approach and calculate the similarity measure of documents using the document feature vectors. These feature vectors may contain presence of words weighted by importance or some extracted features of the document. Some feature selection methods are based on information gain or mutual information [12], least angle regression [10][4] or optimal orthogonal centroid feature selection [14]. Feature extraction techniques like applications of the singular value decomposition [9] or orthogonal locality preserving indexing [2] are also used.

The system presented in this paper is designed for applications in mobile device environments. The most important property of these environments are the limited processor and memory capacity of the devices, but as the communication traffic of a mobile phone usually costs money, the data traffic created during topic comparisons should be limited as well. To conform these requirements, the document topics have to be very small and yet comparable which makes the frequently used huge, weighted document vectors not applicable. To avoid resource consuming transformations, we took the following approach:

A. Dengel et al. (Eds.): KI 2008, LNAI 5243, pp. 38–45, 2008.

We assign a list of topic-specific keywords to every possible topic using a labeled training set. A document is represented by the identifier of the topic, which has the most common keywords with it, and a simple binary vector indicating the presence or absence of the given keywords in the document.

To achieve this, we have to select the most topic-specific keywords. Weighting of the keywords is not possible due to the binary vector and keywords with negative weight cannot be used at all because there are hundreds of keywords never appearing in one topic, but we have to keep the keyword list small to minimize communication traffic. This means that most feature selection approaches like mutual-information and information-gain based ones cannot be applied. [8] proposes a method for keyword selection based on word weights in document vector centroids, but it may produce keywords with negative weight as well.

Following the "topic by example" approach, the proposed system is a part of a project which aims to support searching for documents which are similar to the ones already present on the users mobile device [6][7]. This procedure is executed in background and the user is notified if a new document of interest is accessible.

If there are few keywords, many documents will not contain any of them making them lost for the user. To avoid this, the user has the option to ask for further keywords to retrieve more documents. This feature creates a document search cascade which has multiple levels for searching for further documents if required.

The remaining part of the paper is organized as follows: Section 2 presents the document topic representation and comparison, Section 3 presents the architecture of cascades, Section 4 presents the measurement results and finally, conclusions are summarized in Section 5.

2 Document Topic Representation and Comparison for Mobile Devices

In this section, the keyword selection method and the document similarity measure for comparison is described. As the document search is executed in background and the user is notified if a similar document was found, the rate of misclassifications is of key importance in this application. It is much more important than finding all similar documents.

For performance measurements, we will use the common measures precision, recall and F-measure: If there is a specific target topic, from which we want to select as many documents as possible, c is the number of correctly selected documents, f the number of false selections, and t is the number of documents in the target topic, then precision is defined as $P = c/(c + f)$, recall is $R = c/t$ and F-measure is $F = 2PR/(P + R)$. Our aim to avoid misclassifications means the requirement of a high precision even if the recall will be low.

In the description of the algorithms, we use the following notations:

- $T(d)$ means that the document d belongs to the topic T.
- $K_T = \{w_1, w_2, ..., w_n\}$ is the keyword list (set of keywords w_i) for the topic T. T might refer to a set of topics as well, in which case K_T contains all keywords of the topics in the topic set.

- A *keyword* is a word, which is present in at least one keyword list. This means that they are words used in the topic representations.
- S_T is the selector, which aims to select documents from the topic T. We use the selector expression instead of classifier, because it selects documents for one topic, and leaves the remaining ones to be selected by other selectors. Documents not selected by any selectors will have unidentified topic.
- $S_{\underline{d}}$ is a selector based on a document vector $\underline{d}$ and selects documents which have common keywords with the document represented by $\underline{d}$.
- $S_T(d)$ means that the document d is selected by the selector S_T.

2.1 Creating Topic Specific Keyword Lists

In the following techniques, we will require topic-specific keyword lists. These are created using the Precision based Keyword Selection (PKS) algorithm described and evaluated in [3] in details. It creates keyword lists conforming the following definitions:

Definition 1 (Individual precision). *Individual precision of a keyword w is the precision of a selector, for which $S_w(d) \Leftrightarrow w \in d$, that is, it selects documents containing w.*

Definition 2 (Topic specific keyword). $w \in K_T$ *exactly if* $P(T|w) \geq minprec$, *that is, if the conditional probability of the topic given w is present in a document, is higher than a predefined minimal limit minprec.*

As the probability $P(T|w)$ is the expected individual precision of w, the minimal limit in the definition is called minimal precision limit, *minprec*. The PKS algorithm selects words with high individual precision and the *minprec* limit is set to allow the maximal F-measure for the selection of documents in the given topic.

2.2 Searching for Similar Documents

All documents are represented by the identifier of the best matching keyword list and a binary vector indicating the presence or absence of keywords. Using the keyword list identifier, these binary vectors are mapped into a global keyword space, where the scalar product returns the number of common keywords even of documents represented with different keyword lists. The number of common keywords is the similarity measure employed by the proposed system. If not noted otherwise, in the following we consider document vectors in the global keyword space.

If $\underline{t}$ and $\underline{d}$ are binary document vectors in the space of all possible keywords than the similarity of the two documents is defined as

$$\text{similarity}(\underline{t}, \underline{d}) = \underline{t} \cdot \underline{d} \tag{1}$$

Searching for similar documents to multiple base documents is performed with the merged base document vector, which represents all local documents together:

Definition 3 (Merged base document vector $\underline{b}$)

$$\underline{b} := \mathrm{sign}\Big(\sum_{d \in B} \underline{d}\Big) \tag{2}$$

where B is the set of base documents.

The minimal similarity measure which a retrieved document must have to the base documents is the *threshold* parameter which is defined by the user. It allows controlling the balance between precision and recall.

3   Similar Document Search Cascade

In this section, we present the definition, operation and training method of a cascade of similar document selectors.

Definition 4 (Accepted and Denied topic set). $\mathbf{A}$ *is the set of topics T, so that $T \in \mathbf{A} \leftrightarrow \exists d \in B : T(d)$. The set of denied topics is $\mathbf{D} = \overline{\mathbf{A}}$.*

The $\mathbf{A}$ set of accepted topics is the set of topics of the base documents. Similar documents are allowed to have these topics. Remote documents belonging to one of the denied topics must not be selected, because they are off-topic for the user.

Definition 5 (Document vector over given keyword list). $\underline{d}^{(K_T)}$ *is the document vector of the document d over the keyword list K_T defined as*

$$\underline{d}^{(K_T)}_w := \underline{d}_w \text{ if } w \in K_T,\ 0 \text{ else} \tag{3}$$

where $\underline{d}_w$ stands for the component of the $\underline{d}$ vector corresponding the keyword w.

In a document vector over a given keyword list, all components for keywords not in the keyword list are set zero. This means that only the keywords in the keyword list are used.

Definition 6 (Exclude vector $\underline{e}$)

$$\underline{e}_w := 1 \text{ if } w \in K_{\mathbf{D}},\ 0 \text{ else} \tag{4}$$

If for a given remote document d, $\underline{d} \cdot \underline{b}^{(K_{\mathbf{A}})} \geq 0$, it has at least one common keyword with a base document. If $\underline{d} \cdot \underline{e} \geq 0$, if has a keyword belonging to a denied topic which means d is considered off-topic.

Definition 7 (Cascade element). *The i-th cascade element is $Cas_i := (S_b^{(i)}; S_e^{(i)})$. $S_b^{(i)} := S_{\underline{b}^{(i)}}$ where $\underline{b}^{(i)} := \underline{b}^{(K_{\mathbf{A}}^{(i)})}$. Similarly, $S_e^{(i)} := S_{\underline{e}^{(i)}}$ where $\underline{e}^{(i)} := \underline{e}^{(K_{\mathbf{D}}^{(i)})}$. If for a given d document $S_b^{(i)}(d)$, the document is selected for the user. If $S_e^{(i)}(d)$, the document may not be selected by either the i-th, or any of the further cascade levels.*

A cascade element is a pair of selectors: $S_b^{(i)}$ compares documents with the base documents and $S_e^{(i)}$ compares them with $\underline{e}$ which is a virtual document representing all denied topics. The set of keywords used in these selectors differs from level to level. The training of the cascade means creating these keyword lists. If $S_b^{(i)}$ selects a document, it is selected for the user. If $S_e^{(i)}$ selects it, it is excluded for the further levels of the cascade and considered off-topic. A cascade is a series of cascade elements.

The aim of a cascade is the following: in every cascade element, the selector is trained assuming that every document arriving to its input was not selected by any selectors of previous levels, nor for allowed, neither for denied topics. This means, that the document had no common keyword with any previously used keyword lists.

As the previous level executed selectors for all allowed topics using a merged keyword list $K_\mathbf{A}$, the current level still has to make sure, that none of the selectors for denied topics would have selected the document either. This check is performed by the selector $S_e^{(i)}$, which excludes those documents which are recognized to be off-topic.

It should be noted that $K_\mathbf{A}^{(i-1)} \subseteq K_\mathbf{A}^{(i)}$ for every $i > 1$: the number of common keywords between $K_\mathbf{A}^{(i-1)}$ and a given document cannot decrease while moving to the next level of the cascade.

The training of the cascade is performed by subsequently using PKS to create keyword lists for the topics based on the documents in the training set, and removing those documents, which have at least one common keyword with any of the created keyword lists. The training of the next cascade level is based on the remaining documents. This has the significant advantage that some topics are easier to identify if documents from some similar topics have already been removed. In such cases further high-quality keywords can be employed which would have been bad keywords if the other topics documents would be still present.

4   Measurements

In this section, we present measurements evaluating the capabilities of the cascade. The measurements were performed using a subset of the commonly used data set 20 Newsgroups [11] containing the topics *hardware.PC*, *hardware.MAC*, *sport.baseball*, *sport.hockey*, *science.electronics* and *science.space*. The baseline measurement employed the TFIDF based feature selection and the linear classifier of the "TMSK: Text-Miner Software Kit" [13] system. The results are presented in Fig. 1.

In this measurement we investigate the performance of a cascade. We created 5 cascade levels, each containing a keyword list for every topic of the data set. The performance is evaluated in terms of recall and precision after every cascade level. We analyzed the effect of the number of base documents (Fig. 2) and the threshold (Fig. 3) on the cascade performance.

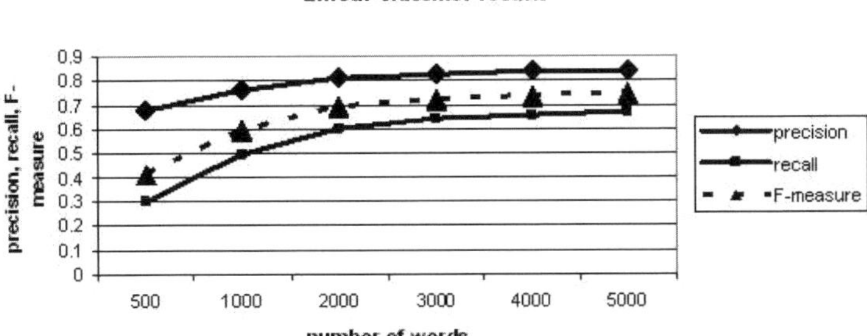

Fig. 1. Results with Linear classifier

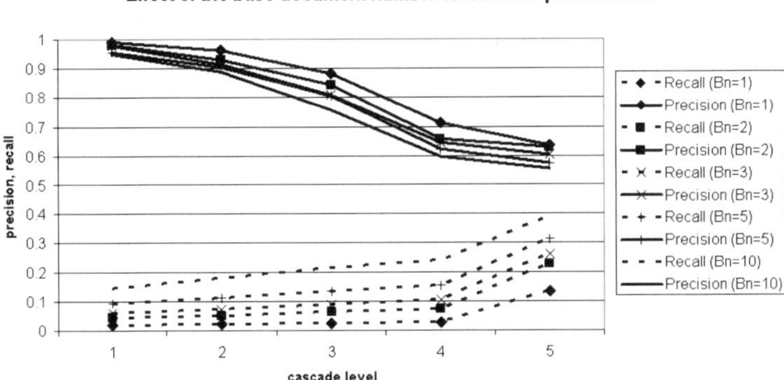

Fig. 2. Effect of the base document number on cascade performance. Threshold is 2, exclusion threshold used by $S_e^{(i)}$ is 1. Results are presented for various number of base documents indicated by Bn in the legend.

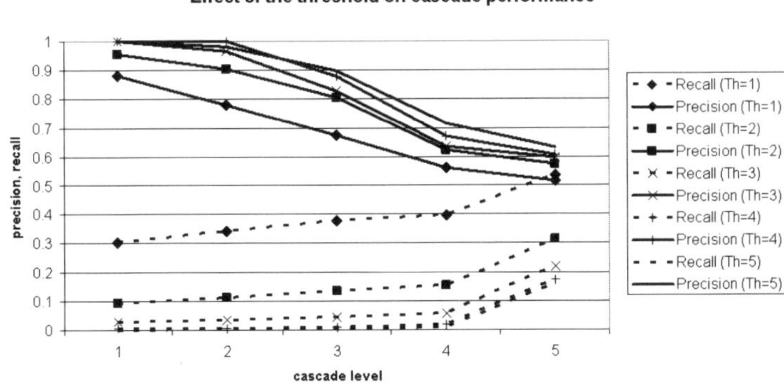

Fig. 3. Effect of the threshold on cascade performance. The number of base documents is 5 and exclusion threshold is 1. Results presented for various threshold values indicated by Th in the legend.

The proposed classifier system is much more biased towards high precision than the linear classifier in the baseline measurements. It achieves higher precision, but usually lower recall. As our system usually used less than 100 keywords in a keyword list, it requires far less keywords to achieve this performance, than the linear classifier. The measurements confirm, that later cascade levels have lower precision and higher recall. This is not surprising because the additional keywords are often of lower quality for topic identification, but more keywords cover more documents.

The number of available base documents is critical for the system performance: few base documents do not provide enough keywords for an acceptable recall. In our measurement, at least 10% was reached in the first cascade level if the number of base documents was at least 5.

Fig. 3 highlights the effect of the threshold. If the user sets a high threshold, the recall is lower and the precision is higher. On the other hand, if threshold is 1, recall reaches 30% in the first cascade level with good precision, and similarly good performance is provided in the second cascade level too.

5   Conclusions

This paper presented a cascade consisting of multiple similar document selectors for applications in mobile device environments to search for documents similar to the users own documents already present on the device. To satisfy the requirements of the low resource environment, the proposed system employs simple and fast algorithms and keeps the communication traffic low by employing very compact topic representations. As the retrieval of less documents is much less annoying then misclassifications, our system concentrates on high precision document selection even with lower recall. The cascade allows the user to request for further searches with additional keywords to find more documents of interest.

Acknowledgements

This work has been fund of the Hungarian Academy of Sciences for control research and the Hungarian National Research Fund (grant number T68370).

References

1. Buntine, W.: Topic-specific scoring of documents for relevant retrieval. In: Proceedings of ICML 2005 Workshop 4: Learning in Web Search, Bonn, Germany, 7 August (2005)
2. Cai, D., He, X.: Orthogonal locality preserving indexing. In: SIGIR 2005: Proceedings of the 28th annual international ACM SIGIR conference on Research and development in information retrieval, pp. 3–10. ACM Press, New York (2005)
3. Csorba, K., Vajk, I.: Supervised term cluster creation for document clustering. Scientific Bulletin of Politehnica University of Timisoara, Romania, Transactions on Automatic Control and Computer Science 51 (2006)

4. Efron, B., Hastie, T., Johnstone, I., Tibshirani, R.: Least angle regression Technical report, Department of Statistics, Stanford University (2002)
5. Fellbaum, C. (ed.): WordNet: An Electronic Lexical Database. MIT Press, Cambridge (1989)
6. Forstner, B., Charaf, H.: Neighbor selection in peer-to-peer networks using semantic relations. WSEAS Transactions on Information Science and Applications 2(2), 239–244 (2005) ISSN 1790-0832
7. Forstner, B., Kelenyi, I., Csucs, G.: Towards Cognitive and Cooperative Wireless Networking: Techniques, Methodologies and Prospects. In: Chapter Peer-to-Peer Information Retrieval Based on Fields of Interest, pp. 311–325. Springer, Heidelberg (2007)
8. Fortuna, B., Mladenić, D., Grobelnik, M.: Semi-automatic construction of topic ontology. In: Proceedings of SIKDD 2005 at multiconference is 2005, Ljubljana, Slovenia (2005)
9. Furnas, G., Deerwester, S., Dumais, S.T., Landauer, T.K., Harshman, R., Streeter, L.A., Lochbaum, K.E.: Information retrieval using a singular value decomposition model of latent semantic structure. In: Chiaramella, Y. (ed.) Proceedings of the 11th Annual International ACM SIGIR Conference, Grenoble, France, pp. 465–480. ACM, New York (1988)
10. Keerthi, S.S.: Generalized lars as an effective feature selection tool for text classification with SVMS. In: Proceedings of the 22nd International Conference on Machine Learning, Bonn, Germany (2005)
11. Lang, K.: NewsWeeder: learning to filter netnews. In: Prieditis, A., Russell, S.J. (eds.) Proceedings of ICML-95, 12th International Conference on Machine Learning, Lake Tahoe, US, pp. 331–339. Morgan Kaufmann Publishers, San Francisco (1995)
12. Oded Maimon, L.R.: The Data Mining and Knowledge Discovery Handbook. Springer, Heidelberg (2005)
13. Weiss, S.M., Indurkhya, N., Zhang, T., Damerau, F.J.: Text Mining, Predictive Methods for Analysing Unstuctured Information. Springer, Heidelberg (2005)
14. Yan, J., Liu, N., Zhang, B., Yan, S., Chen, Z., Cheng, Q., Fan, W., Ma, W.-Y.: Ocfs: optimal orthogonal centroid feature selection for text categorization. In: SIGIR 2005: Proceedings of the 28th annual international ACM SIGIR conference, pp. 122–129. ACM Press, New York (2005)

Limits and Possibilities of BDDs in State Space Search[*]

Stefan Edelkamp and Peter Kissmann

Faculty of Computer Science
TU Dortmund, Germany

Abstract. This paper investigates the impact of symbolic search for solving domain-independent action planning problems with binary decision diagrams (BDDs). Polynomial upper and exponential lower bounds on the number of BDD nodes for characteristic benchmark problems are derived and validated. In order to optimize the variable ordering, causal graph dependencies are exploited.

1   Introduction

Optimal planning is a natural requirement for many applications, and posed in form of challenging benchmarks at international planning competitions. Due to the unfolding of the planning graph, parallel-optimal planners [2, 20, 21, 24, 25] appear to be less effective in large search depth, even though more efficient encodings have been found [23].

The state-of-the-art in optimal sequential planning includes heuristic search planning approaches with admissible search heuristics, such as the flexible abstraction heuristic [14] and explicit-state pattern databases [12]. The corresponding planners refer to a fully-instantiated planning problem together with a minimized state encoding [6, 13] and have been reported to compare well with other techniques [11, 16, 26].

Binary decision diagrams (BDDs) contribute to various successful planning approaches, e.g. [4, 18]. The idea of using BDDs is to reduce the memory requirements for planning state sets as problem sizes increase. For several optimal sequential planning benchmark domains, symbolic breadth-first search with BDDs is competitive with the state-of-the-art [5]. Recent experimental findings additionally show large compression ratios for symbolic pattern databases in many planning benchmark domains [1].

This paper studies the causes for good and bad BDD performance by providing lower and upper bounds for BDD growth in characteristic benchmark domains. The approach applies to step-optimal propositional planning and transfers to planning with additive action costs.

The paper is structured as follows. First, we introduce BDD-based search. In the core of the paper, we then study the impact of BDDs for selected characteristic benchmark problems, for which we provide state space encodings, lower and upper bounds. We consider improvements obtained in bidirectional search. Finally, we give some concluding remarks.

---

[*] Extended version of a poster at the National Conference of Artificial Intelligence (AAAI-08) having the same title [8]. Thanks to DFG for support in ED 74/3 and 74/2.

A. Dengel et al. (Eds.): KI 2008, LNAI 5243, pp. 46–53, 2008.

2 Symbolic Planning

Throughout the paper, we consider optimal planning using the SAS^+ representation of planning tasks [13]. More formally, a SAS^+ planning task with costs is defined as $\mathcal{P} = (\mathcal{S}, \mathcal{V}, \mathcal{A}, \mathcal{I}, \mathcal{G}, \mathcal{C})$ with $\mathcal{S}$ being a set of states, $\mathcal{V} = \{v_1, \ldots, v_n\}$ being a set of state variables (each $v \in \mathcal{V}$ has finite domain D_v), $\mathcal{A}$ being a set of actions given by a pair (P, E) of preconditions and effects, $\mathcal{I}$ being the initial state, and $\mathcal{G}$ being the goal state set in form of a partial assignment. Such partial assignment for $\mathcal{V}$ is a function s over $\mathcal{V}$ with $s(v) \in D_v$, if $s(v)$ is defined. Moreover, cost function $\mathcal{C}$ maps $\mathcal{A}$ to $I\!\!N$. The task is to find a sequence of actions $a_1 \ldots, a_k \in \mathcal{A}$ from $\mathcal{I}$ to $\mathcal{G}$ with minimal $\sum_{i=1}^{k} \mathcal{C}(a_i)$. For step-optimal planning we have $\mathcal{C}(a) = 1$ for all $a \in \mathcal{A}$.

Symbolic planning sometimes refers to analyzing planning graphs [2], or to checking the satisfiability of formulas [20]. In contrast, we refer to the exploration in the context of using BDDs [3]. For constructing the BDDs for the initial and goal states as well as the individual actions, variables are encoded in binary.

Sets of planning states are represented in form of Boolean functions, and actions are formalized as (transition) relations. This allows to compute the successor state set (the image) as a conjunction of the given state set and the transition relation, existentially quantified over the set of the current state variables. This way, all states reached by applying one action to a state in the input set are determined. Iterating the process (starting with the representation of the initial state) yields a symbolic implementation of breadth-first search. Results in international planning competitions show that symbolic bidirectional breadth-first search performs well for many benchmarks[1].

The variable ordering has an exponential impact to the size of the BDDs. In our setting, state variables are encoded in binary and ordered along their causal graph dependencies [13] by greedily minimizing $\sum_{1 \leq i \neq j \leq n, v_i \ depends \ on \ v_j} |i - j|$. Roughly speaking, the more related two variables are, the more likely they will be located next to each other.

In the following we study the impact of BDDs for space improvement in characteristic problems, for which we provide state space encodings, lower and upper bounds.

3 Exponential Lower Bound

Let us consider permutation games on $(0, \ldots, N - 1)$, such as the $(n^2 - 1)$-Puzzle, where $N = n^2$. It is well known that for the $(n^2 - 1)$-Puzzle, half of all $n^2!$ possible permutations are reachable from the initial state. With explicit-state breadth-first search, Korf and Schultze [22] have generated the entire state space of the 15-Puzzle on disk.

The characteristic function f_N of all permutations on $(0, \ldots, N - 1)$ has $N \lceil \log N \rceil$ binary state variables and evaluates to *true*, if every block of $\lceil \log N \rceil$ variables corresponds to the binary representation of an integer and every satisfying path of N integers is a permutation. In his master's thesis, Hung [17] has shown the following result.

---

[1] The fact that BFS performs well wrt. heuristic search is an observation also encountered in explicit-state space search [15]. It is related to the exponential increase in the number of states below the optimum cost for many benchmark domains, even if only small constant errors in the heuristic are assumed.

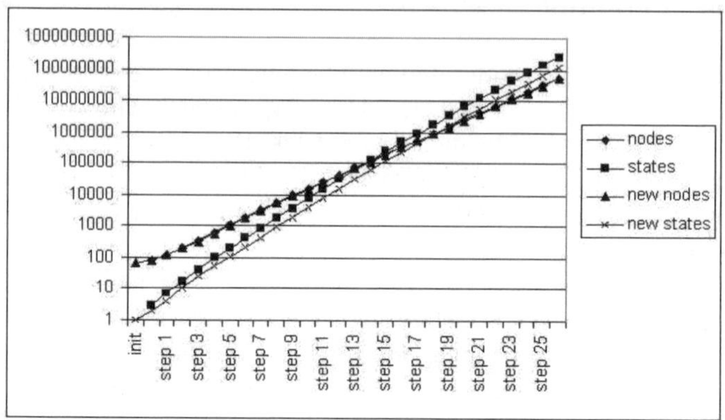

Fig. 1. Symbolic BFS in the 15-Puzzle

Theorem 1. *The BDD for f_N has more than $\lfloor \sqrt{2^N} \rfloor$ nodes for any variable ordering.*

Given that BDD nodes also consume space, the memory savings of BDDs for permutation games like the $(n^2 - 1)$-Puzzle are limited. Fig. 1 validates this hypothesis for the BFS layers of the 15-Puzzle. The growth of BDD nodes is smaller than the growth of states (numbers match the ones by Korf & Schultze), but still the symbolic exploration cannot be finished (within 16 GB RAM). Note that a 15-Puzzle instance can be encoded in 60 bit (4 bits per tile), while a BDD node consumes about 20 bytes. In related research [9] the overall compression ratio for a 6-tile symbolic pattern database of the 35-Puzzle was about factor 2.65.

4   Polynomial Upper Bound

In other search spaces, we obtain an exponential gain using BDDs. In Gripper, there is one robot to transport $2k = n$ balls from one room A to another room B. The robot has two grippers to pick up and put down a ball.

The state space grows exponentially. Since we have $2^n = \sum_{i=0}^{n} \binom{n}{i} \leq n\binom{n}{k}$, the number of all states with k balls in one room is $\binom{n}{k} \geq 2^n/n$. Helmert and Röger [15] have shown that the precise number of all reachable states is $S_n = 2^{n+1} + n2^{n+1} + n(n-1)2^{n-1}$, where $S_n^0 := 2^{n+1}$ corresponds to the number of all states with no ball in a gripper. All states, where the number of balls in each room is even (apart from the two states with all balls in the same room and the robot in the other room), are part of an optimal plan. For larger values of n, therefore, heuristic search planners even with a constant error of only 1 are doomed to failure.

The robot's cycle for the best delivery of two balls has length 6 (picking up the two balls, moving from one room to the other, putting down the two balls, and moving back), such that every sixth BFS layer contains the states on an optimal plan with no ball in a gripper. Yet there are still exponentially many of these states, namely $S_n^0 - 2$.

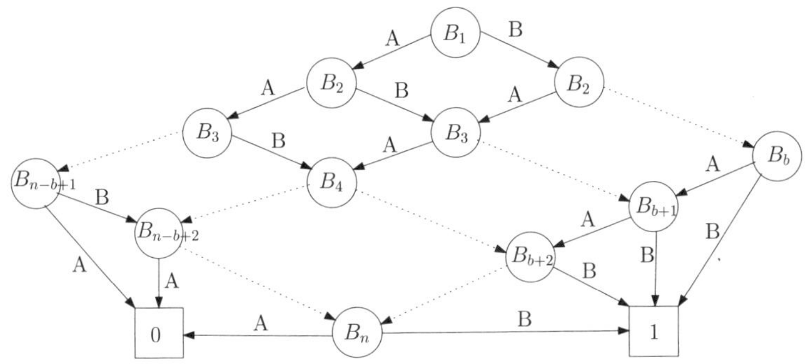

Fig. 2. BDD structure for representing b balls in room B

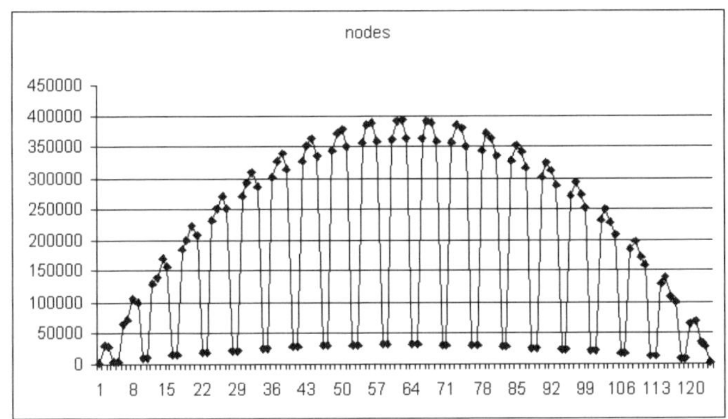

Fig. 3. Sybombolic BFS in Gripper-20

Theorem 2. *There is a binary state encoding and an associated variable ordering, in which the BDD size for the characteristic function of the states on any optimal path in the breadth-first exploration of Gripper is polynomial in n.*

Proof. To encode states in Gripper[2] , $1 + 2 \cdot \lceil \log(n+1) \rceil + 2n$ bits are required: one for the location of the robot (A/B) , $\lceil \log(n + 1) \rceil$ for each of the grippers to denote which ball it currently carries, and 2 for the location of each ball $(A/B/\text{none})$.

According to BFS, we divide the set of states on an optimal path into layers l, $0 \leq l \leq 6k - 1$. If both grippers are empty, we are in level $l = 6d$ and all possible states with $b = 2d$ balls in the right room have to be represented, which is available using $\mathcal{O}(bn)$ BDD nodes (see schema in Fig. 2).

The number of choices with 1 or 2 balls in the gripper that are addressed in the $2\lceil \log(n+1) \rceil$ variables is bounded by $\mathcal{O}(n^2)$, such that intermediate layers with $l \neq 6d$ lead to an at most quadratic growth. Hence, each layer restricted to the states on the optimal plan contains less than $\mathcal{O}(n^2 \cdot dn) = \mathcal{O}(dn^3) = \mathcal{O}(n^4)$ BDD nodes in total.

---

[2] This encoding is found using a static analyzer.

Accumulating the numbers along the path, whose size is linear in n, we arrive at a polynomial number of BDD nodes needed for the entire exploration.

The forward BFS exploration in Gripper for $n = 42$ in Fig. 3 validates the theoretical result (including the reduction/increase of BDD nodes every sixth step). Roughly speaking, *permutation* benchmarks are bad for BDDs, and *counting* benchmarks are good.

5   Symbolic Bidirectional Search

Bidirectional search algorithms are distributed in the sense that two search frontiers are searched concurrently. For explicit-state space search, they solve the one-pair shortest path problem. Since the seed in symbolic search can be any state set and symbolic predecessors are the straight inverse of symbolic successors, symbolic bidirectional search algorithms start immediately from the partial goal assignment. Moreover, bidirectional BFS comes at a low price as it does not rely on any heuristic evaluation function.

In Blocksworld, towers of labeled blocks have to be built. As a full tower can be casted as a permutation, we do not expect polynomially sized BDDs. If we look at benchmark instances (see Fig. 4) bidirectional search improves symbolic BFS from stacking 9 blocks step-optimally to 15 blocks[3].

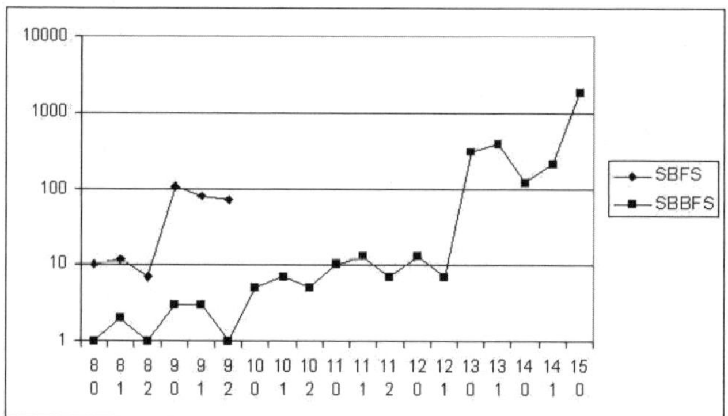

Fig. 4. Time Performance (in seconds) of Symbolic Uni- and Bidirectional BFS in Blocksworld

Fig. 5 gives insights to the exploration for problem 15-0. Backward search (sorted to the left of the plot to avoid an interleaved order) shows that it contains a large number of illegal states, and forward search (to the right of the plot) shows that the number of states exceeds the number of BDD nodes.

We next look at the Sokoban domain, a problem that is well studied in AI research [19]. A level represents a maze of cells, where stones appear to be randomly placed. The player, also located in one of the cells, pushes the stones around the maze so that, at the end, all stones are on a target location. We observe another exponential gap between explicit-state and symbolic representation.

---

[3] All runtimes wrt. Linux PC with 2.6 GHz CPU.

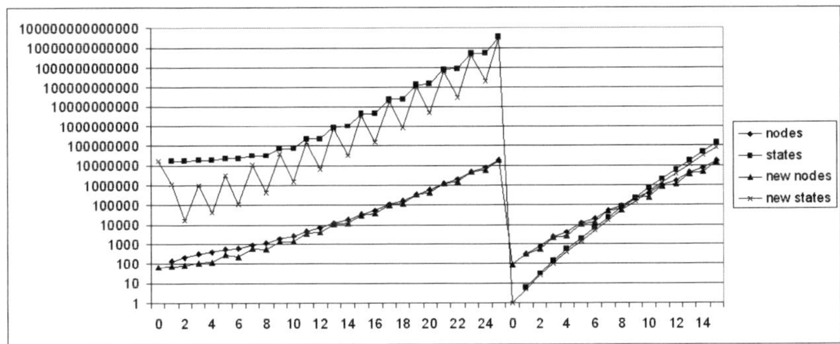

Fig. 5. Symbolic Bidirectional BFS in Blocksworld-15-0

Theorem 3. *If all $\binom{n}{k} \cdot (n - k)$ configurations with k balls in a maze of n cells in Sokoban are reachable, there is a binary state encoding and an associated variable ordering, in which the BDD size for the characteristic function of all reachable states in Sokoban is polynomial in n.*

Proof. To encode states in Sokoban, $2n$ bits are required[4], i.e., 2 bits for each cell (stone/player/none). If we were to omit the player and branch on binary state variables (stone/none), we would observe the same pattern that was shown in Fig. 2, where the left branch would denote an empty cell and the right branch a stone, leaving a BDD of $\mathcal{O}(nk)$ nodes. Integrating the player gives us a second BDD of size $\mathcal{O}(nk)$ with links from the first to the second. Therefore, the complexity for representing all reachable Sokoban positions requires a polynomial number of BDD nodes.

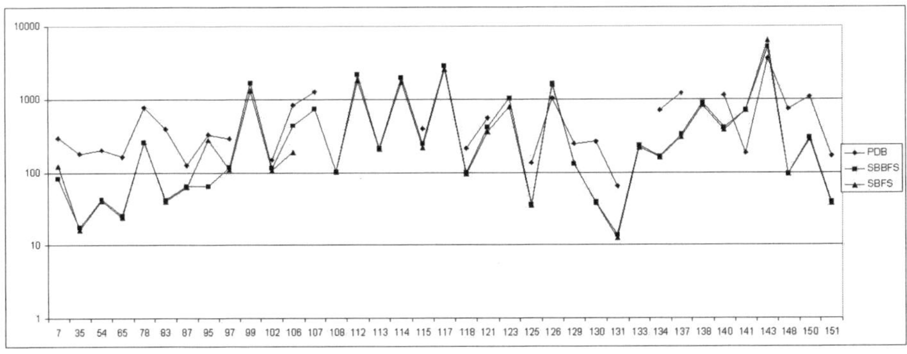

Fig. 6. Results in Sokoban; Unsolved Instances for Symbolic Bidirectional BFS are 93, 105, and 145; Symbolic BFS additionally cannot solve Instance 107

As $\binom{n}{k} \leq (\frac{n}{k})^k$, the number of all reachable states is clearly exponential. In Fig. 6, we compare exploration results of symbolic uni-/bidirectional BFS (SBFS/SBBFS) with

---

[4] This encoding is found using our static analyzer.

explicit-state pattern database heuristic search (PDB) [12]. In all but instances 126, 141 and 143, SBBFS is faster. Moreover, it solves 9 of the 12 unsolved instances (out of the 40 problem instances of [12]) in reasonable time. The discrepancy between uni- and bidirectional search is often small; the former is often slightly better in finding long plans (due to saturation of the reachable set) and sometimes worse on small-sized plans (due to non-saturation).

6 Conclusion and Future Work

The advantage of BDD-based compared to SAT-based and constraint-based planning is that the number of variables does not increase with the search depth. Moreover, heuristics are often of limited help in many planning benchmarks.

We gave insights on when and why BDDs work well for compressing the planning state space and its exploration. State spaces that have permutation character are less appropriate than state spaces with selection of elements in a set (balls in a room, locations in a maze). In difference to other posterior state space compressing techniques, see e.g. [10], the compression is on-the-fly, during the exploration of the state space. We obtained promising results for a selection of planning benchmarks, improving the best known results for Sokoban problems.

Optimal planning problems with discrete action costs and soft constraints are the main challenges in the deterministic part of the international planning competition in 2008. The net-benefit to optimize is the total benefit of satisfying the goals, minus the total action cost to achieve them. This results in a plan metric that is a linear expression over indicator variables for the violation of the constraints added to the action cost total. For computing net-benefits with BDDs, we contributed a symbolic breadth-first branch-and-bound search algorithm together with several search refinements [7], to which the above studies apply.

References

1. Ball, M., Holte, R.C.: The compression power of symbolic pattern databases. In: ICAPS 2008 (to appear, 2008)
2. Blum, A., Furst, M.L.: Fast planning through planning graph analysis. In: IJCAI, pp. 1636–1642 (1995)
3. Bryant, R.E.: Symbolic manipulation of boolean functions using a graphical representation. In: DAC, pp. 688–694 (1985)
4. Cimatti, A., Roveri, M., Traverso, P.: Automatic OBDD-based generation of universal plans in non-deterministic domains. In: AAAI, pp. 875–881 (1998)
5. Edelkamp, S.: Symbolic shortest path planning. In: ICAPS Workshop on Heuristics for Domain-independent Planning: Progress, Ideas, Limitations, Challenges (2007)
6. Edelkamp, S., Helmert, M.: Exhibiting knowledge in planning problems to minimize state encoding length. In: Biundo, S., Fox, M. (eds.) ECP 1999. LNCS, vol. 1809, pp. 135–147. Springer, Heidelberg (2000)
7. Edelkamp, S., Kissmann, P.: GAMER: Bridging planning and general game playing with symbolic search. In: Proceedings of the 6th International Planning Competition (to appear, 2008)

8. Edelkamp, S., Kissmann, P.: Limits and possibilities of BDDs in state space search. In: AAAI, pp. 1452–1453 (2008)
9. Edelkamp, S., Kissmann, P., Jabbar, S.: Scaling search with symbolic pattern databases. In: MOCHART 2008 (to appear, 2008)
10. Felner, A., Meshulam, R., Holte, R.C., Korf, R.E.: Compressing pattern databases. In: AAAI, pp. 638–643 (2004)
11. Grandcolas, S., Pain-Barre, C.: Filtering, decomposition and search-space reduction in optimal sequential planning. In: AAAI, pp. 993–998 (2007)
12. Haslum, P., Botea, A., Helmert, M., Bonet, B., Koenig, S.: Domain-independent construction of pattern database heuristics for cost-optimal planning. In: AAAI, pp. 1007–1012 (2007)
13. Helmert, M.: A planning heuristic based on causal graph analysis. In: ICAPS, pp. 161–170 (2004)
14. Helmert, M., Haslum, P., Hoffmann, J.: Flexible abstraction heuristics for optimal sequential planning. In: ICAPS, pp. 176–183 (2007)
15. Helmert, M., Röger, G.: How good is almost perfect? In: AAAI, pp. 944–949 (2008)
16. Hickmott, S.L., Rintanen, J., Thiébaux, S., White, L.B.: Planning via petri net unfolding. In: IJCAI, pp. 1904–1911 (2007)
17. Hung, N.N.W.: Exploiting symmetry for formal verification. Master's thesis, Faculty of the Graduate School, University of Texas at Austin (1997)
18. Jensen, R., Hansen, E., Richards, S., Zhou, R.: Memory-efficient symbolic heuristic search. In: ICAPS, pp. 304–313 (2006)
19. Junghanns, A.: Pushing the Limits: New Developments in Single-Agent Search. PhD thesis, University of Alberta (1999)
20. Kautz, H., Selman, B.: Pushing the envelope: Planning propositional logic, and stochastic search. In: ECAI, pp. 1194–1201 (1996)
21. Kautz, H., Selman, B., Hoffmann, J.: Satplan: Planning as satisfiability. In: Proceedings of the 5th International Planning Competition (2006)
22. Korf, R.E., Schultze, T.: Large-scale parallel breadth-first search. In: AAAI, pp. 1380–1385 (2005)
23. Robinson, N., Gretton, C., Pham, D.N., Sattar, A.: A compact and efficient SAT encoding for planning. In: ICAPS (to appear, 2008)
24. Vidal, V., Geffner, H.: Branching and pruning: An optimal temporal POCL planner based on constraint programming. Artificial Intelligence 170(3), 298–335 (2006)
25. Xing, Z., Chen, Y., Zhang, W.: MAXPLAN: Optimal Planning by Decomposed Satisfiability and Backward Reduction. In: Proceedings of the 5th International Planning Competition (2006)
26. Zhou, R., Hansen, E.: Breadth-first heuristic search. In: ICAPS, pp. 92–100 (2004)

Interactive Dynamic Information Extraction

Kathrin Eichler, Holmer Hemsen, Markus Löckelt, Günter Neumann,
and Norbert Reithinger

Deutsches Forschungszentrum für Künstliche Intelligenz - DFKI,
66123 Saarbrücken and 10178 Berlin, Germany
`FirstName.SecondName@dfki.de`

Abstract. The IDEX system is a prototype of an interactive dynamic Information Extraction (IE) system. A user of the system expresses an information request for a topic description which is used for an initial search in order to retrieve a relevant set of documents. On basis of this set of documents unsupervised relation extraction and clustering is done by the system. In contrast to most of the current IE systems the IDEX system is domain-independent and tightly integrates a GUI for interactive exploration of the extraction space.

1   Introduction

Information extraction (IE) involves the process of automatically identifying instances of certain relations of interest, e.g. produce(⟨company⟩, ⟨product⟩, ⟨location⟩), in some document collection and the construction of a database with information about each individual instance (e.g., the participants of a meeting, the date and time of the meeting). Currently, IE systems are usually domain-dependent and adapting the system to a new domain requires a high amount of manual labor, such as specifying and implementing relation–specific extraction patterns manually or annotating large amounts of training corpora.

For example, in a hand–coded IE system a topic expert manually implements task–specific extraction rules on the basis of her manual analysis of a representative corpus. Note that in order to achieve a high performance the topic expert usually also has to have a very good expertise in Language Technology (LT) in order to specify the necessary mapping between natural language expressions and the domain–specific concepts. Of course, nowadays, there exists a number of available LT components, which can be used to preprocess relevant text documents and determine linguistic structure. However, this still requires a fine–grained and careful analysis of the mapping between these linguistic structures and the domain–knowledge.

This latter challenge is relaxed and partially automatized by means of corpus–based IE systems. Here, the task–specific extraction rules are automatically acquired by means of Machine Learning algorithms, which are using a sufficiently large enough corpus of topic–relevant documents. These documents have to be collected and costly annotated by a topic–expert. Of course, also here existing LT core technology can be used to pre–compute linguistic structure. But still,

A. Dengel et al. (Eds.): KI 2008, LNAI 5243, pp. 54–61, 2008.

the heavy burden lies in a careful analysis of a very large corpus of documents with domain specific knowledge in order to support effective training of the ML engines.

One important issue for both approaches is, that the adaptation of an IE system to a new extraction task or domain have to be done offline, i.e., before the specific IE system is actually created. Consequently, current IE technology is highly statically and inflexible with respect to a timely adaptation of an IE system to new requirements in form of new topics.

1.1 Our Goal

The goal of our IE research is the development of a core IE technology to produce a new IE system automatically for a given topic on-demand and in interaction with an information analyst. The pre–knowledge about the information request is given by a user online to the IE core system (called IDEX) in the form of a topic description . This information request is used for an initial search in order to retrieve a relevant set of documents. This set of documents (i.e., the corpus) is then further passed over to Machine Learning algorithms which extract and collect (using the IE core components of IDEX) a set of tables of instances of possible relevant relations in an unsupervised way. These tables are presented to the user (who is assumed to be the topic–expert), who will investigate the identified relations further for his information research. The whole IE process is dynamic, since no offline data is required, and the IE process is interactive, since the topic expert is able to navigate through the space of identified structure, e.g., in order to identify and specify new topic descriptions, which express his new attention triggered by novel relationships he was not aware beforehand.

In this way, IDEX is able to adapt much better to the dynamic information space, in particular because no predefined patterns of relevant relations have to be specified, but relevant patterns are determined online. In the next section, we further motivate the application potential and impact of the IDEX approach by an example application, before more technical details of the system are described.

1.2 Application Potential

Consider, e.g., the case of the exploration and the exposure of corruptions or the risk analysis of mega construction projects. Via the Internet, a large pool of information resources of such mega construction projects is available. These information resources are rich in quantity, but also in quality, e.g., business reports, company profiles, blogs, reports by tourist, who visited these construction projects, but also Web documents, which only mention the project name and nothing else. One of the challenges for the risk analysis of mega construction projects is the efficient exploration of the possible relevant search space. Developing manually an IE system is often not possible because of the timely need of the information, and, more importantly, is probably not useful, because the needed (hidden) information is actually not known. In contrast, an unsupervised and dynamic IE system like IDEX can be used to support the expert in the exploration of the search space through pro–active identification and clustering of structured entities. Named entities like, for example, person names and locations, are often useful indicators for relevant text passages, in particular, if the names stand in relationship. Furthermore, because the found relationships are visualized using advanced graphical user interfaces, the user can select specific names and their associated relationships to other names, to the documents they occur in or she can search for paraphrases of sentences.

2   IDEX — System Overview

The next Fig. 1 shows the main components of the IDEX system. The system consists of two main parts: IDEXEXTRACTOR, which is responsible for the information extraction, and IDEXVISOR, which realizes the graphical user interface.

Processing in IDEX is started by sending topic or domain relevant information in form of a search engine query to a web crawler by the user of the IDEX system. The set of documents retrieved are further processed by the IE core components, which perform Named Entity extraction, identification and clustering of interesting relations. All identified and extracted information units together with their textual and linguistic context are stored in different SQL tables, which are maintained by a standard SQL DB server. These DB tables are the input for the IDEXVISOR, which dynamically creates different visual representations of the data in the tables, in order to support flexible search and exploration of all entities by the user. Since all extracted information units are linked to each other and with their information sources, the user can simply bop around the different entities. For example, the user might decide to firstly investigate the extracted named entities, and then might jump to the positions of the original text sources in order to check the sentences in which interesting pairs of names appear. He can then decide to inspect the internal structure of the sentences in order to test, whether they belong to an interesting "hidden" semantic relationship. The user might then decide to specify a new topic in form of a more fine-grained search engine query in order to initiate a more focused web crawl.

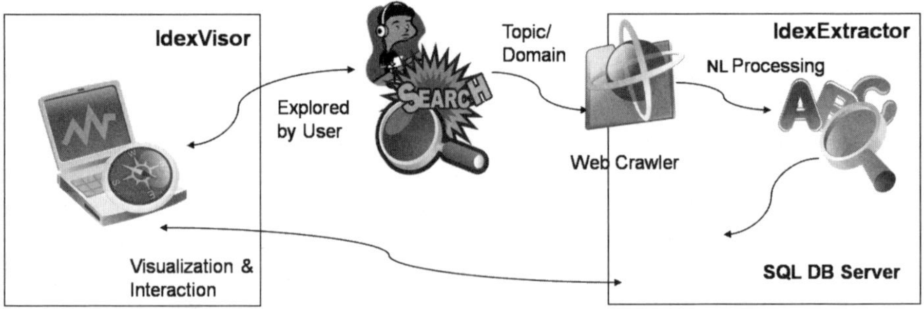

Fig. 1. Architecture of the IDEX System

2.1 IdexExtractor — Unsupervised Relation Extraction

Preprocessing After specifying the topic the documents (HTML and PDF) are automatically crawled from the web using the Google search engine and converted into plain text files. We apply LingPipe [1] for sentence boundary detection, named entity recognition (NER) and coreference resolution. As a result of this step database tables are created, containing references to the original document, sentences and detected named entities (NEs).

Relation extraction. We define a sentence to be of potential relevance if it at least contains two NEs. In the first step, so-called skeletons (simplified dependency trees) are extracted. To build the skeletons, the Stanford parser [2] is used to generate dependency trees for the potentially relevant sentences. For each NE pair in a sentence, the common root element in the corresponding tree is identified and the elements from each of the NEs to the root are collected. In the second step, information based on dependency types is extracted for the potentially relevant sentences. We focus on verb relations and collect for each verb its subject(s), object(s), preposition(s) with arguments and auxiliary verb(s). We consider only those relations to be of interest where at least the subject or the object is an NE. Relations with only one argument are filtered out.

Relation clustering. Relation clusters are generated by grouping relation instances based on their similarity. Similarity is measured based on the output from the different preprocessing steps as well as lexical information. WordNet [3] information is used to determine if two verb infinitives match or if they are in the same synonym set. The information returned from the dependency parser is used to measure the token overlap of the two subjects and objects, respectively. In addition, we compare the auxiliary verbs, prepositions and preposition arguments found in the relation. We count how many of the NEs match in the sentences in which the two relations are found, and whether the NE types of the subjects and objects match. Manually analyzing a set of extracted relation instances, we defined weights for the different similarity measures and calculated a similarity score for each relation pair. We then defined a score threshold and clustered relations by putting two relations into the same cluster if their similarity score exceeded this threshold value.

2.2 IdexVisor — Interactive Exploration of the Extraction Space

Using the IDEXVISOR-Frontend, a user can access different visualizations of the extracted data and he can navigate through the data space in a flexible and dynamic manner. IDEXVISOR is a platform independent application. It can be configured dynamically with respect to the underlying structure of the data base model used by IDEXEXTRACTOR using an XML–based declarative configuration.

The Frontend is placed between the user and a MySQL DB server. IDEXVISOR follows the *Model-View-Controller*-Approach (MVC), as is illustrated in Fig. 2. The results of IDEXEXTRACTOR are available in form of a number of tables. Each table represents one aspect of the extracted data, e.g., a table contains

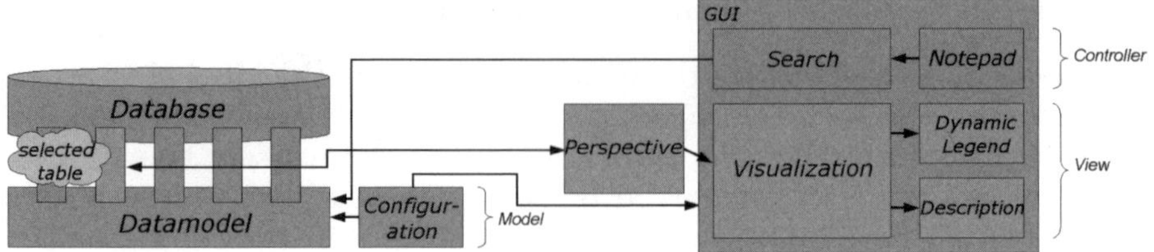

Fig. 2. Architecture of IDEXVISOR

all sentences of all documents with information about the textual content, the document link, the language and a pointer to the topic the sentence belongs to. The configuration of the Frontend specifies meta–information about the part of the data that has to be visualized. With help of the *View*, the selected data is visualized together with additional information, e.g., a dynamic help comment or a caption. Using the *Controller* a user can navigate through the selected data, he can search for specific information or he can specify a queries for a search engine. The visualization offers a set of *perspectives* that offer different viewpoints on subsets of the data that are accessible via separate tabs on the GUI. For example, one perspective shows named entities, and their inferred type (person, geographical location, or organization), clustered according to the documents they occur in; another clusters entities according to whether they are syntactical arguments (subject, object, preposition etc.) to the same predicate. The perspectives are intended to be used in combination to find the answers to questions; query terms can therefore be transferred from one perspective to another.

3   Experiments

We have performed two different experiments in order to test the two subsystems. In the first experiment we evaluated the quantity and quality of the output of IDEXEXTRACTOR. In the second experiment, a number of test users measured the performance of IDEXVISOR.

3.1   Experiments with IdexExtractor

We built a test corpus of documents related to the topic "Berlin Hauptbahnhof" by sending queries describing the topic (e.g., "Berlin Hauptbahnhof", "Berlin central station") to Google and downloading the retrieved documents specifying English as the target language. After preprocessing these documents, our corpus consisted of 55,255 sentences from 1,068 web pages, from which 10773 relations were automatically extracted and clustered.

Clustering. From the extracted relations, the system built 306 clusters of two or more instances, which were manually evaluated by two authors of this paper. 81

of our clusters contain two or more instances of exactly the same relation, mostly due to the same sentence appearing in several documents of the corpus. Of the remaining 225 clusters, 121 were marked as consistent (i.e., all instances in the cluster express a similar relation), 35 as partly consistent (i.e., more than half of the instances in the cluster express a similar relation), 69 as not useful. The clusters marked as consistent can be grouped into three major types 1) relation paraphrases 2) different instances of the same pattern 3) relations about the same topic (NE). Of our 121 consistent clusters, 76 were classified as being of the type 'same pattern', 27 as being of the type 'same topic' and 18 as being of the type 'relation paraphrases'. As many of our clusters contain two instances only, we are planning to analyze whether some clusters should be merged and how this could be achieved.

Relation extraction. In order to evaluate the performance of the relation extraction component, we manually annotated 550 sentences of the test corpus by tagging all NEs and verbs and manually extracting potentially interesting verb relations. We define 'potentially interesting verb relation' as a verb together with its arguments (i.e., subject, objects and PP arguments), where at least two of the arguments are NEs and at least one of them is the subject or an object. On the basis of this criterion, we found 15 potentially interesting verb relations. For the same sentences, the IDEX system extracted 27 relations, 11 of them corresponding to the manually extracted ones. This yields a recall value of 73% and a precision value of 41%. There were two types of recall errors: First, errors in sentence boundary detection, mainly due to noisy input data (e.g., missing periods), which lead to parsing errors, and second, NER errors, i.e., NEs that were not recognised as such. Precision errors could mostly be traced back to the NER component (sequences of words were wrongly identified as NEs) (see [4] for details). To judge the usefulness of the extracted relations, we applied the following soft criterion: A relation is considered useful if it expresses the main information given by the sentence or clause, in which the relation was found. According to this criterion, six of the eleven relations could be considered useful. The remaining five relations lacked some relevant part of the sentence/clause (e.g., a crucial part of an NE, like the 'ICC' in 'ICC Berlin').

Possible enhancements. With only 15 manually extracted relations out of 550 sentences, we assume that our definition of 'potentially interesting relation' is too strict, and that more interesting relations could be extracted by loosening the extraction criterion, for example, by extraction of relations where the NE is not the complete subject, object or PP argument, or by extraction of relations with a complex VP. Further details are presented in [4].

3.2 Experiments with IdexVisor

Seven users (average age 33; 4 males/3 females) evaluated the IDEXVISOR-Frontend. After an introduction of the functionality of the system and a demonstration of a complex search query, the users tried to answer the following four corpus–related questions via interaction with the system:

1. *Find out information about a person "Murase". The complete name, whether the person owns a company, and if so, what is the name of it.*
2. *Find one or more documents in which Hartmut Mehdorn, the CEO of the Deutsche Bundesbahn, and Wolfgang Tiefensee, the Minister of Transport occur together.*
3. *Who built the Reichstag and when? During your search also use the synonym perspective of the system in order to find alternative predicates.*
4. *How often is Angela Merkel mentioned in the corpus?*

The users were then asked about different aspects of the interaction of the system. For each question they could give a real number as answer (cf. Fig. 3) and a short text, where they could describe what they like, what they missed, and could suggest possible improvements. Here are the results of the verbal answers:

Overall usability: All users were able to answer the questions, but with different degree of difficulties in the interaction. Users stated that (a) switching between perspectives was perceived cumbersome, (b) the benefits not obvious, and (c) the possibility was not salient to them, although it was explained in the short introduction to the system. Different types of search queries were not recognized. Parts of the user interface were overlooked or actually not recognized. It was also said, that the short introduction time (10 minutes) was not enough to complete understood the system. The search speed was judge generally as "fast".

Possible Improvements: The synonyms should not be represented as a separate perspective, but should be integrated automatically with the search. The clustering according to semantic similarity should be improved. The major critical point for the representation of the relations was that the text source was only shown for some nodes and not all.

Question	Possible Answers	⊘
How did you like the introduction ?	1=useless/5=helpful	4,42
How useful is the system?	1=useless/5=helpful	4,14
Do you think you might use such a system in your daily work?	1=no/5=yes	4,14
How do you judge the computed information?	1=useless/5=very informative	3,71
How do you judge the speed of the system?	1=very slow/5=very fast	4,42
How do you judge the usability of the system?	1=very laborious/5=very comfortable	3,42
Is the graphical representation of the results useful?	1=totally not/5=very useful	3,57
Is the graphical representation appealing?	1=totally not/5=very appealing	3,71
Is the navigation useful in the system ?	1=totally not/5=very useful	3,57
Is the navigation intuitive in the system?	1=totally not/5=very intuitive	3,57
Did you have any problems using the system?	1=heavy/5=no difficulties	4,28

Fig. 3. Results of the evaluation of IDEXVISOR

4   Related Work

The tight coupling of unsupervised IE and interactive information visualization and navigation is, to best of our knowledge, novel. The work on IDEXVISOR has been influenced by work on general interactive information visualization techniques, such as [5]. Our work on relation extraction is related to previous work on domain-independent unsupervised relation extraction, in particular Shinyama and Sekine [6] and Banko et al. [7]. Shinyama and Sekine [6] apply NER, coreference resolution and parsing to a corpus of newspaper articles to extract two-place relations between NEs. The extracted relations are grouped into pattern tables of NE pairs expressing the same relation, e.g., hurricanes and their locations. Clustering is performed in two steps: they first cluster all documents and use this information to cluster the relations. However, only relations among the five most highly-weighted entities in a cluster are extracted and only the first ten sentences of each article are taken into account. Banko et al. [7] use a much larger corpus, namely 9 million web pages, to extract all relations between noun phrases. Due to the large amount of data, they apply POS tagging only. Their output consists of millions of relations, most of them being abstract assertions such as (executive, hired by, company) rather than concrete facts. Our approach can be regarded as a combination of the two approaches: Like Banko et al. [7], we extract relations from noisy web documents rather than comparably homogeneous news articles. However, rather than extracting relations from millions of pages we reduce the size of our corpus beforehand using a query in order to be able to apply more linguistic preprocessing. Unlike Banko et al. [7], we concentrate on relations involving NEs, the assumption being that these relations are the potentially interesting ones.

References

1. LingPipe, `http://www.alias-i.com/lingpipe/`
2. Stanford Parser, `http://nlp.stanford.edu/downloads/lex-parser.shtml`
3. WordNet, `http://wordnet.princeton.edu/`
4. Eichler, K., Hemsen, H., Neumann, G.: Unsupervised relation extraction from web documents. In: Proceedings of the 6th edition of the Language Resources and Evaluation Conference (LREC 2008), Marrakech, Morocco, May 28–30 (2008)
5. Heer, J., Card, S.K., Landay, J.A.: prefuse: a toolkit for interactive information visualization. In: CHI 2005, Human Factors in Computing Systems (2005)
6. Shinyama, Y., Sekine, S.: Preemptive information extraction using unrestricted relation discovery. In: Proc. of the main conference on Human Language Technology Conference of the North American Chapter of the Association of Computational Linguistics, Association for Computational Linguistics, pp. 304–311 (2006)
7. Banko, M., Cafarella, M.J., Soderland, S., Broadhead, M., Etzioni, O.: Open information extraction from the web. In: Proc. of the International Joint Conference on Artificial Intelligence (IJCAI) (2007)

Fusing DL Reasoning with HTN Planning

Ronny Hartanto[1,2] and Joachim Hertzberg[2]

[1] Bonn-Rhein-Sieg Univ. of Applied Sciences,
53757 Sankt Augustin, Germany
[2] University of Osnabrück, 49069 Osnabrück, Germany

Abstract. We describe a method for cascading Description Logic (DL) representation and reasoning on the one hand, and HTN action planning on the other. The planning domain description as well as the fundamental HTN planning concepts are represented in DL and can therefore be subject to DL reasoning; from these representations, concise planning problems are generated for HTN planning. We show that this method yields significantly smaller planning problem descriptions than regular representations do in HTN planning. The method is presented by example of a robot navigation domain.

1 Introduction

AI action planning algorithms have gained much in efficiency in recent years, as can be seen, for example, in planning system contests like IPC. Still, planning is computationally intractable in theory, dependent on the problem domain description size. Now consider an application domain like robot planning, where a planner should generate a high-level plan for execution by a robot working in, say, a six-storey office building, supposed to do delivery and manipulation tasks. No single plan in this scenario would typically be of impressive length, involving in fact only a tiny fraction of all the rooms, locations, or objects that the robot has to know about of its environment: Only small number of individuals is normally *relevant* for a single planning problem at hand. Yet, the planner has no chance of knowing or estimating what is relevant until it has finished a plan, as it is unable to reason about the domain in any other way than by generating a plan – bogged down by the full environment representation.

Here is the idea, which is described in this paper, for delivering to the planner only a substantially filtered version of the environment representation to work on for any individual planning problem. The planner type used is an HTN planner. The environment is primarily represented in terms of a Description Logic (DL) theory; basic HTN planning notions are represented in DL, too. The problem description for a concrete planning problem is generated algorithmically from the DL representation by expanding in terms of the DL ontology the concepts (methods, operators, state facts) that are part of the planning start and goal formulations in DL. In most cases, this process extracts only a small fraction of the full set of domain individuals to send to the planner, and it does so efficiently.

The rest of the paper is organised as follows: The next section sketches the DL and HTN planning background. Sec. 3 presents our approach and the two

A. Dengel et al. (Eds.): KI 2008, LNAI 5243, pp. 62–69, 2008.

extraction algorithms. After that, Sec. 4 describes an implementation, including some experiments and results. Sec. 5 concludes.

2 Description Logics and HTN Planning

Description Logics (DL) [1] is a family of state-of-the-art formalisms in knowledge representation. It is inspired by the representation and reasoning capabilities of classical semantic networks and frame systems. Yet, it is a well-defined, decidable subset of first order logic; some DL exemplars are even tractable.

A DL knowledge base or *ontology* consists of two components, namely, TBox and ABox: The TBox (*terminology*) defines objects, their properties and relations; the ABox (*assertions*) introduces individuals into the knowledge base. Typical reasoning tasks in DL are consistency check, entailment, satisfiability check, and subsumption.

Planning systems [2] also define a family of knowledge representation formalisms, yet one of a different ontological slicing than DLs. They model state facts, actions, events and state transition functions as a planning domain. The reasoning performed by a planning system is mainly to search the domain for a set and ordering of actions whose execution would lead to a goal state.

Apart from algorithmic differences, planning systems differ in the assumptions that they make about planning domains and their representations. Hierarchical Task Network (HTN) planning [2, Ch. 11] is one of the planning methods that use heuristics not only in the their algorithms, but also in the action representation: Standard procedures of elementary actions in the domain are modeled in a hierarchical way into compound tasks, which have to get disassembled during planning. In HTN planning, the objective then is to perform some set of tasks, rather than achieving a set of goals. HTN planning has been used widely in practical applications, owing to this option of pre-coding procedural domain knowledge into the task hierarchy [2].

However, note that there is no way to do DL-type reasoning in an HTN formalism, as these deal with incommensurable aspects of representing a domain. Yet, it turns out, and we will give examples later, that HTN planning can profit from DL reasoning in the domain in the sense that DL reasoning is able to determine that some subset of objects is irrelevant for a concrete planning problem. Therefore, we will propose in next section our approach to cascading HTN planning and DL reasoning.

3 General Approach

The basic idea of our approach is to split tasks between DL reasoning and HTN planning. The planning domain as well as basic HTN planning concepts are modeled in DL; DL reasoning is then used for extracting automatically a tailored representation for each and every concrete planning problem, and handing only this to the planning system. Note that the DL model can be much larger than the proper planning domain. It may include any information that we want our system to have; in particular, it may span information that we would otherwise model in separate planning domain descriptions for planning efficiency reasons. HTN planning is dedicated to finding a solution within the representation extracted by DL, which is reduced by facts, operators, and events found irrelevant by the previous DL reasoning.

Next, we describe how the necessary HTN planning concepts, independent of a domain, are represented in DL; Sec. 3.2 explains the algorithms for extracting the planning problem from the DL domain representation.

3.1 Modelling HTN Planning in DL

In order to permit the DL reasoning to generate an HTN planning problem, the DL ontology must include HTN planning itself, i.e., how actions, methods, and tasks are defined. In this section, we recapitulate some HTN planning definitions, which are required to build a complete planning problem. Then we show how the HTN planning definitions are modelled into DL formalisms. Note that space limits do not permit to define all HTN concepts from scratch here. We refer to [2, Ch. 11] for details.

Definition 1. *An* HTN planning domain *is a pair* $\mathcal{D} = (O, M)$*, where* O *is a set of operators and* M *is a set of methods.* [2, Def. 11.11]

Definition 2. *An* HTN planning problem *is a four-tuple* $\mathcal{P} = (s_0, w, O, M)$*, where* s_0 *is the initial state,* w *is the initial task network and* $\mathcal{D} = (O, M)$ *is an* HTN planning domain. [2, Def. 11.11]

Definition 3. *An* HTN method *is a four-tuple*

$$m = (name(m), task(m), subtasks(m), constr(m)),$$

where $name(m)$ *is a unique method name and its variable symbols,* $task(m)$ *is a nonprimitive task, and* $(subtasks(m), constr(m))$ *is a task network.* [2, Def. 11.10]

A task network, stated simply, is a pre-defined sub-plan, consisting of a partial order of sub-tasks. See [2, Ch. 11.5] for details.

An HTN planning algorithm, stated in a nutshell, is supposed to deliver a sequence of actions $\pi = \langle a_1, ..., a_n \rangle$ for a given planning problem $\mathcal{P} = (s_0, w, O, M)$ such that a_1 is applicable in s_0, and π is accomplishes w in the sense that it is a properly instantiated version of w.

To make these notions available for modeling individual domains, we model them on a generic level. This concerns the following four concepts in the DL ontology: *Planning-Domain, Planning-Problem, Method,* and *Operator*. This is shown in Fig. 1. The figure also shows that we introduce an additional *Planning* concept, which is itself a *Thing*; the purpose of this concept is set the generic planning concepts apart from all domain-dependent ones.Fig. 1 also specifies the four concepts in DL syntax. The *hasOperator* role domain is *Planning-Domain, Method,* and *Operator*, and role range is *Operator*. The *hasMethod* role domain is *Planning-Domain* and *Method*, and the role range is *Method*.

$Planning\text{-}Domain \sqsubseteq Planning \sqcap$
  $\exists hasMethod.Method \sqcap$
  $\exists hasOperator.Operator$
$Planning\text{-}Problem \sqsubseteq Planning \sqcap$
  $\exists hasDomain.Planning\text{-}Domain$
$Method \sqsubseteq Planning \sqcap$
  $\exists hasMethod.Method \sqcap$
  $\exists hasOperator.Operator \sqcap$
  $\geqslant 1\ useState$
$Operator \sqsubseteq Planning \sqcap$
  $\exists hasOperator.Operator \sqcap$
  $\geqslant 1\ useState$
$Planning\text{-}Domain \sqcap Planning\text{-}Problem \sqcap Method \sqcap Operator \sqsubseteq \bot$

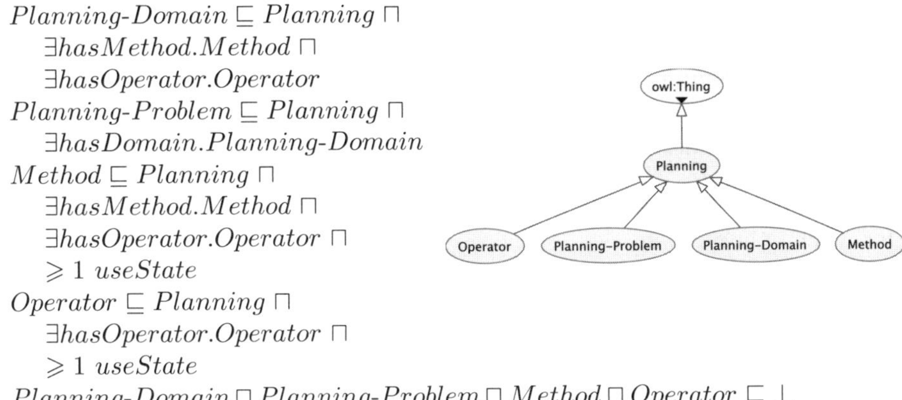

Fig. 1. HTN planning ontology

Planning-Domain corresponds to Def. 1. However, *Planning-Problem* consists only of the planning domain, although, according to Def. 2, it should be a four-tuple. The trick here lies in the *Method* and *Operator* concepts: together, they implicitly assemble the task network. Both concepts have an additional property, namely, "useState", which has the purpose of generating the initial state s_0 for $\mathcal{P}$. The task network w, which is the goal of the planning problem, will be given later by the user. Thereby the planning problem $\mathcal{P}$ is complete.

3.2 Planning Problem Generator Algorithms

The planning concepts that are described in the previous section represent the TBox in the DL formalism. Instances of these concepts must be asserted in the ABox with respect to the corresponding concepts, before a usable planning problem can be deduced from it.

Let $\mathbb{D}$ denote *Planning-Domain*, $\mathbb{O}$ *Operator*, and $\mathbb{M}$ *Method*. Let d_i represent domain instances, so $\mathbb{D} = \{d_1, ..., d_n\}$ denotes all domain instances in the knowledge base. The d_i are independent domains, but a domain could be a subset of another domain.

A domain d consists of operators $\mathbb{O}$ and methods $\mathbb{M}$, such that $d = \{\mathbb{O}, \mathbb{M}\}$. A non-primitive operator is an operator that is defined in terms of other operator(s). This is visible in the *Operator* concept, which uses the *hasOperator* property. Therefore, an operator is formalized as $o_x = \{o_1, ..., o_n | o_x \neq o_1, ...o_x \neq o_n\}$. Analogously, a method is formalized as $m_x = \{m_1, ..., m_n, o_1, ...o_p | m_x \neq m_1, ..., m_x \neq m_n\}$.

One could define the planning goal or *initial task network* in two ways: First, by choosing a domain $d \in \mathbb{D}$ and then selecting the goal from d; second, by choosing from $\mathbb{M}$. Goals are represented as methods in the knowledge base.

Using the first way, three of the four planning problem elements in Def. 2 are given, namely, $d_x = \{m_1, ..., m_n, o_1, ...o_p\}$ and w. Algorithm 1, *extractInitialState*, takes domain d as parameter and assembles the required state s_0

from the knowledge base. Its complexity is linear in the size of the sets of methods/operators and of the ontology. In detail, let $m = (|\mathbb{M}| + |\mathbb{O}|)$ and $n = (|Thing| - |Planning|)$, then the run time is $O(mn)$.

Algorithm 1. extractInitialState($input$)

Require: $input = d$
Ensure: $output = s_0$ ($hashtable$)
  for each m in $input$ **do**
    for all $state$ in $m.useState$ **do**
      $output \Leftarrow +state$
    end for
  end for
  for each o in $input$ **do**
    for all $state$ in $o.useState$ **do**
      $output \Leftarrow +state$
    end for
  end for

Using the second way, only the goal or *initial task network* w is defined. It may consist of several methods, hence $w = \{m_1, .., m_n\}$. Algorithm 2, *generateDomain*, recursively derives the planning domain from w, starting from the methods in w and unfolding them recursively. However the complexity of this algorithm is linear in $n = (|\mathbb{M}| + |\mathbb{O}|)$, since it is called only once for each method and operator. Calling Algorithm 1 with the generated planning domain will return the *initial state* s_0.

Algorithm 2. generateDomain($input$)

Require: $input = m$ or o
Ensure: $output = d$
  $output \Leftarrow +input$
  if $input$ is a method **then**
    for each m from $input.hasMethod$ **do**
      if output does not contain m **then**
        $output \Leftarrow +generateDomain(m)$
      end if
    end for
  end if
  for each o from $input.hasOperator$ **do**
    if output does not contain o **then**
      $output \Leftarrow +generateDomain(o)$
    end if
  end for

Either way, a complete planning problem for the HTN planning can be assembled from the knowledge base. Our system ensures that the generated planning problem is a valid HTN planning problem. Note that, in addition to running the two algorithms, the ABox needs to be filled with the recent domain facts.

4 Implementation

4.1 Formalisms and Systems Used

We have used the Web Ontology Language (OWL) to implement the DL formalism. OWL is the current standard, used by the W3C organisation, for representation in the Semantic Web. It has three different varieties, namely, OWL-Lite, OWL-DL, and OWL-Full [3]. OWL-DL is a higly expressive DL that is related to $\mathcal{SHOIN}(\mathbf{D})$ [4]. OWL-Lite is $\mathcal{SHIF}(\mathbf{D})$, and as such constitutes a subset of OWL-DL. OWL-Full is a superset of OWL-DL [4].

We have chosen OWL-DL for representing knowledge. The choice of OWL-DL as opposed to OWL-Lite or OWL-Full has several motivating factors. OWL-Lite's inference process is exponential (ExpTime) in the worst case, yet the expressive power is still below that of the OWL-DL [4]. In contrast, OWL-Full's power goes beyond that of DL, the language is even undecideable. OWL-DL's expressiveness is sufficient for our purposes, yet is still NExpTime [4].

Several inference engines can be used with OWL-DL, such as Pellet (an open source OWL-DL reasoner in java) [5], FaCT++ (new generation of FaCT - Fast Classification of Terminologies) [6] and RacerPro (RACER - Renamed ABox and Concept Expression Reasoner) [7]. In our work, we use Pellet.

For the planning part, several HTN planner implementations are available, such as Nonlin, SIPE-2 (System for Interactive Planning and Execution), O-Plan (Open Planning Architecture), UMCP (Universal Method Composition Planner) and SHOP (Simple Hierarchical Ordered Planner).

Four SHOP variants are available, namely, SHOP, JSHOP, SHOP2 and JSHOP2. SHOP2 won one of the top four prize at the 2002 International Planning Competition. We use JSHOP2, the Java variant of SHOP2. The Java implementation is a major reason for choosing JSHOP2, because we need to integrate the planning with reasoning into a coherent system. In addition, JSHOP2 compiles each domain description into separate domain-specific planners, thus providing a significant performance increase in the speed of planning [8].

4.2 Experiment

As an example, we use a mobile robotic navigation domain. The goal in this problem is to find a sequence of actions to bring the robot to the target destination. Domain concepts need to be defined accordingly in the DL system. These consists of two basic concepts, namely, *Actor* and *Fixed-Object*. In our case, *Robot* is an actor; *Building*, *Room*, *Corridor* are fixed objects.

We use the following concepts for modeling opportunities for action for the robot: *Door*, *OpenDoor*, *DriveableRoom* and *DriveableRoomInBuilding*. Fig. 2 sketches the ontology. Once these concepts are defined, we can assert data for the system. As usual, DL reasoning can deduce inferred instances. In this case, *DriveableRoom*, *DriveableRoomInBuilding* and *OpenDoor* are concepts, which are automatically filled by applying the DL reasoner on the model.

The navigation domain has two methods, namely, m_{navi1} and m_{navi2}. They serve the same purpose, but differ in parameters. m_{navi1} expects three

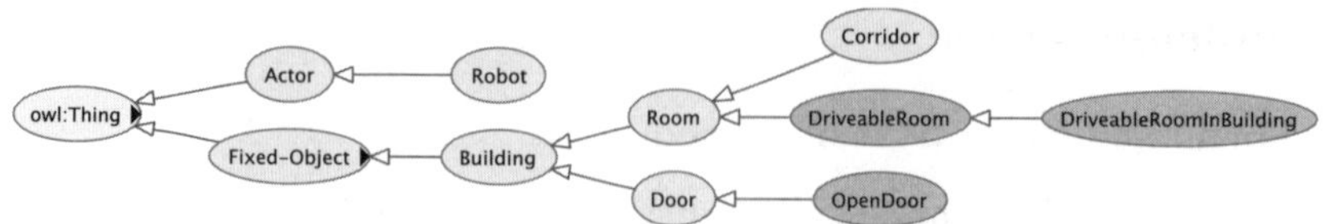

Fig. 2. Actor & Fixed-Object Ontology. Concepts for which instances are only inferred are shown in dark (orange).

parameters: actor, current-location, and destination. m_{navi2} is an abstraction of m_{navi1}, requiring only two parameters: actor and destination.

Either methods contains additional information for the planning generator as defined in Fig. 1. The complete methods and domain are:

$$m_{navi1} = \{o_{drive-robot}, o_{visit}, o_{unvisit}, s_{room}, s_{robot-at}, s_{adjacent-room}\},$$
$$m_{navi2} = \{m_{navi1}, s_{robot-at}\}$$
$$d_{navi-domain} = \{m_{navi1}, m_{navi2}, o_{drive-robot}, o_{visit}, o_{unvisit}\}.$$

The ontology allows to distinguish objects of some type from those objects of the same type that have a certain required property; for example, there are *Rooms* and *DrivableRooms*, i.e., those which are known from sensor data or other information to be currently accessible. (Note that the robot cannot open doors by itself in our tiny example model!) This distinction is exploited when a goal is to be instantiated: We can generate a goal instance sufficiently specialized to filter out irrelevant objects. For example, by specializing s_{room} and $s_{adjacent-room}$ with $s_{driveableroom}$ and $s_{adjacent-driveableroom}$, the generated planning problem will have only states for the rooms with an open door. Getting even more special within the ontology in Fig. 2, one could restrict rooms to *DriveableRoomInBuilding*, i.e., those ones accessible that are located in the current building.

In our DL domain model we had 12 methods, 8 operators and 3 planning domains for the test. A user can choose which domain or goal to use for planning, as described above. Choosing $d_{navi-domain}$ or m_{navi2} will always produce a correct planning domain, which has only 2 methods and 3 operators. However, m_{navi1} produces a smaller planning domain with 1 method and 3 operators. Even with one method less, m_{navi1} is still a valid planning domain.

Table 1 gives an impression of how specializing alters the number of states s_0, based on the number of rooms to consider in some concrete planning problem. As expected, using a more specific concept reduces the number of generated states s_0, which can improve the planning efficiency by reducing the search space.

Table 1. Generated states for different specializations of the *Fixed-Object* concept

Condition	Number of states
s_{room}	16
$s_{room-in-building}$	8
$s_{driveable-room}$	8
$s_{driveable-room-in-building1}$	5
$s_{driveable-room-in-building2}$	3

5 Conclusions

In this paper, we propose a novel approach for integrating DL reasoning and HTN planning. Domain concepts are represented in DL rather than in some special-purpose planning domain representation format like PDDL. Using DL inference, the information provided for the HTN planner is filtered during the process of generating the planning problem from the DL representation, allowing more compact planning problems to be generated, and thereby reducing the planner's search space. Modeling in DL sub-concepts directly related to action execution, such as *DriveableRoom*, is the basis for this filtering process.

In our approach, different planning domains may coexist in the same knowledge base without affecting the planning performance. In our mobile robot domain, additional information is stored in the KB to be used later by the execution layer, such as robot-specific commands for physically executing primitive operators.

References

1. Baader, F., et al.: The Description Logic Handbook: Theory, Implementation, and Applications. Cambridge University Press, Cambridge (2003)
2. Ghallab, M., Nau, D., Traverso, P.: Automated Planning: Theory and Practice. Morgan Kaufmann, San Francisco (2004)
3. Dean, M., et al.: OWL web ontology language reference. W3C Recommendation (2004), http://www.w3.org/TR/owl-ref/
4. Horrocks, I., Patel-Schneider, P.F., van Harmelen, F.: From SHIQ and RDF to OWL: The Making of a Web Ontology Language. Journal of Web Semantics 2003 (2003)
5. Sirin, E., et al.: Pellet: A Practical OWL-DL Reasoner. Journal of Web Semantics 5(2) (2007)
6. Tsarkov, D., Horrocks, I.: FaCT++ Description Logic Reasoner: System Description. In: Furbach, U., Shankar, N. (eds.) IJCAR 2006. LNCS (LNAI), vol. 4130, pp. 292–297. Springer, Heidelberg (2006)
7. Haarslev, V., Möller, R.: Racer: An OWL Reasoning Agent for the Semantic Web. In: Proceedings of the International Workshop on Applications, Products and Services of Web-based Support Systems, Ê Halifax, Canada, October 13, pp. 91–95 (2003)
8. Ilghami, O., Nau, D.S.: A General Approach to Synthesize Problem-Specific Planners. Technical report, Department of Computer Science, ISR, and IACS University of Maryland (2003)

Multi-value Classification of Very Short Texts

Andreas Heß, Philipp Dopichaj, and Christian Maaß

Lycos Europe GmbH, Gütersloh, Germany
{andreas.hess,philipp.dopichaj,christian.maass}@lycos-europe.com

Abstract. We introduce a new stacking-like approach for multi-value classification. We apply this classification scheme using Naive Bayes, Rocchio and kNN classifiers on the well-known Reuters dataset. We use part-of-speech tagging for stopword removal. We show that our setup performs almost as well as other approaches that use the full article text even though we only classify headlines. Finally, we apply a Rocchio classifier on a dataset from a Web 2.0 site and show that it is suitable for semi-automated labelling (often called tagging) of short texts and is faster than other approaches.

1 Introduction

Multi-value text classification is an interesting and very practical topic. In many applications, a single label only is not enough to appropriately classify documents. This is especially true in many applications on the web. As opposed to traditional documents, some texts on the web, especially on Web 2.0 sites, are very short, for example pin-board entries, comments to blog posts or captions of pictures or videos. Sometimes these texts are mere snippets, being at most one or two sentences long. Yet, in some Web 2.0 applications, labelling or tagging such short snippets does not only make sense but could be the key to success. Therefore we believe it is important to investigate how multi-value text classification algorithms perform when very short texts are classified. To test this, we classified news articles from the well-known Reuters data-set based only on the headlines and compared the results to older approaches in literature that used the full text. We applied the same algorithm to a dataset from Web 2.0 site Lycos iQ. An empirical evaluation shows that text classification algorithms perform well in both setups.

The remainder of this paper is organised as follows: first, we present a new stacking approach for multi-value classification. By comparing the performance of classifiers trained only on the short headlines of the well-known Reuters news articles benchmark to results achieved with similar classifiers using the full article text we show that classification of very short texts is possible and the loss in accuracy is acceptable. Second, we present an application of text classification for tagging short texts from a Web 2.0-site. We demonstrate that presenting suggestions to the user can greatly improve the quality of tagging.

A. Dengel et al. (Eds.): KI 2008, LNAI 5243, pp. 70–77, 2008.

2   Stacking for Multi-value Classification

In contrast to a standard single-value classification where each instance is assigned exactly one class label, multi-value (also called multi-label) classification allows for assigning an arbitrary number of labels to instances. A classical example where multi-value classification makes sense is the labelling of texts that have more than one topic. Class labels for such texts could be either of different granularity or they could even be orthogonal. For example, a news article about the presidential elections in the United States could be labelled as 'politics', 'election' and 'USA'. The labels 'politics' and 'election' are based on a topic, but 'election' is a more detailed description of the content. As opposed to the other labels, 'USA' refers to location.

Although multi-value classification is a natural solution for many machine-learning problems, most algorithms can only handle single-value classification. Therefore it is common practise in multi-value classification that single-value classification algorithms are adapted by means of some combination method; see [6] for a recent survey. The most common approach is 'one-vs-all' classification: for each class label, a binary classifier that decides whether an instance is a member of this specific class is trained. This approach has a strong disadvantage: the datasets the binary classifiers are trained on are imbalanced. Consider e. g. a kNN-classifier with $k = 3$. If an instance that is to be classified is equidistant to three instances that have a single class label each and are of three different classes, all three binary classifiers would classify the new instance as a negative example. Therefore, no prediction is made, although this is most probably not correct, and a more intuitive classification would be to assign all three class labels. This argument against the one-vs-all scheme holds for other classifiers as well. Support Vector Machines are known to be sensitive towards imbalanced datasets. Godbole and Sarawagi [3] exploit the relations between classes that exist if class labels are not independent from each other by using a stacking approach to add the predictions of other binary classifiers to train another set of classifiers in a second pass. Many classification algorithms output a ranked list of predicted class labels with confidence values. Well-known algorithms in this group are the Naive Bayes, Rocchio and kNN classifiers. When using such an algorithm, another scheme for multi-value classification is thresholding: selecting class labels that are within the top n predictions or have a confidence score higher than a certain threshold. Different methods of thresholding have been discussed by Yang [9]. The *SCut*-method applies a different threshold for each class. In the *RTCut*-method, rank and confidence score are combined to a single value before thresholding is applied.

2.1   Description of Our Algorithm

We propose a multi-value classification scheme which we call MVS (multi-value classification stacking) that is similar to *RTCut*: we use a classifier with confidence scores and rank, but instead of creating an artificial score, we train a binary meta-classifier for each class on the confidence score and rank as computed by the base classifier. The meta-classifiers decide whether a specific class label predicted by the base classifier should be included in the final set of predictions. In our implementation, we used JRip as implemented in WEKA [7] as meta-classifier. To train our algorithm, we first train the base classifier. Second, we train one binary meta-classifier per class label. The training examples for the meta-classifiers are created as follows: We classify each instance in the training set with the base classifier and iterate over the top n predictions. For each prediction, we check whether the predicted class label q is a true class label of the instance. If this is the case, we add the rank and confidence score of the prediction as a positive example to the training set for the meta-classifier for class q. Otherwise, we add it as a negative example. Finally, we train the meta-classifiers for each class on their respective training sets. Algorithm 2.1 illustrates the training phase of our classification scheme.

Algorithm 1. MVS: Training Phase

Require: $T_{1..t}$, training instances
Require: B, base classifier
Require: $M_{1..l}$, meta-classifiers (one per class label)
  Train B on T
  $N_{1..t} \Leftarrow$ set of instances for meta-classifiers (initially empty)
  for $j = 1$ to t **do**
    $C \Leftarrow$ true class labels of T_j
    $P \Leftarrow$ top n predictions of B for T_j
    for $l = 1$ to n **do**
      $q \Leftarrow$ class label of P_l
      if $q \in C$ **then**
        add P_l as positive example to N_q
      else
        add P_l as negative example to N_q
      end if
    end for
  end for
  for $m = 1$ to l **do**
    Train M_m on N_m
  end for

The classification phase of our scheme is straightforward: first, we classify the instance using the base classifier and iterate over the top n predictions. For each prediction, we use the respective meta-classifier to determine whether the prediction is true or false. It should be noted that for some classification algorithms our MVS scheme reduces the overall complexity compared to one-vs-all. Consider for example a Rocchio classifier: When trained as a multi-class classifier, we need to compute the centroid for each class. When Rocchio is used in a one-vs-all setup, we need to compute the centroid for each *negative* class as well. Another advantage of our scheme is that it can be combined with ensemble learning. In a variety of tasks, ensembles of several classifiers have been shown to be more effective (e.g., [1]). The intention is that two or more diverse classifiers (that are assumed to have independent classification errors) are combined so that classification errors by one classifier will be compensated for by the correct predictions of another classifier. One classical ensemble learning scheme is stacking [8]: a meta-learner is trained on the output of two or more base classifiers. The basic version of our scheme can be regarded as stacking with only one base classifier. It is straightforward to extend our scheme to use more than one base classifier: the meta-classifiers simply use the output of more than one base classifier as features.

2.2   Empirical Evaluation

We decided to implement the MVS scheme with the widely used Naive Bayes, Rocchio and kNN classifiers. The Rocchio algorithm has the known disadvantage of becoming inaccurate when classes are not spheres of similar size in vector space, and it does not handle non-spherical classes (e. g. multi-modal classes that consist of more than one cluster) very well. However, Rocchio classification has been shown to work well for text classification when the texts are short and of similar length. Since a human reader is usually able to recognise the topic of a newspaper article just by looking at the headline, we experimented with the categorisation of very short texts. For stopword removal and further dimensionality reduction we used a part-of-speech tagger and selected only verbs, nouns and proper nouns for inclusion in the feature set.

Table 1. Performance results (F1 in percent) of different setups on the top 10 classes of the Reuters-21578 dataset. The first six classifiers were trained on the headlines only; the last three classifiers were trained on the full text and are listed for comparison, these results were reported in [2] (Naive Bayes and Rocchio) and [4] (kNN). Classifiers were (from left to right) N-1/N-2: Naive Bayes, one-vs-all/MVS; R-1/R-2: Rocchio, one-vs-all/MVS; K-1/K-2: kNN, one-vs-all/MVS; S-3: MVS with kNN, Naive Bayes and Rocchio combined; N-x: Naive Bayes, full text, one-vs-all; F-x: Findsim (similar to Rocchio), full text, one-vs-all; K-x: kNN, full text, one-vs-all.

	Headlines						Full text		
	N-1	N-2	R-1	R-2	K-1	K-2	S-3	N-x	F-x	K-x
earn	93	94	73	91	84	80	95	96	93	97
acq	81	84	72	60	65	63	71	88	65	92
money-fx	51	45	46	60	51	59	61	57	47	78
grain	50	61	57	74	68	68	76	79	68	82
crude	55	70	43	73	54	63	73	80	70	86
trade	45	60	25	69	47	56	64	64	65	77
interest	50	56	31	68	60	61	63	65	63	74
ship	38	53	31	52	34	47	57	85	49	79
wheat	37	57	46	63	40	55	58	70	69	77
corn	32	55	13	61	38	47	61	65	48	78
Mavg(10)	53	64	44	67	54	60	68	75	64	82

We tested our algorithm on the well known Reuters-21578 collection. We used the well-known *ModApte*-split to separate training and test data. Unlabelled instances were kept. Table 1 shows the results. In preliminary experiments, we used a thresholding approach similar to *SCut* instead of MVS. These settings performed consistently worse and are not presented here. With all classifiers tested, the MVS scheme clearly outperforms the traditional one-vs-all-setup. When comparing the performance of our setup to the results presented in [2], we can conclude that classification of news articles based only on headlines is possible with only a small, acceptable loss in accuracy compared to similar classifiers trained on the full article text. The Rocchio algorithm in the MVS setting trained on the headlines even outperformed the Findsim classifier (a variation of Rocchio) trained on the full text. In general, we observe that Rocchio performs surprisingly well, which we assume is due to the fact that the text are very short and equally long, a situation were Rocchio has been shown to perform well. A stacking-approach as described above where kNN, Naive Bayes and Rocchio have been combined performs best for most classes, however, on some classes, the individual classifiers performed better, strongly affecting macro-averaged F1. We conclude that apparently the meta-classifier tended to overfit and the rules it produced are not optimal. This problem could probably be solved by validating the rules on a hold-out set.

3 Semi-automated Tagging

Although text classification is an old and well-researched topic in information retrieval and machine learning it has not been widely used for automatic tagging in Web-2.0 applications yet. An exception is AutoTag [5], a system that uses a k-nearest-neighbour classifier for automated tagging of blog posts. AutoTag [5] uses a search engine to locate similar blog posts. The search query is derived from the text that is to be classified using statistical query rewriting techniques. In the next step, tags from the search results are aggregated and re-ranked using information about the user. Yet, this method of predicting tags for posts has a disadvantage: rewriting the query at classification time is computationally costly. Given that many annotations are plausible and the user is involved in the classification process, it is not necessary that the algorithm predicts the exact set of true annotations. Opposed to the labelling in the Reuters benchmark, it is therefore acceptable that a classifier outputs a ranked list with suggested tags. Considering the high number of classes and the need for an incremental learning algorithm, using vector space classification algorithms such as kNN or Rocchio is a logical choice.

3.1 Experimental Setup

To evaluate our approach, we used a corpus of 116417 questions from the question-and-answer-community web site Lycos iQ that were tagged with 49836 distinct tags. We used the same setup of the Rocchio classifier as described above in section 2. Preliminary experiments showed that using Naive Bayes classification is not viable due to the high number of classes. Also, the Rocchio classifier performed faster that Naive Bayes. For comparison, we included a kNN classifier with $k = 10$ that queries the index for the ten nearest questions (postings) in the database and aggregates the tags from the results. This approach is close to AutoTag [5], but because we perform stopword removal it is not needed to rewrite the query on classification time.

3.2 Empirical Evaluation

Given the nature of the two algorithms, we expect that Rocchio classification will be faster; a factor that is very important in an interactive setting, when users are not willing to accept long response times. We measured the classification time per instance for both approaches on an Intel Core 2 machine with 1.86 GHz and 1 GB RAM. As expected, Rocchio classification was much faster then kNN. The classification time for each instance was 155 ms for kNN and 57 ms for Rocchio.

It is important to note that the tags assigned by users should not be regarded as a gold standard. Tags are not drawn from an ontology, taxonomy or controlled vocabulary, but are free text entered by users and thus prone to spelling mistakes. Also, inexperienced users tend to assign either no tags at all, only very few tags or they tag inconsistently. Given the large number of users, we also expect that users use different synonyms to denote the same concept. Due to these ambiguities and inconsistencies we expect that the accuracy of any automated approach is considerably lower than its true usefulness. In tests kNN only achieved a 26% precision for its top prediction, Rocchio reached 32% precision.

In order to circumvent the problem of noise in the test set, we distributed questionnaires and had test persons check the plausibility of tags suggested by our semi-automated approach. To reduce the workload for the test persons and because it outperformed the kNN-classifier in the automated tests, we decided to test only the Rocchio-style approach. For comparison, we also had the test persons check the precision of the user-assigned tags, since we assumed many nonsensical or inconsistent tags among them. Every test person was given one or two chunks of 100 out of a random sample of 200 questions that were either machine-tagged or hand-tagged. Every question was checked by four persons to average out disagreement about the sensibility of tags.

As expected, we could observe that there was a big disagreement among the test persons and the users who originally tagged the questions as well as between the test persons themselves. As explained above, the total 200 questions that were evaluated were split in two sets of 100 questions, yielding four different questionnaires (two for the original user-assigned tags and two for machine-annotated tags) and each chunk of 100 questions was checked by four persons. Each test person was checking at most two sets of questions. To highlight the huge difference of the several test persons, we report the individual results in the table below. For the human-annotated tags, we evaluated precision, defined as the number of useful tags divided by the total number of assigned tags. For the machine-assigned tags, we also report the fraction of questions with at least one correctly predicted tag.

For all manual tests, we evaluated the algorithms with five suggested tags only. We believe that in a real-world semi-automated setting, we cannot assume that an inexperienced user is willing to look at more than five tags. The questions that were manually tagged had mostly three tags each, some of them only two and very few questions had more than three tags.

Table 2. Evaluation on Lycos iQ dataset. Results are shown for tags *assigned* by the users and for the tags *suggested* by our system.

Test	TP	TP+FP	avg. Prec.
assigned tags	1535	1856	0.83
suggested tags	1866	3360	0.56

Test	Person 1	Person 2	Person 3	Person 4
Set 1, assigned tags, prec.	0.89	0.89	0.93	0.96
Set 2, assigned tags, prec.	0.52	0.73	0.73	0.87
Set 1, suggested tags, prec.	0.41	0.52	0.53	0.71
Set 2, suggested tags, prec.	0.51	0.54	0.59	0.65
Set 1, at least one correct	0.84	0.84	0.86	0.87
Set 2, at least one correct	0.87	0.87	0.91	0.91

As expected, different persons disagreed significantly on both the human-annotated and the machine-annotated tags (see table 2). It is interesting to note when looking at the second set of questions, that, although the human annotations on this set were rated worse than those from the first set, the tags suggested by our algorithm were on average rated slightly better. Since we envision a semi-automated scenario with human intervention, we see this as a confirmation that automatically suggested tags can help to improve the quality of tagging.

When looking at macro-averaged precision, it is obvious that a classification system is still not good enough for fully automated tagging. However, it is important to note that even the human-annotated questions were rated far below 100% correct by the test persons. More than half of the suggested tags were rated as useful by the test persons. We believe that this is certainly good enough for a semi-automated scenario, were users are presented a small number of tags to choose from. In absolute numbers, interestingly, the automatic classifier produced more helpful tags than were assigned by users, even compared to the number of all user-assigned tags, not just the ones perceived as helpful by the test persons. We believe that this confirms our hypothesis that users will assign more tags when they are supported by a suggestion system. However, this can only be finally answered with a user study done with a live system.

Finally, the high number of questions where at least one of the predictions by the algorithm was correct underlines our conclusion that semi-automated tagging is good enough to be implemented in a production environment. In almost nine out of ten cases there was at least one helpful tag among the suggestions.

4 Conclusion

In this paper, we have made two contributions: first, we introduced a new mode for adapting single-label classifiers to multi-label classification we called MVS. This scheme has the advantage of being more accurate and at the same time faster than the traditional one-vs-all classification and is easily extensible to using multiple base classifiers in an ensemble. Second, we introduced part-of-speech tagging as a method for stopword removal and showed that multi-value text classification is possible at acceptable accuracy even if the texts are very short. We applied this on the real-world task of tagging for Web 2.0 and have shown that it performs well enough to be used in a semi-automatic setting. In future work, we want to extend our research in various directions. Our experiments with the Reuters dataset left some important questions open. For example, we are currently ignoring relations between classes, an approach that proved successful [3]. Also, more experiments on different datasets and classifiers are needed.

Acknowledgements. The research presented in this paper was partially funded by the German Federal Ministry of Economy and Technology (BMWi) under grant number 01MQ07008. The authors are solely responsible for the contents of this work. We thank our colleagues at Lycos Europe who gave valuable feedback.

References

1. Dietterich, T.G.: Ensemble methods in machine learning. In: Proc. of the First Int. Workshop on Multiple Classifier Systems (2000)
2. Dumais, S., Platt, J., Heckerman, D., Sahami, M.: Inductive learning algorithms and representations for text categorization. In: CIKM 1998: Proc. of the 7th International Conf. on Information and Knowledge Management. ACM, New York (1998)
3. Godbole, S., Sarawagi, S.: Discriminative methods for multi-labeled classification. In: Proc. of the 8th Pacific-Asia Conf. on Knowledge Discovery and Data Mining (PAKDD) (2004)
4. Joachims, T.: Text categorization with support vector machines: Learning with many relevant features. In: Proc. European Conf. on Machine Learning (ECML). Springer, Heidelberg (1998)
5. Mishne, G.: Autotag: a collaborative approach to automated tag assignment for weblog posts. In: Proc. of the 15th Int. World Wide Web Conference. ACM Press, New York (2006)
6. Tsoumakas, G., Katakis, I.: Multi-label classification: An overview. International Journal of Data Warehousing and Mining 3(3), 1–13 (2007)
7. Witten, I.H., Frank, E.: Data Mining: Practical Machine Learning Tools with Java Implementations. Morgan Kaufmann, San Francisco (1999)
8. Wolpert, D.H.: Stacked generalization. Neural Netw. 5(2) (1992)
9. Yang, Y.: A study of thresholding strategies for text categorization. In: Proc. of the 24th Int. ACM SIGIR Conf. (2001)

Analysis and Evaluation of Inductive Programming Systems in a Higher-Order Framework*

Martin Hofmann, Emanuel Kitzelmann, and Ute Schmid

University of Bamberg, Germany
`lastname.surname@uni-bamberg.de`

Abstract. In this paper we present a comparison of several inductive programming (IP) systems. IP addresses the problem of learning (recursive) programs from incomplete specifications, such as input/output examples. First, we introduce conditional higher-order term rewriting as a common framework for inductive program synthesis. Then we characterise the ILP system GOLEM and the inductive functional system MAGICHASKELLER within this framework. In consequence, we propose the inductive functional system IGOR II as a powerful and efficient approach to IP. Performance of all systems on a representative set of sample problems is evaluated and shows the strength of IGOR II.

1 Introduction

Inductive programming (IP) is concerned with the synthesis of *declarative* (logic, functional, or functional logic) programs from incomplete specifications, such as input / output (I/O) examples. Depending on the target language, IP systems can be classified as inductive logic programming (ILP), inductive functional programming (IFP) or inductive functional logic programming (IFLP).

Beginnings of IP research [1] addressed inductive synthesis of functional programs from small sets of positive I/O examples only. Later on, some ILP systems had their focus on learning recursive logic *programs* in contrast to learning classifiers (FFOIL [2], GOLEM [3], PROGOL [4], DIALOGS-II [5]). Synthesis of functional logic programs is studied with the system FLIP [6]. Now, induction of functional programs is covered by the analytical approaches IGOR I [7] and IGOR II [8] and by the search-based approach ADATE [9] and MAGICHASKELLER [10]. Analytical approaches work example-driven and are guided by the structure of the given I/O pairs, while search-based approaches enumerate hypothetical programs and evaluate them against the I/O examples.

At the moment, neither a systematic empirical evaluation of IP systems under a common framework nor a general vocabulary for describing and comparing the different approaches in a systematic way exists. Both are necessary for further progress in the field exploiting the strengths and tackling the weaknesses of current approaches.

We present conditional combinatory term rewriting as a uniform framework for describing IP systems and characterise and compare some systems in it. Then, we introduce IGOR II, which realises a synthesis strategy which is more powerful and not less efficient as the older approaches, and evaluate their performance on a set of example problems and show the strength of IGOR II. We conclude with some ideas on future research.

---

* Research was supported by the German Research Community (DFG), grant SCHM 1239/6-1.

A. Dengel et al. (Eds.): KI 2008, LNAI 5243, pp. 78–86, 2008.

2 A Unified Framework for IP

2.1 Conditional Constructor Systems

We sketch term rewriting, conditional constructor systems and an extension to higher-order rewriting as, e.g., described in [11]. Let Σ be a set of function symbols (a signature), then we denote the set of all terms over Σ and a set of variables $\mathcal{X}$ by $\mathcal{T}_\Sigma(\mathcal{X})$ and the (sub)set of ground (variable free) terms by $\mathcal{T}_\Sigma$. We distinguish function symbols that denote datatype *constructors* from those denoting (user-)defined functions. Thus, $\Sigma = \mathcal{C} \cup \mathcal{F}, \mathcal{C} \cap \mathcal{F} = \emptyset$ where $\mathcal{C}$ contains the constructors and $\mathcal{F}$ the defined function symbols. Induced programs are represented in a functional style as sets of recursive rewrite rules over a signature Σ, so called *constructor (term rewriting) systems (CS)*.

The lefthand side (lhs) $l := F(p_1, \ldots, p_n)$ of a rewrite rule $l \to r$ consists of a defined function symbol F and a pattern $p_i \in \mathcal{T}_\mathcal{C}(\mathcal{X})$ which is built up from constructors and variables only. We call terms from $\mathcal{T}_\mathcal{C}(\mathcal{X})$ *constructor terms*. This allows for *pattern matching* known from functional languages such as HASKELL. Consequently, all variables of the righthand side (rhs) must also occur in the lhs, i.e. they must be *bound* (by the lhs). If no rule applies to a term the term is in *normal form*. If we apply a defined function to ground constructor terms $F(i_1, \ldots, i_n)$, we call the i_i *inputs* of F. If such an application normalises to a ground constructor term o we call o *output*.

In a *conditional constructor system (CCS)*, each rewrite rule may be augmented with a *condition* that must be met to apply the rule. Conditional rules are written: $l \to r \Leftarrow v_1 = u_1, \ldots, v_n = u_n$ (cf. Fig. 1(1)). So, a condition is an ordered conjunction of equality constraints $v_i = u_i$ with $v_i, u_i \in \mathcal{T}_\Sigma(\mathcal{X})$. A constraint $v_i = u_i$ holds if, after instantiating the lhs and evaluating all $v_j = u_j$ with $j < i$, (i) v_i and u_i evaluate to the same normal form, or (ii) v_i is a pattern that matches u_i and binds the variables in v_i.

To lift a CCS into the higher-order context and extend it to a (conditional) combinatory rewrite system ((C)CRS) [11] we introduce meta-variables $\mathcal{X}_M = X, Y, Z, \ldots$ which are assumed to be different from any variable in $\mathcal{X}$. Meta-variables occur as $X(t_1, \ldots, t_n)$ and allow for generalisation over functions with arity n. To preserve the properties of a CS, we need to introduce an abstraction operator $[-]-$ to bind variables locally to a context. The term $[A]t$ is called abstraction and the occurences of the variable A in t are bound. For example the recursive rule for the well known function map would look like $map([A]Z(A), cons(B, C)) \to cons(Z(B), map([A]Z(A), C))$ and would match a term like $map([A]square(A), cons(1, nil))$.

2.2 Target Languages in the CCRS Framework

To compare all systems under equal premises, we have to fit the different occurences of declarative languages into the CCRS framework[1]. Considering functional target languages, the underlying concepts are either based on abstract theories, as e. g. equational theory [6], constructor term rewriting systems [8], or concrete functional languages as

---

[1] Note the subset relationship between that CS, CCS, and CCRS. So, if the higher-order context is of no matter we use the term CCS, otherwise CCRS.

(1) CCRS

```
multlast([])   -> []
multlast([A]) -> [A]
multlast([A,B|C]) -> [D,D|C]
        <= [D|C] = multlast([B|C])
```

(2) Functional (Haskell)

```
multlast([])      = []
multlast([A])     = [A]
multlast([A,B|C]) =
      let [D|C] = multlast([B|C])
      in [D,D|C]
```

(3) Logic (Prolog)

```
multlast([], []).
multlast([A], [A]).
multlast([A,B|C],[D,D|C]) :-
          multlast([B|C],[D|C]).
```

Fig. 1. Equivalent programs of *multlast* (overwriting a list with last element)

ML [9] or HASKELL[10]. Applying the CCRS framework to IFP or IFLP systems is straight forward, since they all share the basic principles and functional semantics.

In addition to pattern matching and functional operational semantics of CS, CCS can express constructs such as `if`-, `case`-, and `let`-*expressions* in a rewriting context. An `if`-expression is modelled by a condition $v = u$ (case (i) in Sect. 2.1). A `case`-expression is modeled following case (ii), where $v \in \mathcal{T}_C(\mathcal{X})$ and $v \notin \mathcal{X}$. If $v \in \mathcal{X}$, case (ii) models a local variable declaration as in a `let`-expression. Fig. 1 shows a CCRS for a HASKELL program containing a `let`-expression.

In the context of IP, we only consider logic target programs which represent functions, i.e., programs where the output is uniquely determined by the input. Such programs usually are expressed as "functional" predicates such as *multlast* in Fig. 1(3). A *functional predicate* is a relation $r(V_1, \ldots, V_n)$, where for any ground "input" values $i_1, \ldots, i_{n-1}$ of $V_1, \ldots, V_{n-1}$, there is a single "output" value o for V_n such that $\langle i_1, \ldots, i_{n-1}, o \rangle$ belongs to r. The evaluation binds o to V_n using the *input parameters* $V_1, \ldots, V_{n-1}$. If a predicate does not have an output variable it is a "boolean" predicate in the usual sense and evaluates to *true* or *false* if all input parameters are bound. Note that input and output parameters do not have to be variables but may also be constructor terms (containing variables) as known from PROLOG.

In the context of IP, ILP systems require all variable bindings to be directly or indirectly determined by the bindings of the input variables. A Horn clause $h \leftarrow b_1, \ldots, b_n$ is transformed straight forward to a conditional rewrite rule $h' \rightarrow V_h \Leftarrow V_{b_1} = b'_1, \ldots, V_{b_n} = b'_n$, where h' and b'_i are the predicates h and b_i stripped off their output parameters $V_h, V_{b_i} \in \mathcal{T}_C(\mathcal{X})$. For boolean predicates holds $V_h, V_{b_i} \in \{true, false\}$. Transforming Horn clauses containing functional predicates into CCSs is a generalisation of representing Horn clauses as conditional identities as shown in [12].

2.3 IP in the CCRS Framework

Let us now formalise the IP problem in the CCRS setting. Given a CCRS, both, the set of defined function symbols $\mathcal{F}$ and the set of rules R be further partitioned into disjoint subsets $\mathcal{F} = \mathcal{F}_T \cup \mathcal{F}_B \cup \mathcal{F}_I$ and $R = E^+ \cup E^- \cup BK$, respectively. $\mathcal{F}_T$ are the function symbols of the functions to be synthesised, also called *target functions*. $\mathcal{F}_B$ are the symbols of predefined functions that can be used for synthesis. These can either be built in or defined by the user in BK (see below). $\mathcal{F}_I$ is a pool of function variables that can be used for defining invented functions on the fly. E^+ is the set of *positive examples* or *evidence* and E^- the set of *negative examples*, both containing a finite number of I/O pairs as unconditional rewrite rules $F(t_1, \ldots, t_n) \rightarrow r$, where $F \in \mathcal{F}_T$ and $t_1, \ldots, t_n, r \in \mathcal{T}_{\mathcal{C}}(\mathcal{X})$. However, the rules in E^- are interpreted as inequality constraints. BK is a finite set of rules $F(t_1, \ldots, t_n) \rightarrow r \Leftarrow v_1 = u_1 \wedge \ldots \wedge v_n = u_n$ defining auxiliary concepts that can be used for synthesising the target function, where $F \in \mathcal{F}_B$, $t_i \in \mathcal{T}_C(\mathcal{X} \cup \mathcal{X}_M)$ for $i = 1 \ldots n$, and $r, u_i, v_i \in \mathcal{T}_B(\mathcal{X} \cup \mathcal{X}_M)$.

With such a given CCRS, the IP task can be now described as follows: find a finite set R_T of rules $F(t_1, \ldots, t_n) \rightarrow r \Leftarrow v_1 = u_1 \wedge \ldots \wedge v_n = u_n$ (or program for short) where $F \in \mathcal{F}$, $t_1, \ldots, t_n \in \mathcal{T}_C(\mathcal{X} \cup \mathcal{X}_M)$, and $r, u_i, v_i \in \mathcal{T}_\Sigma(\mathcal{X} \cup \mathcal{X}_M)$, such that it covers all positive examples ($R_T \cup BK \models E^+$, *posterior sufficiency or completeness*) and none of the negative examples ($R_T \cup BK \not\models E^-$, *posterior satisfiability or consistency*). In general, this is done by discriminating between different inputs using patterns on the lhs or conditions modelling `case`-expressions and computing the correct output on the rhs. Constructors, recursive calls, functions from the background knowledge, local variable declarations, and invented functions can be used for this. An invented function is hereby a function which symbol occurs only in $\mathcal{F}_I$, i.e. is neither a target function nor defined in BK and is defined by the synthesis system on the fly.

However, there is usually an infinite number of programs satisfying these conditions, e.g. E^+ itself, and therefore two further restrictions are imposed: A restriction on the terms constructed, the so called restriction bias and a restriction on which terms or rules are chosen, the preference bias.

The *restriction bias* may allow only a specific subset of the terms defined for u_i, v_i, t_i, r in a rule $F(t_1, \ldots, t_n) \rightarrow r \Leftarrow u_1 = v_1 \wedge \ldots \wedge u_n = v_n$. It may restrict nested or mutual recursion, allow for or prohibit abstraction and meta-variables, i.e. higher-order context, or demand the rhs to follow a certain program scheme.

The *preference bias* imposes a partial ordering on terms, lhss, rhss, conditions or whole programs defined by the CCS framework and the restriction bias. A correct program is optimal w.r.t. this ordering and satisfying completeness and consistency.

3   Systems Description in the CCRS Framework

So lets put on the CCRS glasses and have a closer look at the systems. The scope of this paper allows us only to consider GOLEM, MAGICHASKELLER, and IGOR II as representatives of ILP, higher-order search-based and analytic approaches. They were chosen as the most powerfull or most suitable to exemplarily illustrate the strength and weaknesses of their kind. Where appropriate, we will refer to other systems to stress differences or similarities.

GOLEM. [3] is, as e.g. FOIL/FFOIL, one of the classic ILP systems. It uses a bottom-up, or example driven approach based on Plotkin's framework of relative least general generalisation (*rlgg*). This avoids searching a large hypothesis space for consistent hypothesis as, but rather constructs a unique clause covering a subset of the provided examples relative to the given background knowledge.

$\mathcal{C}$ and $\mathcal{F}_B$ are unrestricted, but $\mathcal{F}_T$ is restricted to a singleton set and $\mathcal{F}_I$ is always empty (no function invention). E^+ and E^- are both sets of unconditional rules $F(i_1, \ldots, i_n) \rightarrow o$, where i_i is the ith input and o the output. E^- is very important to prune the search space. $F \in \mathcal{F}_T$, $i_i \in \mathcal{C}$, and $o \in \mathcal{C} \cup \{true, false\}$. For BK, full PROLOG syntax, i. e. in CCRS unrestricted conditional rewrite rules are allowed.

It's restriction bias is quite similar to that of FFOIL, however, predicates can now take constructor terms and not only variables as arguments. Thus, l and v_i are proper functional heads, $r \in \mathcal{T}_C(\mathcal{X})$, and $u_i \in \mathcal{T}_C(\mathcal{X}) \cup \{true, false\}$. It synthesises in first-order, so abstraction and meta-variables are not allowed. The preference bias is defined as the clause covering most of the positive and no negative examples in a lattice over clauses constructed by computing the rlgg of two examples relative to the background knowledge. However, such a search space explodes and makes search nearly intractable.

Therefore, to generate a single clause, GOLEM first randomly picks pairs of positive examples, computes their rlggs and chooses the one with the highest coverage, i.e., with the greatest number of positive examples covered. By randomly choosing additional examples and computing the rlgg of the clause and the new examples, the clause is further generalised. After removing irrelevant literals in a postprocessing step, this is repeated using the clause with the highest coverage until generalisation does not yield a higher coverage. To generate further clauses GOLEM uses the sequential covering approach. It generates one clause that covers some positive and no negative examples, removes the covered examples from the training set and generates the next clause until every positive example is covered by some clause.

MAGICHASKELLER. [10] is a comparable new search-based synthesiser which generates HASKELL programs. Exploiting type-constraints, it searches the space of λ-expressions for the smallest program satisfying the user's specification. The expressions are created from user provided functions and data-type constructors via function composition, function application, and λ-abstraction (anonymous functions in HASKELL).

Generally, $\mathcal{C}$ and $\mathcal{F}_B$ are unrestricted, so for BK fully-fledged higher-order functions are allowed. It is noteworthy that this has a direct impact on the synthesiseable functions, since functions in BK immediately define the search space. The system itself is not able to detect recursion, but depends on functions to iterate over or through the defined data types. Therefore to be successful, it needs in addition to the type constructors a paramorphism, i. e. a function that decomposes a given data type, probably applying some function to the primitive part and applying the paramorphism to the rest (i. e. an extended map-function for lists). Only one target-function can be learnt at a time and no function invention is possible. So, $\mathcal{F}_T$ is a singleton set and $\mathcal{F}_I$ is always empty. E^- is empty, too, but E^+ is defined as constraints expressed in a boolean function, so it is possible to define allowed and prohibited outputs of the target function.

It's restriction bias, similar to other search-based approaches (cf. ADATE), is determined by the data types and functions defined in its BK library. So only functions that can be constructed out of these can be synthesised. The system's preference bias can be characterised as a breadth-first search over the length of the candidate programs guided by the type of the target function. Therefore it prefers the smallest program constructable from the provided functions that satisfies the user's constraints.

Forecast. As far as one can already say, GOLEM, typically for ILP systems, is hampered by a greedy sequential covering strategy. Consequently, partial rules are never revised and lead to local optima, and dependencies between rules become lost. Nevertheless, it is more flexible in discriminating the inputs on the lhss, because it, contrarily to FFOIL, allows for constructor terms. However, random sampling is too unreliable to balance out the greedy search and assure for an optimal partition of the inputs, especially when the data structures are more complex or programs with many rules are needed.

MAGICHASKELLER is a promising example of including higher-order features into IP and how functions like map or $filter$ can be applied effectively, when used advisedly, as some kind of program pattern or scheme. Nevertheless, it exhibits the usual pros and cons common to all search-based approaches: The more extensive the BK library, the more powerfull the synthesised programs are, the greater is the search space and the longer are the runs. However, contrarily to GOLEM, it is not mislead by partial solutions and shows again that only a complete search can be satisfactory for IP.

4   IGOR II

In contrast to GOLEM and MAGICHASKELLER, IGOR II is a system specialised to learn *recursive programs*. In order to do this reliably, partitioning of input examples, i.e., the introduction of patterns and predicates, and the synthesis of expressions computing the specified outputs, are strictly separated. Partitioning is done systematically and completely instead of randomly (GOLEM) or by a greedy search (FFOIL). All subsets of a partition are created in parallel, i.e., IGOR II follows a "simultaneous" covering approach. Also the search for expressions is complete. A complete search is tractable even for relative complex programs because construction of hypotheses is data-driven. IGOR II combines analytical program synthesis with search.

IGOR II induces several dependent target functions in one run, no restrictions apply to $\mathcal{F}_T$, $\mathcal{F}_B$ and $\mathcal{C}$. Auxiliary functions are invented if needed, but $\mathcal{F}_I$ is restricted that the domain of each invented function is equal to the domain of the "calling" function, in particular it is not possible to introduce accumulator variables by invention of a auxiliary function. E^- is empty and both E^+ and BK are given as unconditional example equations which may contain variables. In order to achieve confluence it is assured that the induced lhss for one target function do not overlap, i. e., they can be regarded as set and imply a unique partition of the inputs. Rhss of induced rules are restricted in that invented functions cannot be applied at the root. Conditions are restricted to alternative (i) (Sect. 2.1), i.e., simulation of `let`-expressions is not possible.

Fewer case distinctions, most specific patterns, and fewer recursive calls or calls to background functions are preferred. Thus, the initial hypothesis is a single rule per target function. Initial rules are least general generalisations (lggs) [13] of the example equations, i.e., patterns are lggs of the example inputs, rhss are lggs of the outputs w.r.t. the substitutions for the pattern, and conditions are empty. Successor hypotheses have to be computed, if unbound variables occur in rhss. Three ways of getting successor hypotheses are applied: (i) Partitioning of the inputs by replacing one pattern by a set of disjoint more specific patterns or by adding a predicate to the condition. (ii) Replacing the rhs by a (recursive) call of a defined function, where finding the argument of the function call is treated as a new induction problem, i.e., a help function is invented. (iii) Replacing the rhs *subterms* in which unbound variables occur by a call to new subprograms. In cases (ii) and (iii) help functions are invented. This includes the generation of

Table 1. System specific runtimes on different problems in seconds

	lasts	*last*	*member*	*odd/even*	*multlast*	*isort*	*reverse*	*weave*	*shiftr*	*mult/add*	*allodds*
I/O size (G/M/I)	$3/2/2_\vee$	$4/3/3_\vee$	$4/3/3_\vee$	$//3_\vee$	$4/3/3_\vee$	$4/3/3_\vee$	$/3/3_\vee$	$4/4/4_\vee$	$4/4/4_\vee$	$//4_\vee$	$4/3/3_\vee$
GOLEM	1.062	< 0.000	0.033	—	< 0.000	0.714	—	$0.049^\perp$	0.298	—	$0.016^\perp$
MAGICH.	7.620	0.040	0.540	—	0.230	—	0.100	31.480	3.620	—	2.100
IGOR II	5.695	0.007	0.152	0.019	0.023	0.105	0.103	0.200	0.127	⊙	⊙

— not tested   × stack overflow   ⊙ time out   ⊥ wrong   ∨ variabilised.

I/O-examples from which they are induced. For case (ii) this is done as follows: Function calls are introduced by matching the currently considered outputs with the outputs of any defined function. If all current outputs match, then the matched defined function can be called. The argument of the call must map the currently considered inputs to the corresponding inputs of the matched defined function. For case (iii), the example inputs of the new defined function also equal the currently considered inputs. The outputs are the corresponding subterms of the currently considered outputs. The search ends when the best hypothesis regarding the preference bias has no unbound variables.

5   Empirical Results

As problems we have chosen some of those occurring in the accordant papers and some to bring out the specific strengths and weaknesses. They have the usual semantics on lists: *multlast* was introduced above, *lasts* applies *last* on a list of lists, *isort* is insertion-sort, *allodds* checks for odd numbers, *shiftr* makes a right-shift and *weave* alternates elements from two lists into one. For *odd/even* and *mult/add* both functions need to be learnt at once. The functions in *odd/even* are mutually recursive, *lasts*, *multlast*, *isort*, *reverse*, *mult/add*, *allodds* suggest to use function invention, but only *reverse* is explicitly only solvable with. *lasts*, *allodds* and *odd/even* split up in more than two rules.

Because GOLEM usually performs better with more examples, whereas MAGIC-HASKELLER and IGOR II do better with less, each system got as much examples as necessary up to certain complexity, but then all examples completely, so no specific cherry-picking was allowed. For synthesising *isort* all systems had a function to insert into a sorted list, and the predicate < as background knowledge. The definition of the background knowledge was extensional (except for MAGICHASKELLER), IGOR II was allowed to use variables and for GOLEM additionally the accordant negative examples were provided. MAGICHASKELLER had paramorphic functions to iterate over a data type in BK. Table 1 shows the runtimes of the different systems on the example problems and the data type size up to which examples were provided.

Due to GOLEM's random sampling, the best result of ten runs was chosen. Despite its randomisation, it exceeds other ILP and IFLP systems due to its capability of introducing `let`-expressions (cf. *multlast*) where IGOR II needs function invention to balance this weak-point. So `let`-constructs can be considered as "poor man's function invention" showing to be quite usefull and promise to help pushing the boundaries of learnable problems even further. On *reverse* and *allodds* MAGICHASKELLER demonstrates the power of higher order functions. Although it does not invent auxiliary functions, *reverse* was solved using its paramorphism which provides some kind of accumulator. The time increase with *weave* shows the limits of search-based approaches.

6 Conclusions and Further Work

Based on a uniform description of some well-known IP systems and as result of our empirical evaluation of IP systems on a set of representative sample problems, we could show that the analytical approach of IGOR II is highly promising. It can induce a large scope of recursive programs, including mutual recursion and incorporates a straightforward technique for function invention. Background knowledge, in form of example equations, can be included in the inference process in a natural way. As consequence of IGOR II's generalisation principle, induced programs are guaranteed to terminate and to be the least generalisations. Although construction of hypotheses is not restricted by some greedy heuristics, induction is highly time efficient. Furthermore, it needs only a small set of positive I/O examples together with the data type specification of the target function and no further information such as schemes.

The most challenging problem will be to allow function invention for the outmost function without prior definition of the positions of recursive calls and to include the introduction of `let`-expressions and higher-order functions (e. g. map, reduce, filter).

References

1. Biermann, A.W., et al.: Automatic Program Construction Techniques. Free Press, NY (1984)
2. Quinlan, J.R.: Learning first-order definitions of functions. Journal of Artificial Intelligence Research 5, 139–161 (1996)
3. Muggleton, S., Feng, C.: Efficient induction of logic programs. In: Proceedings of the 1st Conference on Algorithmic Learning Theory, Ohmsma, Tokyo, Japan, pp. 368–381 (1990)
4. Muggleton, S.: Inverse entailment and Progol. New Generation Computing, Special issue on Inductive Logic Programming 13(3-4), 245–286 (1995)
5. Flener, P.: Inductive logic program synthesis with Dialogs. In: Muggleton, S. (ed.) Proceedings of the 6th International Workshop on ILP, pp. 28–51. Stockholm University (1996)
6. Hernández-Orallo, J., et al.: Inverse narrowing for the induction of functional logic programs. In: Freire-Nistal, et al. (eds.) Joint Conference on Declarative Programming, pp. 379–392 (1998)
7. Kitzelmann, E., Schmid, U.: Inductive synthesis of functional programs: An explanation based generalization approach. Journal of Machine Learning Research 7, 429–454 (2006)
8. Kitzelmann, E.: Data-driven induction of recursive functions from input/output-examples. In: Kitzelmann, E., Schmid, U. (eds.) Proceedings of the ECML/PKDD 2007 Workshop on Approaches and Applications of Inductive Programming (AAIP 2007), pp. 15–26 (2007)
9. Olsson, R.J.: Inductive functional programming using incremental program transformation. Artificial Intelligence 74(1), 55–83 (1995)
10. Katayama, S.: Systematic search for λ-expressions. In: Trends in Functional Programming, pp. 111–126 (2005)

11. Terese,: Term Rewriting Systems. Cambridge Tracts in Theoretical Computer Science, vol. 55. Cambridge Univ. Press, Cambridge (2003)
12. Baader, F., Nipkow, T.: Term Rewriting and All That. Cambridge Univ. Press, UK (1998)
13. Plotkin, G.: A further note on inductive generalization. In: Machine Intelligence, vol. 6, Edinburgh Univ. Press (1971)

High-Level Expectations for Low-Level Image Processing

Lothar Hotz[2], Bernd Neumann[1], and Kasim Terzic[1]

[1] Cognitive Systems Laboratory, Department Informatik, Universität Hamburg
22527 Hamburg, Germany
{terzic,neumann}@informatik.uni-hamburg.de
[2] HITeC e.V. c/o Department Informatik, Universität Hamburg
22527 Hamburg, Germany
hotz@informatik.uni-hamburg.de

Abstract. Scene interpretation systems are often conceived as extensions of low-level image analysis with bottom-up processing for high-level interpretations. In this contribution we show how a generic high-level interpretation system can generate hypotheses and initiate feedback in terms of top-down controlled low-level image analysis. Experimental results are reported about the recognition of structures in building facades.

1 Introduction

In recent years, growing interest in artificial cognitive systems has brought about increased efforts to extend the capabilities of computer vision systems towards higher-level interpretations [2,14,19,17,7,6,10]. Roughly, a high-level interpretation can be defined as an interpretation beyond the level of recognised objects. Typical examples are monitoring tasks (e.g. detecting a bank robbery), analysing traffic situations for a driver assistance system or interpreting aerial images of complex man-made structures. While existing approaches to high-level interpretation differ in many respects, they have in common that prior knowledge about spatial and temporal relations between several objects has to be brought to bear, be it in terms of probabilistic models [17], frame-based models called aggregates, logic-based conceptual descriptions [14], Situation Graph Trees [13] or Scenarios [5]. In the following, we will use the term "aggregate" for meaningful multiple-object units of a high-level scene interpretation.

In extending vision to high-level interpretations, one of the challenges is to exploit expectations derived from high-level structures for improved low-level processing. There exists much work addressing expectation-guided image analysis [13,18,12,4], but to our knowledge few vision systems with a generic architecture have been proposed which allow to feed back expectations from aggregates at arbitrarily high levels of abstraction to image analysis procedures at the level of raw images. Nagel [1] has been one of the first to demonstrate with concrete experiments in the street traffic domain that high-level hypotheses about intended vehicle behaviour could in fact be used to influence the tracking unit and thus improve tracking under occlusion.

A. Dengel et al. (Eds.): KI 2008, LNAI 5243, pp. 87–94, 2008.

In this contribution we show how a scene interpretation system based on aggregates, introduced as generic high-level conceptual units in [9,15,14], can generate feedback in a generic manner. This is demonstrated by experiments with the fully implemented scene interpretation system SCENIC. One of the core mechanisms of SCENIC is the capability of part-whole reasoning, i.e. the capability of establishing an aggregate instantiation based on evidence for any of the aggregate parts. This allows to generate strong expectations about further evidence for parts of this aggregate and to feed back these expectations to lower levels. In SCENIC, this feedback process has been extended to control image-analysis processes below the level of recognised objects which provide the input to the high-level interpretation system.

In Section 2 we will describe the generic high-level reasoning facilities which may lead to object hypotheses supported only by the high-level context. The middle layer, called Match Box in SCENIC, is described in Section 3. It has the task to mediate between hypotheses and evidence, including the initiation of goal-oriented low-level image analysis. Experiments and a summary are given in Section 4 and 5, respectively.

2   Top-Down Expectation Generation

In this section we describe techniques developed in SCENIC to generate high-level hypotheses about a scene, possibly containing incomplete information, and propagate consequences top-down to influence low-level processing. It is shown that this can be achieved by navigating in a highly structured interpretation space with a fixed set of interpretation steps. Expectations are generated by passing information obtained from evidence upwards and downwards along taxonomical and compositional relations, and laterally to parts of the same aggregate.

High-level scene interpretation in SCENIC is based on conceptual knowledge about aggregates and their parts, embedded in compositional and taxonomical hierarchies (illustrated in Fig. 1). With `Scene` as the root of the compositional hierarchy, the conceptual knowledge base implicitly represents all possible scene interpretations. Each concept is described by attributes with value ranges or value sets, constraints between attributes and relations to other concepts. Spatial attributes representing the potential position and size of the scene objects in terms of ranges are of primary importance. They are represented as constraints which are automatically adjusted as new evidence or related hypotheses restrict the attribute values. Aggregate concepts have a generic structure:

aggregate name	contains a symbolic ID
parent concepts	contains IDs of taxonomical parents
external properties	provide a description of the aggregate as a whole
parts	describe the subunits out of which an aggregate is composed
constraints	specify which relatins must hold between the parts

The conceptual knowledge base is completely described in terms of aggregate structures (including primitive objects which are aggregates without parts). Note

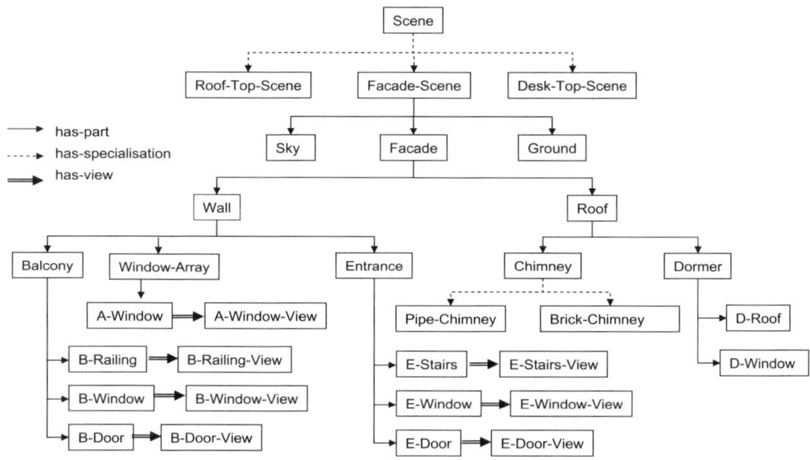

Fig. 1. Structure of facade knowledge base used for high-level interpretation in SCENIC Evidence can be related to object concepts by means of view concepts which specify properties of possible object views. It is the task of the middle layer (Section 3) to assign evidence to view concepts.

that this does not allow constraints between parts of different aggregates. This restriction is intended to channel information flow exclusively along the structure of the compositional and taxonomical hierarchies.

The interpretation process obeys the following basic algorithm:

```
Repeat
    Check for goal completion
    Check for new evidence
    Determine possible interpretation steps and update agenda
    Select from agenda one of
        {  evidence matching,
           aggregate instantiation,
           aggregate expansion,
           instance specialization,
           parameterization,
           constraint propagation }
    Check for conflict
end
```

As elucidated in [9,15,14], the interpretation steps allow to construct all partial models of the knowledge base consistent with the evidence. Typically, we are interested in a single interpretation which meets a given goal, for example, to instantiate the concept **Scene** and its parts down to the primitive objects of the compositional hierarchy such that the views of primitive objects optimally match evidence. Note that this allows to hypothesise objects without evidence - indispensable for realistic scene interpretation with occlusions, deficient low-level evidence, model limitations etc. Within the logic-based framework presented so far, there is no preference between consistent interpretations. A probabilistic framework providing a preference measure will be presented in a forthcoming publication.

We now describe the information flow, based on generic interpretation steps, which leads to top-down expectations and possible control of subsequent low-level operations. As an exemplary situation, we consider the recognition of a window-array based on evidence of a single window and causing expectations about more windows in the vicinity and at the same height.

Step 1 (evidence matching): Low-level evidence is assigned to `A-Window-View`, and a corresponding instance is created incorporating evidence properties (e.g. location and shape). This step is performed by the Match Box.

Step 2 (physical-object instantiation): The view instance leads to an instantiation of the corresponding physical-object concept `A-Window`. Image properties are transformed into physical-object properties using constraints between the physical-object concept and its views. This step is performed automatically and hence not subject to selection from the agenda as other interpretation steps.

Step 3 (aggregate instantiation): The `A-Window` is tentatively interpreted as part of the aggregate `Window-Array`. Constraints between external properties of the aggregate and its parts are set up and give rise to the spatial range where additional windows may be expected.

Step 4 (aggregate expansion): An additional `A-Window` instance is created as part of the `Window-Array`, meeting the constraints between the aggregate and its parts. While the size is closely restricted by the evidence for the first window, its location obeys loose restrictions specified in the conceptual description of the `Window-Array`. At this time, the instance is hypothesised by part-whole reasoning and not supported by evidence.

Step 5 (view instantiation): The window hypothesis leads to the instantiation of a corresponding view hypothesis with properties specified in image coordinates. In particular, the range of possible window locations is now expressed as an image area and the window size is given in pixel size. This step is performed automatically analog to Step 2.

Step 6 (evidence matching): The hypothesised view, restricted by constraints in Steps 4 and 5, is matched to evidence. This allows goal-oriented low-level image analysis meeting the restrictions of the view hypothesis. A decision has to be made whether the hypothesis has to be refuted because of missing or conflicting evidence.

In this report, we emphasise the use of feedback for low-level image analysis to resolve cases of insufficient evidence. But also other kinds of analysis have to be invoked such as occlusion reasoning, illumination and shadow analysis. All this is the task of the Match Box.

3   Middle Layer

The middle layer, called Match Box in SCENIC, acts as an interface between image analysis and symbolic interpretation. This corresponds to the three-layer architecture often proposed for cognitive systems and robotics [16,3]. The main tasks in the context of a vision system are:

- Matching low-level evidence to high-level concepts,
- Confirming or refuting high-level hypotheses, and
- Initiating low-level activities from high-level hypotheses.

The combination of these tasks allows the interpretation system to operate in a feedback loop as originally proposed by [11] almost three decades ago.

Matching low-level evidence to high-level concepts is basically object classification. There are, however, several differences which complicate the task. First, the object classes for choice are not predefined but are influenced by restrictions resulting from the high-level context generated in preceding interpretation steps, typically including a ROI (region of interest) and size restrictions. Second, there is the problem of evidence assignment: Several pieces of evidence may qualify for the range restrictions of a hypothesis, and one has not only to choose between classes but also between evidences. The SCENIC Match Box currently operates in a simplified setting, where preclassified evidence is matched to views by selecting pairings with maximal spatial overlap.

Confirming and refuting high-level hypotheses is an additional aspect of the task. Hypotheses may be false due to bad high-level interpretation choices, hence it is necessary to compare theses hypotheses with relevant evidence. Currently, the Match Box distinguishes between two evidence qualities, "primary evidence" by low-level analysis with a high significance threshold, and "secondary evidence" with a lower significance threshold. The idea is to initially interpret the image based on primary evidence and invoke a refined image analysis to obtain secondary evidence only if required.

To compare a hypothesis with evidence, the Match Box has access to the view concepts of the knowledge base. If there is evidence within the ROI matching the evidence types of the hypothesis, the hypothesis can be confirmed. If there is conflicting evidence, it must be refuted. If there is no primary evidence, low-level image analysis is initiated to obtain secondary evidence. Here, the Match Box has a repertoire of image processing modules (IPMs) at its disposal with parameters which can be set corresponding to expectations.

4   Experiments

The experimental scene interpretation system SCENIC has been applied to numerous images of a database of ca. 600 building facades assembled in the EU-funded project eTRIMS (eTraining for the Interpretation of Man-made Scenes). The thrust of the project is to develop learning methods for structured objects, and the conceptual knowledge base of SCENIC actually comprises several learnt concepts, including `Window-Array`, `Balcony` and `Entrance` [8].

To demonstrate the feedback cycle, a rectified facade image has been processed by a low-level image analysis procedure trained to discover T-style windows. Fig. 2 shows the resulting primary evidence. As can be expected from bottom-up image analysis of a natural scene, the results also contain several false positives and false negatives.

Fig. 2. Evidence created with an image processing module (IPM) trained to recognise windows

Fig. 3. Left: Hypothesised window-arrays with four additional window hypotheses. Right: Result after restarting the low-level IPM with a hypothesised region of interest.

The Match Box receives this evidence and creates window views from evidence items with high confidence values. It then passes these views to the interpretation system.

The interpretation system has been initiated to interpret a facade scene by instantiating the corresponding root node of the conceptual knowledge base. The initialisation also includes information relating image coordinates to world coordinates. The first interpretation phase has the goal to instantiate all obligatory descendants of the instantiated root node, in this case `Facade` and `Wall`.

In the second phase, the interpretation system creates physical-object instances corresponding to the view instances received from the Match Box. Furthermore, all aggregates which follow uniquely from the available instances are created in a bottom-up manner. In the experiment, instances of the concept `Window-Array` are created, when three windows exist which fulfil the `Window-Array` constraints. The concept `Window-array` specifies the following information:

```
Number of parts is [3 to inf].
All parts have type window.
All parts have similar y-position.
Any two neighbouring parts have similar distance.
All parts have similar height.
```

While the preceding steps have been obligatory, the next phase deals with uncertain interpretation decisions. The aggregates now trigger the creation of hypotheses for not yet instantiated optional parts. In the experiment, additional windows are searched at appropriate positions inferred from the established parts to complement the existing rudimentary window arrays. First, established parts integrated, then new parts are hypothesised if there are remaining gaps. Fig. 3 (left) shows the resulting window arrays with three gaps filled by hypothesised windows.

The interpretation system now creates view hypotheses for the newly hypothesised physical windows and asks the Match Box to confirm or refute these view hypotheses. As described in Section 3, the Match Box tries to match the views to existing evidence. In the experiment, there is insufficient evidence to confirm the hypotheses, so the Match Box initiates low-level image analysis of the ROIs with parameters set for weak evidence. The feedback results are shown in Fig. 3 (right): three new correct window evidences at places where no window has been recognised in the first run (first three from left), and one new false window evidence in the wall area on the right. This confirms all window hypotheses generated by high-level interpretation.

5   Summary and Future Work

In this paper we have presented a generic approach for combining low-level image analysis and high-level interpretation so that feedback to low-level analysis can be achieved. To this end, a middle layer has been introduced which mediates between low-level evidence and high-level hypotheses. The combination of low-level, middle-layer and high-level techniques has been implemented in the system SCENIC, experimental results have been presented demonstrating the use of feedback for images with building facades. Feedback has been shown to allow a coarse-to-fine strategy, with fine-grained analysis only where required. Feedback also allows to invoke procedures specialised for the kinds of views which are to be verified. For example, texture analysis can be invoked for the verification of a wall background or an occluding tree.

In ongoing work, we integrate a probabilistic model for the compositional concept hierarchy to provide a preference measure for interpretation steps. Future work will also deal with time-yarying scenarios, where high-level expectations often concern future events and are particularly valuable for scene interpretation.

References

1. Arens, M., Ottlik, A., Nagel, H.-H.: Using Behavioral Knowledge for Situated Prediction of Movements. In: Biundo, S., Frühwirth, T., Palm, G. (eds.) KI 2004. LNCS (LNAI), vol. 3238, pp. 141–155. Springer, Heidelberg (2004)
2. Borg, M., Thirde, D., Ferryman, J., Fusier, F., Valentin, V., Bremond, F., Thonnat, M.: A Real-Time Scene Understanding System for Airport Apron Monitoring. In: Proc. of IEEE International Conference on Computer Vision Systems ICVS 2006 (2006)
3. Coradeschi, S., Saffiotti, A.: An Introduction to the Anchoring Problem. Robotics and Autonomous Systems 43(2-3), 85–96 (2003)
4. Dickmanns, E.: Expectation-based Dynamic Scene Understanding. In: Blake, A., Yuille, A. (eds.) Active Vision (1993)
5. Georis, B., Mazière, M., Brémond, F., Thonnat, M.: Evaluation and Knowledge Representation Formalisms to Improve Video Understanding. In: Proc. of IEEE International Conference on Computer Vision Systems ICVS 2006 (2006)
6. Gerber, R., Nagel, H.-H.: Occurrence Extraction from Image Sequences of Road Traffic Scenes. In: van Gool, L., Schiele, B. (eds.) Proceedings Workshop on Cognitive Vision, ETH Zurich, Switzerland, pp. 1–8 (2002)
7. Gong, S., Buxton, H.: Understanding Visual Behaviour. Image and Vision Computing 20(12), 825–826 (2002)
8. Hartz, J., Neumann, B.: Learning a knowledge base of ontological concepts for high-level scene interpretation. In: International Conference on Machine Learning and Applications, Cincinnati (Ohio, USA) (December 2007)
9. Hotz, L., Neumann, B.: Scene Interpretation as a Configuration Task. Künstliche Intelligenz 3, 59–65 (2005)
10. Howarth, R.J., Buxton, H.: Conceptual descriptions from monitoring and watching image sequences. Image and Vision Computing 18(2), 105–135 (2000)
11. Kanade, T.: Region Segmentation: Signal vs. Semantics. In: Proc. International Joint Conference on Pattern Recognition (IJCPR 1978), Kyoto, Japan (1978)
12. Mohnhaupt, M., Neumann, B.: On the Use of Motion Concepts for Top-Down Control in Traffic Scenes. In: Proc. ECCV-1990, pp. 598–600. Springer, Heidelberg (1990)
13. Nagel, H.-H.: Natural Language Description of Image Sequences as a Form of Knowledge Representation. In: Burgard, W., Christaller, T., Cremers, A.B. (eds.) KI 1999. LNCS (LNAI), vol. 1701, pp. 45–60. Springer, Heidelberg (1999)
14. Neumann, B., Möller, R.: On Scene Interpretation with Description Logics. In: Christensen, H.I., Nagel, H.-H. (eds.) Cognitive Vision Systems. LNCS, vol. 3948, pp. 247–275. Springer, Heidelberg (2006)
15. Neumann, B., Weiss, T.: Navigating through Logic-based Scene Models for High-level Scene Interpretations. In: 3rd International Conference on Computer Vision Systems - ICVS 2003, pp. 212–222. Springer, Heidelberg (2003)
16. Russel, S., Norvig, P.: Artificial Intelligence - A Modern Approach. Prentice-Hall, Englewood Cliffs (2003)
17. Sage, K., Howell, J., Buxton, H.: Recognition of Action, Activity and Behaviour in the ActIPret Project. Künstliche Intelligenz 3, 30–33 (2005)
18. Tenenbaum, J.M., Barrow, H.G.: Experiments in Interpretation Guided Segmentation. Artificial Intelligence Journal 8(3), 241–274 (1977)
19. Vincze, M., Ponweiser, W., Zillich, M.: Contextual Coordination in a Cognitive Vision System for Symbolic Activity Interpretation. In: Proc. of IEEE International Conference on Computer Vision Systems ICVS 2006 (2006)

Automatic Bidding for the Game of Skat[*]

Thomas Keller and Sebastian Kupferschmid

University of Freiburg, Germany
{tkeller, kupfersc}@informatik.uni-freiburg.de

Abstract. In recent years, researchers started to study the game of Skat. The strength of existing Skat playing programs is definitely the card play phase. The bidding phase, however, has been treated quite poorly so far. This is a severe drawback since bidding abilities influence the overall playing performance drastically. In this paper we present a powerful bidding engine which is based on a k-nearest neighbor algorithm.

1 Introduction

Although mostly unknown in the English-speaking world, the game of Skat is the most popular card game in continental Europe, surpassed in world-wide popularity only by Bridge and Poker. With about 30 million recreational players and about 40,000 people playing at a competitive level, Skat is mostly a German phenomenon, although national associations exist in twenty countries on all six inhabited continents. It is widely considered the most interesting card game for three players.

Despite its popularity, only in recent years have researchers started to study the game of Skat. This is not due to lack of challenge, as Skat is definitely a game of skill – significant experience is required to reach a tournament playing level. Like Bridge, Skat consists of two playing phases: after the bidding, which determines the alliance and the trump for the game, the actual card playing begins.

The best performing Skat playing programs we are aware of are the implementations of Kupferschmid and Helmert [1] and Schäfer [2]. The strength of these programs is the card playing phase. Both programs also have a bidding module, but in both cases their performance is rather poor. The problem with poor bidding ability is that it influences the overall performance of a Skat player dramatically. The reason for this is that a victory can easily be turned into a loss if the wrong trump was chosen. In this paper, we investigate how a k-nearest neighbor algorithm can be used to build a more sophisticated bidding engine for the game of Skat.

The paper is structured as follows. Section 2 briefly introduces the rules of Skat. Section 3 focuses on the bidding phase and provides a general architecture for a bidding engine. Afterward a k-nearest neighbor based algorithm for the bidding is presented and applied to discarding in Section 5. Further improvements are described afterwards. Section 7 concludes.

---

[*] This work was partly supported by the German Research Foundation (DFG) as part of the Transregional Collaborative Research Center "Automatic Verification and Analysis of Complex Systems" (SFB/TR 14 AVACS). See http://www.avacs.org/ for more information.

A. Dengel et al. (Eds.): KI 2008, LNAI 5243, pp. 95–102, 2008.

2 Skat

In this section we provide a brief overview of the skat rules. For a more detailed description, we refer to the official rules [3].

Skat is a three-player imperfect information game played with 32 cards, a subset of the usual Bridge deck. At the beginning of a game, each player is dealt ten cards, which must not be shown or communicated to the other players. The remaining two cards, called the *skat*, are placed face down on the table. Like in Bridge, each hand is played in two stages, *bidding* and *card play*.

The bidding stage determines the alliance for the hand and proceeds as follows: two players announce and accept increasing bids until one of them passes. The winner of the first bidding phase now competes in the same way with the third player. The successful bidder of the second bidding phase, henceforth called the *declarer*, plays against the other two. In this process, the maximum bid a player can announce depends on the kind of game (grand, ♣, ♠, ♡, ◇, null) the player wants to play and a factor determined by the jacks the player holds. To avoid confusion about "game" and "kind of game", we henceforth call the latter "trump". The declarer decides on the trump. Before declaring the game, the declarer may pick up the skat and then discard any two cards from his hand, face down. These cards count towards the declarer's score.

Card play proceeds as in Bridge, except that the trumps and card ranks are different. In grand games, the four jacks are the only trumps. In suit games, the four jacks and the seven other cards of the selected suit are trumps. There are no trumps in null games. Non-trump cards are grouped into suits as in Bridge. Each card has an associated *point value* between 0 and 11, and the declarer must score more points than the opponents (i. e. at least 61 points) to win. Null games are an exception and follow *misère* rules: the declarer wins iff he scores no trick. Note that we do not deal with null games in this paper since they are rather trivial, from a bidder's perspective, because there are efficient rule-based criteria to decide whether a null game can be won or not.

3 Bidding and Discarding

From a player's perspective, the problem of bidding is to decide if there is a trump to win the game. This decision is only based on the cards the player holds at the beginning of the game, as this is the only information a player has available. To estimate a game's outcome as precisely as possible is crucial for a favourable playing performance, but very difficult given limited information as well. Some factors that determine a hand's strength are rather obvious, such as the number of trumps, but in fact there are more subtle intricacies that influence the potential of a player's hand enormously.

Discarding cards is a similar problem. If the declarer picks up the skat, he has to decide which two of his twelve cards to discard. Again some decisions seem rather obvious, e. g., a player should discard a ten when he does not hold the ace of the same suit, but in general the choice is far from simple.

The bidding engine of Kupferschmid's and Helmert's Double Dummy Skat Solver (DDSS) [1] is based on a least mean square error algorithm (cf. [4]), which tries to separate the winnable from the non-winnable games by learning a linear function. The problem with this approach is that it is very unlikely that Skat games can be linearly separated. This is the main reason for DDSS' rather poor bidding behavior.

The bidding system we are presenting in this paper is based on a k-nearest neighbor algorithm (KNN). The advantage of this machine learning algorithm is that it is well suited to approximate difficult separation functions. Our bidding module consists of several different *evaluation functions* $V^t : S \rightarrow \{0, \ldots, 120\}$, one for each $t \in T = \{\text{grand}, \clubsuit, \spadesuit, \heartsuit, \diamondsuit\}$. Such a function maps each set of ten cards (a player's hand) $s \in S$ to a number between 0 and 120 (the possible outcome of the game), estimating the potential of the given cards if trump t is played. With these evaluation functions it is possible to describe our bidding system as $C(s) = \arg\max_{t \in T}\{V^t(s) \mid V^t(s) > th\}$. If none of the estimates $V^t(s)$ is above the threshold th (usually $th = 60$) then $C(s)$ evaluates to $\emptyset$. In case that there are more than one t, the t allowing the highest bid is selected.

With our bidding system it is also possible to solve the discarding problem. As said before, the problem here is to select two cards out of twelve so that the remaining ten cards are as strong as possible. For every of the 66 possible subsets that contain ten cards our bidding system is applied. The cards that yield the highest estimation of the outcome are discarded. A more detailed description, also including some refinements of this procedure is presented in Sec. 5.

4   A KNN Based Bidding System

The k-nearest neighbor algorithm belongs to the group of instance-based learning algorithms, which have in common that a hand is classified by comparing it to instances for which the correct classification is known. These instances form the *knowledge base*. To be able to compare different hands, a metric on the set of instances has to be defined. Depending on the trump t, we first project an instance s to a numeric feature vector $\langle f_1^t(s), f_2^t(s), \ldots, f_8^t(s)\rangle \in \mathbb{R}^8$. The distance to another instance is then defined as the Euclidean distance of the corresponding normalized feature vectors.

It should be mentioned that, due to the imperfect information, it is not possible to design a bidding system recognizing every winnable game as winnable. Even an optimal

Table 1. Features describing a hand s if a game with trump t is played

feature	description	range
$f_1^t(s)$	number of jacks in s	0–4
$f_2^t(s)$	number of non-jack trumps in s	0–7 (grand: 0)
$f_3^t(s)$	number of non-trump aces and 10s. 10s only if the ace of the same suit is in s	0–6 (grand: 0–8)
$f_4^t(s)$	accumulated points of the cards in s	0–90
$f_5^t(s)$	number of suits not in s	0–3
$f_6^t(s)$	number of non-trump 10s without the ace of the same suit in s	0–3
$f_7^t(s)$	valuation of jacks in s	0–10*
$f_8^t(s)$	valuation of non-jack trumps in s	0–17 (grand: 0)**

$*\ \clubsuit J : 4, \spadesuit J : 3, \heartsuit J : 2, \diamondsuit J : 1$
$**\ A : 5, 10 : 4, K : 3, Q : 2, 9, 8 \text{ and } 7 : 1$

bidder will sometimes lose a game. The task of developing adequate features lies in finding features that are significant to describe a hand, but still leave the feature space as small as possible. Only jacks and non-jack trumps, the most basic factors for the potential of a hand, are represented twice in the features, both as a simple quantity function and a valuation function as can be seen in Table 1. The other features describe other important aspects of a hand.

To classify a given hand, the distance between that hand and each instance in the knowledge base is computed and the k sets of nearest neighbors are determined. For a trump t the estimated score of a game is the averaged score of all elements of those sets. Note that more than k instances are used for classification, since each set of neighbors contains all instances with the same distance.

4.1 Construction of the Knowledge Base

Unlike most other algorithms, KNN does not have a training phase. Instead the computation takes place directly after a query occurs. The only thing needed before using the algorithm is the knowledge base, which is substantial for KNN. In this work, DDSS was used to generate a knowledge base. Therefore we randomly generated 10,000 pairwise distinct games. For each game g and each trump t DDSS was used to approximate the correct game theoretic value. Note that, since Skat is an imperfect information game, it is not possible to efficiently calculate the exact game theoretic value.

To judge the quality of our training and evaluation games it is necessary to know how DDSS works. It is a Monte Carlo based Skat playing program. Its core is a fast state-of-the-art algorithm for perfect information games. In order to decide which card should be played in a certain game position, the following is done. The Monte Carlo approach repeatedly samples the imperfect information (i. e., the cards of the other players). For each of the resulting perfect information games, the game theoretic value is determined. The card which yields the best performance in most of these perfect information games is then played in the imperfect game. Ginsberg [5] first proposed to use a Monte Carlo approach for imperfect information games. However, the problem with this approach is that even if all possible card distributions are evaluated, the optimal strategy cannot be computed. This was proven by Basin and Frank [6]. Another problem is that a Monte Carlo player will never play a card in order to gain additional information rather than points. Nevertheless, DDSS has high playing abilities.

For the generation of the knowledge base, the number of samples per card was set to 10. This corresponds to the strength of an average Skat player. The higher the number of samples the better is the playing performance. But since the runtime is exponential in the number of samples, we were forced to use a rather small number of samples.

To evaluate the performance of our bidding system, another 5,000 games were randomly generated and calculated by DDSS in the same way. Approximately 60 % of these games were calculated as not winnable. This is a reasonable percentage because we are regarding so called *hand* games[1].

---

[1] A game is a hand game if the declarer does not pick up the skat.

4.2   Choosing the Right Number of Neighbors

In this section, we address how to determine an optimal number of neighbors k for this task. Figure 1 shows the average square error depending on the used k. Because of the sweet spot between 8 and 30, we set k to 15. With this k more than 90 % of the games that were calculated as not winnable in the evaluation set were classified correctly. Just 50 % of the winnable games were classified as winnable, but the identification of the correct trump worked in more than 80 % of those cases. Altogether, 73.2 % of all games were classified correctly which is already a pretty good result.

It may be observed that too many games are classified as not winnable. This is especially a problem as all the games are regarded as *hand* games, and thus the possible improvement by picking up the skat is not regarded yet. A bidder should rather overestimate a bit and speculate on a slight improvement than bid too carefully. However, the bidding system is already well suited to classify *hand* games.

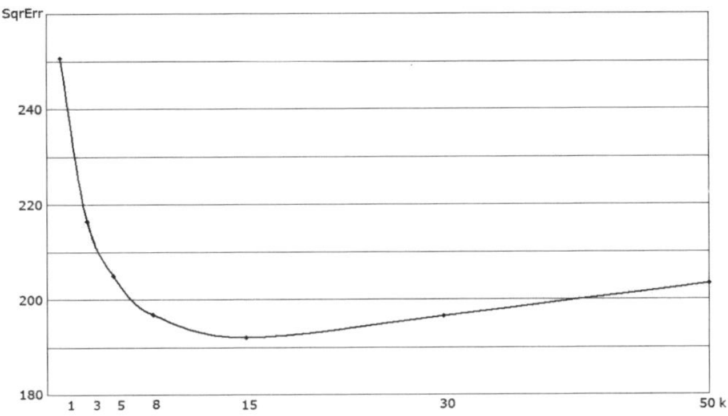

Fig. 1. Average square errors for different k

4.3   Failed Improvement Attempts

In the literature, a couple of improvements to KNN can be found. The most basic improvement is to weigh each nearest neighbor depending on its distance to the query s. The evaluation showed, that the results could not be improved this way. The reason for this is the high degree of incomplete information in Skat. Seemingly, very different hands can describe very similar games due to the cards in the skat or the distribution of the unknown cards. By lowering the influence of further distanced instances on the result, this kind of game is not included strong enough in the result.

Since our knowledge base consists of randomly generated games, there are densely populated areas containing lots of instances with similar feature vectors and sparse areas. A query in a dense area is then classified by more neighbors than a query in a sparse part – even with the same k. Therefore, we created a knowledge base that is uniformly distributed over the feature space – and obtained slightly worse results. The reason for this is twofold. First, the probability for certain features differs significantly, e. g., it is more likely to hold two jacks than holding four. So, dense areas in our knowledge base are those areas that come up with a higher probability, and most of our evaluation games should fall into dense populated areas as well. Now, most of our evaluation games were classified with less instances and thus a little worse in average, while just a few instances with a rarely upcoming distribution were classified by more instances. The second reason lies in the nature of the game. Most of the distributions that are less probable to occur have either a very high or a very low score – games with either a lot or a few trumps and jacks are usually easy to dismiss or clearly winnable. In those cases, it is not so important to calculate a very accurate estimate of the possible score. The quality of a bidding system is rather given by its ability to decide whether close games can be won.

5   Discarding with KNN

As outlined in Section 3, discarding can also be implemented with KNN. For a fixed trump, all 66 possibilities to discard any two cards out of twelve must be examined. If the trump is not yet decided, there are even 330 possibilities, not considering null games. Although KNN is a pretty fast algorithm, we drop some of these possibilities. First, only those trumps are considered that are still playable given the last bid of the bidding auction. Second, it never makes sense to discard a jack or a trump card. Hence, those possibilities are excluded as well. The remaining possibilities are all classified with our bidding system, yielding a discarding algorithm that only needs 0.1 seconds for discarding and game announcement.

For the discarding system, the knowledge base was changed slightly. On the one hand, the score of each instance was lowered by the value of the skat. On the other hand, the result of the evaluation functions $V^t(s)$ was then increased by the value of the discarded cards. In bidding, an estimation of the points in the skat is necessary because those points belong to the declarer, but are unknown while bidding. That estimation was given naturally by the structure of the knowledge base. For discarding on the other hand, no estimation is necessary as the real point value of the discarded cards is known.

A detailed evaluation of this process was impossible: to discard all 66 possibilities for the 5,000 evaluation games and play the resulting game with every trump would haven taken too long with DDSS. Therefore, we randomly chose 100 games and calculated all possibilities just for those. Two representative games are shown in Fig. 2, both the calculated optimum and the bidders result.

For the upper game, our approach discards exactly the cards calculated as optimal. The lower one does not match the calculated optimum, but it seems that this optimum is more questionable than our result. First, it is definitely better to discard $\diamond$A instead of $\diamond$10, because the declarer will have a higher score – in both cases the remaining card will win the same tricks. Regarding the second row, most skilled players try to get rid of one suit completely as proposed by our algorithm. The questionable calculation is caused by the small number of samples, as already stated by the authors of DDSS. Note that similar errors in our knowledge base distort the bidding systems' results as well.

However, the performance of our discarding system is even better in all regarded games than the bidding system's performance so far. This is mainly due to the changes in the knowledge base.

trump: ♡	hand									skat	
bidder	♣J	♠J	◇J	♣7	♡10	♡Q	♡9	♡8	◇A	◇Q	♠10	♠Q
DDSS	♣J	♠J	◇J	♣7	♡10	♡Q	♡9	♡8	◇A	◇Q	♠10	♠Q

trump: ♠	hand									skat	
bidder	♠J	◇J	♣Q	♣7	♠Q	♠9	♠7	◇A	◇10	◇9	♡Q	◇K
DDSS	♠J	◇J	♣7	♠Q	♠9	♠7	♡Q	◇A	◇K	◇9	♣Q	◇10

Fig. 2. Discarding with KNN

6 Skat Aware Bidding

The knowledge base for bidding can also be improved in a similar way. It was already mentioned, that the bidder estimates the scores based on *hand* games that are randomly generated. Of course, this does not adequately describe Skat, as most games simply cannot be won without picking up the skat, which causes an underestimation. Therefore, the knowledge base was changed as follows. For all games the system picked up the skat, discarded by use of the discarding system, and the resulting games were then calculated with DDSS. This new score simply replaced the former one in the knowledge base – the features were left unchanged as they describe a hand during bidding sufficiently.

Now complete games were simulated: we assume that the bidder has to bid until its highest possible bid is reached and thus cannot play trumps with a lower value. The games the bidder chose to bid on were calculated (after discarding) by DDSS. The rest of the games (the ones the bidder classified as ∅) were discarded as well and then calculated by DDSS. If they were not won by DDSS we regarded them as not winnable. Additionally, different values for the threshold th were tested. That value can serve as a risk modulator, the lower it is the riskier the system bids. Figure 3 shows the results of the evaluation.

The overall results are very similar to the ones of our first presented implementation, with 75 %, they are just slightly higher. However, the change to a more realistic scenario made the task a bit harder – the system was not allowed to announce every trump, so

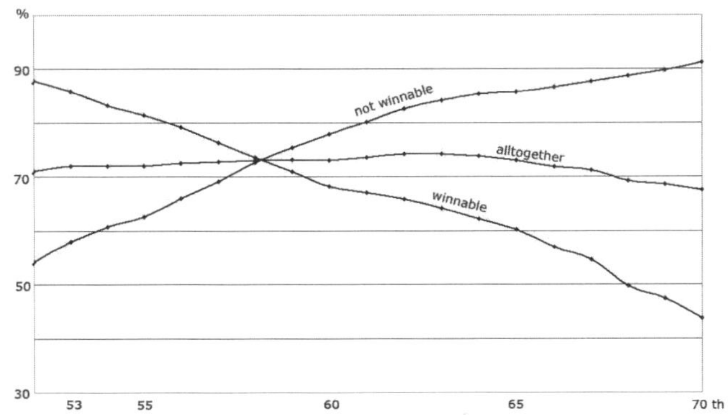

Fig. 3. Evaluation results for bidding

an improvement was definitely achieved. Also, the underestimation has turned into the preferred overestimation, as the optimal threshold lies between 61 and 64. In general the range in which good results are obtained is rather large. Everywhere between 52 and 67, more than 70 % of the games are classified correctly, while the percentage of correctly classified winnable or not winnable games differs significantly in this range. This is a perfect precondition for a regulating system for the desired risk.

7   Conclusion

Regarding the dimension of imperfect information in a game of Skat, the presented results are impressive. We implemented an efficient bidding and discarding system. First comparisons with Schäfer's bidding system [2] revealed that our implementation detects more winnable games, without playing too risky. In general, KNN seems very well suited for tasks with complex separation functions like bidding in skat. Our final system classifies 75 % of all games correctly.

The results may even be improved with a knowledge base created by DDSS using more samples. This is one of the possible improvements planned for the future. Applying the results of the last section to discarding might further improve the discarding systems behavior as well. We also plan to investigate how our system can be further improved by combining it with a Monte Carlo approach. Therefore, the features have to be expanded on the opponent's hands as well (that are known for the instances of the knowledge base), and the query has to be sampled. Since KNN is a time efficient algorithm, the additional computing time should be neglectable.

References

1. Kupferschmid, S., Helmert, M.: A Skat player based on Monte-Carlo simulation. In: van den Herik, H.J., Ciancarini, P., Donkers, H.H.L.M(J.) (eds.) CG 2006. LNCS, vol. 4630, pp. 135–147. Springer, Heidelberg (2007)
2. Schäfer, J.: Monte-Carlo-Simulationen bei Skat. Bachelor thesis, Otto-von-Guericke-Universität Magdeburg (2005)
3. International Skat Players Association: International Skat and Tournament Order 2007 (2008), `http://www.skatcanada.ca/canada/forms/rules-2007.pdf`
4. Mitchell, T.M.: Machine Learning. McGraw-Hill, New York (1997)
5. Ginsberg, M.L.: GIB: Steps toward an expert-level Bridge-playing program. In: Proceedings of the 16th International Joint Conference on Artificial Intelligence, pp. 584–589 (1999)
6. Frank, I., Basin, D.A.: Search in games with incomplete information: A case study using Bridge card play. Artificial Intelligence 100, 87–123 (1998)

Automobile Driving Behavior Recognition Using Boosting Sequential Labeling Method for Adaptive Driver Assistance Systems

Wathanyoo Khaisongkram[1], Pongsathorn Raksincharoensak[1],
Masamichi Shimosaka[2], Taketoshi Mori[2], Tomomasa Sato[2], and Masao Nagai[1]

[1] Tokyo University of Agriculture and Technology, 2-24-16 Koganei, Tokyo, Japan
{wathan,pong,nagai}@cc.tuat.ac.jp
[2] The University of Tokyo, 7-3-1 Hongo, Bunkyo-ku, Tokyo, Japan

Abstract. Recent advances in practical adaptive driving assistance systems and sensing technology for automobiles have led to a detailed study of individual human driving behavior. In such a study, we need to deal with a large amount of stored data, which can be managed by splitting the analysis according to the driving states described by driver maneuvers and driving environment. As the first step of our long-term project, the driving behavior learning is formulated as a recognition problem of the driving states. Here, the classifier for recognizing the driving states is modeled via the boosting sequential labeling method (BSLM). We consider the recognition problems formed from driving data of three subject drivers who drove on two roads. In each problem, the classifier trained through BSLM is validated by analyzing the recognition accuracy of each driving state. The results indicate that even though the recognition accuracies of braking and decelerating states are mediocre, accuracies of the following, cruising an stopping states are exceptionally precise.

Keywords: Automobile driving behavior, adaptive driving assistance system, driving-state recognition, boosting sequential labeling method.

1   Introduction

As active safety has been a common target of transport technology trends, studies of practical cooperative safety systems that suit individual maneuvering behavior has become popularized among transport system researchers. This gives rise to the development of *adaptive driver assistance systems*, i.e., the systems capable of providing assistance that fits driver's needs and ongoing traffic conditions. To realize such systems, a serious study on individual human driving behavior is necessary, from which our research goal originates.

There has been a large body of literature relevant to driver assistant systems and driving behavior learning [1,2,3,4,5,6,7,8]. In some of these works [7], ongoing driver's maneuvers are usually categorized into mutually exclusive classes, which, in this research, shall be called *driving states*. Definitions of these driving states are based on driver's operations and traffic environment. In this paper, we assume

A. Dengel et al. (Eds.): KI 2008, LNAI 5243, pp. 103–110, 2008.

that the driving behavior comprises three stages: the situation-recognizing stage that perceives ongoing events, the decision-making stage that selects the proper driving states and the actuating stage that implements the maneuvers. In order to recognize the driving behavior, our objective is to develop a classifier that can recognize the driving states efficiently using the data on vehicle motions and environment. In other words, this classifier is intended to model the situation-recognizing and decision-making stages of the driving behavior.

As the driving-state-recognition problem involves many features, it is preferable to adopt non-generative models like maximum entropy Markov models (MEMMs) or conditional random fields (CRFs) [9,10]. Here, we employ the boosting sequential labeling method, BSLM [11], whose concept is partially based on the CRFs framework, which has been proved to be superior to MEMMs [9].

In the next section, a recognition problem of the driving behavior is formulated. Definitions of the driving states and a driving behavior model are given. In Section 3, a short overview on the framework of BSLM and how it is applied to our problem are presented. Subsequently, the computational results are given and discussed in Section 4. In the last section, we state the concluding remarks and perspectives.

2   Problem Formulation of Driving Behavior Recognition

In this research, the experimental vehicle is equipped with a drive recorder comprises such measuring and capturing devices as driver pedal maneuvers, laser radar, multi-view on-vehicle camera system etc. By means of these devices, driving data and videos showing ongoing operations of the driver and surroundings can be used in the subsequent driving behavior analysis. In this phase, we confine the interesting driving operations to only those in longitudinal direction. Hence, measured driving data of which we make use in this work includes

- R: range, distance from the preceding vehicle [m].
- $\dot{R}$: rate of change of R [m/s].
- V: velocity [km/h].
- a_x: longitudinal acceleration [m/s^2].
- X: GPS longitude coordinate [deg].
- Y: GPS latitude coordinate [deg].

The measurement sampling frequency is 30 Hz. In addition, the time headway T_{hw} and the inverse of time to collision T_{tc}^{-1} are also employed in the driving behavior analysis, which can be computed as $T_{\mathrm{hw}} = R/V$ and $T_{\mathrm{tc}}^{-1} = -\dot{R}/R$. Figure 1 shows the graphical definitions of the measured data.

2.1   Definitions of Driving States and Their Labels

The driving states are exhaustively and exclusively categorized into five cases: following, braking, cruising, decelerating and stopping states. Five labels are assigned to these states as $\mathbb{F}$, $\mathbb{B}$, $\mathbb{C}$, $\mathbb{D}$ and $\mathbb{S}$, respectively. These driving states are defined as follows.

1. $\mathbb{F}$: There exists a preceding vehicle and either the driver is approaching the preceding vehicle or the driver intends to maintain R by hitting the acceleration pedal or hit no pedal at all.
2. $\mathbb{B}$: There exists a preceding vehicle and the driver slows down the host vehicle by hitting the brake pedal.
3. $\mathbb{C}$: A preceding vehicle is absence or is so distant that the driver does not intend to pursue, and thus, cruises with desired speed by hitting the acceleration pedal or hit no pedal at all.
4. $\mathbb{D}$: The conditions are the same as in $\mathbb{C}$ except that the driver hits the brake pedal just as in the case that the driver wants to stop at the traffic light.
5. $\mathbb{S}$: The host vehicle is not moving.

Exemplary situations associated to these states are demonstrated in Figure 2. Distinguishing these states manually is meticulous and requires detailed information on vehicle motions, driver's actual movements and traffic environment.

Here, a state flow model, which produces state sequences, is described by possible transitions between states and their probabilities. The model is depicted in Figure 3 where each arrow represent each state transition. For the normalization condition to hold, we need $\sum_j a_{ij} = 1$ for each i.

2.2   Automobile Driving Behavior Model

The driving behavior model is assumed to be described by a hierarchical model consisting of three stages. The first one is the situation-recognizing state which represents perceptions of the driver. The second stage is the decision-making stage in which the driver selects the driving states based on the foregoing driving state, the current vehicle motions and traffic environment. The third stage is an actuating stage in which the driver decides what and how maneuvers should be executed. This concept is described by a block-diagram in Figure 4.

To recognize the driving behavior, we start by developing a driving-state classifier which resembles the situation-recognizing and the decision-making stages. This can be accomplished by identifying the state flow model in Figure 3, or equivalently, computing the joint probability distribution over the state sequence given the data sequence. Suppose $\underline{\mathbf{x}} \in \mathrm{R}^8$ is a vector-valued random variable of the data sequence, i.e., $\underline{\mathbf{x}} = [R, \dot{R}, V, a_x, X, Y, T_{hw}, T_{tc}^{-1}]^T$, and y is a random variable of the corresponding state sequence, over the set $\Lambda = \{\mathbb{F}, \mathbb{B}, \mathbb{C}, \mathbb{D}, \mathbb{S}\}$. The classifier can be viewed as the joint conditional probability $p(y|\underline{\mathbf{x}})$. In the next section, we state how the BSLM framework is adopted to compute $p(y|\underline{\mathbf{x}})$, and how we measure accuracy of recognized results.

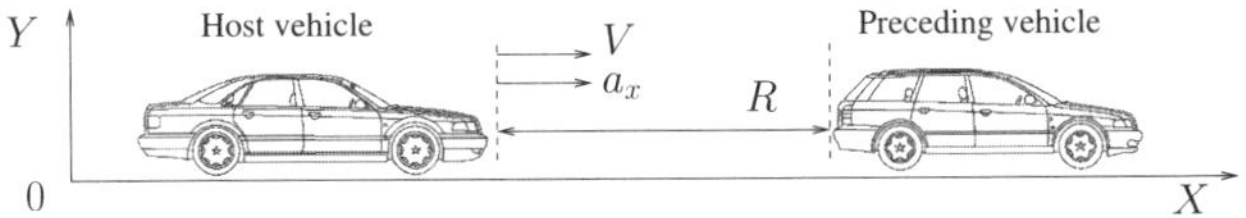

Fig. 1. Graphical definitions of the measured driving data

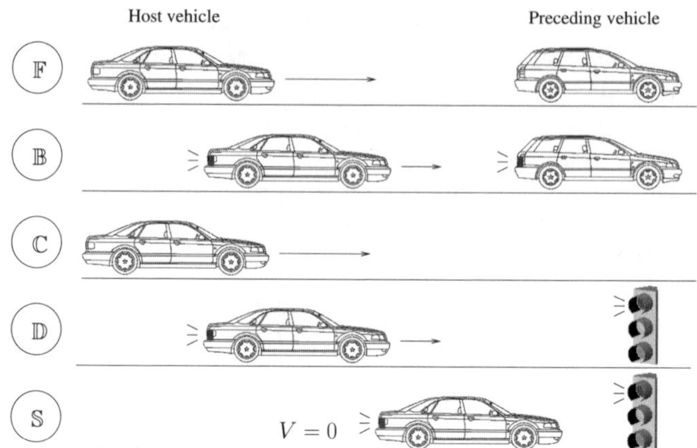

Fig. 2. Pictorial definitions of the considered driving states

3 Driving-State Recognition Via BSLM

In BSLM [11] (literature currently available only in Japanese), $p(y|\underline{\mathbf{x}})$ is constructed directly from pairs of an observed data sequence $\underline{\mathbf{x}}$ and a state sequences y, which is manually-labeled using knowledge of $\underline{\mathbf{x}}$, without explicitly calculating the probability $p(\underline{\mathbf{x}})$. Its joint conditional probability $p(y|\underline{\mathbf{x}})$ takes the form

$$p(y_t = \lambda|\underline{\mathbf{x}}_{1:t}) \propto \exp\left[F_\lambda(\underline{\mathbf{x}}_t) + G_\lambda(y_{t-1})\right] \tag{1}$$

where $\lambda \in \Lambda = \{\mathbb{F}, \mathbb{B}, \mathbb{C}, \mathbb{D}, \mathbb{S}\}$, y_t is the label of y at the tth sampling instant and $\underline{\mathbf{x}}_{1:t}$ is the sub-sequence of $\underline{\mathbf{x}}$ from the first to the tth sampling instant. The functions $F_\lambda : \mathrm{R}^8 \mapsto \mathrm{R}$ and $G_\lambda : \Lambda \mapsto \mathrm{R}$ are obtained through a classifier training process. Basically, this requires a training data set, denoted by D, of some training data pairs, say N pairs. The nth training data pair is formed by a pair of the labeled state sequence $y_{1:T_n}^{(n)}$ and the observed data sequence

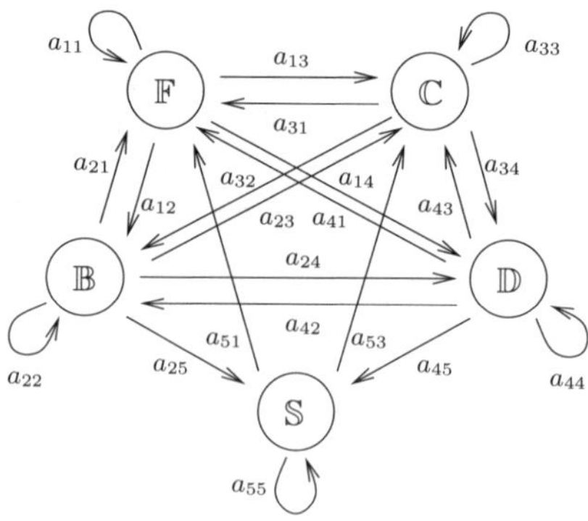

Fig. 3. The state-flow model showing the transitions of the driving states

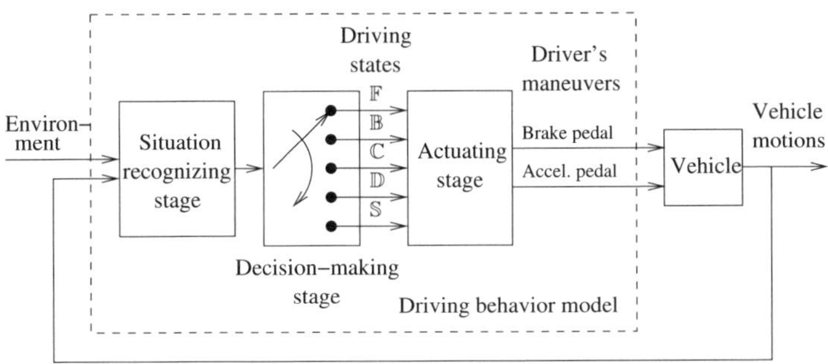

Fig. 4. The three-stage human driving behavior model

$\underline{\mathbf{x}}_{1:T_n}^{(n)}$ such that $D = \{\underline{\mathbf{x}}_{1:T_n}^{(n)}, y_{1:T_n}^{(n)}\}_{n=1}^{N}$. The purpose of the data training process is to find F_λ and G_λ that maximize a log-likelihood objective function $J = \sum_{n=1}^{N} \sum_{t=1}^{T_n} \ln p(y_t^{(n)}|\underline{\mathbf{x}}_{1:t}^{(n)})$. Via boosting, F_λ and G_λ are updated with the *weak learners* f_λ and g_λ as $F_\lambda \leftarrow F_\lambda + \nu f_\lambda$ and $G_\lambda \leftarrow G_\lambda + \nu g_\lambda$ for the prescribed number of iterations, say, $I_{\max}$. Here, $\nu \in (0,1]$ and $I_{\max} \in \mathbb{N}$ are adjustable algorithm parameters. For example, let $N = 1$. Based on the 2nd-order Taylor series expansion of J, given some specific forms of f and g, both of them must minimize $\sum_t w_{t,\lambda}(\phi - z_{t,\lambda})^2$ where ϕ represents either f_λ or g_λ, and for each λ,

$$
\begin{aligned}
w_{t,\lambda} &= \tilde{p}(y_t = \lambda)[1 - \tilde{p}(y_t = \lambda)], \\
z_{t,\lambda} &= \begin{cases} 1/\tilde{p}(y_t = \lambda), & \text{if} \quad y_t = \lambda, \\ -1/[1 - \tilde{p}(y_t = \lambda)], & \text{if} \quad y_t \neq \lambda. \end{cases}
\end{aligned} \tag{2}
$$

In this work, we formulate the forms of the learners f_λ and g_λ as follows:

$$
f_\lambda(\underline{\mathbf{x}}_t) = \begin{cases} \alpha_\lambda, & \text{if} \quad \underline{\mathbf{x}}_t(i) > \theta, \\ \alpha'_\lambda, & \text{if} \quad \underline{\mathbf{x}}_t(i) \leq \theta, \end{cases} \quad \text{and} \quad g_\lambda(y_{t-1}) = \frac{\sum_t w_{t,\lambda} z_{t,\lambda} \tilde{p}(y_{t-1})}{\sum_t w_{t,\lambda} \tilde{p}(y_{t-1})} \tag{3}
$$

where $\tilde{p}(y_{t-1}) = p(y_{t-1}|\underline{\mathbf{x}}_{1:t-1})$, $\underline{\mathbf{x}}_t(i)$ is the ith element of $\underline{\mathbf{x}}_t$. Let $\mathcal{H}$ be the Heaviside step function; parameters $\alpha_\lambda, \alpha'_\lambda$ in (3) are computed as

$$
\alpha_\lambda = \frac{\sum_t w_{t,\lambda} z_{t,\lambda} \mathcal{H}[\underline{\mathbf{x}}_t(i) - \theta]}{\sum_t w_{t,\lambda} \mathcal{H}[\underline{\mathbf{x}}_t(i) - \theta]} \quad \text{and} \quad \alpha'_\lambda = \frac{\sum_t w_{t,\lambda} z_{t,\lambda}(1 - \mathcal{H}[\underline{\mathbf{x}}_t(i) - \theta])}{\sum_t w_{t,\lambda}(1 - \mathcal{H}[\underline{\mathbf{x}}_t(i) - \theta])}, \tag{4}
$$

by seeking for θ (e.g., a binary search) and varying the index ith to minimize $\sum_t w_{t,\lambda}(f_\lambda(\underline{\mathbf{x}}_t) - z_{t,\lambda})^2$. In short, starting from $F_\lambda \equiv 0$ and $G_\lambda \equiv 0$, we compute $\tilde{p}(y_t)$ at each t to obtain $w_{t,\lambda}, z_{t,\lambda}$ and $g_\lambda(y_{t-1})$; these are then used to find the optimal $f_\lambda(\underline{\mathbf{x}}_t)$, which is in turn used to update F_λ and G_λ of the next iteration.

In practice, we first precondition raw data to eliminate noises and errors and then label the data sequences $\underline{\mathbf{x}}$ to obtained the corresponding state sequences y for the classifier training via BSLM. The state-labeling process is conducted at each sampling instant, one after another, by considering all the measured data and videos showing driver's actions as well as traffic around the host vehicle.

To validate the classifier and examine the recognition accuracy, $p(y|\underline{x})$ is used to recognitze a state sequence when the actual human-labeled state sequence, which shall be called *ground truth*, is known. The recognized state sequence is then compared with the ground truth at each instant, and the accuracy of each state are computed. In so doing, driving data from three subject drivers on two road sections are collected. We name the subject drivers as S1, S2 and S3, and the road sections as R1 and R2. The drivers' ages range from 23 to 28 years old with driving experience of three to ten years. The distances of R1 and R2 are about 2 and 3.5 [km] with the speed limits of 40 and 60 [km/h], respectively. For each driver on each road section, eight sets of data (one set is associated to one course of driving) are collected and labeled. Among these sets, seven of them form a training data set D; the left one is the validation data set.

We evaluate the recognition accuracy using a measure called *F-measure* which is a harmonic mean of two other measures called *precision* and *recall*. Suppose that we are to compute the recognition accuracy of $\mathbb{F}$. Let N_1 and N_2 be the total numbers of time instants at $\mathbb{F}$ of the ground truth and of the recognized state, respectively. Let N_c be the total number of time instants at $\mathbb{F}$ where both ground truth and recognized state match. The precision and recall are given by $(N_c/N_2) \times 100\%$ and $(N_c/N_1) \times 100\%$, respectively. The F-measure definition is

$$\text{F-measure} = \frac{2(\text{precision} \times \text{recall})}{\text{precision} + \text{recall}}. \tag{5}$$

4   Computational Results and Discussion

Firstly, the detailed results of driving-state recognition of S1 on R2 is presented. Then, we show the accuracy of each exclusive recognition for each driver on each road section.

4.1   Recognition of S1 on R2

In Figure 5, a diagram showing the ground truth and the recognized state transitions during the driving course and some plots of time responses are illustrated. The segment of each state are colored as follows: $\mathbb{F}$-blue, $\mathbb{B}$-red, $\mathbb{C}$-green, $\mathbb{D}$-magenta, $\mathbb{S}$-black. The shorthand notations GT and RSS stand for the ground truth and the recognized state, respectively. The recognition accuracy of each driving state is given in the next section together with the other drivers on different road sections.

As parts of the results, we also obtain $F_\lambda(\underline{x})$, appears in (1), for each $\lambda \in \Lambda$, which can be viewed as relationships between the measured data items and the driving states. The following facts have been observed. Firstly, $\mathbb{F}$ depends much on almost every measured parameters except Y; it depends strongly on a_x and T_{hw}. $\mathbb{B}$ mainly depends on a_x and somewhat depends on V. $\mathbb{C}$ mainly depends on R and somewhat depends on a_x. In contrast, $\mathbb{D}$ mainly depends on a_x but somewhat depends on R and X. finally, $\mathbb{S}$ almost entirely depends on V.

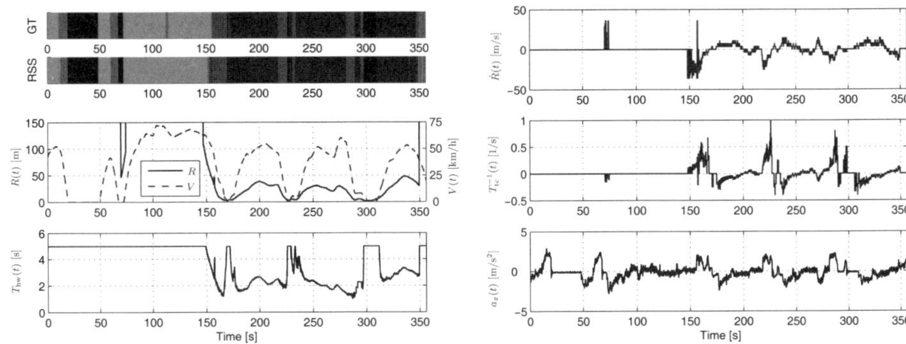

Fig. 5. State transition diagrams and time responses of vehicle motion data. GT and RSS stands for the ground truth and the recognized state sequence, respectively. Different states are displayed with different colors: $\mathbb{F}$-blue, $\mathbb{B}$-red, $\mathbb{C}$-green, $\mathbb{D}$-magenta and $\mathbb{S}$-black.

4.2   Exclusive Recognitions of All Drivers on All Roads

Here, we present the recognition F-measure associated with each driving state in each exclusive validation, i.e., both training and validation data sets are from the same driver on the same road section. The results are given in Table 1.

Table 1. The F-measures of the exclusive validations associated with each driver on each road section for different driving states. The F-measures for $\mathbb{S}$ are all 100%.

| States | F-measure (%) | | | | | |
| | R1 | | | R2 | | |
	S1	S2	S3	S1	S2	S3
$\mathbb{F}$	96.6	90.2	90.3	98.2	97.3	97.3
$\mathbb{B}$	68.5	75.9	70.1	91.7	91.5	90.0
$\mathbb{C}$	96.5	92.1	94.8	96.1	96.9	93.0
$\mathbb{D}$	87.8	51.8	73.5	89.0	86.2	73.8

From this table, recognitions associated to $\mathbb{S}$ has no mistake since it is very obvious by the fact that $V = 0$ in this state. Recognition of $\mathbb{F}$ and $\mathbb{C}$ are relatively accurate compared to $\mathbb{B}$ and especially $\mathbb{D}$. The accuracies of $\mathbb{D}$ are relatively low because it is generally difficult to recognize with currently available measured data. For instance, $\mathbb{D}$ occurs when the drivers want to stop at traffic lights; however, with no knowledge of the timing of traffic signals along the road, we can hardly predict the time that the drivers start hitting the brake pedal to change to $\mathbb{D}$. On the other hand, recognizing $\mathbb{B}$ can be relatively easier than $\mathbb{D}$ since the classifier can observe R and $\dot{R}$. Note that there are more traffic lights on R1 than R2, so the driving behavior is more varying on R1, which further reduces the accuracy of $\mathbb{B}$ and $\mathbb{D}$ recognitions on R1.

5   Conclusions and Perspectives

We have investigated individual human driving behavior by constructing the state classifier, which imitates human decision-making in selecting driving states based on driving data. The probabilistic approach BSLM has been employed and the joint conditional probability is obtained, which represents the state classifier. The validations have revealed that this classifier can effectively recognize $\mathbb{F}$, $\mathbb{C}$ and $\mathbb{S}$, while yielding relatively lower accuracy in recognizing $\mathbb{B}$ and $\mathbb{D}$. In the future, if we could acquire data from traffic signals, $\mathbb{D}$ recognition can be improved. The recognition of $\mathbb{B}$ may also be ameliorated by adding some measured feature into the observed data. Besides, we aim to extend the wider range of subject drivers in further experiments as well.

References

1. Liu, A., Pentland, A.: Towards Real-Time Recognition of Driver Intentions. In: Proc. 1997 IEEE Inteligent Transportation Systems Conf. (1997)
2. Kuge, N., Yamamura, T., Shimoyama, O., Liu, A.: A Driver Behavior Recognition Method Based on a Driver Model Framework. In: Proc. the 2000 SAE World Congress (2000)
3. Liu, A., Salvucci, D.: Modeling and Prediction of Human Driver Behavior. In: Proc. 9th International Conf. Human-Computer Interaction (2001)
4. Amano, Y., Sugawara, T., Nagiri, S., Doi, S.: Driver Support System by the Prediction of Driving Manoeuvre. Trans. the Japan Society of Mechanical Engineers 70(698), 2932–2939 (2004)
5. Kumagai, T., Akamatsu, M.: Human Driving Behavior Prediction Using Dynamic Bayesian Networks. In: Proc. JSAE Annual Congress, pp. 13–18 (2005)
6. McCall, J., Trivedi, M.: Human Behavior Based Predictive Brake Assistance. In: Proc. IEEE Intelligent Vehicle Symp., pp. 13–18 (2006)
7. Gerdes, A.: Driver Manoeuvre Recognition. In: Proc. 13th World Congress on ITS (2006)
8. Plöchl, M., Edelmann, J.: Driver Models in Automobile Dynamics Application. Vehicle System Dynamics 45(7–8), 699–741 (2007)
9. Lafferty, J., McCallum, A., Pereira, F.: Conditional Random Fields: Probabilistic Models for Segmenting and Labeling Sequence Data. In: Proc. 18th International Conf. on Machine Learning, pp. 282–289 (2001)
10. Duda, R., Hart, P., Stork, D.: Pattern Classification, 2nd edn. John Wiley and Sons, New York (2001)
11. Shimosaka, M., Nejigane, Y., Mori, T., Sato, T.: Online Action Recognition Based on Boosted Sequential Classification. Jour. the Robotics Society of Japan 25(6), 108–114 (2007) (in Japanese)

Identifying and Analysing Germany's Top Blogs

Darko Obradovič and Stephan Baumann

DFKI, Knowledge Management Department &
Technical University of Kaiserslautern, Computer Science Department
Kaiserslautern, Germany
{obradovic,baumann}@dfki.uni-kl.de

Abstract. Blogs are popular communication instruments in today's web
and altogether, they form the so-called blogosphere. This blogosphere has
repeatedly been subject to structural analyses, while a compilation of the
definitive top blogs is still missing. Using the German blogosphere as an
example, we developed a method to identify an accurate representation of
the top blogs and used their reciprocal recommendations as the input for
social network algorithms. These revealed interesting positions of blogs
that are not respected by existing ranking services up to date.

1   Introduction

Blogs, defined as dynamic internet pages containing articles in reverse chrono-
logical order, are an interesting phenomenon of the Web 2.0 movement. Their
authors utilise them for various purposes, ranging from private online diaries
over so-called *citizen journalism* up to corporate information delivery. Blogs offer rich possibilities for interaction. Authors can include textual and multimedia
content, link to original content, refer to articles in other blogs or let visitors post
comments to their own articles. Thus blogs can and do link to each other, either
by referring to other blog entries in their articles or comments, or by explicitly recommending other blogs in a link set, the *blogroll*. This linked network of
blogs forms the *blogosphere*, which attracted many researchers eagerly analysing
its structure and its dynamics. This is usually done quantitatively with methods
and tools from the field of *Social Network Analysis* (SNA) or qualitatively with
visualisations and field studies.

One of the findings is the discovery of the *A-List* blogs [1,2,3], described by
[4] as "those that are most widely read, cited in the mass media, and receive
the most inbound links from other blogs", which is the first characteristic of this
subnetwork. The studies have also revealed a second important characteristic,
namely the fact that these blogs also heavily link among each other, but rarely
to the rest of the blogosphere. This rest is often referred to as the *long tail* and
consists of millions of blogs that are only partially indexed. Many services attempt to rank the Top 100 blogs, with Technorati[1] being the most popular one.
When looking through different ranking lists, one will usually find roughly the

---

[1] http://www.technorati.com/

A. Dengel et al. (Eds.): KI 2008, LNAI 5243, pp. 111–118, 2008.

same set of blogs in a very different order, although they are all based on algorithms counting inbound links. This discrepancy is mainly due to the particular fraction of long tail blogs that happen to be indexed by the ranking services. We question such an A-List representation, that is also arbitrarily limited to 100 blogs. This is seconded by [4], who compiled a list of top blogs that had to be listed in two of three popular Top 100 lists and ended up with only 45 blogs, which is way too restrictive according to our interpretation of the A-List. As previously mentioned, several studies described the A-List as a set defined rather by its internal link structure than by a certain number. By now, there has been no study yet dealing with the particular structure and the roles inside the A-List subnetwork, but we can base our research on related work describing the structure, the roles and the mechanics in the blogosphere as a whole [5,6]. Therefore we seek to gather a good representation of the A-List by being the first ones to exploit the second characteristic, that A-List blogs tend to link mostly to other A-List blogs and rarely to blogs in the long tail. This would be a valuable input for research and algorithms that require a list of relevant information sources on the web.

Our subject of analysis will be the German blogosphere. For languages other than English, there exist separate sub-blogospheres that are more cohesively interlinked among each other than the blogosphere as a whole is, and thus provide a better suited data set for researchers. This resulted in language-specific studies, such as a quantitative and structural analysis of the Spanish blogosphere [7], an empirical and comparative study of the German blogosphere with the Chinese one [8] or a qualitative sociological field study of the German blogosphere [9]. In Germany, there exists a very active community, even a conference where A-List bloggers meet each other[2], and many national blog ranking services like the German Blogcharts[3].

The rest of the paper is organised as follows: In Section 2 we derive and present our network representation of the A-List and the analytical methods we use for our analyses. In Section 3 we describe our procedure to construct a complete representation of the A-List, which is then analysed and discussed in Section 4 before we conclude this paper in Section 5.

2   Methodology

2.1   Using the Blogroll

The blogroll of a blog is the author's explicit list of recommendations of other blogs. We choose to use these links instead of references from articles or comments. From a social network point of view, a recommendation link is much more expressive and better to be interpreted than an arbitrary reference whose semantics is unknown, as it could signify an aritcle recommendation, a counterstatement, a notion of similar content or any other relation, including spam

---

[2] http://re-publica.de/
[3] http://www.deutscheblogcharts.de/

from comments. In contrast, a blogroll link is a definitve positive relation to another blog by the original author. Of course, in certain cases the blogroll might be outdated, but we expect this to be rather an exception than the rule in an active A-List blog. Nevertheless there are some doubts about the expressiveness of blogroll links to be aware of, e.g. we know from random interviews with community members that bloggers might use their blogroll more for identity management than for real recommendations.

2.2 Data Set

We pursue the goal to deliver an ongoing and comprehensive list of German A-List blogs and thus need an automatically generated data set. Previous blog studies all used large one-time generated data sets for a nonrecurring analysis. E.g. [4] used a manually aquired sample of 5,517 blogs and 14,890 links. We instead need a systematically constructed and fast to aquire set of german top blogs with a certain confidence that it really represents the A-List.

We initially use the set of blogs from an existing ranking service, ignoring their positions in that list. We choose the German Blogcharts as our sample, which is a weekly updated list of the 100 best german blogs in Technorati's ranking. This filtering by language is done manually by the maintainer of the site. We implemented a set of scripts to scrape the actual blogs from the Blogcharts website and to crawl the entries from their individual blogrolls, if present. We encountered three pitfalls in this task. First, we had to develop a sufficiently good heuristic for locating the blogroll entries, as their inclusion on the blog page is not standardised in a way we could rely on. The second pitfall is the existence of multiple URLs for one blog. We check every single blogroll entry with an HTTP request in order to not insert blog links to synonymous or redirected URLs another time into our database. This would cause a split of one blog into two separate nodes and thus distort our network and our results. This is a common problem, e.g. Technorati often ranks a blog multiple times and thus leads to biased results[4]. And the last pitfall is the reachability of a blog. Blogs that are not reachable during our crawl are ignored with respect to their own blogroll links, but remain in the data set and can be recommended by other blogs of course. The data used in this paper has been collected in late September 2007 and contains 7,770 links. Crawling the blogrolls costs us several hours, as we have to verify each new link as described above along with eventual timeouts.

In Section 1 we postulated the hypothesis, that language-specific subsets of the blogosphere are more interconnected than the anglophone international blogosphere is. Assuming that the A-List phenomenon is valid for any language, and there are no indicators against this assumption, we are going to compare the interconnectedness of the Technorati Top 100 samples of German and international blogs. We therefor crawl the corresponding networks as described above and compare the relevant measures in Table 1, i.e. the number of links, the density and the number of isolated blogs with respect to weak connectivity. Notably,

---

[4] Check the two links `http://shopblogger.de/blog` and `http://shopblogger.de` via the Technorati API for an example.

Table 1. Network structure comparison

	International	German Initial	German Extended
Blogs	100	100	185
Links	201	419	1453
Density	0.0203	0.0423	0.0427
Isolated Blogs	32	6	5

all metrics clearly indicate that the german Top 100 blogs are significantly better interconnected than the international ones.

2.3   Social Network Analysis

We use well established SNA methods [10] for the analysis of our A-List network. The blogs are the vertices and the blogroll links are the edges. Most SNA measures assume a certain meaning for transitivity and thus use metrics based on path lengths, such as radius, diameter and distance. In our case, transitivity does not play any role. We argue that a transitive recommendation has no influence in terms of authority or trust. As we are also interested in the roles of the A-List blogs, we find the Kleinberg algorithm [11] to be best suited in our case and elect it to be the main SNA metric for our analyses. It reveals important vertices in two aspects that mutually depend on each other and can be expressively mapped to roles of blogs in the blogosphere:

Authorities. These are blogs that are recommended by many good hubs, and therefore are considered to provide good content.
Hubs. These are blogs that recommend many good authorities and therefore are considered to provide good recommendations for finding good content.

3   Generating the A-List Data Set

Our idea is to use the Top 100 data set described in Section 2.2 as a starting point for finding other blogs that also belong to the A-List. The 100 blogs most frequently referenced by the whole blogosphere are assumed to belong to the A-List, and in Section 2.2 we have observed that a blog of this set in average recommends 4.19 other blogs from this set. Our conclusion is, that any other german blog that is also recommended as many times from within that set, belongs to the A-List and not to the rarely referenced long tail blogs. Extending the A-List according to that rule should result in an equally structured network, maintaining all the A-List characteristics.

We already have the required blogroll links of the Top 100 blogs at hand, but how to determine whether a referenced website really is a blog, and whether it is a german one or not? This task is not trivial, and [4] decided to verify the blog question manually, as automatic blog detection yields too many false classifications. Thanks to the fact that our study restricts itself to A-List blogs, which are definitely indexed by Technorati, we can use its API[5] to verify blogs by URL. Once verified that a recommendation link really is a blog, we need to check whether it is in german language or not. We therefor utilise a stop word-based algorithm to identify the language of the plain text that we extracted from the blog's start page. We found this method to be extremely reliable due to the typically rich textual content on blog sites.

We extend the data set iteratively according to the average number of recommendation links as described above. In each iteration, we evaluate the required number of recommendation links inside our data set and include all german blogs that are recommended at least that often. We always round down the required link number, in order to not exclude any blogs close to the limit. We then process the blogrolls of these newly added blogs, integrate their recommendation links and proceed to the next iteration step, as long as we find new blogs to be included. As described in Section 1, theory expects this procedure to only slightly extend the data set inside the A-List, as the blogosphere as a whole is only sparsely connected compared to the A-List. Although we include new blogs based solely on inbound links, based on the findings of previous studies, we expect the new blogs to integrate well and to link back often into the existing data set. The results of the extension procedure are shown in Table 2, listing the iteration number, the required number of links for being included, the number of candidate URLs we had to check and the number of German blogs that were finally included in that step.

Table 2. Extension Results

Step	Required	Candidates	Included
1	4(.19)	68	47
2	5(.88)	25	22
3	6(.76)	13	13
4	7(.28)	4	3
5	7(.40)	0	0

We end up with an extended data set of 185 blogs, which we expect to be a more accurate representation of the German A-List than before. In order to verify this, we compare the network properties of this extended data set with the initial Top 100 data set. although there exist some retentions on the expressiveness of network density when comparing networks with a different number of vertices (see [10], pp. 93ff.), Table 1 reveals that our extended data set, even while ignoring the network size difference, has a slightly higher density and contains less isolated blogs, i.e. is at least as well connected as the initial data set or even more tightly connected. This means that we received a large number of recommendation links from the newly included blogs back into the initial data set and thus, remembering once more the typical properties of an A-List network,

---

can conclude that our extended data set is definitely a more complete and better representation of the German A-List than the intial one.

4   Analysis

We used our sample set and Kleinberg's algorithm [11] to identify authorities and hubs. Table 3 and Table 4 list the top 10 authorities and hubs of the extended data set along with their coefficients and the data set they first appeared in.

Table 3. Top 10 Authorities

Rank	Blog URL	Coefficient	Data Set
1.	`http://wirres.net`	0.269	initial
2.	`http://don.antville.org`	0.228	initial
3.	`http://schwadroneuse.twoday.net`	0.177	extended
4.	`http://www.ankegroener.de`	0.176	initial
5.	`http://www.whudat.de`	0.174	initial
6.	`http://www.dasnuf.de`	0.172	extended
7.	`http://stefan-niggemeier.de/blog`	0.170	initial
8.	`http://ivy.antville.org`	0.169	extended
9.	`http://blog.handelsblatt.de/indiskretion`	0.166	initial
10.	`http://www.spreeblick.com`	0.164	initial

Table 4. Top 10 Hubs

Rank	Blog URL	Coefficient	Data Set
1.	`http://don.antville.org`	0.297	initial
2.	`http://schwadroneuse.twoday.net`	0.251	extended
3.	`http://www.whudat.de`	0.229	initial
4.	`http://www.mequito.org`	0.214	extended
5.	`http://blog.franziskript.de`	0.203	extended
6.	`http://www.nerdcore.de/wp`	0.195	initial
7.	`http://wortschnittchen.blogger.de`	0.184	extended
8.	`http://hinterstuebchen.juliette-guttmann.de`	0.183	extended
9.	`http://www.jensscholz.com`	0.179	extended
10.	`http://www.burnster.de`	0.175	extended

The list of the top authorities is populated by a few more blogs from the initial data set than from the extended one, whereas the list of the top hubs shows exactly the opposite tendency. This is the opposite of the expected outcome of an extension in a random network and hence confirms a second time that we have extended our data set inside the A-List subnetwork.

There are three blogs appearing in both lists. The two measures mutually depend on each other, but ulimately they are not directly related for an individual blog, hence this implies another special role in the blogosphere. These blogs hold a *key position* in the network, because they are recommended by others for having good content and are known to recommend good content themselves. We want to express that role in a new coefficient c_{key_pos} as defined in Formula 1, in which we simply combine the authority and hub coefficients c_{auth} and c_{hub} multiplicatively. The square root is used for reasons of visual expressivity and comparability of the coefficient value.

$$c_{key\_pos} = \sqrt{c_{auth}c_{hub}} \tag{1}$$

As a blog recommendation network lives from both, authorities and hubs, we suggest a measure reflecting the overall importance of a blog in that network. For being important a blog needs to be either a strong authority, a strong hub, hold a key position or partially all of this. We determine the *importance* coefficient c_{imp} as defined in Formula 2 by additively combining the authority, hub and key position coefficients, whereat the authority coefficient is weighted stronger, as authority is the most important attribute.

$$c_{imp} = \frac{2c_{auth} + c_{hub} + c_{key\_pos}}{4} \tag{2}$$

Table 5 lists the 10 most important blogs from the extended data set according to our formula. As expected, we see the blogs listed in both, the authorities and the hubs list, ranked on top of the new list. There also appear some new blogs that replaced previous top blogs, which were either purely strong authorities or purely strong hubs. This final list reflects our opinion on how to best rank blogs inside the A-List, based on our assumptions and findings presented throughout this paper.

Table 5. Top 10 Important Blogs

Rank	Blog URL	Coefficient	Data Set
1.	`http://don.antville.org`	0.254	initial
2.	`http://schwadroneuse.twoday.net`	0.204	extended
3.	`http://www.whudat.de`	0.194	initial
4.	`http://www.ankegroener.de`	0.175	initial
5.	`http://blog.franziskript.de`	0.168	extended
6.	`http://wortschnittchen.blogger.de`	0.168	extended
7.	`http://www.mequito.org`	0.168	extended
8.	`http://blog.handelsblatt.de/indiskretion`	0.160	initial
9.	`http://www.burnster.de`	0.158	extended
10.	`http://www.spreeblick.com`	0.154	initial

5   Conclusion

We had the goal to identify the blogs of Germany's A-List and to analyse more in depth their individual roles and positions in that subset of the blogosphere. We justified our decisions to constrain our study to the German blogosphere, to use blogroll links for the analyses and to construct an extended data set with the German Blogcharts as its seed. Despite all potential pitfalls and eventual shortcomings, our results quantitatively confirmed the decisions we had taken, although the accuracy of our final ranking is difficult to be judged. This definitely requires some feedback from the blogging community.

This list can be effectively used for research and algorithms in the fields of opinion mining, trend detection, news aggregation, recommendation systems. Our next steps are to comparatively analyse it to other national blogospheres and to optimise the extension rules and the importance formula accordingly, as well as to further lower the error rates and to accelerate the crawling procedure.

References

1. Delwiche, A.: Agenda-setting, opinion leadership, and the world of web logs. First Monday 10(12) (2005)
2. Marlow, C.: Audience, structure and authority in the weblog community. In: Proceedings of the International Communication Association Conference (2004)
3. Park, D.: From many, a few: Intellectual authority and strategic positioning in the coverage of, and self-descriptions of, the big four weblogs. In: Proceedings of the International Communication Association Conference (2004)
4. Herring, S.C., Kouper, I., Paolillo, J.C., Scheidt, L.A., Tyworth, M., Welsch, P., Wright, E., Yu, N.: Conversations in the blogosphere: An analysis from the bottom up. In: Proceedings of the 38th Annual Hawaii International Conference on System Sciences (HICSS 2005). IEEE Computer Society, Los Alamitos (2005)
5. Gruhl, D., Guha, R., Liben-Nowell, D., Tomkins, A.: Information diffusion through blogspace. In: Proceedings of the 13th international conference on World Wide Web, pp. 491–501. ACM Press, New York (2004)
6. Kumar, R., Novak, J., Raghavan, P., Tomkins, A.: On the bursty evolution of blogspace. World Wide Web 8(2), 159–178 (2005)
7. Tricas, F., Ruiz, V., Merelo, J.J.: Do we live in an small world? measuring the spanish–speaking blogosphere. In: Proceedings of the BlogTalk Conference (2003)
8. He, Y., Caroli, F., Mandl, T.: The chinese and the german blogosphere: An empirical and comparative analysis. In: Mensch und Computer, pp. 149–158 (2007)
9. Schmidt, J.: Weblogs - Eine kommunikationssoziologische Studie. UVK Verlagsgesellschaft (2006)
10. Scott, J.: Social Network Analysis: A Handbook. SAGE Publications, Thousand Oaks (2000)
11. Kleinberg, J.M.: Authoritative Sources in a Hyperlinked Environment. In: Proceedings of the 9th Annual ACM-SIAM Symposium on Discrete Algorithms, pp. 668–677. AAAI Press, Menlo Park (1998)

Planar Features for Visual SLAM

Tobias Pietzsch

Technische Universität Dresden, International Center for Computational Logic,
01062 Dresden, Germany
`Tobias.Pietzsch@inf.tu-dresden.de`

Abstract. Among the recent trends in real-time visual SLAM, there has
been a move towards the construction of structure-rich maps. By using
landmarks more descriptive than point features, such as line or surface
segments, larger parts of the scene can be represented in a compact form.
This minimises redundancy and might allow applications such as object
detection and path planning. In this paper, we propose a probabilistic map representation for planar surface segments. They are measured
directly using the image intensities of individual pixels in the camera
images. Preliminary experiments indicate that the motion of a camera
can be tracked more accurately than with traditional point features, even
when using only a single planar feature.

1   Introduction

In simultaneous localisation and mapping (SLAM), we are concerned with estimating the pose of a mobile robot and simultaneously building a map of the
environment it is navigating [1]. The problem is formulated in a Bayesian framework where noisy measurements are integrated over time to create a probability
distribution of the state of a dynamical system, consisting of landmark positions
and the robot's pose. Visual SLAM tackles this problem with only a camera
(monocular or stereo, typically hand-held) as a sensor.

Since the seminal work by Davison [2], the majority of existing systems for
visual SLAM build sparse maps representing 3D locations of scene points and
their associated uncertainties. This particular representation has been attractive
because it allows real-time operation whilst providing sufficient information for
reliable tracking of the camera pose. If we want to visual SLAM to move beyond
tracking applications, we need maps which allow geometric reasoning. Planar
structures allow a compact representation of large parts of the environment.
For tasks such as scene interpretation, robot navigation and the prediction of
visibility/occlusion of artificial objects in augmented reality, maps consisting of
planar features [3,4] seem more suitable than wire frame models [5,6].

In this paper, we propose a map representation for planar surface segments
within the framework of [2]. Those planes are measured directly using the image
intensities of individual pixels in the camera images.

Gee et al.[3] presented a visual SLAM system in which planar structural components are detected and embedded within the map. Their goal is to compact

A. Dengel et al. (Eds.): KI 2008, LNAI 5243, pp. 119–126, 2008.

the representation of mapped points lying on a common plane. However these planes are not directly observable in the camera images, but are inferred from classical point feature measurements. Our approach has a closer relationship to the works by Molton et al. [4] and Jin et al. [7]. We also use pixel measurements directly to update estimates of plane parameters. Molton et al. [4] regard feature points as small locally planar patches and estimate their normal vectors. However, this is done outside of the SLAM filter, thus ignoring correlations between normal vectors and rest of the state. While being useful for improving the observability of point features, their approach does not scale to larger planar structures because these correlations cannot be neglected in this case. Jin et al. [7] are able to handle larger planes but rely on the linearity of the image gradient which limits their approach with respect to image motion and tolerable camera acceleration. We solve this issue by casting the measurement update in an iterative framework. Furthermore, our feature model employs an inverse-depth parameterisation which has been shown to be more suitable to linearisation than previous representations [8].

In the following section, we describe Davison's framework [2] for visual SLAM. In Sec. 3, we introduce our feature representation and measurement model. After discussing details of the measurement process in Sec. 4, we present experimental results for simulated and real scenarios in Sec. 5. Sec. 6 concludes the paper.

2   EKF-Based Visual SLAM

The task in visual SLAM is to infer the state of the system, i.e., the pose of the camera and a map of the environment from a sequence of camera images. A commonly used and successful approach is to address the problem in a probabilistic framework, using an Extended Kalman Filter (EKF) to recursively update an estimate of the joint camera and map state [2,5,3,4].

The belief about the state $\mathbf{x}$ of the system is modeled as a multivariate Gaussian represented by its mean vector $\mu_{\mathbf{x}}$ and covariance matrix $\Sigma_{\mathbf{x}}$. The state vector can be divided into parts describing the state of the camera $\mathbf{x}_v$ and of map features $\mathbf{y}_i$.

$$\mu_{\mathbf{x}} = \begin{pmatrix} \mu_{\mathbf{x}_v} \\ \mu_{\mathbf{y}_1} \\ \vdots \\ \mu_{\mathbf{y}_n} \end{pmatrix} \qquad \Sigma_{\mathbf{x}} = \begin{bmatrix} \Sigma_{\mathbf{x}_v \mathbf{x}_v} & \Sigma_{\mathbf{x}_v \mathbf{y}_1} & \cdots & \Sigma_{\mathbf{x}_v \mathbf{y}_n} \\ \Sigma_{\mathbf{y}_1 \mathbf{x}_v} & \Sigma_{\mathbf{y}_1 \mathbf{y}_1} & \cdots & \Sigma_{\mathbf{y}_1 \mathbf{y}_n} \\ \vdots & \vdots & \ddots & \vdots \\ \Sigma_{\mathbf{y}_n \mathbf{x}_v} & \Sigma_{\mathbf{y}_n \mathbf{y}_1} & \cdots & \Sigma_{\mathbf{y}_n \mathbf{y}_n} \end{bmatrix} \tag{1}$$

The state estimate is updated sequentially using the predict-update cycle of the EKF. Whenever a new image is acquired by the camera, measurements of map features can be made and used to update the state estimate, resulting in a decrease of uncertainty in the update step. In the prediction step, a process model is used to project the estimate forward in time. The process model describes how the state evolves during the period of "temporal blindness" between images. Only the camera state $\mathbf{x}_v$ is affected by the process model, because we assume an otherwise static scene. The camera is assumed to be moving with constant linear and angular velocity. The (unknown) accelerations that cause deviation from this assumption are modeled as noise. This results in an increase of camera state uncertainty in the prediction step. Similar to [2] we model the camera state as $\mathbf{x}_v = \begin{pmatrix} \mathbf{r} \, \mathbf{q} \, \mathbf{v} \, \boldsymbol{\omega} \end{pmatrix}^\top$. Position and orientation of the camera with respect to a fixed world frame $\mathcal{W}$ are described by the 3D position vector $\mathbf{r}$ and the quaternion $\mathbf{q}$. Translational and angular velocity are described by $\mathbf{v}$ and $\boldsymbol{\omega}$.

The EKF update step integrates new information from measurements of map features into the state estimate. A generative measurement model

$$\mathbf{z} = \mathbf{h}(\mathbf{x}_v, \mathbf{y}_i) + \boldsymbol{\delta} \tag{2}$$

describes the measurement vector $\mathbf{z}$ as a function of the (true, unknown) state of the camera $\mathbf{x}_v$ and the feature $\mathbf{y}_i$. The result is affected by the zero-mean measurement noise vector $\boldsymbol{\delta}$. The current estimate of the camera and feature state can be used to *predict* the expected measurement. The difference between the predicted and actual measurement is then used in the EKF to update the state estimate.

In [2], a map feature $\mathbf{y}_i = \begin{pmatrix} x \, y \, z \end{pmatrix}^\top$ describes the 3D coordinates of a fixed point in the environment. A measurement of such a feature then consists of the 2D coordinates $\mathbf{z} = (u, v)^\top$ of the projection of this point into the current camera image. To be able to actually make measurements, new features are initialised on salient points in a camera image. With each new feature a small template patch of its surrounding pixels in this initial image is stored. In subsequent images, measurements of this feature are made by searching for the pixel with maximal correlation to this template. This search is carried out in a elliptic region determined by the predicted measurement and its uncertainty.

3   Feature Representation and Measurement Model

The template-matching approach to the measurement process works well as long as the feature is observed from a viewpoint reasonably similar to the one used for initialisation. To improve the viewpoint range from which a feature is observable, we can try to predict the change in feature appearance due to changed viewpoint prior to the correlation search. To do this, we assume that the feature is located on a locally planar scene surface. Given the surface normal and the prior estimate of the camera position the features template patch can be warped to give the expected appearance in the current camera image. Simply assuming a surface normal facing the camera in the initial image already introduces some tolerance to changing viewing distance and rotation about the camera axis. Molton et al. [4] go one step further by estimating the surface normal using multiple feature observations from different viewpoints. Both approaches are aimed at improving the stability of the template matching procedure by removing the effects of varying viewpoint.

However, changes in feature appearance also provide information which can be directly used to improve the state estimate. For instance if we observe that

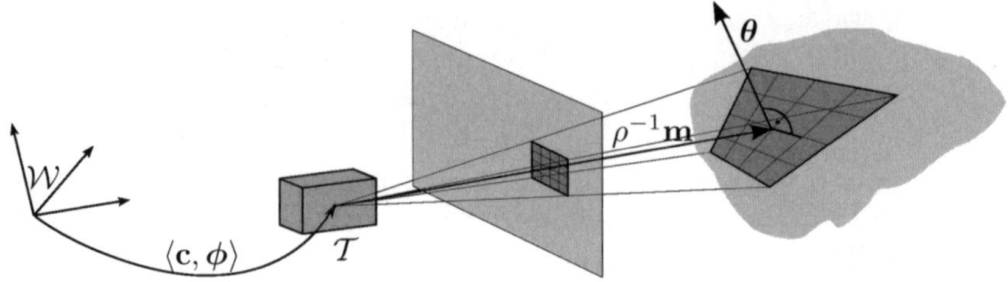

Fig. 1. The relative orientation of world frame $\mathcal{W}$ and template frame $\mathcal{T}$ is given by translation $\mathbf{c}$ and rotation ϕ. The unit vector $\mathbf{m}$ defines a ray to the feature center, ρ is the inverse depth along this ray. The plane normal is described by $\boldsymbol{\theta}$.

the scale of the observed image patch is larger than we expected, this tells us that the camera is closer to the feature than we predicted. Also, the amount of perspective distortion is directly related to the relative orientation between the camera and scene surface. To exploit such information, modeling a feature measurement as a single 2D coordinate as described above is not sufficient.

In the following, we describe a feature representation of planar segments in the scene. A measurement model of such a planar features is developed directly using the raw camera images, i.e., a measurement comprises a set of pixel intensity values. In our system, features represent planar surface segments in the environment. Such a segment is described in terms of its appearance in the image where it was initially observed, and the position of the camera when this initial image was taken, cf. Fig. 1.

For each feature, the initial camera image is stored as the template image $T : \mathbb{R}^2 \to \mathbb{R}$. We assume that the outline of the compact image area corresponding to the planar scene segment is known. Some (arbitrary) pixel within this area is chosen as the *feature center* projection. The unit vector $\mathbf{m}$ is the ray from the camera center through this pixel. Both T and $\mathbf{m}$ are fixed parameters, i.e., not part of the probabilistic state.

In the EKF state vector, the i-th map feature is parametrised as

$$\mathbf{y}_i = \begin{pmatrix} \mathbf{c} \; \phi \; \rho \; \boldsymbol{\theta} \end{pmatrix}^\top . \tag{3}$$

The parameters $\mathbf{c}$, ϕ describe the camera translation and rotation at the initial observation. This initial camera coordinate frame is referred to as the template frame $\mathcal{T}$. The parameter ρ is the inverse depth[1] measured along the ray defined by $\mathbf{m}$. The last component $\boldsymbol{\theta}$ is the normal vector of the feature plane, encoded in polar coordinates. Both $\boldsymbol{\theta}$ and $\mathbf{m}$ are measured with respect to the template frame $\mathcal{T}$.

Having defined the feature representation, we can now proceed to formulate a measurement model of the form (2). It is well-known from the computer vision literature that two images of a plane are related by a homography transformation [9]. Thus, pixel coordinates in the current camera image and the template

---

[1] Inverse depth parameterisation was chosen, because it can cope with distant features and its better suited to approximation by a Gaussian than depth [8].

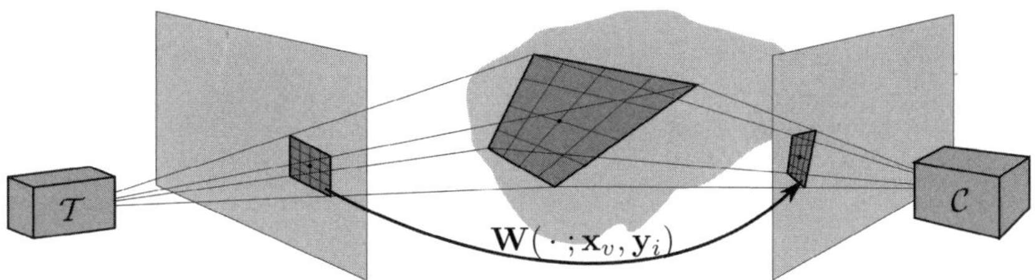

Fig. 2. The camera $\mathbf{x}_v$ and feature state $\mathbf{y}_i$ define a homography warp $\mathbf{W}$ between the template and current image. $\mathbf{W}$ can be used to predict intensities in the current image by looking up the intensities at the corresponding template pixels. The measurement consists of the observed intensities inside the feature outline in the current image.

image can be related through the homography induced by the feature plane. We will denote by $\mathbf{W}(\,\cdot\,;\mathbf{x}_v,\mathbf{y}_i)$ the homography warp function which maps points in the template image T to the corresponding points in the current image C, cf. Fig. 2. The homography is fully determined by the orientation of both the template and current camera coordinate frames, and the position and normal of the feature plane. Therefore, $\mathbf{W}$ is parameterised by the current state of the camera $\mathbf{x}_v$ and the feature state $\mathbf{y}_i$. The exact form of the warp function can be derived similarly to [4]. This is omitted here due to space limitations.

Using the warp function $\mathbf{W}(\,\cdot\,;\mu_{\mathbf{x}_v},\mu_{\mathbf{y}_i})$ parameterised on the prior state estimate, we can project the features outline into the current image to determine where we expect to observe the feature. Let $\mathbf{U} = \{\mathbf{u}_1,\ldots,\mathbf{u}_m\}$ be the set of pixel locations inside the expected projected outline. The measurement vector $\mathbf{z} = (z_1,\ldots,z_m)^\top$ consists of the intensities measured at these locations, i.e., $z_j = C(\mathbf{u}_j)$.

To formulate a generative model, we have to define the function $\mathbf{h}(\mathbf{x}_v,\mathbf{y}_i)$, i.e., we have to express what we expect to measure given a certain camera and feature state. The intensity we expect to measure at a given location $\mathbf{u}_j$ in the current image is the intensity at the corresponding location in the template image. This corresponding location is computed using the inverse of the warp function, $\mathbf{W}^{-1}$. Thus, the generative model for a single pixel is

$$z_j = h_j(\mathbf{x}_v,\mathbf{y}_i) + \delta_j = T(\mathbf{W}^{-1}(\mathbf{u}_j;\mathbf{x}_v,\mathbf{y}_i)) + \delta_j \qquad (4)$$

where $\delta_j \sim \mathcal{N}(0,r_j)$ is zero-mean Gaussian noise with variance r_j, uncorrelated between pixel locations.

4   Measurement Process

The EKF update step involves a linearisation of the measurement model, i.e., (2) is replaced by its first-order Taylor expansion around the prior state estimate. For the proposed measurement model, (4) must be linearised for every intensity measurement z_j. This involves linearisation of the template image function T around the template pixel position predicted as corresponding to $\mathbf{u}_j$. The intensity in the neighbourhood of a template pixel is approximated using the image

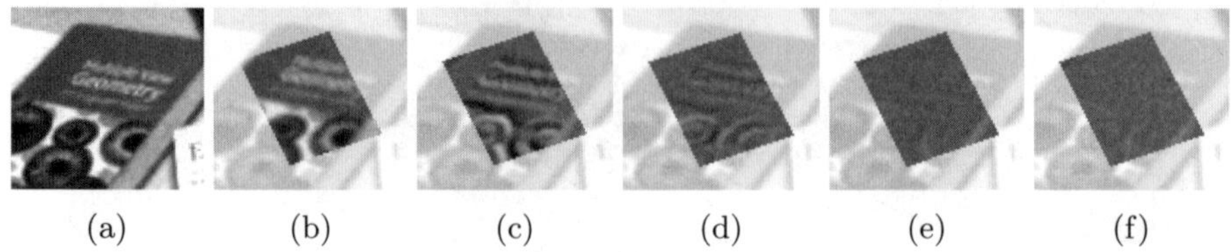

(a)   (b)   (c)   (d)   (e)   (f)

Fig. 3. A real-world example of the measurement update iteration. (a) the current image. (b) the predicted feature template. (c)-(d) show the difference between the warped template and the current image. (c) before the first iteration, i.e., the difference between (a) and (b). (d) after the first iteration with the restricted measurement model. (e,f) after the second and third iteration which are performed using the full planar measurement model.

gradients at that pixel. Image functions in general are highly non-linear. Therefore, the linearised model used in the EKF will provide good approximation only in a small region of state-space around the true state. To deal with this problem, we use an Iterated Extended Kalman Filter (IEKF). The basic idea of the IEKF is that because the posterior estimate is closer to the true state than the prior estimate, linearisation should rather be performed around the posterior estimate. The update step is repeated several times, where in each iteration linearisation is performed around the posterior estimate obtained in the previous iteration. During rapid camera accelerations the prior estimate can be quite far from the true state, rendering the initial linearisation of T useless. To guide the update process to the correct region of convergence the first iteration is done using a restricted model where the measurement consists only of the 2D image coordinates of the projection of the feature center. The measurement is obtained using point-feature-like correlation search for the feature template warped according to the prior estimate. This iterative process is illustrated in Fig. 3.

In our current implementation we use a stereo camera. Conceptually there is not a great difference between a monocular and a stereo setting. The additional intensity measurements obtained from the second camera eye contribute additional pixel measurements (4) with a different warp function $\mathbf{W}'$ which takes into account the baseline offset of the second eye from the camera center.

5   Experimental Results

We compared the performance of the proposed measurement model to that of a point-based model on a artificial stereo image sequence. The sequence was generated using the raytracer POV-Ray. The scene consists of a single large planar object. The camera parameters were chosen to resemble a Point Grey Bumblebee® stereo camera.[2] The camera is initially 1 m away from the planar object. It travels a path where the object is viewed from varying distance (0.4 to 1.7 m) and varying angles ($0°$ to $70°$).

---

[2] This type of camera was used for the real-world experiments.

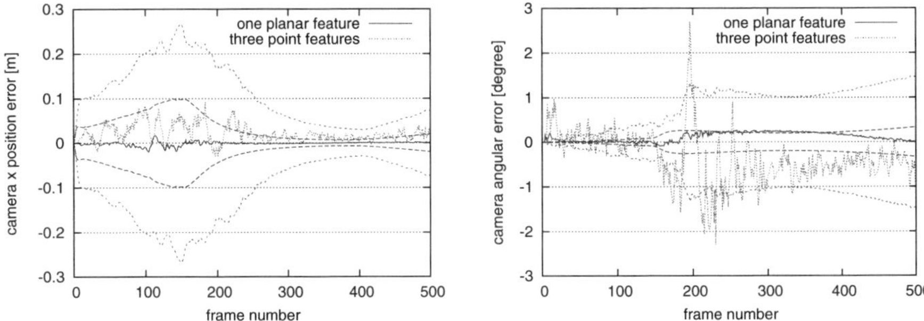

Fig. 4. Comparison of absolute errors in estimated camera orientations for the artificial image sequence. The position error in x direction is shown on the left. The total angular error is shown on the right. The dashed lines indicate the 3σ confidence bounds.

To fully constrain the camera orientation estimate only one (sufficiently large) planar feature is needed. With the point-based model three features are needed. We use state-of-the-art point features with an inverse-depth representation and predictive template warping assuming camera-facing normal. All features where initialised by hand on salient image areas.

In Fig. 4 we compare the errors of the reconstructed camera trajectories with respect to ground truth. The trajectory obtained using the planar feature is more accurate and a lot smoother than the trajectory obtained using point features. Several factors contribute to this result. By using intensities at a large number of pixels[3] as measurements much more information from the images is used than by artificially reducing the measurement to a single 2D coordinate for each of the point features. Also, point feature measurements are made at integer pixel coordinates, although the approach could be extended to compute a sub-pixel interpolation of the correlation search maximum. Iterating the update process as described in Sec. 4 provides sub-pixel accurate registration of the feature template while at the same time keeping the same stability with respect to prior uncertainty as point features.

We have also tested the planar feature model with real image sequences, that were recorded with a Point Grey Bumblebee® stereo camera. Examples from one sequence are shown in Fig. 5. Again, only a single planar feature is used. In this case, the feature template is 41×41 pixels. The feature was selected by hand on a salient image area. The feature normal estimate is initialised with high uncertainty as pointing towards the camera. Convergence to the (visually judged) correct normal occurs after one frame. No ground truth was available for this sequence. To give an impression of the accuracy of the reconstruction the images in Fig. 5 are augmented with a coordinate frame attached to the reconstructed feature plane such that its Z axis coincides with the plane normal.

---

[3] The feature used corresponds to a 71×71 pixel patch in the initial image, which means that ca. $10,000$ intensity measurements are contributed by one stereo image.

Fig. 5. Results for some pictures of a real image sequence using a single planar feature. The projected outline of the feature is shown in red. A coordinate frame attached to the reconstructed feature plane such is shown in black.

6   Conclusion

We presented a representation of plane segments as features in a probabilistic map and a method to directly measure such features in camera images. We have shown in experiments that the motion of a camera can be tracked more accurately than with traditional point features, even when using only a single planar feature. In order to build a fully functional SLAM system one important issue has yet to be addressed, namely the *detection* of planar scene structures which can be used as features. This is the focus of immediate future work.

References

1. Thrun, S., Burgard, W., Fox, D.: Probabilistic Robotics. MIT Press, Cambridge (2005)
2. Davison, A.J.: Real-Time Simultaneous Localisation and Mapping with a Single Camera. In: International Conference on Computer Vision (2003)
3. Gee, A.P., Chekhlov, D., Mayol, W., Calway, A.: Discovering Planes and Collapsing the State Space in Visual SLAM. In: BMVC (2007)
4. Molton, N.D., Davison, A.J., Reid, I.D.: Locally Planar Patch Features for Real-Time Structure from Motion. In: BMVC (2004)
5. Smith, P., Reid, I., Davison, A.: Real-Time Monocular SLAM with Straight Lines. In: BMVC (2006)
6. Eade, E., Drummond, T.: Edge Landmarks in Monocular SLAM. In: BMVC (2006)
7. Jin, H., Favaro, P., Soatto, S.: A semi-direct approach to structure from motion. The Visual Computer 19(6), 377–394 (2003)
8. Montiel, J., Civera, J., Davison, A.J.: Unified inverse depth parameterization for monocular SLAM. In: Robotics: Science and Systems (RSS) (2006)
9. Hartley, R., Zisserman, A.: Multiple View Geometry in Computer Vision. Cambridge Univerity Press, Cambridge (2000)

Extracting and Querying Relations in Scientific Papers

Ulrich Schäfer, Hans Uszkoreit, Christian Federmann, Torsten Marek,
and Yajing Zhang*

German Research Center for Artificial Intelligence (DFKI), Language Technology Lab
Campus D 3 1, Stuhlsatzenhausweg 3, D-66123 Saarbrücken, Germany
{ulrich.schaefer,uszkoreit,christian.federmann,torsten.marek,
yajing.zhang}@dfki.de

Abstract. High-precision linguistic and semantic analysis of scientific texts is an emerging research area. We describe methods and an application for extracting interesting factual relations from scientific texts in computational linguistics and language technology. We use a hybrid NLP architecture with shallow preprocessing for increased robustness and domain-specific, ontology-based named entity recognition, followed by a deep HPSG parser running the English Resource Grammar (ERG). The extracted relations in the MRS (minimal recursion semantics) format are simplified and generalized using WordNet. The resulting 'quriples' are stored in a database from where they can be retrieved by relation-based search. The query interface is embedded in a web browser-based application we call the Scientist's Workbench. It supports researchers in editing and online-searching scientific papers.

1 Introduction

Research in the HyLaP project (in particular here the subproject HyLaP-AM for the research on a personal digital associative memory) focuses on exploring hybrid (e.g. deep and shallow) methods to develop a framework for building a densely interlinked, associative memory on the basis of email and documents on the PC or laptop of a user.

As part of the application scenario, intelligent search based on factual relations extracted from parsed scientific paper texts is provided which we will describe in this paper. Because precision is in focus when facts are to be found in research papers, deep linguistic analysis of the paper contents is performed, assisted by shallow resources and tools for incorporating domain knowledge and ontology information, and for improving robustness of deep processing.

In this paper, we concentrate on two aspects of this rather complex enterprise. First, we explain how we extract the facts (relations) from the papers, and second, we describe the GUI application that has been built and that includes access to the extracted information.

---

* This work has been supported by a grant from the German Federal Ministry of Education and Research (FKZ 01 IW F02).

A. Dengel et al. (Eds.): KI 2008, LNAI 5243, pp. 127–134, 2008.

We start in Section 2 with a description of the hybrid parsing approach to get semantic structures from natural language sentences. In Section 3, we explain how we compute the core relations from the linguistics-oriented output results of the previous step. We present the GUI application Scientist's Workbench and the underlying system architecture in Section 4. Finally, we briefly discuss related work, conclude and give an outlook to future work.

2   Parsing Science

In order to get documents related to the application domain, we downloaded the contents of the ACL Anthology[1], a collection of scientific articles from international conferences, workshops and journals on computational linguistics and language technology. We acquired all papers from the years 2002–2007, resulting in a collection of 4845 PDF documents, along with bibliographical information in BibTeX format, if available.

The texts were then parsed using the hybrid NLP platform Heart of Gold [2] – so far only paper abstracts, parsing the full papers will be addressed in a later project phase. Heart of Gold is an XML-based middleware architecture for the integration of multilingual shallow and deep natural language processing components.

The employed Heart of Gold configuration instance starts with a tokenizer, the part-of-speech tagger TnT [3] and the named entity recognizer SProUT [4]. These shallow components help to identify and classify open class words such as person names, events (e.g. conferences) or locations.

Furthermore, the tagger can guess part-of-speech tags of words unknown to the deep lexicon. For both (unknown words and named entities), generic lexicon entries are generated in the deep parser. Using an XML format, the shallow preprocessing results are then passed to the high-speed HPSG parser PET [5] running the open source broad-coverage grammar ERG [6].

Details of the general approach and further configurations as well as evaluations of the benefits of hybrid parsing are described in [2].

In contrast to shallow parsing systems, the ERG not only handles detailed syntactic analyses of phrases, compounds, coordination, negation and other linguistic phenomena that are important for extracting relations, but also generates a formal semantic representation of the meaning of the input sentence.

Ambiguities resulting in multiple readings per input sentence are ranked using a statistical model based on the Redwoods treebank [7]. We got full parses for 62.5% of the 17 716 abstract sentences in HoG. The average sentence length was 18.9 words for the parsed, 27.05 for the unparsed, and 21.95 for all.

The deep parser returns a semantic analysis in the MRS representation format (Minimal Recursion Semantics; [8]) that is, in its robust variant, specifically suited to represent hybrid, i.e. deep and comparably underspecified shallow NLP semantics. A sample MRS as produced by ERG in Heart of Gold is shown in Figure 1.

---

[1] [1]; http://acl.ldc.upenn.edu

As part of the shallow preprocessing pipeline, named entity recognition is performed to recognize names and domain-relevant terms. To improve recognition in the science domain of language technology and computational linguistics, we enriched the lingware resources of the generic named entity recognizer SProUT [4] by instance and concept information from an existing domain ontology using the OntoNERdIE tool [10].

We used the LT World ontology [9], containing about 1200 concepts and approx. 20 000 instances such as terminology, conferences, persons, projects, products, companies and organizations.

The MRS representations resulting from hybrid parsing, however, are close to linguistic structures and contain more detailed information than a user would like to query and search for. Therefore, an additional extraction and abstraction step is necessary before storing the semantic structures with links to the original text in a database.

3   From MRS to Quriples: Relation Extraction and Abstraction

The produced MRSes contain relations (EPs) with names e.g. for verbs that contain the stem of the recognized verb in the lexicon, such as *show_rel*. We can extract these relations and represent them in *quriples*.

3.1   Quriple Generation

Instead of using simple triples, we decided to use *quriples* to represent all arguments in the sentence. *Quriples* are query-oriented quintuples including subject (SUB), predicate (PRD), direct object (DOBJ), other complement (OCMP) and adjunct (ADJU), where OCMP includes indirect object, preposition complement, etc. ADJU contains all other information which does not belong to any of the other four parts.

Due to semantic ambiguity, the parser may return more than one reading per sentence. Currently up to three readings are considered (the most probable ones according to the trained parse ranking model) by the algorithm, and quriples are generated for each reading respectively. Multiple readings may lead to the same quriple structure, in which case only a single one is stored in the database. We illustrate the extraction procedure using the following example. Its corresponding MRS representation produced by Heart of Gold is shown in Figure 1.

Example 1
We evaluate the efficiency and performance empirically against the corpus.

Generally speaking, the extraction procedure starts with the predicate, and all its arguments are extracted one by one using depth-first until all EPs are exhausted. In practice, we start with the top handle *h1* which in this case contains the *prop-or-ques_m_rel* relation (cf. Figure 1). Its first argument ARG0 indicates the predicate (*evaluate*) which has 3 arguments. ARG0 normally refers to the EP

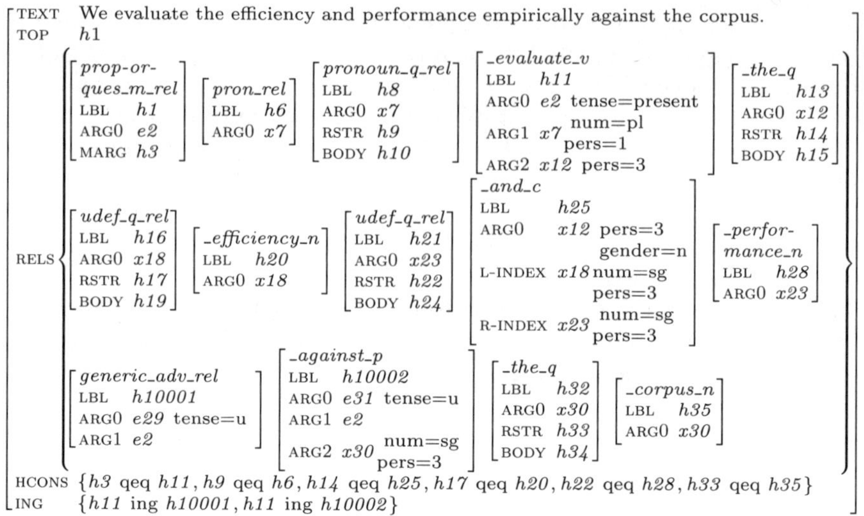

Fig. 1. MRS representation (robust variant) as produced by ERG in Heart of Gold

itself, and the rest ARG1 and ARG2 are the arguments taken by the predicate. Therefore, a transitive verb has altogether three ARGs and a ditransitve verb has four ARGs.

The predicate arguments are processed sequentially. ARG1 of the predicate, $x7$ in the example, has two relations: *pron_rel* and *pronoun_q_rel*, both of which are realized as *we* in the sentence. ARG2 of the predicate i.e. $x12$ in this case has *_and_c* relation where two EPs shown in L-INDEX and R-INDEX respectively are coordinated. The extraction of L-INDEX results in two lexemes: *its* and *efficiency*, and the extraction of R-INDEX results in *performance*.

Finally the adjunct, indicated by the message relation MARG in this case, is extracted. Following the *qeq* (equality modulo quantifiers) constraint under HCONS, handle $h3$ is *qeq* to label $h11$. As the implicit conjunction indicates (marked by ING relation), $h11$ conjuncts with label $h10002$, the *_against_p* relation, introducing the adjunct *against the corpus*. In this way, all EPs can be extracted from the MRS, and the quriples generated are shown in Table 1.

Table 1. Quriples generated for Examples 1 and 2

SUB	We	trained, neural, networks, with, varying, error, rates, depending, corpus, type
PRD	evaluate	classify
DOBJ	the, efficiency, and, performance	unseen, input
OCMP	-	-
ADJU	against, the, corpus	-

Although ARG0 of the top handle refers to the predicate of the sentence, it is not necessarily the main verb of the sentence. We discuss two more cases: conjunction and passive.

Passive: For passive sentences, ARG1 refers to the semantic object and ARG2 refers to the subject. The past participle, but not *be*, is extracted as PRD (cf. right column of Table1).

Example 2
Unseen input was classified by trained neural networks with varying error rates depending on corpus type.

Conjunction: In Example 1, the conjunction relation connects two noun phrases, both of them being DOBJ. Therefore, no new quriple is necessary. However, we decided to distinguish cases where conjunction connects two sentences or verb phrases. In such cases, quriples are generated for each part respectively. Example 3 shows an *AND* relation. However, conjunction relations may also be realized in different lexemes, e.g. *and, but, or, as well as*, etc.

Example 3
The system automatically extracts pairs of syntactic units from a text and assigns a semantic relation to each pair.

For this example, two quriples are generated separately with their own PRD, DOBJ and OCMP (cf. Table 2):

Table 2. Two quriples generated for the conjunction in Example 3

SUB	The, system	The, system
PRD	extract	assign
DOBJ	pairs, syntactic, units	a, semantic, relation
OCMP	from, a, text	to, each, pair
ADJU	automatically	automatically

In MRS, a conjunction with more than two arguments is represented in the coordination relation with an embedded structure, i.e. ARG1 first coordinates with ARG2, where ARG2 includes all the rest arguments. ARG2 then coordinates with ARG3 where ARG1 and ARG2 are excluded. Currently, we only deal with conjunctions on the top level, the embedded structure is not touched.

All fully parsed sentences were processed and quriples were generated. Currently, a relational database is used to save the quriples. It should be pointed out here that the same extraction steps that are used in the offline extraction process can also be used for processing natural language queries (parsed using the same hybrid pipeline as described before) for efficient and robust online search in the built relation database.

However, the query interface implemented so far only consists of a form-based search with input fields for subject, predicate and rest (objects etc.).

3.2   Integration of WordNet

Since the same relation can be very often realized differently using various lexemes, e.g. *present(x,y)* can be realized in *presents/shows/demonstrates ...*, a pure string match using the predicate of a sentence is far from enough. In order to improve the robustness of our system, WordNet [11] is integrated to search for the synsets of predicates.

As an explicit synset class is not defined for verbs in WordNet, for approximation we retrieved verbs from all senses of the current predicate. In this way, synonyms of more than 900 predicates were retrieved and saved in the database as an extra table. Table 3 shows some PRDs and their synonyms.

Table 3. Predicates and their synonyms extracted from WordNet

PREDICATE	SYNONYM
demonstrate	demo, prove, establish, show, manifest exhibit, present
evaluate	measure, judge, value, assess, valuate
assign	put, attribute, specify
result in	result, leave, lead
search for	look, explore, search, research, seek
find out	check, determine, discover, learn, pick up see, find

The integration of WordNet enables the user to conduct the search process not only based on the PRDs themselves, but also take the synonyms of PRDs into consideration. The second optionality can be activated by selecting 'allow predicate synonyms' on the workbench GUI.

4   Implementation of the Scientist's Workbench

The Scientist's Workbench is a web-based GUI application containing a simple editor, hyperlinked result boxes, and a Flash-based ontology browser. Edited text can be submitted to the application server that annotates every term in the input text that is known in the underlying ontology. As no ontology can ever be complete, we also integrated a Heart of Gold service to perform a SProUT named entity analysis of the text data. This ensures that even terms that are not contained within the ontology can be found and annotated. In case of person names, even variants such as 'Dr. Smith' can be analyzed. Different colors are used to distinguish between the possible annotation concepts such as person name, project, etc.

All annotations that have been returned from the application server can be clicked by the user and reveal more information about the annotated term. These *search results* may contain information about emails that a detected person has sent or received, papers a person has authored, meetings that a person has scheduled, and ontology concepts.

For papers we show both associated metadata such as title, authors, etc. and also the quriples that have been found within them. If the user clicks on such a quriple, the original PDF document will be opened in another window and the corresponding text lines will be highlighted in the original layout.

Furthermore, the GUI includes an interface to directly query quriples within the parsed papers. *Search Quriples* allows to specify a subject, a predicate and some 'rest' which will then be used to filter out any matching quriples. The user may choose to use WordNet synsets to allow predicate synonyms when defining the query.

Once the user has defined a query it is sent to the application server which looks up matching quriples from the quriple store and returns them to the GUI.

In order to enable relation queries, we stored all extracted quriples in a relational database, along with information where to find them, i.e. the source document, page and position the quriple was extracted from.

In the results view, the user can choose to view the original document (i.e. the original article in PDF form) and the sentence this relation was extracted from will also be highlighted, so that users immediately find the source instead of having to search through a probably lengthy document themselves.

5   Related Work

Using HPSG combined with shallow domain-specific modeling for high-precision analysis of scientific texts is an emerging research area. Another ERG-based approach to relation and information extraction from scientific texts in the DELPH-IN context[2] is SciBorg [12] (chemistry research papers).

[13] use shallow dependency structure and results from HPSG parsing for extracting protein-protein interactions (PPI) from research papers. The same group has also worked on medical texts: MEDIE[3] is a semantic search engine to retrieve biomedical correlations from MEDLINE articles.

What distinguishes our approach from those, besides concentration on a different scientific field, is the focus on and use of ontology information as integrated part of linguistic analysis, and the interactive editor user interface (Scientist's Workbench application).

6   Summary and Outlook

We have described methods to extract interesting factual relations from scientific texts in the computational linguistics and language technology fields taken from the ACL Anthology.

The coverage of 62.5% full parses of the abstract sentences without any specific adaptations to the domain except for the recognition of instances from the LT World ontology domain is very good and promising. Coverage could be further increased by improving the still rather coarse-grained deep-shallow interface.

A promising fallback solution in case deep parsing fails is using (robust) MRS analyses from a shallow parser as is done in SciBorg [12]. Also, using fragmentary parsing results in cases where a sentence could not be analyzed entirely can help to improve the overall coverage.

---

[2] DEep Linguistic Processing with Hpsg INitiative; `http://www.delph-in.net`

[3] `http://www-tsujii.is.s.u-tokyo.ac.jp/medie/`

Furthermore, various approaches exist to anaphora resolution that could be incorporated and help to improve coverage and quality of the extracted relations.

References

1. Bird, S., Dale, R., Dorr, B., Gibson, B., Joseph, M., Kan, M.Y., Lee, D., Powley, B., Radev, D., Tan, Y.F.: The ACL anthology reference corpus: a reference dataset for bibliographic research. In: Proc. of LREC, Marrakech, Morocco (2008)
2. Schäfer, U.: Integrating Deep and Shallow Natural Language Processing Components – Representations and Hybrid Architectures. PhD thesis, Faculty of Mathematics and Computer Science, Saarland University, Saarbrücken, Germany (2007)
3. Brants, T.: TnT - A Statistical Part-of-Speech Tagger. In: Proc. of Eurospeech, Rhodes, Greece (2000)
4. Drożdżyński, W., Krieger, H.U., Piskorski, J., Schäfer, U., Xu, F.: Shallow processing with unification and typed feature structures – foundations and applications. Künstliche Intelligenz 2004(1), 17–23 (2004)
5. Callmeier, U.: PET – A platform for experimentation with efficient HPSG processing techniques. Natural Language Engineering 6(1), 99–108 (2000)
6. Copestake, A., Flickinger, D.: An open-source grammar development environment and broad-coverage English grammar using HPSG. In: Proc. of LREC, Athens, Greece, pp. 591–598 (2000)
7. Oepen, S., Flickinger, D., Toutanova, K., Manning, C.D.: LinGO redwoods: A rich and dynamic treebank for HPSG. In: Proc. of the Workshop on Treebanks and Linguistic Theories, TLT 2002, Sozopol, Bulgaria, September 20–21 (2002)
8. Copestake, A., Flickinger, D., Sag, I.A., Pollard, C.: Minimal recursion semantics: an introduction. Journal of Research on Language and Computation 3(2–3) (2005)
9. Uszkoreit, H., Jörg, B., Erbach, G.: An ontology-based knowledge portal for language technology. In: Proc. of ENABLER/ELSNET Workshop, Paris (2003)
10. Schäfer, U.: OntoNERdIE – mapping and linking ontologies to named entity recognition and information extraction resources. In: Proc. of LREC, Genoa, Italy (2006)
11. Miller, G.A., Beckwith, R., Fellbaum, C., Gross, D., Miller, K.J.: Five papers on WordNet. Technical report, Cognitive Science Lab, Princeton University (1993)
12. Rupp, C., Copestake, A., Corbett, P., Waldron, B.: Integrating general-purpose and domain-specific components in the analysis of scientific text. In: Proc. of the UK e-Science Programme All Hands Meeting 2007, Nottingham, UK (2007)
13. Sætre, R., Kenji, S., Tsujii, J.: Syntactic features for protein-protein interaction extraction. In: Baker, C.J., Jian, S. (eds.) Short Paper Proc. of the 2nd International Symposium on Languages in Biology and Medicine (LBM 2007), Singapore (2008)

Efficient Hierarchical Reasoning about Functions over Numerical Domains

Viorica Sofronie-Stokkermans

Max-Planck-Institut für Informatik, Campus E 1.4, D-66123 Saarbrücken, Germany

Abstract. We show that many properties studied in mathematical analysis (monotonicity, boundedness, inverse, Lipschitz properties possibly combined with continuity, derivability) are expressible by formulae in a class for which sound and complete hierarchical proof methods for testing satisfiability of sets of ground clauses exist. The results are useful for automated reasoning in analysis and in the verification of hybrid systems.

1 Introduction

Efficient reasoning about functions over numerical domains subject to certain properties (monotonicity, convexity, continuity or derivability) is a major challenge both in automated reasoning and in symbolic computation. Besides its theoretical interest, it is very important for verification (especially of hybrid systems). The task of automatically reasoning in extensions of numerical domains with function symbols whose properties are expressed by first-order axioms is highly non-trivial: most existing methods are based on heuristics. Very few sound and complete methods or decidability results exist, even for specific fragments: Decidability of problems related to monotone or continuous functions over $\mathbb{R}$ were studied in [3,4,1]. In [3,4] Harvey and Seress give a decision procedure for formulae of the type $(\forall f \in \mathcal{F})\phi(f, x_1, \ldots x_n)$, where $\mathcal{F}$ is the class of continuous (or differentiable) functions over $\mathbb{R}$, x_i range over $\mathbb{R}$ and ϕ contains only existential or only universal quantifiers, evaluations of f and comparisons w.r.t. the order on $\mathbb{R}$. Reasoning about functions which satisfy other axioms, or about *several* functions is not considered there. In [1], Cantone et al. give a decision procedure for the validity of universally quantified sentences over continuous functions satisfying (strict) convexity or concavity conditions and/or monotonicity.

In this paper we apply recent methods for hierarchical reasoning we developed in [8] to the problem of checking the satisfiability of ground formulae involving functions over numerical domains. The main contributions of the paper are:

(1) We extend the notion of locality of theory extensions in [8] to encompass additional axioms and give criteria for recognizing locality of such extensions.
(2) We give several examples, including theories of functions satisfying various monotonicity, convexity, Lipschitz, continuity or derivability conditions and combinations of such extensions. Thus, our results generalize those in [1].
(3) We illustrate the use of hierarchical reasoning to tasks such as deriving constraints between parameters which ensure (un)satisfiability.

A. Dengel et al. (Eds.): KI 2008, LNAI 5243, pp. 135–143, 2008.

Illustration. Assume that $f:\mathbb{R}\to\mathbb{R}$ satisfies the bi-Lipschitz condition (BL_f^λ) with constant λ and g is the inverse of f. We want to determine whether g satisfies the bi-Lipschitz condition on the codomain of f, and if so with which constant λ_1, i.e. to determine under which conditions the following holds:

$$\mathbb{R} \cup (\mathsf{BL}_f^\lambda) \cup (\mathbf{Inv}(f,g)) \models \phi, \tag{1}$$

where $\phi : \forall x, x', y, y'(y{=}f(x)\wedge y'{=}f(x') \to \frac{1}{\lambda_1}|y-y'|{\leq}|g(y)-g(y')|{\leq}\lambda_1|y-y'|)$;

$$(\mathsf{BL}_f^\lambda) \qquad \forall x, y(\frac{1}{\lambda}|x-y| \leq |f(x)-f(y)| \leq \lambda|x-y|);$$

$$\mathsf{Inv}(f,g) \qquad \forall x, y(y=f(x) \to g(y)=x).$$

Entailment (1) is true iff $\mathbb{R}\cup(\mathsf{BL}_f^\lambda)\cup(\mathbf{Inv}(f,g))\cup G$ is unsatisfiable, where $G = (c_1{=}f(a_1)\wedge c_2{=}f(a_2)\wedge(\frac{1}{\lambda_1}|c_1{-}c_2| > |g(c_1){-}g(c_2)| \vee |g(c_1){-}g(c_2)| > \lambda_1|c_1{-}c_2|))$ is the formula obtained by Skolemizing the negation of ϕ.

Standard theorem provers for first order logic cannot be used in such situations. Provers for reals do not know about additional functions. The Nelson-Oppen method [7] for reasoning in combinations of theories cannot be used either.

The method we propose reduces the task of checking whether formula (1) holds to the problem of checking the satisfiability of a set of constraints over $\mathbb{R}$. We first note that for any set G of ground clauses with the property that *"if $g(c)$ occurs in G then G also contains a unit clause of the form $f(a) = c$"* every partial model P of G – where (i) f and g are partial and defined exactly on the ground subterms occurring in G and (ii) P satisfies $\mathsf{BL}_f^\lambda \cup \mathbf{Inv}(f,g)$ at all points where f and g are defined – can be completed to a total model of $\mathbb{R} \cup (\mathsf{BL}_f^\lambda) \cup (\mathbf{Inv}(f,g)) \cup G$ (cf. Thm. 7 and Cor. 8). Therefore, problem (1) is equivalent to

$$\mathbb{R} \cup (\mathsf{BL}_f^\lambda \cup \mathbf{Inv}(f,g))[G] \cup G \models \perp,$$

where $(\mathsf{BL}_f^\lambda \cup \mathbf{Inv}(f,g))[G]$ is the set of those instances of $\mathsf{BL}_f^\lambda \cup \mathbf{Inv}(f,g)$ in which the terms starting with g or f are ground terms occurring in G, i.e.

$$\begin{aligned}
(\mathsf{BL}_f^\lambda \cup \mathbf{Inv}(f,g))[G] = {}& \frac{1}{\lambda}|a_1 - a_2| \leq |f(a_1) - f(a_2)| \leq \lambda|a_1 - a_2| \wedge \\
& (c_1 = f(a_1) \to g(c_1) = a_1) \ \wedge \ (c_2 = f(a_1) \to g(c_2) = a_1) \ \wedge \\
& (c_1 = f(a_2) \to g(c_1) = a_2) \ \wedge \ (c_2 = f(a_2) \to d(c_2) = a_2).
\end{aligned}$$

We separate the numerical symbols from the non-numerical ones by introducing new names for the extension terms, together with their definitions $D = (f(a_1) = e_1 \wedge f(a_2) = e_2 \wedge g(c_1) = d_1 \wedge g(c_2) = d_2)$ and replacing them in $(\mathsf{BL}_f^\lambda \cup \mathbf{Inv}(f,g))[G] \cup G$. The set of formulae obtained this way is $\mathsf{BL}_0 \cup \mathbf{Inv}_0 \cup G_0$. We then use – instead of these definitions – only the instances $\mathsf{Con}[G]_0$ of the congruence axioms for f and g which correspond to these terms. We obtain:

$$\mathsf{BL}_0 : \ \frac{1}{\lambda}|a_1 - a_2| \leq |e_1 - e_2| \leq \lambda|a_1 - a_2|$$

$$\mathsf{Inv}_0 : \ (c_1{=}e_1{\to}d_1{=}a_1) \wedge (c_2{=}e_1{\to}d_2{=}a_1) \wedge (c_1{=}e_2{\to}d_1{=}a_2) \ \wedge \ (c_2{=}e_2{\to}d_2{=}a_2)$$

$$G_0 : \ c_1 = e_1 \wedge c_2 = e_2 \wedge (|d_1 - d_2| < \frac{|c_1 - c_2|}{\lambda_1} \vee |d_1 - d_2| > \lambda_1|c_1 - c_2|)$$

$$\mathsf{Con}[G]_0 \ \ c_1 = c_2 \to d_1 = d_2 \wedge a_1 = a_2 \to e_1 = e_2$$

Thus, entailment (1) holds iff $\mathsf{BL}_0 \wedge \mathsf{Inv}_0 \wedge G_0 \wedge \mathsf{Con}[G]_0$ is unsatisfiable, i.e. iff

$$\exists a_1, a_2, c_1, c_2, d_1, d_2, e_1, e_2 (\mathsf{BL}_0 \wedge \mathsf{Inv}_0 \wedge G_0 \wedge \mathsf{Con}[G]_0) \text{ is false.}$$

The quantifiers can be eliminated with any QE system for $\mathbb{R}$. We used REDLOG [2]; after simplification (w.r.t. $\lambda > 1, \lambda_1 > 1$ and some consequences) we obtained:

$$\lambda_1 \lambda^2 - \lambda < 0 \ \vee \ \lambda_1 \lambda - \lambda^2 < 0 \ \vee \ \lambda_1 - \lambda < 0 \ \vee \ (\lambda_1 \lambda - \lambda^2 > 0 \wedge \lambda_1 = \lambda) \ \vee$$
$$(\lambda_1^2 \lambda - \lambda_1 > 0 \ \wedge \ (\lambda_1^2 - \lambda_1 \lambda < 0 \ \vee \ (\lambda_1^2 - \lambda_1 \lambda > 0 \ \wedge \ \lambda_1 = \lambda))) \ \vee$$
$$(\lambda_1^2 \lambda - \lambda_1 > 0 \ \wedge \ \lambda_1^2 - \lambda_1 \lambda < 0 \ \wedge \ \lambda_1 - \lambda > 0) \ \vee \ (\lambda_1^2 \lambda - \lambda_1 > 0 \ \wedge \ \lambda_1^2 - \lambda_1 \lambda < 0).$$

If $\lambda > 1, \lambda_1 > 1$, this formula is equivalent to $\lambda_1 < \lambda$. Hence, if $\lambda > 1, \lambda_1 > 1$ we have:

$$\mathbb{R} \cup (\mathsf{BL}_f^\lambda) \wedge (\mathbf{Inv}(f, g)) \models \phi, \quad \text{iff } \lambda_1 \geq \lambda. \tag{2}$$

The constraints we obtain can be used for optimization (e.g. we can show that the smallest value of λ_1 for which g satisfies the bi-Lipschitz condition is λ).

In this paper we investigate situations where this type of reasoning is possible. In Sect. 2 local extensions are defined, and ways of recognizing them, and of hierarchical reasoning in such extensions are discussed. Section 3 provides several examples from analysis with applications to verification[1].

2 Local Theory Extensions

Let $\mathcal{T}_0$ be a theory with signature $\Pi_0 = (S_0, \Sigma_0, \mathsf{Pred})$, where S_0 is a set of sorts, Σ_0 is a set of function symbols, Pred is a set of predicate symbols. We consider extensions $\mathcal{T}_1$ of $\mathcal{T}_0$ with new sorts and function symbols (i.e. with signature $\Pi = (S_0 \cup S_1, \Sigma_0 \cup \Sigma_1, \mathsf{Pred})$), satisfying a set $\mathcal{K}$ of clauses. We denote such extensions by $\mathcal{T}_0 \subseteq \mathcal{T}_1 = \mathcal{T}_0 \cup \mathcal{K}$. We are interested in disproving ground formulae G in the extension Π^c of Π with new constants Σ_c. This can be done efficiently if we can restrict the number of instances to be taken into account without loss of completeness. In order to describe such situations, we need to refer to "partial models", where only the instances of the problem are defined.

2.1 Total and Partial Models

Partial Π-structures are defined as total ones, with the difference that for every $f \in \Sigma$ with arity n, f_A is a partial function from A^n to A. Evaluating a term t with respect to a variable assignment $\beta : X \to A$ in a partial structure A is the same as for total algebras, except that this evaluation is undefined if $t = f(t_1, \ldots, t_n)$ and either one of $\beta(t_i)$ is undefined, or $(\beta(t_1), \ldots, \beta(t_n))$ is not in the domain of f_A. For a partial structure A and $\beta : X \to A$, we say that $(A, \beta) \models_w (\neg)P(t_1, \ldots, t_n)$ if either (a) some $\beta(t_i)$ is undefined, or (b) $\beta(t_i)$ are all defined and $(\neg)P_A(\beta(t_1), \ldots, \beta(t_n))$ holds in A. (A, β) *weakly satisfies a clause* C (notation: $(A, \beta) \models_w C$) if $(A, \beta) \models_w L$ for at least one literal L in C.

---

[1] For proofs and additional examples cf.
`www.mpi-inf.mpg.de/~sofronie/papers/ki-08-full.pdf`

A weakly satisfies a set of clauses $\mathcal{K}$ $(A \models_w \mathcal{K})$ if $(A, \beta) \models_w C$ for all $C \in \mathcal{K}$ and all assignments β.

2.2   Locality

Let Ψ be a closure operator stable under renaming constants, associating with sets $\mathcal{K}$ and T of axioms resp. ground terms, a set $\Psi_{\mathcal{K}}(T)$ of ground terms. We consider condition (Loc^{Ψ}) (cf. also [5]):

(Loc^{Ψ}) for every ground formula G, $T_1 \cup G \models \perp$ iff $T_0 \cup \mathcal{K}[\Psi_{\mathcal{K}}(G)] \cup G$ has no weak partial model in which all terms in $\Psi_{\mathcal{K}}(G)$ are defined,

where $\mathcal{K}[\Psi_{\mathcal{K}}(G)]$ consists of all instances of $\mathcal{K}$ in which the terms starting with extension functions are in the set $\Psi_{\mathcal{K}}(G) := \Psi_{\mathcal{K}}(\mathsf{st}(\mathcal{K}, G))$, where $\mathsf{st}(\mathcal{K}, G)$ is the set of ground terms occurring in $\mathcal{K}$ or G. (If Ψ is the identity function, we denote $\mathcal{K}[\Psi_{\mathcal{K}}(G)]$ by $\mathcal{K}[G]$ and the locality condition by (Loc).)

2.3   Hierarchical Reasoning

Assume the extension $T_0 \subseteq T_1 = T_0 \cup \mathcal{K}$ satisfies (Loc^{Ψ}). To check if $T_1 \cup G \models \perp$ for a set G of ground Π^c-clauses, note that, by locality, $T_1 \cup G \models \perp$ iff $T_0 \cup \mathcal{K}[\Psi_{\mathcal{K}}(G)] \cup G$ has no weak partial model. We *purify* $\mathcal{K}[\Psi_{\mathcal{K}}(G)] \cup G$ by introducing, bottom-up, new constants c_t (from a set Σ_c) for subterms $t = f(g_1, \ldots, g_n)$ with $f \in \Sigma_1$, g_i ground $\Sigma_0 \cup \Sigma_c$-terms, together with their definitions $c_t \approx t$. Let D be the set of definitions introduced this way. The formula thus obtained is $\mathcal{K}_0 \cup G_0 \cup D$, where $\mathcal{K}_0 \cup G_0$ is obtained from $\mathcal{K}[\Psi_{\mathcal{K}}(G)] \cup G$ by replacing all extension terms with the corresponding constants.

Theorem 1 ([5]). *Assume that* $T_0 \subseteq T_0 \cup \mathcal{K}$ *satisfies* (Loc^{Ψ})*. Let* $\mathcal{K}_0 \cup G_0 \cup D$ *be obtained from* $\mathcal{K}[\Psi_{\mathcal{K}}(G)] \cup G$ *as explained above. The following are equivalent:*

(1) $T_0 \cup \mathcal{K} \cup G$ *is satisfiable;*
(2) $T_0 \cup \mathcal{K}_0 \cup G_0 \cup \mathsf{Con}[G]_0$ *has a (total) model, where* $\mathsf{Con}[G]_0$ *is the set of instances of the congruence axioms corresponding to* D*:*

$$\mathsf{Con}[G]_0 = \{ \bigwedge_{i=1}^{n} c_i \approx d_i \rightarrow c = d \mid f(c_1, \ldots, c_n) \approx c, f(d_1, \ldots, d_n) \approx d \in D\}.$$

Thus, if the extension $T_0 \subseteq T_1$ satisfies (Loc^{Ψ}) then satisfiability w.r.t. T_1 is decidable for all ground clauses G for which $\mathcal{K}_0 \cup G_0 \cup N_0$ is finite and belongs to a fragment $\mathcal{F}_0$ of T_0 for which checking satisfiability is decidable. Theorem 1 also allows us to give parameterized complexity results for the theory extension:

Theorem 2. *Let* $g(m)$ *be the complexity of checking the satisfiability w.r.t.* T_0 *of formulae in* $\mathcal{F}_0$ *of size* m*. The complexity of checking satisfiability of a formula* G *w.r.t.* T_1 *is of order* $g(m)$*, where* m *is a polynomial in* $n = |\Psi_{\mathcal{K}}(G)|$ *whose degree* (≥ 2) *depends on the number of extension terms in in* $\mathcal{K}$*.*

2.4   Recognizing Locality

We can recognize local extensions $\mathcal{T}_0 \subseteq \mathcal{T}_1$ by means of flat and linear[2] clauses as follows.

Theorem 3 ([5]). *Assume that $\mathcal{K}$ is flat and linear and $\Psi_{\mathcal{K}}(T)$ is finite for any finite T. If the extension $\mathcal{T}_0 \subseteq \mathcal{T}_0 \cup \mathcal{K}$ satisfies condition (Comp_w^{Ψ}) then it satisfies $(\mathsf{Loc})^{\Psi}$, where:*

(Comp_w^{Ψ}) *Every weak partial model A of $\mathcal{T}_1$ with totally defined Σ_0-functions, such that the* definition *domains of functions in Σ_1 are finite and such that the set of terms $f(a_1, \ldots, a_n)$ defined in A is closed under Ψ weakly embeds into a total model B of $\mathcal{T}_1$ s.t. $A_{|\Pi_0} \simeq B_{|\Pi_0}$ are isomorphic.*

Theorem 4 (Considering additional axioms.). *Let $\mathcal{A}x_1$ be an additional set of axioms in full first-order logic. Assume that every weak partial model A of $\mathcal{T}_1$ with totally defined Σ_0-functions satisfying the conditions in (Comp_w^{Ψ}) weakly embeds into a total model B of $\mathcal{T}_0 \cup \mathcal{K} \cup \mathcal{A}x_1$. Let G be a set of ground clauses. The following are equivalent:*

(1) $\mathcal{T}_0 \cup \mathcal{K} \cup \mathcal{A}x_1 \cup G \models \perp$.
(2) $\mathcal{T}_0 \cup \mathcal{K}[\Psi_{\mathcal{K}}(G)] \cup G$ *has no partial model in which all ground subterms in* $\mathcal{K}[\Psi_{\mathcal{K}}(G)] \cup G$ *are defined.*

3   Examples of Local Extensions

We give several examples of local extensions of numerical domains. Besides axioms already considered in [8,9,6] (Sect. 3.1) we now look at extensions with functions satisfying inverse conditions, convexity/concavity, continuity and derivability. For the sake of simplicity, we here restrict to unary functions, but most of the results also hold for functions $f : \mathbb{R}^n \to \mathbb{R}^m$.

3.1   Monotonicity and Boundedness Conditions

Any extension of a theory with free function symbols is local. In addition the following theory extensions have been proved to be local in [8,9,6]:

Monotonicity. Any extension of the theory of reals, rationals or integers with functions satisfying $\mathsf{Mon}^{\sigma}(f)$ is local $((\mathsf{Comp}_w)$ holds [8,9])[3]:

$$\mathsf{Mon}^{\sigma}(f) \quad \bigwedge_{i \in I} x_i \leq_i^{\sigma_i} y_i \wedge \bigwedge_{i \notin I} x_i = y_i \rightarrow f(x_1, .., x_n) \leq f(y_1, .., y_n).$$

---

[2] A *non-ground formula* is Σ_1-*flat* if function symbols (also constants) do not occur as arguments of functions in Σ_1. A Σ_1-flat non-ground formula is called Σ_1-*linear* if whenever a universally quantified variable occurs in two terms which start with functions in Σ_1, the two terms are identical, and if no term which starts with a function in Σ_1 contains two occurrences of the same universal variable.

[3] For $i \in I$, $\sigma_i \in \{-, +\}$, and for $i \notin I$, $\sigma_i = 0$; $\leq^+ = \leq, \leq^- = \geq$.

The extension $\mathcal{T}_0 \subseteq \mathcal{T}_0 \cup \mathsf{SMon}(f)$ is local if $\mathcal{T}_0$ is the theory of reals (and $f : \mathbb{R} \to \mathbb{R}$) or the disjoint combination of the theories of reals and integers (and $f : \mathbb{Z} \to \mathbb{R}$) [5]. The extension of the theory of integers with $(\mathsf{SMon}_{\mathbb{Z}}(f))$ is local.

$$\mathsf{SMon}(f) \quad \forall i, j(i<j \to f(i)<f(j)) \qquad \mathsf{SMon}_{\mathbb{Z}}(f) \quad \forall i, j(i<j \to (j-i) < f(j)-f(i)).$$

Boundedness. Assume $\mathcal{T}_0$ contains a reflexive binary predicate $\leq$, and $f \notin \Sigma_0$. Let $m \in \mathbb{N}$. For $1 \leq i \leq m$ let $t_i(x_1, \ldots, x_n)$ and $s_i(x_1, \ldots, x_n)$ be terms in the signature Π_0 and $\phi_i(x_1, \ldots, x_n)$ be Π_0-formulae with (free) variables among $x_1, \ldots, x_n$, such that $\mathcal{T}_0 \models \forall \overline{x}(\phi_i(\overline{x}) \to s_i(\overline{x}) \leq t_i(\overline{x}))$, and if $i \neq j$, $\phi_i \wedge \phi_j \models_{\mathcal{T}_0} \bot$. Let $\mathsf{GB}(f) = \bigwedge_{i=1}^{m} \mathsf{GB}^{\phi_i}(f)$ and $\mathsf{Def}(f) = \bigwedge_{i=1}^{n} \mathsf{Def}^{\phi_i}(f)$, where:

$$\mathsf{GB}^{\phi_i}(f) \quad \forall \overline{x}(\phi_i(\overline{x}) \to s_i(\overline{x}) \leq f(\overline{x}) \leq t_i(\overline{x})) \qquad \mathsf{Def}^{\phi_i}(f) \quad \forall \overline{x}(\phi_i(\overline{x}) \to f(\overline{x}) = t_i(\overline{x}))$$

(i) The extensions $\mathcal{T}_0 \subseteq \mathcal{T}_0 \cup \mathsf{GB}(f)$ and $\mathcal{T}_0 \subseteq \mathcal{T}_0 \cup \mathsf{Def}(f)$ are both local [9,5].
(ii) Any extension of a theory for which $\leq$ is a partial order (or at least reflexive) with functions satisfying $\mathsf{Mon}^\sigma(f)$ and $\mathsf{Bound}^t(f)$ is local [9,5].
$$\mathsf{Bound}^t(f) \qquad \forall x_1, \ldots, x_n(f(x_1, \ldots, x_n) \leq t(x_1, \ldots, x_n))$$
where $t(x_1, \ldots, x_n)$ is a Π_0-term with variables among $x_1, \ldots, x_n$ whose associated function has the same monotonicity as f in any model. Similar results hold for strictly monotone functions.

Injectivity. An extension $\mathcal{T}_0 \subseteq \mathcal{T}_1 = \mathcal{T}_0 \cup \mathsf{Inj}(f)$ with a function f of arity $\mathsf{i} \to \mathsf{e}$ satisfying $\mathsf{Inj}(f)$ is local provided that in all models of $\mathcal{T}_1$ the cardinality of the support of sort i is lower or equal to the cardinality of the support of sort e.

$$\mathsf{Inj}(f) \quad \forall i, j(i \neq j \to f(i) \neq f(j)).$$

3.2   Inverse Conditions

Consider the following inverse condition:

$$\mathsf{Inv}(f, g) \quad \forall x, y(y = f(x) \to g(y) = x).$$

Such conditions often occur in mathematics and are important in verification (e.g. to model direct and inverse links between certain objects).

Theorem 5. *Let $\mathcal{T}_0$ be a theory and $f, g \notin \Sigma_0$. Assume that $\mathcal{T}_0 \subseteq \mathcal{T}_0 \cup \mathcal{K}(f)$ satisfies Comp_w, and that $\mathcal{T}_0 \cup \mathcal{K}(f) \models \mathsf{Inj}(f)$. Then $\mathcal{T}_0 \cup (\mathcal{K}(f) \cup \mathsf{Inv}(f, g)) \cup G \models \bot$ iff $\mathcal{T}_0 \cup (\mathcal{K}(f) \cup \mathsf{Inv}(f, g))[G] \cup G \models \bot$ for all sets G of ground clauses with the property that if $g(c)$ occurs in G then also some $f(a) = c$ occurs in G.*

3.3   Convexity/Concavity

Let f be a unary function, and $I = [a, b]$ a subset of the domain of definition of f. We consider the axiom:

$$\mathsf{Conv}^I(f) \quad \forall x, y, z \left(x, y \in I \wedge x \leq z \leq y \to \frac{f(z)-f(x)}{z-x} \leq \frac{f(y)-f(x)}{y-x} \right).$$

Theorem 6. *Let $\mathcal{T}_0$ be one of the theories: (i) $\mathbb{R}$, (ii) $\mathbb{Z}$ (a theory of integers), (iii) the many-sorted combination of the theories of reals (sort real) and integers (sort int). Let f be a new, unary function (for (iii) assume f has arity int $\to$ real). $\mathcal{T}_0 \cup \mathsf{Conv}_f^I$ and $\mathcal{T}_0 \cup \mathsf{Conc}_f^I$ are local extensions of $\mathcal{T}_0$, where $\mathsf{Conc}^I(f) = \mathsf{Conv}^I(-f)$.*

3.4 Lipschitz Conditions

Consider the following conditions:

$(\mathsf{L}_f^\lambda(c_0))$	$\forall x(\lvert f(x) - f(c_0)\rvert \leq \lambda\lvert x - c_0\rvert)$	Lipschitz condition at c_0
(L_f^λ)	$\forall x, y(\lvert f(x) - f(y)\rvert \leq \lambda\lvert x - y\rvert)$	(uniform) Lipschitz condition
(BL_f^λ)	$\forall x, y(\frac{1}{\lambda}\lvert x - y\rvert \leq \lvert f(x) - f(y)\rvert \leq \lambda\lvert x - y\rvert)$	bi-Lipschitz condition

Such conditions occur in the verification of hybrid systems when specifying (by universal axioms) that the derivative of a function is bounded by a given value.

Theorem 7. $\mathbb{R} \cup (\mathsf{L}_f^\lambda(c_0))$, $\mathbb{R} \cup (\mathsf{L}_f^\lambda)$, *and* $\mathbb{R} \cup (\mathsf{BL}_f^\lambda)$ *are local extensions of* $\mathbb{R}$.

From Thms. 5 and 7 we obtain the result used in the illustration in Sect. 1.

Corollary 8. *The extension $\mathbb{R} \cup \mathsf{BL}_f^\lambda \cup \mathsf{Inv}(f, g)$ of $\mathbb{R}$ has the property that for all sets G of ground clauses such that if $g(c)$ occurs in G then also $f(a) = c$ occurs in G, $\mathbb{R} \cup (\mathsf{BL}_f^\lambda \cup \mathsf{Inv}(f, g)) \cup G \models \bot$ iff $\mathbb{R} \cup (\mathsf{BL}_f^\lambda \cup \mathsf{Inv}(f, g))[G] \cup G$ has no weak partial model in which all subterms of G are defined (and only those).*

3.5 Continuity, Derivability

We consider the following continuity conditions for a function $f : \mathbb{R} \to \mathbb{R}$:

$\mathsf{Cont}_f(c_0)$	$\forall\epsilon(\epsilon>0 \to \exists\delta(\delta>0 \wedge \forall x(\lvert x-c_0\rvert<\delta \to \lvert f(x)-f(c_0)\rvert<\epsilon)))$	continuity at c_0
Cont_f	$\forall x(\mathsf{Cont}_f(x))$	continuity

and the following derivability conditions for a (continuous) function f:

$$\mathsf{Der}(f, f')(c_0) : \qquad \forall\epsilon(\epsilon>0 \to \exists\delta(\delta>0 \wedge \forall x(\lvert x-c_0\rvert<\delta \to \lvert \tfrac{f(x)-f(c_0)}{x-c_0} - f'(c_0)\rvert<\epsilon)))$$

$$\mathsf{Der}^{\leq n}(f, f^1, \ldots, f^n)(c_0) : \qquad \bigwedge_{i=1}^{n} \mathsf{Cont}_{f^{i-1}}(c_0) \wedge \mathsf{Der}(f^{i-1}, f^i)(c_0)$$

$\mathsf{Der}(f, f') := \forall x\mathsf{Der}(f, f')(x)$; $\mathsf{Der}^{\leq n}(f, f^1, \ldots, f^n) = \forall x\mathsf{Der}^{\leq n}(f, f^1, \ldots, f^n)(x)$ (axiomatizing derivability – resp. n-times derivability – at every point, where $n \in \mathbb{N} \cup \{\infty\}$, $f^0 = f$ and f^i is the i-th derivative of f).

Theorem 9. *(1) Any partial function over the reals with a finite domain of definition extends to a total continuous function over the reals.*
(2) $\mathbb{R} \cup \mathsf{Cont}_f(c_0) \cup \mathsf{Der}(f, f')(c_0)$ and $\mathbb{R} \cup \mathsf{Cont}_f \cup \mathsf{Der}(f, f')$ are Ψ-local extensions of $\mathbb{R}$, where $\Psi(T) = T \cup \{f(c) \mid f'(c) \in T\} \cup \{f'(c) \mid f(c) \in T\}$. $\mathbb{R} \subseteq \mathbb{R} \cup \mathsf{Der}^{\leq n}(f, f^1, \ldots, f^n)(c_0)$ and $\mathbb{R} \subseteq \mathbb{R} \cup \mathsf{Der}^{\leq n}(f, f^1, \ldots, f^n)$ are Ψ^n-local extensions, where $\Psi^n(T) = T \cup \{f^k(c) \mid 0 \leq k \leq n$ if $f^i(c) \in T$ for some $0 \leq i \leq n\}$.

Proof. We can use any polynomial interpolation theorem to compute a total model from any partial model (e.g. the Hermite interpolation theorem). $\square$

3.6 Combinations

Analyzing the proofs in the previous sections we notice that the same completion for the partial functions can be used for (i) monotone, strictly monotone, convex/concave, Lipschitz and continuous functions over $\mathbb{R}$. The same completion (possibly different from that in (i)) is used (ii) for Lipschitz, and for (uniformly) continuous and n-derivable functions over $\mathbb{R}$.

Theorem 10. *The following axiom combinations define local extensions of $\mathbb{R}$:*

(1) Arbitrary combinations of $[\mathsf{S}]\mathsf{Mon})(f)$, Conv_f, $\mathsf{L}_f^\lambda[(c_0)]$, BL_f^λ, $\mathsf{Cont}_f[(c_0)]$;

(2) Arbitrary combinations of $\mathsf{L}_f^\lambda[(c_0)]$, BL_f^λ, $\mathsf{Cont}_f[(c_0)]$, $\mathsf{Cont}_f[(c_0)] \wedge$ $\mathsf{Der}(f,f')[(c_0)]$, $\mathsf{Der}^{\leq n}(f, f^1, .., f^n)(c_0)$, *and* $\mathsf{Der}^{\leq n}(f, f^1, .., f^n)$.

However, care is needed when combining $\mathsf{Der}(f, f')$ with boundedness or monotonicity conditions on f', or with convexity/concavity conditions on f or f'.

The types of extensions considered before can be combined up to a certain extent.

Theorem 11. *Let* $\{f_1, \ldots, f_n\}$ *be unary function symbols, and* $\mathcal{K}_1, \ldots, \mathcal{K}_n$ *be systems of axioms such that for every i, $\mathcal{K}_i$ is a set of formulae over the signature of $\mathbb{R}$ augmented with f_i. Assume that for every $i \in \{1, \ldots, n\}$, $\mathcal{K}_i$ is in one of the classes considered in Thm. 10. Then* $\mathbb{R} \subseteq \mathbb{R} \cup \mathcal{K}_1 \cup \cdots \cup \mathcal{K}_n$ *is a local extension.*

The constructions in the previous sections can be relativized to a subinterval I of the domain of definition of the function. We denote this by adding the index I to the corresponding axiom. Locality is preserved for families $\{C_f^I \mid I \in \mathcal{J}\}$ of axioms in the class above relativized over a family of mutually disjoint intervals.

4 Conclusions

We presented a class of extensions of numerical domains with additional functions for which sound and complete proof methods exist, which allow to reduce testing satisfiability of quantifier-free formulae, hierarchically, to a satisfiability problem in the "base", numerical domain.[4] These results can be applied for automated reasoning in mathematical analysis as well as in verification. An example we considered in the frame of AVACS involved train control systems [6]. We used hierarchical reasoning to determine constraints between the parameters of such control systems which guarantee safety. The new results we present here open new possibilities for efficient verification, since Lipschitz conditions, as well as continuity and derivability conditions occur naturally in the verification of (parametric) hybrid systems (Lipschitz conditions can model e.g. boundedness of derivatives). For tests we used an implementation (cf. also [5]) of the method for hierarchical reasoning in local theory extensions described in [8,5]. All tests and experiments are very encouraging. In the future we will also consider problems involving satisfiability tests for formulae with (alternations) of quantifiers.

---

[4] Note that the hierarchical reduction method we use is sound even for non-local extensions. Locality guarantees completeness for proving validity of universal sentences.

Acknowledgments. We thank Thomas Sturm for advise in using REDLOG. This work was partly supported by the German Research Council (DFG) as part of the Transregional Collaborative Research Center "Automatic Verification and Analysis of Complex Systems" (SFB/TR 14 AVACS, www.avacs.org).

References

1. Cantone, D., Cincotti, G., Gallo, G.: Decision algorithms for fragments of real analysis. I. Continuous functions with strict convexity and concavity predicates. Journal of Symbolic Computation 41, 763–789 (2006)
2. Dolzmann, A., Sturm, T.: Redlog: Computer algebra meets computer logic. ACM SIGSAM Bulletin 31(2), 2–9 (1997)
3. Friedman, H., Serres, A.: Decidability in elementary analysis, I. Adv. in Mathematics 76(1), 94–115 (1989)
4. Friedman, H., Serres, A.: Decidability in elementary analysis, II. Adv. in Mathematics 70(2), 1–17 (1990)
5. Ihlemann, C., Jacobs, S., Sofronie-Stokkermans, V.: On local reasoning in verification. In: Ramakrishnan, C.R., Rehof, J. (eds.) TACAS 2008. LNCS, vol. 4963, pp. 265–281. Springer, Heidelberg (2008)
6. Jacobs, S., Sofronie-Stokkermans, V.: Applications of hierarchical reasoning in the verification of complex systems. Electronic Notes in Theoretical Computer Science 174(8), 39–54 (2007)
7. Nelson, G., Oppen, D.C.: Simplification by cooperating decision procedures. ACM Transactions on Programming Languages and Systems (1979)
8. Sofronie-Stokkermans, V.: Hierarchic reasoning in local theory extensions. In: Nieuwenhuis, R. (ed.) CADE 2005. LNCS (LNAI), vol. 3632, pp. 219–234. Springer, Heidelberg (2005)
9. Sofronie-Stokkermans, V., Ihlemann, C.: Automated reasoning in some local extensions of ordered structures. Journal of Multiple-Valued Logics and Soft Computing 13(4–6), 397–414 (2007)

A Drum Machine That Learns to Groove

Axel Tidemann[1] and Yiannis Demiris[2]

[1] IDI, Norwegian University of Science and Technology
tidemann@idi.ntnu.no
[2] BioArt, ISN, EEE, Imperial College London
y.demiris@imperial.ac.uk

Abstract. Music production relies increasingly on advanced hardware and software tools that makes the creative process more flexible and versatile. The advancement of these tools helps reduce both the time and money required to create music. This paper presents research towards enhancing the functionality of a key tool, the drum machine. We add the ability to *learn how to groove* from human drummers, an important human quality when it comes to drumming. We show how the learning drum machine overcomes limitations of traditional drum machines.

Keywords: Imitation Learning, Machine Learning, Music.

1 Introduction

Digital audio workstations allow users to quickly program drums, but the drawback is the machine-like sound of the programmed drums. Our research goal is to make a drum machine that has the groove of a human drummer, by modeling the user-specific variations of human drummers. The groove of a drummer is made up of its large-scale variations (i.e. playing a break) and the small-scale variations (i.e. varying tempo and how hard a beat is played). The approach of drumming software so far has been to add random noise to the produced drum patterns, to make them sound more human. Our approach is to learn drum patterns from human drummers and *model* the variations that make a drummer groovy to make an *intelligent* drum machine that sounds like a human drummer.

2 Background

Saunders et al. [1] and Tobudic and Widmer [2] focus on variations of tempo and dynamics when modeling the style of pianists, albeit with different techniques (string kernels and clustering of similar phrases, respectively). Pachet uses Markov models in his Continuator system [3] to model the probability that one note will follow the other. The Continuator learns the tonal signature of the pianist, and has been implemented in a real-time system. Mantaras and Arcos focus on imitating expressiveness, such as joyful or sad, by using case-based reasoning [4]. Raphael [5] focuses on the problem that arises when soloists do not

A. Dengel et al. (Eds.): KI 2008, LNAI 5243, pp. 144–151, 2008.

have the opportunity to play with a real orchestra, but must do with a static recording. His system dubbed "Music Plus One" lets soloists practice along with an orchestra played by a computer, and the system then learns to follow the variations in tempo introduced by the soloist.

Current drumming software (e.g. FXpansion BFD, Toontrack EZdrummer, DigiDesign Strike, Reason Drum Kits, Native Instruments Battery) consist of huge sample libraries (typically in the gigabytes range), where meticulous precision has gone into sampling the different dynamics when playing drums (i.e. from soft to hard). Still, the user must program the drum tracks. The user can typically select which pattern to play and adjust some parameters to add random noise to both onset time and velocity, to make the patterns sound more human. The lack of intelligent ways to generate human-like drum patterns in these applications forms the motivation for our research.

3   Architecture

Our architecture is called "Software for Hierarchical Extraction and Imitation of drum patterns in a Learning Agent" (SHEILA), see figure 1. SHEILA learns drum patterns and playing style from human drummers. The drum patterns and the models of how they are played are stored in a library. The repetitive nature of playing drum patterns makes it suitable for machine learning techniques to extract user-specific variations. After the learning process, SHEILA can be used as an *intelligent and groovy* drum machine that has the ability to imitate the style of the drummers that served as teachers.

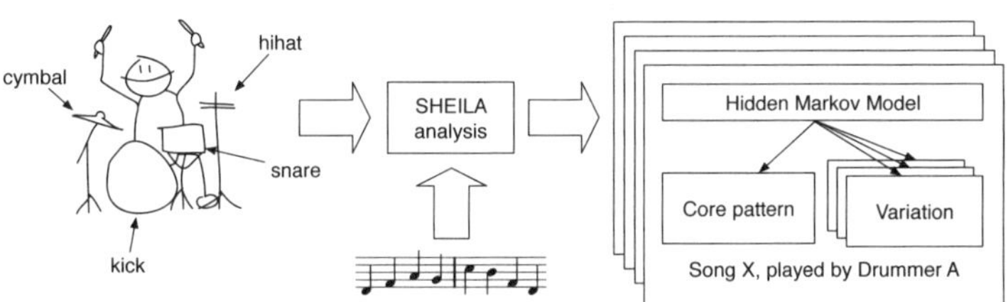

Fig. 1. The SHEILA architecture. Drum patterns are learned, accompanied with melody. In addition to learning drum patterns, SHEILA models the *style* of each drummer that served as a teacher. After learning, SHEILA uses the acquired models to imitate a drum pattern in the style of a given drummer.

3.1   Modeling the Groove

There are two inputs to SHEILA: drum patterns and melody, presented in the MIDI[1] format. SHEILA models both small- and large-scale variations of drum patterns. The *large-scale variations* are discovered the following way: the most

---

[1] Musical Instrument Digital Interface, a protocol for electronic music equipment to communicate in real time.

frequently played patterns are defined as *core patterns*. To find the core patterns, the MIDI drum pattern sequence is transformed into a string, and examined for *supermaximal repeats*, a technique used in computational biology to find sequences of genes [6]. A supermaximal repeat is a recurring pattern that is not a substring of any other pattern. The same approach is taken on the melody MIDI sequence; the supermaximal repeats are used to divide the song into different parts (e.g. the verse/chorus/bridge of the song), thus the melodic input serves as musical context. For each of these parts, the most commonly played pattern is then considered to be the core pattern of that part, yielding several core patterns for a song. Patterns that differ from the core pattern within the part are defined as the large-scale variations of the core pattern. The sequence of core patterns and variations within similar parts is modeled using a Hidden Markov Model (HMM). The *small-scale variations* are defined as the variations in timing and velocity for each beat that occurs when a drummer plays a pattern. We model the small-scale variations by calculating the mean (μ) and standard deviation (σ) of both the onset time (i.e. how much the beat differs from the metronome) and velocity (i.e. how hard a note is played) of each beat across similar patterns. Our pilot study [7] showed that the Gaussian distribution was an appropriate model. One of the leading software samplers on the market, FXpansion BFD, use the same approach to model human variations in its "Humanize panels", however the user must specify the μ and σ using a graphical interface[2]. Each entry in the SHEILA library consists of a core pattern and variations of the core pattern and the HMM modeling their sequence. Each beat in all the patterns is assigned the normal distribution parameters μ and σ of both velocity and onset time.

3.2 Imitating the Groove

The user of SHEILA presents the desired pattern in the MIDI format to SHEILA. SHEILA presents a list of drummers who have played the desired pattern, along with the name of the song this pattern was played on (indicating what the imitation will sound like, presuming the user knows the song). The user then selects which drummer SHEILA should imitate when playing the desired pattern, and for how many bars. The HMM for each core pattern is used to generate output sequences (i.e. core patterns and variations). This is how *large-scale variations* are introduced. To create the actual beats, random numbers determining the velocity and onset time are drawn from the normal distribution assigned for each beat, introducing *small-scale variations* to the imitated pattern.

The generated onset times are averaged locally based on a 10 beat window (five beats before and five beats after the current beat). A human drummer will vary the onset time of the beats, but this variation will not occur randomly within a close timeframe. The local averaging makes beats played at the same time sound more coherent, whereas the onset time is still allowed to drift over time. Averaging over ten beats was found to be the value that sounded most

---

[2] See page 118 of the user manual (accessed 2008-04-08),
   www.fxpansion1.com/resourceUploads/BFD_Manual_English.pdf

natural; less than 10 yielded too random-sounding onset times, whereas more than 10 made the onset times sound too rigid.

The combination of large- and small-scale variations in the imitated drum patterns will result in a pattern that is different from the original, but still in the *style* of drummer X. The output is in the MIDI format, ready to be imported into music production software with high quality drum samples.

4   Experimental Setup

SHEILA was implemented in MatLab. *vmatch*[3] was used to find the super-maximal repeats. Propellerheads Reason 3.0[4] loaded with Reason Drum Kits was used for recording MIDI signals and for generating sound from MIDI files. Recording MIDI was done with a Roland TD-3[5] velocity sensitive electronic drum kit. Five male amateur drummers recorded drum tracks to a melody written by the authors, and were told to play specific patterns for the verse (shown in figure 2), chorus and bridge, and a specific break at each 8th bar of the verse. The drummers were free to introduce large-scale variations according to what felt natural to them. The overall structure of the song was verse/chorus/verse/chorus/bridge/chorus/chorus. The tempo was 120 beats per minute, the length of the song was 2:30 minutes.

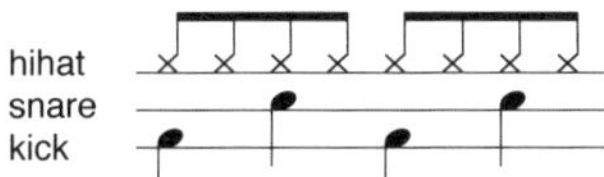

Fig. 2. One of the three core patterns played in the experiment

5   Results

Figures 3 and 4 show how SHEILA models drummers A and E playing the pattern shown in figure 2. [7] provides more examples. The figures show the mean and standard deviation for both velocity and onset time, and how SHEILA models the unique style of each drummer. Most easily observable is the difference in accentuation of the hihat between drummers A and E. Accents are defined as periodic variations in velocity. This is demonstrated by the strong regular variations in velocity that can be seen in the hihat bar plots of drummer A, whereas for drummer E these variations are not so prominent.

The onset time reveals the temporal qualities of the drummers, i.e. how aggressive (ahead of the metronome) or relaxed (behind the metronome) he is. Figure 4 shows how the onset time profiles are different for drummers A and E. The plots show how the playing style of drummer A is slightly more aggressive

---

[3] www.vmatch.de

[4] www.propellerheads.se

[5] www.roland.com

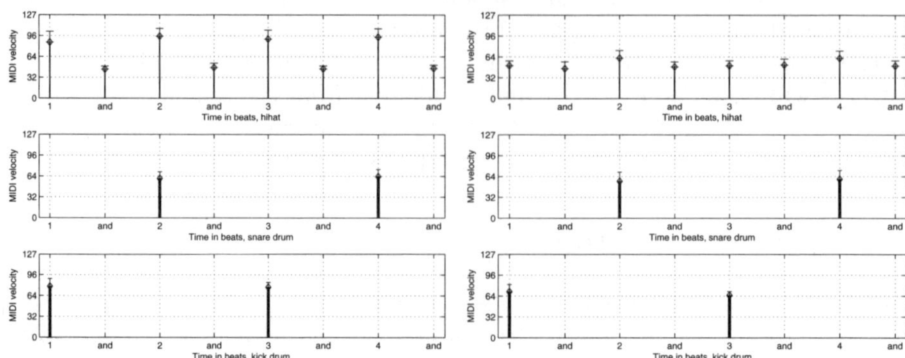

Fig. 3. The plots show the velocity profiles of drummer A (left) and drummer E (right) playing the pattern in figure 2. The MIDI resolution [0−127] is on the Y axis. The X axis represent the beats in a bar. The bar plots show how the velocity varies periodically over time (standard deviation shown as the error bar). This is what makes up the drummer's *groove* in terms of small-scale variations, including the onset times (shown below). The accentuation (i.e. variation of velocity) on the hihat is more pronounced for drummer A compared to drummer E, providing a visual confirmation that the drummers have a different groove when playing the same pattern. The acquired models makes SHEILA capable of imitating the playing style of the drummers.

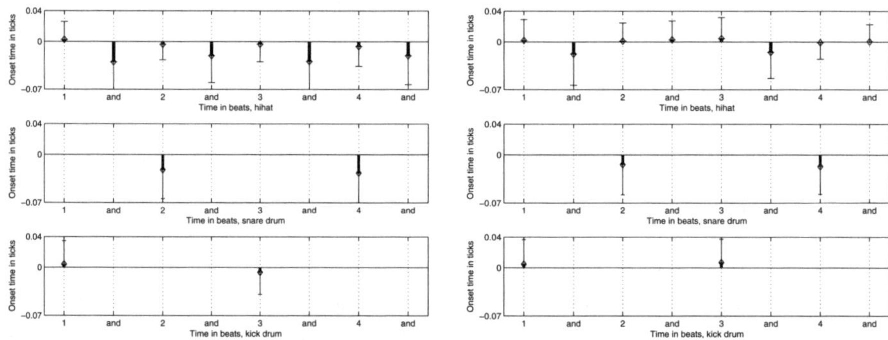

Fig. 4. The plots show the onset time profiles for drummer A (left) and E (right) playing the pattern in figure 2. The Y axis shows the onset time, there are 120 ticks in range [0. − 0.99] between each quarter note. The X scale is the same as in figure 3. The differences in the onset time profiles is further evidence that the drummers have different grooves when playing the same pattern.

than that of drummer E. The reader can listen to available MP3[6] files to get a better sense of these terms.

---

[6] http://www.idi.ntnu.no/∼tidemann/sheila/ki08/

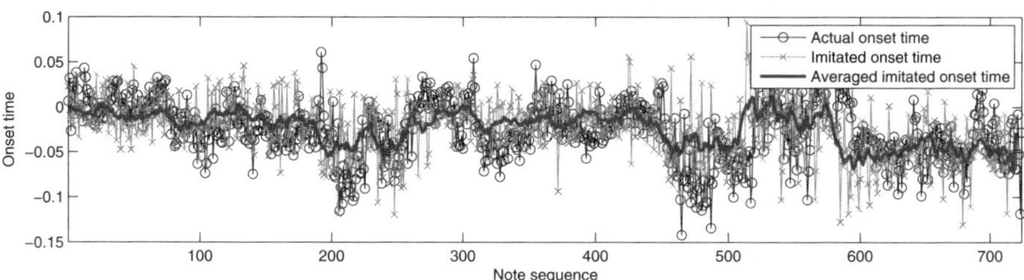

Fig. 5. The circle plots show the tempo drift of drummer A throughout the song. The cross plots show the generated onset times; they are locally averaged since they are drawn from a normal distribution. The local averaging will make beats that are close temporally sound more coherent, whereas the tempo can still drift over time, as shown by the thick line, adding realism to the generated patterns.

Table 1. How often the imitated SHEILA output was correctly classified as being imitated from the corresponding human drummer. Each drummer that participated in the experiment evaluated the imitation, and tried to match which drummer served as a teacher.

Drummer	A	B	C	D	E
Classification	80%	60%	80%	80%	80%

After the learning process, SHEILA was set to imitate the same song used in the training phase to evaluate the quality of the imitation. Since both the large-scale and small-scale variations are introduced randomly using the acquired models (HMMs and normal distributions, respectively), the output will have the human quality that it is different each time a drum track is generated. Figure 5 shows how the local averaging of onset times allows the tempo to drift over time, making the output sound more human.

Each of the drummers participating in the experiment was presented both the original drum tracks and the imitations in random order and asked to classify which drummer SHEILA imitated (each song lasting 2:30 minutes, the available MP3 files include both originals and imitations). All the tracks used the same set of samples and were mixed the same way, i.e. the sonic qualities of the drumming were the same from track to track. Only the playing style of generated tracks were different. It should be noted that normally other qualities will help identify a drummer (e.g. the sound of the drums, the genre of music, the band the drummer is playing in, etc.). We have removed all these factors, making the recognition task harder than what first might be expected. Table 1 shows that the imitated drum tracks were often perceived as being similar to the teacher (average of 76% correct classification).

6   Discussion and Conclusion

This paper presents a more sophisticated version of SHEILA than the pilot study in [7], which does not model large-scale variations, does not employ the local averaging of onset time for added realism, and was trained on fewer and simpler datasets. The were two reasons for using amateur drummers playing on a MIDI drumkit: 1) budget (i.e. professional drummers were too expensive) and 2) simplicity of data gathering. We have bypassed the data acquisition stage of human drummers, that consist of seeing and hearing drums being played. We believe these simplifications allow focus on the core problem, which is learning drum patterns.

Our software drummer learns how a human plays drums by modeling small- and large-scale variations. The intelligent drum machine is capable of mimicking its human teacher, not only by playing the learned patterns, but also by always varying how patterns are played. The imitative qualities of SHEILA was shown to be good by evaluation from the different drummers that served as teachers. Our drum machine is an example of how a tool can be augmented by applying AI techniques in an area where incorporating human behaviour is of essence. Creating music is an area where human qualities such as "feel" and "groove" are important, but hard to define rigorously. By enhancing our software drummer with the ability to learn and subsequently imitate such human qualities, the drum machine evolves from being a static rhythm producer to become closer to a groovy human drummer. The added advantage is the ease and cost-effective way of creating human-like drum tracks.

7   Future Work

A more biologically plausible approach of learning drum patterns would be to make SHEILA listen to music (i.e. using audio files as input) and extract melody lines and drum patterns from low-level audio data. For reasons of simplicity, this was not done in the current implementation, but possible solutions are described in [8].

We also aim to visualize SHEILA, by providing imitative capabilities of upper-body movements. SHEILA could then be used in a live setting, playing with other musicians. Our previous work on dance imitation using motion tracking and imitating arm movements [9] forms the basis for this area of research. This approach uses multiple forward and inverse models for imitation [10].

Acknowledgements

The authors would like to thank the drummers who participated in the experiment (Inge Hanshus, Tony André Søndbø, Daniel Erland, Sven-Arne Skarvik).

References

1. Saunders, C., Hardoon, D.R., Shawe-Taylor, J., Widmer, G.: Using string kernels to identify famous performers from their playing style. In: Boulicaut, J.-F., Esposito, F., Giannotti, F., Pedreschi, D. (eds.) ECML 2004. LNCS (LNAI), vol. 3201, pp. 384–395. Springer, Heidelberg (2004)
2. Tobudic, A., Widmer, G.: Learning to play like the great pianists. In: Kaelbling, L.P., Saffiotti, A. (eds.) IJCAI, pp. 871–876. Professional Book Center (2005)
3. Pachet, F.: Enhancing Individual Creativity with Interactive Musical Reflective Systems. Psychology Press (2006)
4. de Mantaras, R.L., Arcos, J.L.: AI and music from composition to expressive performance. AI Mag. 23(3), 43–57 (2002)
5. Raphael, C.: Orchestra in a box: A system for real-time musical accompaniment. In: IJCAI workshop program APP-5, pp. 5–10 (2003)
6. Gusfield, D.: Algorithms on strings, trees, and sequences: computer science and computational biology. Cambridge University Press, New York (1997)
7. Tidemann, A., Demiris, Y.: Imitating the groove: Making drum machines more human. In: Olivier, P., Kay, C. (eds.) Proceedings of the AISB symposium on imitation in animals and artifacts, Newcastle, UK, pp. 232–240 (2007)
8. Poliner, G.E., Ellis, D.P.W., Ehmann, A.F., Gomez, E., Streich, S., Ong, B.: Melody transcription from music audio: Approaches and evaluation. IEEE Transactions on Audio, Speech and Language Processing 15(4), 1247–1256 (2007)
9. Tidemann, A., Öztürk, P.: Self-organizing multiple models for imitation: Teaching a robot to dance the YMCA. In: Okuno, H.G., Ali, M. (eds.) IEA/AIE 2007. LNCS (LNAI), vol. 4570, pp. 291–302. Springer, Heidelberg (2007)
10. Demiris, Y., Khadhouri, B.: Hierarchical attentive multiple models for execution and recognition of actions. Robotics and Autonomous Systems 54, 361–369 (2006)

Believing Finite-State Cascades in Knowledge-Based Information Extraction

Benjamin Adrian[1] and Andreas Dengel[1,2]

[1] Knowledge Management Department, DFKI,
Kaiserslautern, Germany
[2] CS Department, University of Kaiserslautern
Kaiserslautern, Germany
`firstname.lastname@dfki.de`

Abstract. Common information extraction systems are built upon regular extraction patterns and finite-state transducers for identifying relevant bits of information in text. In traditional systems a successful pattern match results in populating spreadsheet-like templates formalizing users' information demand. Many IE systems do not grade extraction results along a real scale according to correctness or relevance. This leads to difficult management of failures and missing or ambiguous information.

The contribution of this work is applying *belief* of Dempster-Shafer's *A Mathematical Theory of Evidence* for grading IE results that are generated by probabilistic FSTs. This enhances performance of matching uncertain information from text with certain knowledge in knowledge bases. The use of *belief* increases precision especially in modern ontology-based information extraction systems.

1   Introduction

Common information extraction (IE) systems are built upon regular extraction patterns and finite-state transducers (FST) for identifying relevant bits of information in text. In traditional IE systems, a pattern match is a binary decision. It results in populating spreadsheet-like templates formalizing users' information demand. By reason of binary decisions, many IE systems do not grade extraction results along a real scale according to correctness or relevance. This leads to difficult management of failures and missing or ambiguous information.

In this work we use Dempster-Shafer's *A Mathematical Theory of Evidence* [1] in an ontology-based IE (OBIE) system for grading uncertain, extracted bits of information from text with respect to certain background knowledge. We show the resulting increase of precision by evaluating an instance resolution task.

At first, we list related work about similar applications and applied techniques. We provide a short overview about common finite-state approaches in IE and expose the capabilities of Dempster-Shafer's *belief* function in FSTs. An evaluation along our OBIE system iDocument [1] proves the increase of precision. Finally we summarize and conclude the resulting advantages and give an outlook of future perspectives.

----

[1] `http://idocument.opendfki.de`

A. Dengel et al. (Eds.): KI 2008, LNAI 5243, pp. 152–159, 2008.

2   Related Work

Common IE tasks are well explained in [2]. A reference implementation of IE provides Gate [3]. OBIE is accounted in [4]. A survey of IE patterns is given in [5]. Using FST for parsing purpose in natural language is explained in [6], [7]. FSTs are used in IE systems such as Fastus [8] or SMES [9]. The use of weighted, probabilistic FST is explained in [10], [11], and specialized in Markov models [12]. The mathematical concept of *belief* is considered in *A Mathematical Theory of Evidence* [1]. In [13], *belief* is used in Markov models. According to [14] and [15], *belief* has already been successfully used in knowledge-based systems.

The contribution of this work is applying *belief* for grading IE results generated by probabilistic FSTs. This enhances performance of matching uncertain information from text with certain knowledge in knowledge bases.

3   Knowledge-Based Information Extraction

Most relevant techniques in IE systems are regular pattern matchers as described in [5], combined with additional linguistic or domain knowledge for instance corpora, grammars, gazetteers, ontologies, or dictionaries as done in Gate [3].

We explain knowledge-based IE (KBIE) as a special form of a traditional IE process that is well explained in [2]. KBIE extracts relevant bits from text with the help of an underlying knowledge base. If this knowledge base uses formal ontologies for knowledge representation, we speak of ontology-based IE (OBIE). Concerning OBIE systems, the knowledge base separates: ontologies for structuring information domains, rules for expressing integrity and regularities, frame-like instances describing concrete knowledge, and finally relations for expressing behavior between instances. Figure 1 shows the architecture and process of a KBIE system as being implemented in iDocument. The extraction process is based on cascading extraction tasks and comprises the following steps:

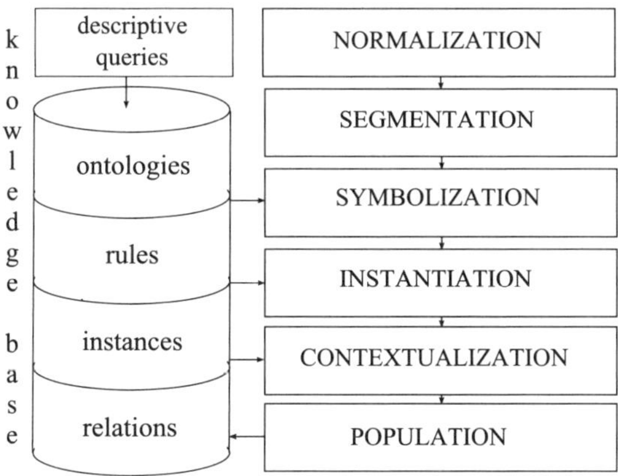

Fig. 1. Architecture and process of a knowledge-based IE system

1. **Normalization:** Extracts metadata about and plain text data from documents that are coded in textual or binary file formats.
2. **Segmentation:** Partitions plain text into hierarchical ordered segments, namely: paragraphs, sentences, and tokens.
3. **Symbolization:** Recognizes known tokens or token sequences as symbols and unknown noun phrases as named entities.
4. **Instantiation:** Resolves and disambiguates symbols as occurrences of instances, relations, and properties. Entities that cannot be resolved to existing instances are classified as new instance.
5. **Contextualization:** Resolves recognized relations between and properties of instances with respect to existing contexts such as the document, the user's question, or the knowledge base.
6. **Population:** Evaluates extracted instances, properties, and relations. Valid information is populated into the knowledge base.

The following overview introduces the use of FST in IE systems as done for instance in Fastus [8], a *Cascaded Finite-State Transducer for Extracting Information from Natural-Language Text.*

4   Finite-State Machines as Grounding IE-Engine

The modeling of finite-state machines (FSM) is a well known technique for processing textual input streams. It is based on states, transitions, and actions. Two groups of FSMs exist, namely recognizers and transducers. Recognizing machines accept certain tokens with a function λ that maps read tokens into the boolean token set $O := \{false, true\}$. Transducing machines (called finite-state transducer) read tokens and create a new formal language with λ and an output alphabet O as shown below in Fig.2.

We set the focus on a special type of FSTs called Moore Machine. The basic nature of a Moore Machine is the fact that it generates output based on its

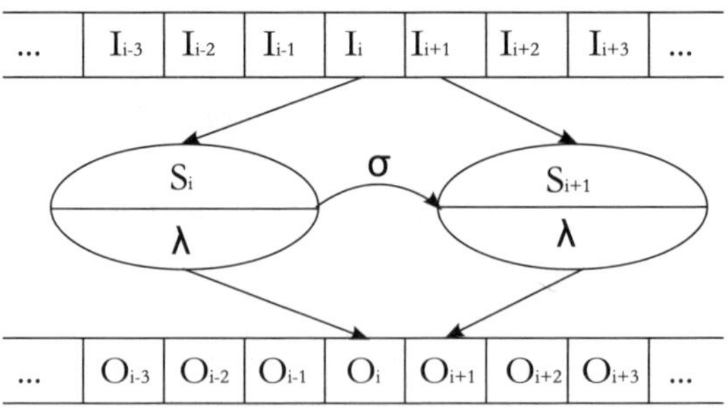

Fig. 2. A simple extract of a finite-state machine

current state. Thus, transitions to other states depend on the input. We describe a Moore Machine as 6-tuple $(S, S_0, I, O, \sigma, \lambda)$ with:

S as finite set of states and $S_0 \in S$ as initial state,
I as input alphabet and O as output alphabet,
$\sigma : S \times I \to S$ as transition function, and finally
$\lambda : S \to O$ as output function.

Fig.2 outlines a Moore machine between a token input stream I and output stream O. The theory of FST allows composing a series of FSTs [10] to finite-state cascades.

In OBIE systems, domain ontologies are used for defining patterns about instances, classes, properties, and relations. Cascading FST match the input for patterns inside the knowledge base. Thus, the output function $\lambda : S \to O$ of a Moore machine is extended to $\lambda_{KB} : S \to O$. KB is a proxy in λ that matches input with patterns inside the knowledge base. In means of KBIE, a positive pattern match is the identification of evidence about a pattern inside the knowledge base. The generated output of λ is called hypotheses. The following section extends FSTs with a facility for grading hypotheses in O.

5   Adding Belief to Finite-State Transducers

In KBIE, input is matched by finite-state cascades and evaluated by an underlying knowledge base. A possible evaluation may be for instance the question: *Is the phrase 'Benjamin Adrian' the name of a known person?* Let X be the variable for this question, then ranking the results with Bayes probabilities may e.g., result in a probability that X is true of $P(X) := 0.6$. In terms of Bayes, this infers that *Benjamin Adrian* is not a person's name with a certainty of $P(\neg X) := 0.4$. This conclusion is not suitable in KBIE, as positive matches between bits of extracted information and formalized knowledge express an unsure similarity with an amount of *certainty*. Therefore the degree $1 - certainty$ should describe the degree of *uncertainty*.

According to this problem, Dempster-Shafer's theory introduces two metrics called *belief* and *plausibility* [1]. It extends Bayes' probability with the main difference that the negative probability $(1 - P(s))$ of an event s to happen does not describe the complement probability of s not to happen with a probability $(P(\neg s))$. This results in $(P(\neg s) \neq 1 - P(s))$. In terms of *belief* and *plausibility*, the degree of $(1 - P(s))$ counts the remaining *uncertainty* about $P(s)$ to happen.

Given a fixed set of hypotheses H about a finite set of states S, the amount of *belief* in a hypothesis is constituted by combining all hypotheses about a certain state $s_i \in S$. *Plausibility* results from all hypotheses about different states $s_j \in S$ whose *uncertainty* indirectly support evidence for s_i.

We introduce *belief* to FSTs by adding a *belief* function bel to the output function $\lambda(S)$. In terms of KBIE it results in $bel(\lambda_{KB}(S))$. We call this type of Moore machine a believing FST.

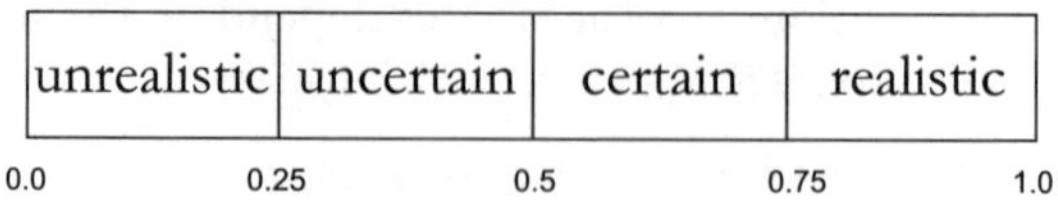

Fig. 3. A scale of belief values

We recommend the condition only to use positive pattern matches for simplifying believing FSMs. This condition ensures that *belief* values about state complements $bel(\neg S)$ are always $bel(\neg S) := 0$. Consequently the general *plausibility pl* of a state S has the value $pl(S) := 1$.

We define the range $]bel, pl]$ between *belief* and *plausibility* in a hypothesis as *uncertainty*. As figured out in Fig. 3, we define four ranges of belief regarding the amount of *certainty* to be used in hypotheses. *Belief* values less than 0.25 are called *uncertain*. *Belief* values less than 0.5 are defined as *unrealistic*. *Realistic belief* values are less than 0.75. Higher values are defined to be *certain*.

If more than one hypothesis holds evidence about a certain state, it is necessary to perform a combination. Based on Bernoulli's rule of combination and our simplified theory of evidence, the combination of two hypotheses functions h_i, h_j adding belief values bel_i and bel_j to a certain state $s \in S$ can be expressed as follows:

$$comb(bel_i, bel_j) = bel_i + (1 - bel_i) \cdot bel_j \tag{1}$$

The propagation of *belief* values in hypotheses h_{i-1}, h_i during a finite-state cascade is done using this combination rule. Consequently, the degree of *belief* of hypotheses H_i in step i is influenced by the degree of *belief* of H_{i-1} and $\lambda_{KB}(H_i)$.

$$
\begin{aligned}
bel(H_i) &= comb(bel(H_{i-1}), bel(\lambda_{KB}(H_i))) \\
&= bel(H_{i-1}) + (1 - bel(H_{i-1})) \cdot bel(\lambda_{KB}(H_i))
\end{aligned}
\tag{2}
$$

While considering an infinite cascade of hypotheses H_i with $1 \leq i < \infty$ and $bel(\lambda_{KB}(H_i)) > 0$, the limit of $bel(H_i)$ results in $\lim_{i \to \infty} : bel(H_i) = 1$. This behavior is not acceptable. If we assume that $bel(\lambda_{KB}(H_i))$ is a random variable with a mathematical expectation of $E(\lambda_{KB}(H_i))$, the limit of $bel(H_i)$ should not lead to higher evidences in form of $\lim_{i \to \infty} : bel(H_i) > E(bel(\lambda_{KB}(H_i)))$. Therefore we introduce a moderation parameter β resulting in a β-combination rule:

$$
\begin{aligned}
bel(H_i) &= comb_\beta(H_{i-1}, bel(\lambda_{KB}(H_i))) \\
&= bel(H_{i-1}) \cdot \beta + (1 - bel(H_{i-1})) \cdot bel(\lambda_{KB}(H_i))
\end{aligned}
\tag{3}
$$

The desired behavior of $\lim_{i \to \infty} : bel(H_i) = E(bel(\lambda_{KB}(H_i)))$ can be reached if β is set to:

$$\beta = \frac{1}{2 - E(bel(\lambda_{KB}(H_i)))} \tag{4}$$

If we estimated a uniform distribution of belief between 0 and 1 ($E = 0.5$), it leads to $\beta = 2/3$.

6   Evaluation

We developed an OBIE prototype called iDocument along the KBIE architecture of Fig.1. Each extraction task of iDocument is implemented as one FST inside the finite-state cascade. Thus, extraction tasks create hypotheses about successful pattern matches. In iDocument, hypotheses are graded using *belief* values. We evaluate the beneficial use of believing FST by comparing results of an instance resolution task in four *belief* threshold settings.

In order to evaluate OBIE systems, an existing knowledge base is needed that describes one information domain. A document corpus with content concerning the information domain has to be used as source for each test run. Finally, for grading test results according to reference values, a ground truth has to exist that contains manually extracted instances from each of the document.

We took a real world organizational repository written in RDFS of the KM Department at DFKI as knowledge base. The document corpus contains 20 research publications written by employees. As referencing gold standard, we used manually created instance annotations about each document inside the corpus. To keep the evaluation simple, we let the annotators only consider already existing instances of the knowledge base.

The evaluation plan covered four routines. Every routine used one *belief* range of Fig.3 as hypotheses threshold, namely *uncertain, unrealistic, realistic,* and *certain.* All hypotheses being generated in a test run had to reach the *belief* threshold for being considered inside the finite-state cascade. Results were compared with the gold standard by using precision, and recall scales, and the harmonic mean of both called balanced F-score [16].

Fig.4 summarizes the evaluation results. The three curves represent values measured in percentage value of precision, recall, and balanced F-score. Each value is computed by taking averages of all 20 test runs. Concerning the four test routines, Fig.4 separates precision, recall, and F-score results in four *belief* threshold ranges. The scale begins with 75%.

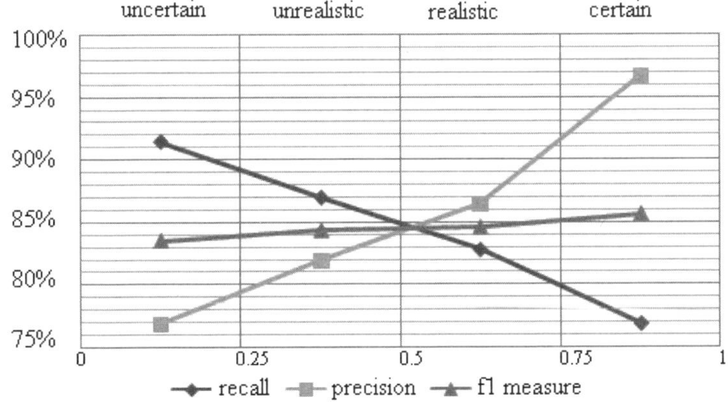

Fig. 4. Evaluation of iDocument

The balanced F-score results describe the general IE performance of our approach regarding an equal weight of precision and recall. It can be seen that it starts with a value of 83% and slightly increases in a linear way regarding belief thresholds to reach 86% in the certain setting. As an explanatory remark, the absolute performance of iDocument's instance resolution capability was not designed to be competitive to state-of-the-art IE systems. It is acceptable for showing benefits of *belief*. The interpretation of F_1 distributions reveals that iDocument's general performance increases, if only hypotheses with high *belief* ratings are accounted.

A progression analysis of precision and recall ratios reveals an antiproportional relation regarding increasing *belief* threshold. Heightening *belief* thresholds, leads to higher precision but lower recall ratings of extraction results. The linear increasing balanced F-score concludes that the rise of precision is stronger than the decrease of recall. In detail, precision increases by relying on certain hypotheses from 77% to 97%. In this case the amount of recall decreases from 91% to 77%. The break-even point of precision, recall, and F-score is near 0.5, which accounts the quality of *belief* distributions in grading hypotheses.

7 Conclusion

This paper explains how and why to use *belief* as grading purpose in KBIE systems. It extends probabilistic weighted FST with *belief* scales. The resulting system uses believing FSTs that generate extraction hypotheses graded with discrete *belief* values. We separated the span of *belief* into four labeled ranges (see Fig. 3) with intermediate steps of 0.25, namely *uncertain, unrealistic, realistic,* and *certain.* We analyzed the limit of *belief* values in finite-state cascades and concluded the β-combination rule (consider Eq. 3) with an optimal β for a given *belief* distribution (see Eq.4). In iDocument, we estimated a mathematical expectation of *belief* values around 0.5 leading to $\beta = 2/3$. The evaluation of iDocument according to *belief* thresholds reveals an increasing F-score of 3%. Heightening *belief* thresholds to be certain, leads to higher precision ($+20\%$) but lower recall (-14%) ratings of extraction results. This knowledge can be used for fine tuning future IE systems. We recommend using certain thresholds in automatic IE systems for ensuring a high degree of quality. Semi-automatic IE should run with between realistic or unrealistic thresholds near 0.5. Finally, manual IE (e.g. recommendation systems) can be run by regarding even uncertain hypotheses for providing a high amount of completeness.

8 Outlook

In future work, we plan to increase the amount of recall in certain threshold settings. This can be done by analyzing more semantic patterns between document and knowledge base. This work was supported by "Stiftung Rheinland-Pfalz für Innovation".

References

1. Shafer, G.: A Mathematical Theory of Evidence. Princeton University Press, Princeton (1976)
2. Appelt, D.E., Israel, D.J.: Introduction to information extraction technology. In: A tutorial prepared for IJCAI 1999, Stockholm, Schweden (1999)
3. Cunningham, H., Maynard, D., Bontcheva, K., Tablan, V.: Gate: A framework and graphical development environment for robust NLP tools and applications. In: Proc. of 40th Anniversary Meeting of the ACL (2002)
4. Bontcheva, K., Tablan, V., Maynard, D., Cunningham, H.: Evolving gate to meet new challenges in language engineering. Nat. Lang. Eng. 10(3-4), 349–373 (2004)
5. Muslea, I.: Extraction patterns for information extraction tasks: A survey. In: Califf, M.E. (ed.) Machine Learning for Information Extraction: Papers from the AAAI Workshop, pp. 94–100. AAAI, Menlo Park (1999)
6. Roche, E.: Parsing with finite state transducers. In: Roche, E., Schabes, Y. (eds.) Finite–State Language Processing, ch. 8. The MIT Press, Cambridge (1997)
7. Abney, S.: Partial parsing via finite-state cascades. Nat. Lang. Eng. 2(4), 337–344 (1996)
8. Appelt, D., Hobbs, J., Bear, J., Israel, D., Tyson, M.: FASTUS: A finite-state processor for information extraction from real-world text. In: Proc. Int. Joint Conference on Artificial Intelligence (1993)
9. Neumann, G., Backofen, R., Baur, J., Becker, M., Braun, C.: An information extraction core system for real world german text processing. In: Proc. of the 5. conf. on Applied natural language processing, pp. 209–216. Morgan Kaufmann Publishers Inc, San Francisco (1997)
10. Mohri, M.: Weighted finite-state transducer algorithms: An overview. Formal Languages and Applications 148, 620 (2004)
11. Vidal, E., Thollard, F., de la Higuera, C., Casacuberta, F., Carrasco, R.C.: Probabilistic finite-state machines-part i and i. IEEE Trans. Pattern Anal. Mach. Intell. 27(7), 1013–1039 (2005)
12. Brants, T.: Cascaded markov models. In: Proc. of the 9th conf. on European chapter of the ACL, pp. 118–125. ACL, Morristown, NJ, USA (1999)
13. Wilson, N., Moral, S.: Fast markov chain algorithms for calculating dempster-shafer belief. In: Wahlster, W. (ed.) ECAI, pp. 672–678. John Wiley and Sons, Chichester (1996)
14. Liu, L., Shenoy, C., Shenoy, P.P.: Knowledge representation and integration for portfolio evaluation using linear belief functions. IEEE Transactions on Systems, Man, and Cybernetics, Part A 36(4), 774–785 (2006)
15. Shenoy, P.P.: Using dempster-shafer's belief-function theory in expert systems, pp. 395–414 (1994)
16. Makhoul, J., Kubala, F., Schwartz, R., Weischedel, R.: Performance measures for information extraction. In: Proc. DARPA Workshop on Broadcast News Understanding, pp. 249–252 (1999)

A Methodological Approach for the Effective Modeling of Bayesian Networks

Martin Atzmueller and Florian Lemmerich

University of Würzburg,
Department of Computer Science VI
Am Hubland, 97074 Würzburg, Germany
{atzmueller,lemmerich}@informatik.uni-wuerzburg.de

Abstract. Modeling Bayesian networks manually is often a tedious task. This paper presents a methodological view onto the effective modeling of Bayesian networks. It features intuitive techniques that are especially suited for inexperienced users: We propose a process model for the modeling task, and discuss strategies for acquiring the network structure. Furthermore, we describe techniques for a simplified construction of the conditional probability tables using constraints and a novel extension of the Ranked-Nodes approach. The effectiveness and benefit of the presented approach is demonstrated by three case studies.

1   Introduction

Modeling Bayesian networks manually is often a difficult and costly task: If a complex domain needs to be completely captured by a domain specialist, then both the network structure and the conditional probability tables need to be specified. A naive approach is often hard to implement – especially for domain specialists that are not familiar with Bayesian networks. In such cases, the user can be supported significantly by techniques that allow the approximate specification and adaptation of the network structure.

In this paper, we present a methodological view on the effective modeling of Bayesian networks. The presented process features intuitive modeling strategies that are especially suited for inexperienced users: Specifically, we utilize *idioms* that specify common patterns for modeling. These can be reused as templates in a semi-automatic way. Furthermore, we provide the ability to include *background knowledge* given by set-covering models or subgroup patterns. These can then be mapped to probabilistic constraints that are utilized for instantiating the (conditional) probability tables of the network. For these we also provide easy-to-apply modeling methods, e.g., a novel *extension* of the common *Ranked-Nodes* approach. The presented approach has been completely implemented as a plugin of the VIKAMINE system (http://vikamine.sourceforge.net).

The rest of the paper is organized as follows: In Section 2 we briefly discuss some technical background issues. Next, Section 3 summarizes the process of modeling Bayesian networks. Section 4 describes the modeling of the graph structure, while Section 5 presents techniques for an easier assignment of the probability tables, including the extended Ranked-Nodes approach. In Section 6 the presented concepts are evaluated by a series of three case studies. Finally, we conclude the paper with a summary in Section 7, and point out interesting directions for future work.

A. Dengel et al. (Eds.): KI 2008, LNAI 5243, pp. 160–168, 2008.

2 Background

In the following, we briefly introduce the concepts of Bayesian networks, subgroup patterns, and set-covering models. A **Bayesian network** [1] is a standard representation of uncertain knowledge and describes the complete probability distribution of a domain.

Definition 1. *A Bayesian network $B = (V, E, C)$ consists of a set of nodes V, a set of directed edges E between the nodes, and a set of conditional probability tables C.*

The nodes correspond to random variables of a domain. An edge $e = (X, Y)$ connecting the nodes X and Y implies that Y is directly influenced by X. The necessarily acyclic *graph structure* of B is defined by the nodes V and the edges E. A conditional probability table $c(v) \in C$ specifies a probability distribution of the values of its node v given each possible value-combination of the parents $parents(v)$.

The focus of **subgroup patterns** [2] are the dependencies of a given (dependent) target variable with a set of explaining variables. Essentially, a subgroup is a subset of all cases of a database with an *interesting* behavior, e.g., with a high deviation of the relative frequency of a specified value v_s of the target variable. Subgroup patterns are usually described as a conjunction of variable-value-pairs called *features*.

Definition 2. *A subgroup description $sd(s) = \{f_1, \ldots, f_n\}$ is given by a set of features $f_i = (V_i, v_{ij})$ for a variable V_i and a value v_{ij} of this variable. A subgroup s is given by the set of all cases of a database that fulfill each feature of its description.*

Another common knowledge representation are **set-covering models** [3]. They describe relations between *diagnoses* and *features* (see above) caused by some diagnoses.

Definition 3. *A set-covering relation scr describes that a diagnosis d evokes features $f_1, \ldots, f_N$, and is noted as follows: $scr : d \rightarrow f_1, \ldots, f_n$. A set-covering model consists of a set of set-covering relations between the given diagnoses and features.*

An extension assigns a *covering strength* to each relation. The higher the covering strength, the stronger the diagnosis indicates the feature of the set-covering relation.

3 Semi-automatic Process Model

For the acquisition of Bayesian networks we propose a four-step iterative process:

1. **Select Variables:** The relevant variables are usually elicitated in cooperation with domain experts. Additionally, the variables can be divided into weakly interacting groups (*modules*) corresponding to smaller subproblems for easier reuse.
2. **Construct Graph Structure:** The connections between the random variables are modeled semi-automatically based on the causal structure of the domain. We discuss effective strategies for this potentially difficult step in Section 4.
3. **Specify Conditional Probability Tables:** The entries of the tables are assigned semi-automatically using either expert or statistical knowledge about the domain. We propose effective and intuitive techniques for this important step in Section 5.
4. **Evaluate:** The evaluation considers whether the network corresponds to the domain specification. It can be performed manually by human experts, although explicitly

formalized (probability) constraints can also be checked semi-automatically. Necessary extensions and improvements can then be identified for the next iteration.

4 Constructing the Graph Structure

There are several different approaches to model the structure of a Bayesian network (see Figure 1). We distinguish *constructive* and *descriptive* methods: Constructive methods identify and add the edges directly in the graph structure (and also create new nodes on demand). The descriptive methods mainly consider probabilistic constraint knowledge, e.g., known subgroups or set-covering models.

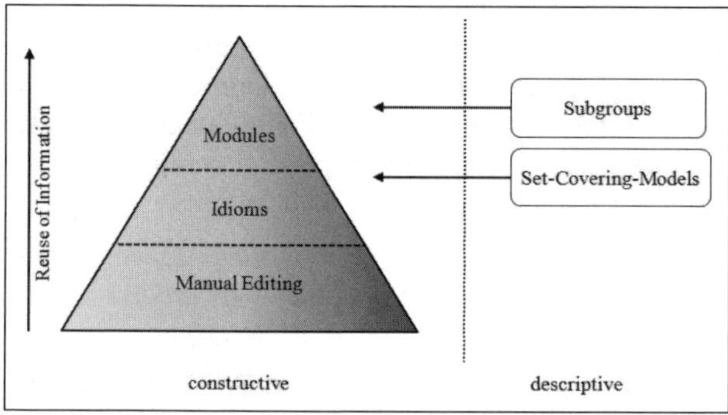

Fig. 1. Approaches for modeling the graph structure

Constructive Methods. For small domains the network structure can often be manually constructed. While a correlation between two variables can be expressed by different network configurations, the edges should indicate direct causality between the random variables where possible. As experience has shown, this matches the domain experts way of thinking (cf. [4]), reduces the network size and leads to more understandable networks. However, identifying causalities one by one is not sufficient for all types of relations and provides no reuse of constructed patterns. On the other hand, large parts of the network can potentially be reused with modular and object-oriented networks, e.g., [5].

Between these approaches lies the concept of *idioms*: Idioms provide small generic fragments of a graph structure, which can be used as templates by human experts. In addition the according network structure is well documented. Neil, Fenton and Nielsen [6] propose several types of idioms: The *Definitional/Synthesis idiom*, for example, is used, if a variable is completely defined by others, cf. Section 6. Then, often a new "hidden" variable is introduced as the target variable.

Descriptive Methods. For the descriptive techniques we apply set-covering models and subgroup patterns: Set-covering models can be transparently mapped onto the graph structure. Each set-covering relation $d \rightarrow f_1, \ldots, f_n$ implies causality from a diagnosis variable to the variables of the findings. Therefore we add edges from d to each f_i.

The integration of known subgroups in the graph structure of a Bayesian network is a much more complicated task. Subgroup patterns neither imply causality nor specify which of the describing variables V_i has a direct or indirect influence on the target V_T. Therefore, we can conclude little more from a single subgroup pattern $(V_1, v_{1k}), \ldots, (V_n, v_{nl}) \rightarrow (V_T, v_t)$ than the existence of a path between each V_i and

V_T. Unfortunately, for each V_i in such a subgroup the number of possibilities to connect a describing variable to the network $(2^{|V|})$ grows exponentially with the number of nodes of the network. For this reason we propose to build a skeletal structure by expert knowledge first. Subgroup patterns are considered afterwards, such that the probability tables can be easily adjusted by connecting the describing independent variables with the target variable directly.

5 Assignment of the Probability Tables

In the following, we discuss effective strategies for determining the conditional probability tables of the Bayesian network. To combine qualitative and constraint knowledge we propose the following steps that we discuss in more detail below:

1. **Manual Acquisition:** First, we utilize qualitative expert knowledge to acquire initial entries for as many probability tables as possible. These can be marked as fixed in order to prevent further changes during the optimization step.
2. **Constraint Integration:** Next, we extract constraints for the probabilities in the Bayesian Network using the given constraint knowledge.
3. **Optimization:** Finally we use an optimization algorithm with the probability tables entries as free parameters to minimize the error on the given constraints.

Manual Acquisition. The most simple way of acquiring the entries of the probability tables for a target node T is to elicit them by experts. However, if each variable has k values, there are roughly $k^{|parents(T)|+1}$ entries to specify, usually a costly task. Techniques based on *causal independence* can decisively reduce the number of necessary elicitations from human experts to about $k^2 \cdot |parents(T)|$, cf. [1].

A recent approach that needs even fewer – $O(|parents(T)|)$ – inputs from experts to construct a complete probability table, is the *Ranked-Nodes* method presented in [7], which is applicable for discrete and ordered variables: We assume the parent nodes to be possible causes for T, each weighted by an human expert. Parents with a higher weight will have a stronger influence on T. We start the construction process by mapping each variables domain bijectively to the continuous interval $[0; 1]$, keeping the value ordering. Let $dom(V_i) = \{v_{i1}, \dots, v_{ik}\}$ denote the ordered value range of the variable V. Then, the value v_{ij} is assigned to the interval $I(v_{ij}) = [\frac{j-1}{k}; \frac{j}{k}]$; the mean of this interval is $m(v_{ij}) = \frac{2j-1}{2k}$. In a next step we will compute a *continuous* probability distribution for the (discrete) variable T for each configuration of its parents. For this purpose we use the doubly truncated normal distribution $TNormal$, which is defined as:

$$TNormal(x, \mu, \sigma) = \begin{cases} \frac{N(x,\mu,\sigma)}{\int_0^1 N(x,\mu,\sigma)dz}, & \text{if } 0 \leq x \leq 1 \\ 0 & \text{otherwise.} \end{cases}$$

where $N(x, \mu, \sigma)$ is the normal distribution with expected value μ and standard deviation σ, with

$$\mu = \frac{\sum_{i=1}^n w_i \cdot m(v_{ij})}{\sum w_i}, \sigma = \sqrt{\frac{1}{\sum_i^n w_i}}.$$

These two parameters are determined considering the weights w_i and the current configurations values v_{ij} of the parent nodes: In this way, σ ensures a lower deviation for strongly influenced nodes, while μ leads to a higher target value for higher values of the parent nodes and also considers the influence of the weights of the parents. In a final step the continuous distribution is transformed into one column of the probability table by integrating over the intervals $I(v_{ij})$ corresponding to the target nodes values. It is easy to see, that the complete computation can be done automatically with appropriate tool support, given only the weights of the parent nodes.

The weighted average function for μ can be substituted by other functions to model different interactions of the parent nodes, cf. [7]. In our implementation of the Ranked-Nodes approach we provide a graphical user interface, which also allows the expert to quickly add offsets to the values of μ and σ or invert the influence of a parent by temporarily reversing the node ordering.

Although the Ranked-Node approach proved to be very useful in our studies, its applicability is limited to specific cases. Therefore we developed an adaptation that allowed us to transfer the basic concept to target nodes with unordered domains: We replace the target variable with k boolean variables, each corresponding to one value of the original target node. As boolean nodes are ordered by definition, the standard Ranked-Nodes algorithms can now be applied with the boolean nodes as targets. As result we get a vector of pseudo-probabilities for the values of T, which can easily be normalized to obtain a sum of 1, i.e., a valid probability distribution of the node T.

To prevent very high standard deviations for boolean nodes without any cause, we propose to adapt the algorithm slightly by computing $\sigma = (\sum_j^k \sum_i^n w_{ij})^{-\frac{1}{2}}$, taking into account *all* weights specified for any boolean node substituting the target node.

Constraint Integration. One difficulty in constructing the probability tables is the integration of formalized background knowledge, e.g., subgroups and set-covering models. We therefore transform such knowledge into constraints concerning conditional and joint probabilities in the Bayesian network. This transformation is straight forward for subgroup descriptions: The conditional probability of the target given the describing features, and the joint probability of the describing features can be directly integrated. Given a set-covering relation $d \rightarrow f_1, \ldots, f_n$ with covering strength cs, n constraints can be derived, e.g., $p_{min}(cs) < P(f_i|D) < p_{max}(cs)$, where $p_{min}(cs)$ and $p_{max}(cs)$ must be specified once by the user. Additional constraints can be elicited manually, e.g., utilizing knowledge about qualitative influences, cf. [8], or using case cards, cf. [9]. A constraint can also be weighted with a priority. The constraints can then be used for evaluating and adapting the (conditional) probability tables.

Optimization. To adjust the probability tables to the acquired constraints we apply an optimization algorithm to minimize the global error, which is computed as follows:

$$\sum_{Constraint\ c} ((goal\_value(c) - current\_value(c))^2 \cdot priority(c))$$

Each table entry of the network, that is not marked as fixed explicitly, is used as a free parameter. We propose a hill climbing approach, where in each step one parameter is modified by a fixed step size. Because computing the global error is quite costly, parameters, which can not improve at least a single constraint, are filtered out first. Besides first choice hill climbing and steepest ascent hill climbing we used two variations: (a) Adaptive step size hill climbing, where the step size is dependent on the global error: *step size = max (basic step size · global error, minimum step size)*, and (b) Optimal step size hill climbing: Once an improving modification is found, increasing step sizes are considered until an optimum step size for an adaptation of this parameter is reached. Due to local minima, a randomized multistart approach is necessary. The efficiency of these variations is evaluated in the case studies in the next section.

6   Case Studies

In a first case study we tested the usefulness of idioms and construction techniques for probability tables in the well-known restaurant domain, cf. [1]. We start the modeling process with the identification of idioms. For example, the variable "Fri/Sat", which is true, if the event is on a friday or saturday is completely determined by the variable "Day of the Week". Thus, we have found a Definitional/Synthesis idiom here and add the appropriate edge to the graph. As another example, the variables "Patrons" and "Reservation" causally influence ("produce") the variable "Wait Estimate". This corresponds to the so called Product-Process idiom and again we can add the respective edges. The complete resulting network includes 12 nodes and probability tables with up to 64 entries, see Figure 2.

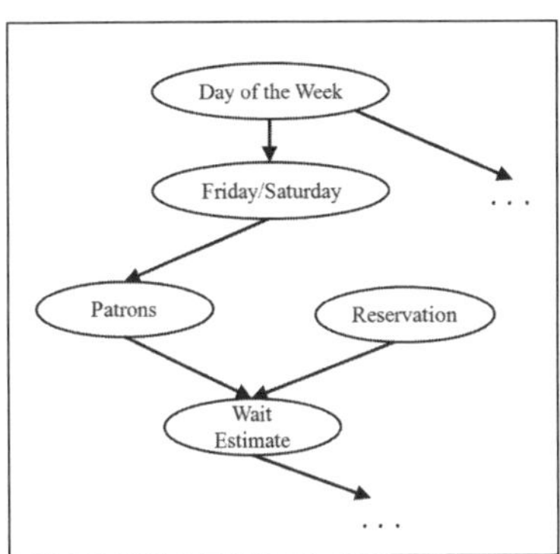

Fig. 2. Part of the graph structure for the restaurant domain case study

A decisive advantage of building the network using the idioms was its implicit documentation, since the resulting graph structure is explained very well by the applied idioms. We expect this advantage to scale up strongly when modeling a large domain in collaboration with several knowledge engineers.

The modeling of the probability tables showed, that using advanced methods like Ranked-Nodes or techniques based on causal independence can save a large amount of time. For example, we could determine the probabilities of "Wait Estimate" (see Figure 3) using our implementation by only weighting the influence of "Patrons" with 20, the inverted Influence of "Reservation" with 10 and adding a μ-offset of -0.1.

In several cases we could use the result of a Ranked-Nodes algorithm directly, while it provided a baseline for manual adaptations in other cases. Even for smaller tables the Ranked-Nodes algorithm was helpful to create a plausible probability table very fast. However, we also had to experience, that in some cases Ranked-Nodes was not applicable at all, as the probability distribution in the target node did not fit to the truncated normal distribution implied by the algorithm.

In a second case study we compared different hill climbing algorithms. Therefore we provided a complete graph structure with 13 nodes in the Wal-Mart domain and

	Reservation	False			True		
	Patrons	None	Some	Full	None	Some	Full
WaitEstimate	0-10 mins	0.433	0.097	0.0070	0.718	0.33	0.056
	10-30 mins	0.464	0.439	0.131	0.264	0.506	0.362
	30-60 mins	0.099	0.394	0.493	0.018	0.155	0.464
	≥ 60 mins	0.0040	0.07	0.369	0.0	0.0090	0.118

Fig. 3. Screenshot of a conditional probability table generated using the Ranked-Nodes method

extracted a set of 11 constraints from given knowledge, e.g., the conditional probability $P(Eggs|Bread \wedge Milk) = 0.5$ from an exemplary subgroup pattern. Then we applied our constraint solver using first choice hill climbing (FC-HC), steepest ascent hill climbing (SA-HC), FC-HC with adaptive step size and FC-HC with optimum step-size search. The mean number of iterations and considered states of the test runs until constraint satisfying probability tables were found can be seen in the following table and in Figure 4.

Algorithm	Iterations	Considered States	States/Iteration
SA-HC	124	9672	78
FC-HC	324	2048	6.3
Adapt-SS	263	1881	7.2
Opt-SS	31	489	15.8

Although SA-HC needed less iterations in comparison to FC-HC, it had to consider almost five times more states and thus was significantly slower. The adaptive step size variation accomplished slight improvements. As Figure 4 shows, the adaption progress starts quite fast, but degenerates to the speed of the unmodified FC-HC, when the step size is determined by the specified minimum step size due to low global errors. For this task, the hill climbing algorithm with search for the optimum step size performed best. We explain this by the fact, that – compared to the other algorithms – this variation can modify parameters by a larger step size in one iteration.

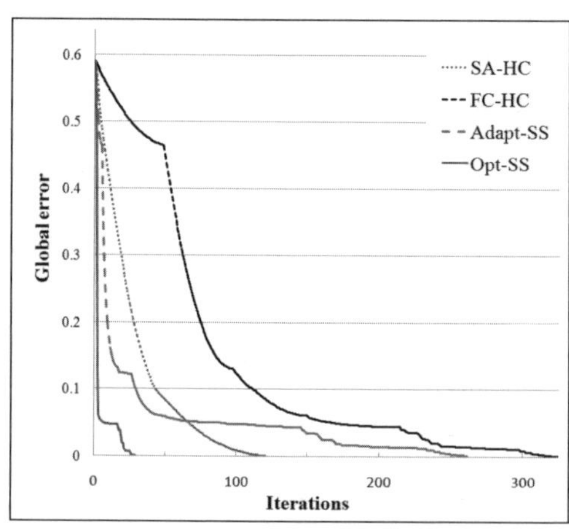

Fig. 4. Global error after iterations for different hill climbing algorithms

In our third case study we tested the integration of set-covering models and evaluated the efficiency of the implemented constraint satisfaction solver in a larger example. To do this we built a Bayesian network from an imported set of 30 set-covering relations. The constructed graph consisted of three layers. The buttom layer contained all findings, the middle

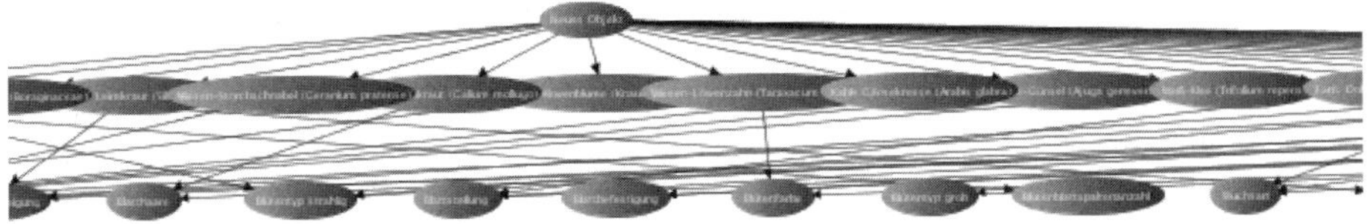

Fig. 5. Part of the generated graph structure of the third case study

layer all diagnoses. The top layer consisted of one single node, which was connected with all diagnoses to disallow multiple diagnoses to apply at the same time. For each set-covering relation we added edges from the diagnosis to all findings. Findings had up to 5 incoming edges which lead to probability tables with up to 32 entries. A part of the resulting graph can be seen in Figure 5. Starting with random table entries the hill climbing algorithm with search for optimal step size found a global optimum very fast, needing less than 100 iterations on average. We assume, that the simple, layered structure of the network is very appropriate for the applied algorithm.

7   Conclusions

In this paper, we have presented an effective approach for modeling Bayesian networks that is especially appealing for inexperienced users. We have proposed an incremental process model that includes all necessary steps from the initial setup to the evaluation. Furthermore, we have discussed techniques for determining the graph structure of the network. Additionally, we have presented novel methods for a simplified construction of the conditional probability tables, i.e., an extension of the common Ranked-Nodes approach and techniques for the constraint-based integration of domain knowledge. The effectiveness and benefit of the approach was demonstrated by three case studies.

In the future, we are planning to consider additional methods for including domain knowledge into the modeling process in order to further increase the effectiveness of the approach. Another interesting option for future work is given by specialized refinement methods with respect to the structure of the Bayesian network.

References

1. Russell, S., Norvig, S.: Artificial Intelligence: A Modern Approach, 2nd edn. Prentice–Hall, Englewood Cliffs (2003)
2. Wrobel, S.: An Algorithm for Multi-Relational Discovery of Subgroups. In: Proc. 1st Europ. Symp. Principles of Data Mining and Knowledge Discovery, pp. 78–87. Springer, Berlin (1997)
3. Puppe, F.: Knowledge Reuse among Diagnostic Problem-Solving Methods in the Shell-Kit D3. Intl. Journal of Human-Computer Studies 49, 627–649 (1998)
4. van der Gaag, L.C., Helsper, E.M.: Experiences with Modelling Issues in Building Probabilistic Networks. In: Gómez-Pérez, A., Benjamins, V.R. (eds.) EKAW 2002. LNCS (LNAI), vol. 2473, pp. 21–26. Springer, Heidelberg (2002)

5. Koller, D., Pfeffer, A.: Object–Oriented Bayesian Networks. In: Proceedings of the Thirteenth Conference on Uncertainty in Artificial Intelligence (UAI 1997), pp. 302–313 (1997)
6. Neil, M., Fenton, N., Nielsen, L.: Building Large-Scale Bayesian Networks. Knowledge Engineering Review (1999)
7. Fenton, N., Neil, M.: Ranked Nodes: A Simple and Effective Way to Model Qualitative Judgements in Large–Scale Bayesian Nets. IEEE Transactions on Knowledge and Data Engineering (2005)
8. Lucas, P.: Bayesian Network Modelling through Qualitative Patterns. Artificial Intelligence 163(2), 233–263 (2005)
9. Helsper, E., van der Gaag, L., Groenendaal, F.: Designing a Procedure for the Acquisition of Probability Constraints for Bayesian Networks. In: Motta, E., Shadbolt, N.R., Stutt, A., Gibbins, N. (eds.) EKAW 2004. LNCS (LNAI), vol. 3257, pp. 280–292. Springer, Heidelberg (2004)

Plan Repair in Hybrid Planning

Julien Bidot*, Bernd Schattenberg, and Susanne Biundo

Institute for Artificial Intelligence
Ulm University, Germany
`firstname.lastname@uni-ulm.de`

Abstract. We present a domain-independent approach to plan repair in a formal
framework for hybrid planning. It exploits the generation process of the failed plan
by retracting decisions that led to the failed plan fragments. They are selectively
replaced by suitable alternatives, and the repaired plan is completed by following
the previous generation process as close as possible. This way, a stable solution
is obtained, i.e. a repair of the failed plan that causes minimal perturbation.

1 Introduction

For many real-world domains, hybrid approaches that integrate hierarchical task decomposition with action-based planning turned out to be most appropriate [1]. On the one
hand, human expert knowledge can be represented and exploited by means of tasks and
methods, which describe how abstract tasks can be decomposed into pre-defined plans
that accomplish them. On the other hand, flexibility to come up with non-standard solutions, to overcome incompleteness of the explicitly defined solution space, or to deal
with unexpected changes in the environment results from the option to insert tasks and
primitive actions like in partial-order-causal-link planning (POCL). Furthermore, hybrid
planning enables the generation of abstract solutions as well as of plans whose prefixes
provide courses of primitive actions for execution, while other parts remain still abstract, ready for a refinement in later stages. With these capabilities hybrid approaches
meet essential requirements complex real-world applications, such as mission or project
planning, impose on AI planning technology. However, the problem of how to deal with
execution failures in this context has not been considered in detail yet. In general, there
are various alternatives for achieving this including contingency planning, replanning,
and repair.

In this paper, we introduce an approach to plan repair in hybrid planning. The motivation is twofold. Firstly, real-world planning problems often result in complex plans
(task networks) with a large number of causal, temporal, and hierarchical dependencies among the tasks involved. In order to keep the plans manageable, constructs such
as conditionals can therefore only sparingly be used and need to be left for those plan
sections which are most likely affected by uncertainty during execution. Replanning is
not appropriate in this context either. Given the complexity of the planning domain, it
is not reasonable to build a new plan from scratch in order to address just a single and

* This work has been supported by a grant of Ministry of Science, Research, and the Arts of
Baden-Württemberg (Az: 23-7532.24-14-1).

A. Dengel et al. (Eds.): KI 2008, LNAI 5243, pp. 169–176, 2008.

exceptional execution failure. Secondly, plan stability is essential in this context. This means, the modified plan should be as similar to the failed plan as possible and should only differ at positions that definitely need to be altered to compensate for the failure. The reason is that large parts of the plan may be unaffected by the failure, some parts may have been executed already, others may require commitments in terms of resource allocations or activities from third parties that have already been requested or even carried out. In a word, perturbation of the plan at hand should be minimized in order to avoid any amount of unnecessary cancellation actions and confusion.

Most plan repair methods known from the literature (cf. Section 5) take aim at non-hierarchical plans and use local search techniques to modify a failed plan by removing and inserting plan steps. In contrast to these approaches, our plan repair method covers hierarchical task decomposition and relies on the plan generation process instead of operating on the failed plan itself.

We present a general *refinement-retraction-and-repair algorithm* that revises the plan generation process of the failed plan by first retracting those development steps that led to the failed plan fragments. In a second step, the repaired plan is produced by following the previous generation process as close as possible. This means to replace the failure-critical development steps by alternatives and redo the uncritical ones. The rationale behind this procedure is as follows: it is realistic to assume that all initial goals persist and that the underlying domain model is adequate and stable. Thus, an execution failure is taken as caused by exceptional conditions in the environment. Nevertheless, such a failure can disturb several dependencies within the complex plan structure and may require revisions that go beyond adding and removing plan steps and include even the decomposition of abstract tasks. Information gathered during the original plan generation process is reused to avoid decisions that inserted the failure-critical plan elements and explore the possible alternatives instead, thereby enabling an efficient and minimally invasive plan repair.

In the following, we first describe the hybrid planning framework. We then introduce the plan repair problem and provide some extensions to the formal framework that allow us to explicitly address failure-affected plan elements. After that, the generic plan repair algorithm is presented. It is based on a hybrid planning system that records information about plan generation processes. The plan repair algorithm exploits information about the generation of the original plan and uses least-discrepancy heuristics to efficiently search for stable repair plans. Finally, we review related work and conclude with some remarks.

2   A Hybrid-Planning Framework

The hybrid planning framework is based on an ADL-like representation of states and *primitive tasks*, which correspond to executable basic actions. States as well as preconditions and effects of tasks are specified through formulae of a fragment of first-order logic. *Abstract tasks*, which show preconditions and effects as well, and *decomposition methods* are the means to represent human expert knowledge in the planning domain model. Methods provide *task networks*, also called *partial plans* that describe how the corresponding task can be solved. Partial plans may contain abstract and primitive tasks.

With that, hierarchies of tasks and associated methods can be used to encode the various ways a complex abstract task can be accomplished. Besides the option to define and exploit a reservoir of pre-defined solutions this way, hybrid planning offers first-principles planning based on both primitive and abstract actions.

Formally, a domain model $D = \langle \mathtt{T}, \mathtt{M} \rangle$ consists of a set of task schemata $\mathtt{T}$ and a set $\mathtt{M}$ of decomposition methods. A partial plan is a tuple $P = \langle TE, \prec, VC, CL \rangle$ where TE is a set of *task expressions* (plan steps) $te = l{:}t(\overline{\tau})$ with t being the task name and $\overline{\tau} = \tau_1, \ldots, \tau_n$ the task parameters; the label l serves to uniquely identify the steps in a plan. $\prec$ is a set of *ordering constraints* that impose a partial order on plan steps in TE. VC are *variable constraints*, i.e. co-designation and non-co-designation constraints $v\doteq\tau$ resp. $v\neq\tau$ on task parameters. Finally, CL is a set of *causal links* $\langle te_i, \phi, te_j \rangle$ indicating that formula ϕ, which is an effect of te_i establishes (a part of) the precondition of te_j. Like in POCL, causal links are the means to establish and maintain causal relationships among the tasks in a partial plan.

A *planning problem* $\pi = \langle D, P_{\text{init}} \rangle$ consists of a domain model D and an initial task network P_{init}. Please note that purely action-based planning problems given by state descriptions s_{init} and s_{goal}, like in POCL planning are represented by using distinguished task expressions te_{init} and te_{goal}, where s_{init} are the effects of te_{init} and te_{goal} has preconditions s_{goal}. The solution of a planning problem is obtained by transforming the initial task stepwise into a partial plan P that meets the following *solution criteria*: (1) all preconditions of the tasks in P are supported by a causal link, i.e. for each precondition ϕ of a task te_j there exists a task te_i and a causal link $\langle te_i, \phi, te_j \rangle$ in P; (2) the ordering and variable constraints of P are consistent; (3) the ordering and variable constraints ensure that none of the causal links is threatened, i.e. for each causal link $\langle te_i, \phi, te_j \rangle$ and each plan step te_k that destroys the precondition ϕ of te_j, the ordering constraints $te_i \prec te_k$ and $te_k \prec te_j$ are inconsistent with the ordering constraints $\prec$. If all task expressions of P are primitive in addition, P is called an *executable solution*. The transformation of partial plans is done using so-called *plan modifications*, also called *refinements*. Given a partial plan $P = \langle TE, \prec, VC, CL \rangle$ and domain model D, a plan modification is defined as $\mathtt{m} = \langle E^{\oplus}, E^{\ominus} \rangle$, where $E^{\oplus}$ and $E^{\ominus}$ are disjoint sets of elementary additions and deletions of so-called *plan elements* over P and D. The members of $E^{\ominus}$ are elements of P, i.e. elements of TE, $\prec$, VC or CL, respectively. $E^{\oplus}$ consists of new plan elements, i.e. task expressions, causal links, ordering or variable constraints that have to be inserted in order to refine P towards a solution. This generic definition makes the changes explicit that a modification imposes on a plan. With that, a planning strategy, which has to choose among all applicable modifications, is able to compare the available options qualitatively and quantitatively [1,2]. The application of a modification $\mathtt{m} = \langle E^{\oplus}, E^{\ominus} \rangle$ to a plan P returns a plan P' that is obtained from P by adding all elements in $E^{\oplus}$ to P and removing those of $E^{\ominus}$. We distinguish various classes of plan modifications.

For a partial plan P that has been developed from the initial task network of a planning problem, but is not yet a solution, so-called *flaws* are used to make the violations of the solution criteria explicit. Flaws list those plan elements that constitute deficiencies of the partial plan. We distinguish various flaw classes including the ones for unsupported preconditions of tasks and inconsistencies of variable and ordering constraints.

It is obvious that particular classes of modifications are appropriate to address particular classes of flaws while others are not. This relationship is explicitly represented by a *modification trigger function* α, which is used in the algorithm presented in [3], that relates flaw classes to suitable modification classes.

The resolution procedure is as follows: 1) the flaws of the current plan are collected; 2) relevant modifications are applied (generation of new plans); 3) the next plan to refine is selected, and we go to 1. This loop repeats, until a flawless plan is found and returned.

3   Formal Representation of the Plan-Repair Problem

After a (partly) executable solution for a planning problem has been obtained, it is supposed to be given to an execution agent that carries out the specified plan. We assume that a dedicated *monitoring component* keeps track of the agent's and the world's state and notices deviations from the plan as well as unexpected behaviour of the environment. It is capable of recognizing failure situations that cause the plan to be no longer executable: actions have not been executed, properties that are required by future actions do not hold as expected, etc. The monitor maps the observation onto the plan data structure, thereby identifying the *failure-affected elements* in the plan. This includes the break-down of causal links, variable constraints that cannot be satisfied, and the like. Note that this failure assessment is not limited to the immediately following actions in the executed plan fragments, as the plan's causal structure allows us to infer and anticipate causal complications for actions to be executed far in the future.

With such a failure description at hand, the system tries to find an alternative fail-safe plan for the previously solved problem, taking into account what has already been executed and is, therefore, viewed as *non-retractable decisions*. Remember that we assume the planning domain model to be valid and the failure event to be an exceptional incident. We hence translate the execution failure episode into a *plan-repair problem* that we will solve relying on the same domain model that we used for the previous problem.

As a prerequisite, we define the notions of failure descriptions and non-retractable decisions. In order to express exceptional incidents we augment a domain model with a set of particular primitive task schemata that represent non-controllable environmental processes: for every fluent, i.e. for every property that occurs in the effects of a task, we introduce two *process schemata* that invert the truth value of the respective fluent. Every failure-affected causal link will impose the plan-repair problem to contain a process, the effect of which falsifies the annotated condition. We assume that failures can be unambiguously described this way – if a disjunctive failure cause is monitored, the problem has to be translated into a set of alternative plan-repair problems, which is beyond the scope of this paper.

Regarding the representation of non-retractable decisions, every executed plan element, i.e. either an executed task or a constraint referring to such, has to occur in the repair plan. To this end, we make use of meta-variables of plan elements such as task expressions over given schemata in D, causal links between two task expression meta-variables, and ordering or variable constraints on appropriate meta-variables. Meta-variables are related with actual plan elements by *obligation constraints*, which serve in this way as a structural plan specification. We thereby distinguish two types of constraints: existence obligations, which introduce meta-variables, and assignment obligations, which are equations on meta-variables and plan elements. A set of obligation constraints OC is *satisfied* w.r.t. a plan P, if (1) for every existence obligation in OC, a respective assignment obligation is included as well and (2) the assignment obligations instantiate a plan schema that is consistent with P.

Given a planning problem $\pi = \langle D, P_{\text{init}} \rangle$ and a plan $P_{\text{fail}} = \langle TE_{\text{fail}}, \prec_{\text{fail}}, VC_{\text{fail}}, CL_{\text{fail}} \rangle$ that is a partially executed solution to π that failed during execution, a plan-repair problem is given by $\pi^r = \langle D^r, P^r_{\text{init}} \rangle$ that consists of the following components: D^r is an extended domain model specification $\langle \text{T}, \text{M}, \text{P} \rangle$ that concurs with D on the task schemata and method definitions and provides a set of process schemata P in addition. The initial plan $P^r_{\text{init}} = \langle TE_{\text{init}}, \prec_{\text{init}}, VC_{\text{init}}, CL_{\text{init}}, OC_{\text{init}} \rangle$ of the plan-repair problem is an obligation-extended plan data structure that is obtained from P_{init} and P_{fail} as follows. The task expressions (incl. te_{init} and te_{goal}), ordering constraints, variable constraints, and causal links are replicated in P^r_{init}. The obligation constraints OC_{init} contain an existence obligation for every executed plan element in P_{fail}. Finally, if broken causal links have been monitored, appropriate existence obligations for process meta-variables are added to OC_{init} and causally linked to the intended link producers. E.g., let $\langle te_a, \varphi, te_b \rangle$ be a broken causal link, then OC_{init} contains existence obligations for the following meta-variables: v_a stands for a task expression of the same schema as te_a, v_φ represents a process that has a precondition φ and an effect $\neg\varphi$, and finally $v_{CL} = \langle te_a, \varphi, te_\varphi \rangle$.

An obligation-extended plan $P^r = \langle TE, \prec, VC, CL, OC \rangle$ is a solution to a plan-repair problem $\pi^r = \langle D^r, P^r_{\text{init}} \rangle$, if and only if the following conditions hold: (1) $\langle TE, \prec, VC, CL \rangle$ is a solution to the problem $\langle D^r, P_{\text{init}} \rangle$ where P_{init} is obtained from P^r_{init} by removing the obligations. (2) The obligations in P^r are satisfied. (3) For every ordering constraint $te_a \prec te_b$ in $\prec$, either OC contains an assignment obligation for te_a or none for te_b.

Solution criterion (1) requires the repaired plan to be a solution in the sense of Section 2, and in particular implies that P^r is an executable solution to the original problem as well. (2) ensures that the non-retractable decisions of the failed plan are respected and that the environment's anomaly is considered. Criterion (3) finally verifies that the repair plan stays consistent w.r.t. execution; i.e., in the partial order, not yet executed tasks do not occur before executed ones.

The problem specification and solution criteria imply appropriate flaw and plan modification classes for announcing unsatisfied obligations and documenting obligation decisions. A specific modification generation function is used for introducing the appropriate processes. With these definitions, the repair mechanism is properly integrated into the general hybrid planning framework and a simple replanning procedure can easily be realized: The obligation-aware functions are added to the respective function sets in the algorithm presented in [3], which in turn is called by **simplePlan**$(P^r_{\text{init}}, \pi^r)$. Note that the obligation extension of the plans is transparent to the framework. With the appropriate flaw and modification generators, the algorithm will return a solution to the repair problem, and while the new plan is developed, the obligation assignments will be introduced opportunistically. Apparently, the solution obtained by replanning

Require: Planning problem $\pi^r = \langle D^r, P^r_{\mathrm{init}} \rangle$ and plan P_{fail}

```
 1: planRepair(P_fail, π^r, M):
 2:   P_rh ← retract(P_fail, M) {Retract modifications}
 3:   while P_rh ≠ fail do
 4:     P_safe ← autoRefine(P_rh, M) {Remake modifications}
 5:     P_new ← simplePlan(P_safe, π^r) {Call the algorithm of [3]}
 6:     if P_new ≠ fail then
 7:       return P_new
 8:     else
 9:       P_rh ← retract(P_new, M) {Retract modifications}
10: return fail
```

Algorithm 1. The refinement-retraction-and-repair algorithm

cannot be expected to be close to the previous solution. Furthermore, all knowledge of the previous successfull plan generation episode and in particular that concerning the un-affected portions of the plan, is lost. The following section will address these issues.

4   The Plan-Repair Algorithm

In this section, we describe a repair algorithm that allows us to find efficiently an executable solution plan to π^r, which is as close to a failed one (P_{fail}) as possible, by reusing the search space already explored to find P_{fail}.

In contrast to replanning, our plan-repair procedure starts from P_{fail}. Our objective w.r.t. search is to minimize the number of modifications to retract, remake, and newly add (minimal invasiveness). To achieve this efficiently we focus on the modifications M that our planning system applied to obtain P_{fail} from the initial planning problem. We thus partition the modifications M into the following two subsequences: $\mathtt{M_{fail}}$ is the sequence of plan modifications that are related to the failure-affected elements of P_{fail}; $\mathtt{M_{rest}} = \mathtt{M} \setminus \mathtt{M_{fail}}$ is the sequence of plan modifications that are not related to the failure-affected elements of P_{fail}. During our refinement-retraction-and-repair procedure, all modifications of $\mathtt{M_{fail}}$ have to be retracted from P_{fail}.

Once an execution-time failure is detected by our monitoring system, we repair the current failed plan as described by Algorithm 1. It first retracts hybrid planning modifications within the *retraction horizon* (line 2). The retraction horizon corresponds to the first partial plan P_{rh} that is encountered during retraction of modifications in M in which no failure-affected elements occur. The retraction step chooses the modifications to retract given M. At this step of the algorithm the retraction horizon comprises all the modifications made from the last one inserted in M until the first modification in M that is related to one or more failure-affected plan elements. In line 3, we have a consistent partial plan, which we name P_{rh}. The next step of Algorithm 1 consists in refining the current partial plan (P_{rh}) automatically by remaking every possible modifications in $\mathtt{M_{rest}}$ (line 4); we have then a consistent partial plan that we name P_{safe}. We then call line 5 algorithm $\mathtt{simplePlan}$ presented in [3] to refine P_{safe} into an executable solution P_{new}. If no consistent plan is found, then the algorithm retracts modifications (line 9). Each time function $\mathtt{retract}$ is called (line 9) after backtrackings of $\mathtt{simplePlan}$, the retraction horizon includes one more modification included in $\mathtt{M_{rest}}$. Each time function $\mathtt{autoRefine}$ is called after backtrackings of $\mathtt{simplePlan}$ (line 4), we remake one less modification of $\mathtt{M_{rest}}$; i.e., we do not remake the modifications of $\mathtt{M_{rest}}$ that are beyond the original retraction horizon. This algorithm runs iteratively until a plan (P_{new}) without flaws is found (line 7), or when no modification can be retracted any longer (line 9). The retraction step (lines 2 and 9) has a linear cost, which depends on the number of modifications to retract. Note that P_{fail}, P_{rh}, P_{safe}, and P_{new} represent each a repair plan P^r, i.e. a partial plan with obligation constraints.

During the retraction of modifications related to committed plan elements, the corresponding assignment obligations are no longer satisfied. During the refinement phase of Algorithm 1 (line 4), obligation flaws related to execution-time failures are addressed by automatically inserting environment processes.

There is no combinatorial search from P_{fail} to P_{safe} (Algorithm 1, lines 2-4).

Our retraction-refinement search algorithm can be extended to find a new plan P_{new} that is similar to P_{fail}. For achieving this, our procedure uses least-discrepancy heuristics when refining P_{safe} (Algorithm 1, line 5). The heuristics guide search with respect to the modifications that have led to plan P_{fail}: function f^{select} prefers the modifications that are similar to the modifications in $\mathtt{M_{rest}}$; i.e., the modification-selection strategy prefers modifications that belong to the same class and that correspond to the same flaw class.

5 Related Work

Plan repair is rarely addressed in the context of hierarchical planning. However, there is a number of studies in the field that are relevant to this objective.

Fox et al. [4] demonstrate with an empirical study that a plan-repair strategy can produce more stable plans than those produced by replanning, and their system can produce repaired plans more efficiently than replanning. Their implementation uses local-search techniques and plan-oriented stability metrics to guide search. To what extent our approach may benefit from using such metrics is subject to future work.

The planning system of Yoon et al. [5] computes a totally-ordered plan before execution; each time it is confronted with an unexpected state, it replans from scratch without reusing previous planning effort.

Nebel and Köhler [6] report a theoretic and practical study about plan reuse and plan generation in a general situation. Using the non-hierarchical propositional STRIPS planning framework, the authors show that modifying a plan is not easier than planning from scratch. Recent work in hierarchical case-based planning includes the approach based on SHOP presented by Warfield et al. [7].

Kambhampati and Hendler [8] present some techniques for reusing old plans. To reuse an old plan in solving a new problem, the old plan, along with its annotations, has to be found in a library of plans and is then mapped into the new problem. A process of annotation verification is used to locate applicability failures and suggest refitting tasks.

Drabble et al. [9] propose plan-repair mechanisms to integrate several pre-assembled repair plans into an ongoing and executing plan when action effects fail for a limited number of reasons.

Van der Krogt and de Weerdt [10] endow a partial-order causal link planning system with capabilities of repairing plans. In this context, plan repair consists of adding and removing actions. During the execution of a plan, their monitoring system may observe uncontrollable changes in the set of facts describing a state, which makes the plan fail.

6 Conclusion

We presented a novel approach to plan repair that is capable of dealing with hierarchical and partial-order plans. Embedded in a well-founded hybrid planning framework, it uses the plan generation process of the failed plan to construct a repair. In a first phase, all plan refinements that introduced failure-affected plan elements – possibly including even task decompositions – are retracted from the failed plan. After that, additional obligation constraints are inserted into the resulting partial plan in order to ensure that already executed plan steps will be respected by the repair plan. The repair procedure constructs a solution by replacing the failure-critical plan modifications and replaying as many of the uncritical ones as possible, thereby achieving stability of the repair plan. The flexibility of our hybrid planning framework, where plan deficiencies and modifications are explicitly defined, enables the repair of various plan and execution failures including both the removal of plan steps or entire plan fragments that became obsolete and the insertion of extra tasks coming up.

References

1. Biundo, S., Schattenberg, B.: From abstract crisis to concrete relief–a preliminary report on combining state abstraction and HTN planning. In: Proc. of ECP 2001, pp. 157–168 (2001)
2. Schattenberg, B., Bidot, J., Biundo, S.: On the construction and evaluation of flexible plan-refinement strategies. In: Hertzberg, J., Beetz, M., Englert, R. (eds.) KI 2007. LNCS (LNAI), vol. 4667, pp. 367–381. Springer, Heidelberg (2007)
3. Schattenberg, B., Weigl, A., Biundo, S.: Hybrid planning using flexible strategies. In: Furbach, U. (ed.) KI 2005. LNCS (LNAI), vol. 3698, pp. 258–272. Springer, Heidelberg (2005)
4. Fox, M., Gerevini, A., Long, D., Serina, I.: Plan stability: Replanning versus plan repair. In: Proc. of ICAPS 2006, pp. 212–221 (2006)
5. Yoon, S., Fern, A., Givan, R.: FF-Replan: A baseline for probabilistic planning. In: Proc. of ICAPS 2006, pp. 352–359 (2006)
6. Nebel, B., Köhler, J.: Plan reuse versus plan generation: A theoretical and empirical analysis. Artificial Intelligence 76, 427–454 (1995)
7. Warfield, I., Hogg, C., Lee-Urban, S., Muñoz-Avila, H.: Adaptation of hierarchical task network plans. In: FLAIRS 2007, pp. 429–434 (2007)
8. Kambhampati, S., Hendler, J.A.: A validation-structure-based theory of plan modification and reuse. Artificial Intelligence 55, 193–258 (1992)
9. Drabble, B., Tate, A., Dalton, J.: Repairing plans on-the-fly. In: Proceedings of the NASA Workshop on Planning and Scheduling for Space (1997)
10. van der Krogt, R., de Weerdt, M.: Plan repair as an extension of planning. In: Proc. ICAPS 2005, pp. 161–170 (2005)

Visual Terrain Traversability Estimation Using a Combined Slope/Elevation Model

Tim Braun, Henning Bitsch, and Karsten Berns

University of Kaiserslautern

Abstract. A stereo vision based terrain traversability estimation method for offroad mobile robots is presented. The method models surrounding terrain using either sloped planes or a digital elevation model, based on the availability of suitable input data. This combination of two surface modeling techniques increases range and information content of the resulting terrain map.

1   Introduction

Autonomous navigation in unstructured offroad terrain has a whole range of interesting applications, but still remains a challenging and largely unsolved problem for robotic vehicles. A key capability within this domain is the estimation of *terrain traversability*, the analysis of sensor data in order to determine how well surrounding terrain can be crossed. Unfortunately, this task is very difficult, as 'traversability' is a complex function of the vehicles' own mechanical capabilities, surface geometry and terrain type.

Outdoor traversability estimation usually starts with the construction of a three-dimensional point cloud outlining the terrain surface. Then, the point cloud is abstracted into a model that allows reasoning about traversability. A popular model is the digital elevation map (DEM) [1]. DEMs are grid maps storing the overall terrain height of a certain metrical patch (e.g. 1m x 1m). By considering height differences between neighboring cells, traversability scores can be computed. However, details such as the 'real' slope of the patch are lost and a distinction between patches containing abrupt steps and those containing continuous slopes becomes impossible. A more accurate terrain model can be produced by fitting planes [2] or triangular meshes. Unfortunately, this requires more points to work reliably than the construction of a DEM patch. Therefore, this model has been predominantly applied to indoor applications requiring shorter ranges and thus obtaining higher density point clouds.

In this paper, we propose a method for terrain traversability estimation in outdoor, offroad environments that adaptively combines both terrain model types. More precisely, areas exhibiting a high point density are described using fitted planes that retain the information about predominant slopes. In areas unsuitable for planar approximation (due to a lack of points or non-planar surfaces), the method falls back to elevation map modeling. We show that this strategy increases the range of a vision-based traversability estimation system compared to

A. Dengel et al. (Eds.): KI 2008, LNAI 5243, pp. 177–184, 2008.

a pure plane-fitting approach. At the same time, it models nearby terrain more accurately than a pure DEM approach, allowing to estimate terrain traversability more precisely.

The remaining paper is structured as follows: Section 2 presents related approaches for terrain traversability estimation. Section 3 then presents our proposed method, whose performance is evaluated in section 4 based on simulations and real experiments. Section 5 summarizes the results and provides an outlook.

2   Related Work

Established techniques for outdoor traversability estimation typically measure the terrain layout using either 3D LIDAR range sensors [3] or stereo systems [4] [5]. After data acquisition, derived properties such as height variances [4] or classified terrain type [5] are used for traversability estimation. While LIDAR systems can quickly produce accurate 3D point clouds distributed along the terrain surface, they do not provide visual information useful for terrain classification such as terrain color or texture. In contrast, stereo cameras capturing high resolution color images can support appearance-based terrain type inference. However, these systems have to derive the distance information for 3D point cloud generation algorithmically, which increases computational load and limits range accuracy. Even worse, stereo reconstruction depends heavily on suitable input data - on problematic image areas (e.g. having low contrast), stereo matching may fail. In order to combine their benefits, several researchers have proposed to use both sensor types simultaneously and perform data fusion [1]. However, consistency over long ranges requires an extremely accurate mutual sensor registration and a mechanically rigid construction to prevent misalignments caused by vibrations during robot operation. Thus, sensor fusion has not been widely adapted for sensor systems with moving part such as pan/tilt units.

3   Proposed Method

Our proposed method for terrain traversability estimation is based on a panning stereo camera system as exclusive information source. As outlined in section 2, this *allows later extension toward terrain type classification* based on visual appearance (e.g. using texture, feature or color analysis). Also, the use of a single camera system avoids registration issues that plague sensor fusion approaches, while the panning degree of freedom can be used to increase the field of view.

Since image acquisition is not the focus of this paper, we assume that a strategy suitable for capturing well-exposed images in outdoor settings (CMOS sensors, HDR image acquisition) is used. Furthermore, we also assume the presence of a canonical stereo system with horizontal epipolar lines. In practice, an offline calibration is performed. The algorithm can be subdivided into the three sucessive steps **Point Cloud Generation**, **Terrain Modeling** and **Traversabiliy Analysis**. Each will now be discussed in sequence.

3.1 Point Cloud Generation

Initially, a 3D point cloud outlining the terrain surface is built using a simple sparse stereo approach. As first step, well localizable image points are identified in both camera images using the 'good features to track' algorithm [6]. Then, left and right image interest points are matched with each other using a highly efficient pyramidal Lukas-Kanade feature tracker [7]. Bad matches are discarded.

We implement an additional filtering step for the remaining point pairs to eliminate outliers. For a point pair $p = (x, y)$, $p' = (x', y')$ with p coming from the left and p' from the right image, the following two criteria must be fulfilled: 1. $d_{min} < x - x' < d_{max}$, 2. $\left| \arctan \frac{y-y'}{x-x'} \right| < \delta$. d_{min} and d_{max} denote minimal and maximal acceptable disparities and are selected according to the spatial range interesting for traversability analysis. δ determines the tolerated vertical skew between matched points. Although this value is theoretically 0 for perfectly calibrated systems, experiments have shown that it should be chosen greater to accommodate slight misalignments that creep in during robot operation. After valid interest point pairs (p_i, p'_i) have been determined, they are projected onto points $P_i = (X_i, Y_i, Z_i)$ in three-dimensional space using standard stereo reconstruction techniques. The 3D coordinate system is chosen so that the z-axis is parallel to the gravitational vector and the x-axis points into the direction of the robot's heading.

As a result of the sparse stereo approach and the additional filtering, the amount of incorrect points in the 3D point cloud is substantially reduced. However, uniformly colored image areas such as smooth roads are sampled much less frequently than vegetated terrain having high contrast. Luckily, the target offroad environment exhibits few objects with low contrast and these objects can typically be approximated well using height planes, which require only few points for construction (see section 3.2).

3.2 Terrain Modeling

In order to allow piecewise modeling of the surrounding terrain, the obtained point cloud must be segmented spatially. One common choice is to use segments having similiar areas on the ground plane, resulting in a grid map structure (fig. 1a). Another possibility is a *camera-centric* approach which segments the point cloud into parts that project onto equal areas (given perfectly flat terrain) on the *input images* (fig. 1c). Both choices have drawbacks. While the environment-centric segmentation approximates the terrain with constant resolution, grid cells further away are covered by a very small part of the image and can thus only be described by few points. In contrast, expected point density is uniform for camera-centric segmentations, but the metrical footprint of the segments increases dramatically with distance, making a planar approximation unrealistic for larger ranges.

To strike a viable compromise, we propose a mixture of environment and camera-centric segmentation as outlined in fig. 1b. Using a cylindrical coordinate system originating at the robot, space is partitioned into $2\pi/\phi$ **sectors** S with

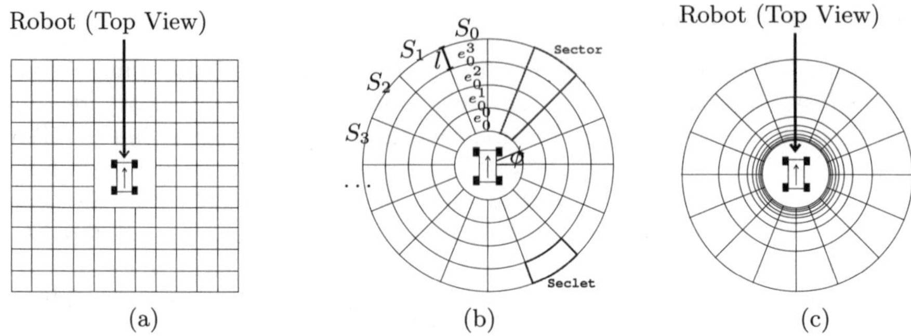

Fig. 1. (a) Environment-Centric Map, (b) Seclet Map, (c) Camera-Centric Map

angular extend ϕ. Each sector S_i is further subdivided into **seclets** $e_i^0 \ldots e_i^n$ of metrical length l. This segmentation ensures that each seclet has at least a constant projected width in the image, while simultaneously the growth in metrical area with increasing distance is moderated. Also, traversability in a given direction γ can later be computed easily by sequential inspection of all seclets in the sector S_i with $i\phi \leq \gamma < (i+1)\phi$ (see sec. 3.3).

After assigning the reconstructed 3D points to their corresponding seclets, each seclet is considered in turn. Seclets with a point set $M = \{P_0, \ldots, P_n\}$ larger than a threshold κ_s are considered as candidates for approximation with a sloped planar surface. To guard against outliers, approximation is performed using a RANSAC-based plane fitting approach extensively described in [2]. In short, this approach proceeds by constructing plane candidates from three randomly picked points of M. Then, all plane-point distances of the points in M are computed. Points with a distance smaller than a threshold d_{plane} are considered **inliers** and added to the inlier set R, all other points are classified as outliers. After several repetitions of this process, the plane supported by the most inliers is selected. Finally, its pose is recomputed using a least-squares fitting of *all* inliers.

With the least-square fitted plane available, it is checked whether a planar approximation is reasonable at all given the seclet's contained points. For this, the ratio between the number of RANSAC inliers and the cardinality of M is calculated. If the ratio falls below a level r_{min}, it is assumed that the seclet's point set actually describes a distinctively non-planar surface and a linear representation would be overly simplistic. Consequently, the sloped plane is discarded.

If a seclet was not modeled using a sloped plane due to either the $|M| > \kappa_s$ or the $|R|/|M| \geq r_{min}$ constraints, the algorithm falls back to using an elevation plane containing only a height value and no slope information (in this case, its plane normal is set to $(0, 0, 1)$). The elevation plane's height is chosen pessimistically as the *maximum* of all valid heights (z-values) in M, which is again determined using the RANSAC algorithm to guard against outliers. The elevation plane RANSAC model is one dimensional. One point P is randomly chosen from M, and its support is defined by all points P' satisfying $Z - d_{height} < Z' \leq Z$.

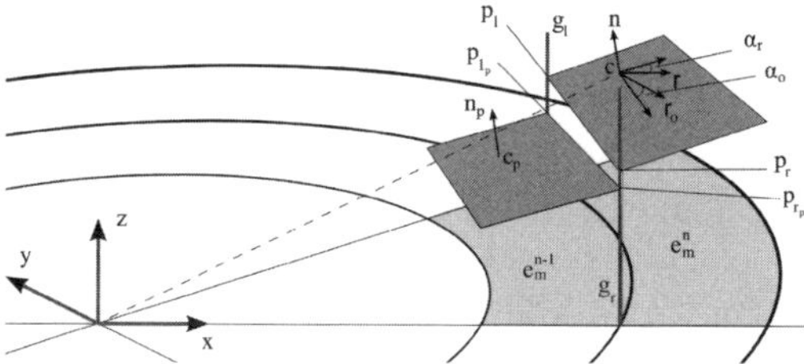

Fig. 2. Parameters for Traversability Analysis

Again, the point with the highest support is selected, and it's z coordinate determines the elevation plane's height. This approach effectively discards sporadic outliers that have high z-values without an underlying 'foundation' of additional points. Nevertheless, the elevation plane model also requires a minimum number of points to work reliably. Therefore, if fewer points than κ_e with $\kappa_e < \kappa_s$ are available, the seclet is completely excluded from terrain modeling.

3.3 Traversability Analysis

The result of the previous steps is a seclet map approximating the geometrical shape of the terrain with seclets containing either sloped planes, elevation planes or no plane information at all. Before traversability can be reasoned upon this data, some assumptions and notations need to be introduced. Firstly, all planes are considered load-bearing and are defined by their normal vector n and their center of gravity (cog) c lying inside the plane (see fig. 2). Secondly, the vehicles mechanical capabilities are defined by the maximum declination α_{max} and the maximum possible step h_{max} it can climb. Thirdly, we assume that the robot's possible driving options can be represented as straight lines from the center of the map towards the cog of a seclet lying in the desired direction. Thus, every map sector contains a single path to consider, and a seclet e_m^n has e_m^{n-1} as predecessor and e_m^{n+1} as successor seclet along this path. If both e_m^n and e_m^{n-1} contain valid planes described by (n, c) and (n_p, c_p), it is possible to compute transitional steps between the two seclets. This is accomplished by inspecting the z distance of the planes at the edges of the common seclet border. At each edge, a vertical straight line g_l and g_r is intersected with the two planes. The resulting intersection point pairs p_{l_p}, p_l and p_{r_p}, p_r provide two step heights.

Given these preconditions, a given seclet e_m^n is considered traversable, if the following conditions are satisfied:

$$\alpha_r \leq \alpha_{max} \wedge \alpha_o \leq \alpha_{max} \tag{1}$$

$$n_p \bullet n \leq \cos(\alpha_{max}) \tag{2}$$

$$\max(|p_{l_p} - p_l|, |p_{r_p} - p_r|) \leq h_{max} \tag{3}$$

Condition 1 ensures, that the maximum absolute inclination in direction of the path (α_r) and orthogonal to it (α_o) remains below α_{max}. The vectors r and r_o

(with $r \perp r_o$) are the projections of the path vector onto the x/y plane, and α_r and α_o represent the plane's declination. In case there is no usable predecessor e_m^{n-1} available, only this condition is considered. Otherwise, Condition 2 limits the allowed difference in orientation between seclets (especially in direction of r) and can be directly calculated using the inner product of n and n_p. Condition 3 is used to detect steps and torsion between two adjacent seclets and becomes especially important for traversability analysis if at least one plane is an elevation plane. As long as the maximum step heights are less or equal h_{max}, the seclets junction is considered traversable.

4   Experiments

To test the performance of the proposed terrain traversability estimation method, a series of test images with a resolution of 1600 x 1200 pixels generated either in a 3D simulation environment or by a real stereo system has been analyzed. For all experiments, the following parameters have been used: $d_{min} = 23$, $d_{max} = 255$, $\phi = 10°$, $l = 1m$, $n = 20$ for the seclet map, $\kappa_s = 35$, $\kappa_e = 10$, $d_{plane} = 15cm$, $r_{min} = 0.6$, $d_{plane} = 50cm$ for terrain modeling. During testing, the stereo system was mounted on a pan/tilt unit on top of the mobile robot RAVON [8]. Unfortunately, only an exisiting system with a very small baseline of $18cm$ and focal length of 1902 $pixels$ was available at the time of the experiments. This limits the usable range of the terrain analysis to about $15m$. At this distance, the depth resolution δz is equal to $0.6m/pixel$, but the sub-pixel accuracy of the used stereo algorithm still suffices to for plane generation.

4.1   Simulation

The first experiment was conducted in a simulated environment without optical inadequacies. In figure 3a the scene (containing two distinct obstacles) can be seen. For each seclet, the computed model plane is indicated by circles drawn around the center of gravity. Green planes are traversable, red marks impassable seclets. Figure 3b depicts the corresponding traversability map using the same

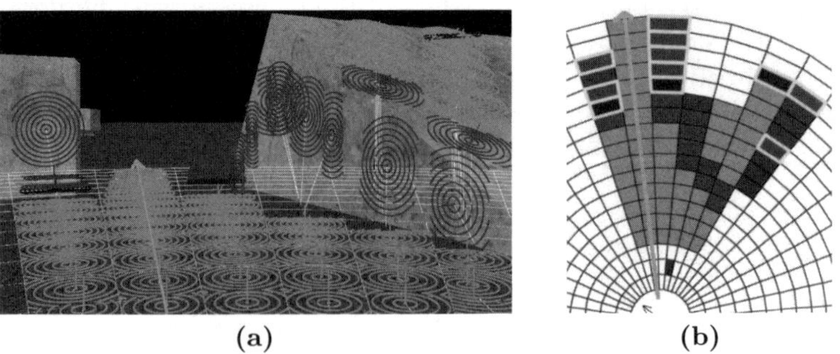

(a)   (b)

Fig. 3. Simulated Terrain Experiment

color code. The darker green/red areas with yellow borders indicate seclets with elevation instead of slope planes. White and dark grey seclets are not used, as they contain zero or less than κ_e points. Based on this map, the best traversable driving direction is marked by a cyan arrow.

As can been seen, most of the seclets provided sufficient points for slope plane generation, resulting in a good terrain approximation and an accurate traversability analysis. Most of the rocks top surface on the right side is marked traversable as the declination does not exceed α_{max}. Only the first top surface seclet in each sector covering the rock is considered impassable as its predecessor is vertical. As expected, elevation planes were primarily used further away from the cameras where fewer points were recorded.

4.2 Real Images

Figures 4a and 4c show two typical outdoor scenes. While the bush is captured quite well on the left side of figure 4a, the tree on the right side is erroneously placed too close to the robot due to imprecise calibration. However, the traversabiliy map shown in 4c can nevertheless be used to correctly determine a well traversable direction. Figure 4c depicts a scene with more complex geometry. This leads to the creation of several elevation planes in areas with non-planar surfaces, and again a correct overall traversability estimate is produced.

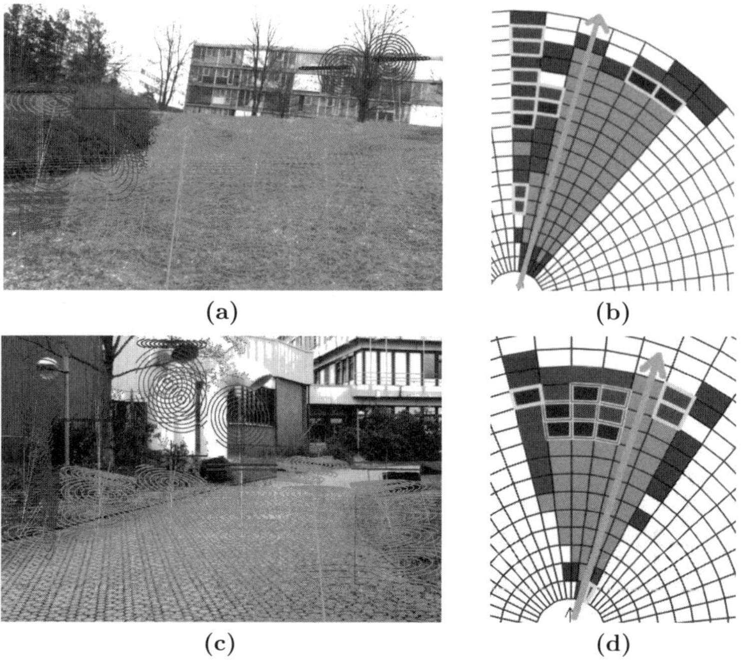

(a)   (b)

(c)   (d)

Fig. 4. Real Terrain Experiments

Algorithm runtime has not been focused during design or implementation, as traversability analysis is done while robot movement is suspended. Current execution times are about 15 seconds per image. About 80% of the time is spent during feature extraction and matching, while RANSAC computations and traversability analysis are negligible. Speed could therefore be substantially increased using GPU computation or code optimization.

5   Conclusions and Future Work

In this paper, an approach for terrain traversability estimation in unstructured offroad terrain was presented that combines outlier-robust elevation map modeling and planar fitting methods. A novel seclet map structure was proposed as a suitable spatial segmentation and algorithms for the generation of hybrid sloped/elevation plane seclets have been presented along with a set of constraints usable for binary traverability decisions. It was shown exemplarily that this approach can produce usable results in both simulated and real environments, and that the combination of sloped and elevation planes increases information content in the terrain model. Future work includes the usage of slope information to calculate continuous traversability scores, an increased baseline of the stereo system, the consideration of overhanging obstacles (tree branches) and the incorporation of a more sophisticated sparse stereo approach exploiting the epipolar constraint already during feature matching.

References

1. Cremean, L., et al.: Alice: An information-rich autonomous vehicle for high-speed desert navigation: Field reports. J. Robot. Syst. 23(9), 777–810 (2006)
2. Weingarten, J., Gruener, G., Siegwart, R.: A fast and robust 3d feature extraction algorithm for structured environment reconstruction. In: Proceedings of the 11th International Conference on Advanced Robotics (ICAR) (July 2003)
3. Vandapel, N., et al.: Natural terrain classification using 3-d ladar data. In: IEEE International Conference on Robotics and Automation (ICRA) (2004)
4. Howard, A., et al.: Towards learned traversability for robot navigation: From underfoot to the far field. Journal of Field Robotics 23 (2006)
5. Kim, D., Oh, S.M., Rehg, J.M.: Traversability classification for UGV navigation: A comparison of patch and superpixel representations. In: IROS 2007 (2007)
6. Shi, J., Tomasi, C.: Good features to track. In: IEEE Conference on Computer Vision and Pattern Recognition (1994)
7. Bouguet, J.Y.: Pyramidal implementation of the lucas kanade feature tracker: Description of the algorithm. Intel Corp., Microprocessor Research Labs (2002)
8. Schäfer, H., et al.: Ravon - a robust autonomous vehicle for off-road navigation. In: European Land Robot Trial (May 2006)

Symbolic Classification of General Two-Player Games[*]

Stefan Edelkamp and Peter Kissmann

Technische Universität Dortmund, Fakultät für Informatik
Otto-Hahn-Str. 14, D-44227 Dortmund, Germany

Abstract. In this paper we present a new symbolic algorithm for the classification, i. e. the calculation of the rewards for both players in case of optimal play, of two-player games with general rewards according to the Game Description Language. We will show that it classifies all states using a linear number of images concerning the depth of the game graph. We also present an extension that uses this algorithm to create symbolic endgame databases and then performs UCT to find an estimate for the classification of the game.

1 Introduction

In General Game Playing (GGP), games are not known beforehand. Thus, for writing an algorithm to play or solve general games no specialized knowledge about them can be used. One language to describe such games is the Game Description Language (GDL) [11]. Since 2005, an annual GGP Competition [9] takes place, which was last won 2007 by Yngvi Björnsson's and Hilmar Finnsson's CADIAPLAYER [6]. GDL is designed for the description of general games of full information satisfying the restrictions to be finite, discrete, and deterministic.

When the games end, all players receive a certain reward. This is an integer value within $\{0, \ldots, 100\}$ with 0 being the worst and 100 the optimal reward. Thus, each player will try to get a reward as high as possible (and maybe at the same time keep the opponent's reward as low as possible). The description is based on the Knowledge Interchange Format (KIF) [8], which is a logic based language. It gives formulas for the initial state (*init*), the goal states (*terminal*), the preconditions for the moves (*legal*), further preconditions to get certain effects (*next*), the rewards in each terminal state for each player (*goal*) and some domain axioms (constant functions).

In this paper, we focus on non-simultaneous two-player games. Opposed to the competition's winners [4,13,6], we are not mainly interested in playing the games, but in classifying (solving) them. That is, we want to get the rewards for both players in case of optimal play (optimal rewards) for each of the states reachable from the initial state. Thus, when the classification is done, we can exploit the information to obtain a perfect player. Our implementation for the classification of these games originally worked with a variation of the Planning Domain Definition Language (PDDL) [12] that we called GDDL [5]. In the meantime, we implemented an instantiator for the KIF files, which results in a format quite close to instantiated PDDL but with *multi-actions* to represent the moves. These multi-actions consist of a global precondition, which is the disjunction of all corresponding *legal* formulas, and several precondition/effect pairs where the

---

[*] Thanks to DFG for the financial support of the authors.

A. Dengel et al. (Eds.): KI 2008, LNAI 5243, pp. 185–192, 2008.

preconditions can be arbitrary formulas while the effects are atoms. These pairs are the result of instantiating the *next* formulas and are interpreted as follows. A multi-action consisting of the global precondition *global* and the precondition/effect pairs pre_1, eff_1, ..., pre_n, eff_n might also be written as $global \land (pre_1 \Leftrightarrow eff_1) \land \ldots \land (pre_n \Leftrightarrow eff_n)$. If *global* holds in the current state, we can take this action. It will create the current state's successor by applying the precondition/effect pairs. An effect has to hold in the successor iff the corresponding precondition holds in the current state. All variables not in any of the effects are set to *false*.

The algorithm to classify two-player turn-taking games with general rewards proposed in [5] is quite inefficient. In this paper, we present a new one that is much faster. We also show how to adapt it to build endgame databases that can be used for the estimation of the optimal rewards of the reachable states using the UCT algorithm [10].

2   Symbolic Search for General Games

Symbolic search is concerned with checking the satisfiability of formulas. For this purpose, we use Binary Decision Diagrams (BDDs) [3], so that we work with state sets instead of single states. In many cases, this saves a lot of memory. E. g., we are able to calculate the complete set of reachable states for Connect Four. The precise number of states reachable from the initially empty board is 4,531,985,219,092, compared to Allis's estimate of 70,728,639,995,483 in [2]. In case of explicit search, for the same state encoding (two bits for each cell (player 1, player 2, empty) plus one for the current player, resulting in a total of 85 bits per state), nearly 43.8 terabytes would be necessary. When using BDDs, 84,088,763 nodes suffice to represent all states.

In order to perform symbolic search, we need BDDs to represent the initial state (*init*), the goal states (*goal*), the transition relation (*trans*) and the conditions for each player p to achieve a specified reward r (*reward* $(p, r), r \in \{0, \ldots, 100\}$). We construct one BDD $trans_a$ for each action a resulting in the transition relation $trans = \bigvee_a trans_a$. Furthermore, we need two sets of variables, one, S, for the predecessor states and the other, S', for the successor states.

To calculate the successors of a state set *state*, we build the (weak) *image*, defined as $image\,(trans, state) = \bigvee_a \exists S.\ trans_a\,(S, S') \land state\,(S)$. The predecessor calculation, the (weak) *pre-image*, works similar: $preImage\,(trans, state) = \bigvee_a \exists S'.\ trans_a\,(S, S') \land state\,(S')$. If we need all the states whose successors hold in *state*, we use the *strong pre-image*, which is defined as $strongPreImage\,(trans, state) = \bigwedge_a \forall S'.\ trans_a\,(S, S') \Rightarrow state\,(S') = \neg preImage\,(trans, \neg state)$. Note that the image and pre-image calculations result in states in the other set of variables. Thereto, after each image or pre-image we perform a replacement of the variables to the ones used before.

3   The Algorithm UCT

In order to calculate an estimate for the current state often Monte-Carlo sampling can be applied. It performs a random search from the current state to a goal state and updates the estimated values of intermediate nodes.

The algorithm UCT [10] (<u>U</u>pper <u>C</u>onfidence <u>B</u>ounds applied to <u>T</u>rees) is an algorithm to work for trees, such as game trees. Starting at the root of the tree, it initially works similar to Monte-Carlo search. If not all actions of the current node were performed at least once, it chooses an unused action randomly. Once this is done for all actions in a node (state s in depth d), it chooses the action that maximizes $Q_t(s, a, d) + c_{N_{s,d}(t), N_{s,a,d}(t)}$, with $Q_t(s, a, d)$ being the estimated value of action a in state s at depth d and time t, $N_{s,d}(t)$ the number of times s was reached up to time t in depth d, $N_{s,a,d}(t)$ the number of times action a was selected when s was visited in depth d up to time t and $c_{n_1,n_2} = 2C_p\sqrt{\ln(n_1)/n_2}$ with appropriate constant C_p. The value c_{n_1,n_2} trades off between exploration and exploitation in that actions with a low estimate will be taken again if the algorithm runs long enough: If an action is chosen, this factor will decrease as the number of times this action was chosen increases; for the actions not chosen, this factor will increase as the number of visits to this state increases.

UCT is successfully used in several game players. In [7], the authors show that in Go UCT outperforms classical α-β-search and give three reasons: UCT can be stopped at any time and still perform well, it is robust by automatically handling uncertainty in a smooth way and the tree grows asymmetrically.

4   Classification of General Two-Player Games

In [5], we proposed a first symbolic algorithm to classify general non-simultaneous two-player games, which has a bad runtime-behavior. In the following, we will present a greatly improved version and a symbolic variant of UCT that uses this algorithm to construct endgame databases.

4.1   Improved Symbolic Classification Algorithm

When classifying a game, we are interested in the optimal rewards of all reachable states. Thereto, we construct a game tree by symbolic breadth-first search (SBFS) with the initial state as its root and each action corresponding to an edge to a succeeding node. As the rewards are only applied in goal states, the internal nodes initially have no known rewards, whereas those of the leaves can be easily determined.

To determine the optimal rewards of each internal node, we propagate these values towards the root until all necessary nodes are classified. As we are concerned with two-player games, this propagation is quite simple. All states whose successors are already classified are given rewards (cf. Fig. 1) according to certain rules. This we repeat iteratively, until finally all states are classified.

Our symbolic algorithm works exactly like that. For it we need a 101×101 matrix (*bucket*) with one BDD for each theoretically possible reward combination. Furthermore, we need a BDD to represent all *classified* states, i. e. states that are in the matrix, a BDD *classifyable* to store those unclassified states whose successors are already classified, and a BDD for the remaining *unclassified* states.

The algorithm's pseudocode is shown in Algorithm 1. First, we determine all reachable states by performing SBFS. Next, we initialize the matrix by setting each possible *bucket* (i, j) to the conjunction of the reachable goal states achieving reward i for player 0 and j for player 1. The *classified* states are the disjunction of all the states in the buckets and the *unclassified* ones the remaining reachable states.

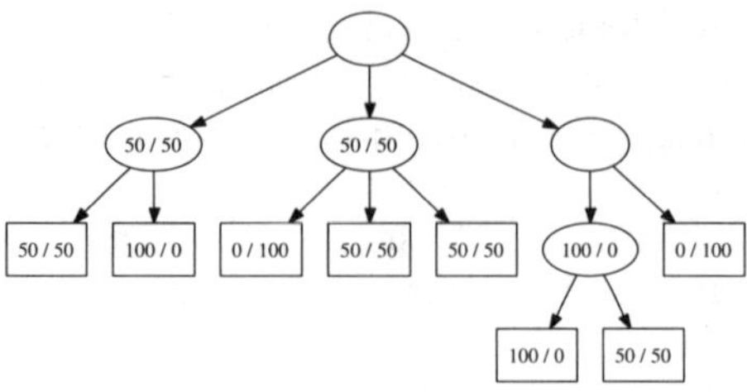

Fig. 1. Example of a game tree. The classification has performed one step.

Algorithm 1. Improved algorithm for general rewards (*classifyGame*)

Input: *reward, goal*
reach ← *reachable* ();
forall $i, j \in \{0, \ldots, 100\}$ **do**
  bucket $(i, j) = reach \wedge goal \wedge reward\,(0, i) \wedge reward\,(1, j)$
classified ← $\bigvee_{0 \leq i,j \leq 100} bucket\,(i, j)$;
unclassified ← *reach* $\wedge \neg classified$;
while *unclassified* $\neq \bot$ **do**
  foreach *player* $\in \{0, 1\}$ **do**
    classifyable ← *strongPreImage* (*trans, classified*) $\wedge$ *unclassified* $\wedge$ *control* (*player*);
    if *classifyable* $\neq \bot$ **then**
      bucket ← *classifyStates* (*classifyable, player, bucket*);
      classified ← *classified* $\vee$ *classifyable*;
      unclassified ← *unclassified* $\wedge \neg classifyable$;

While there are still unclassified states, we continue working on those states, whose successors are all classified. They are found by computing the strong pre-image. Of these states, we only take those where the current player has *control* and classify them. For this, we have two different functions, dependent on the way we wish to compare two states. Either we first maximize the reward for the current player and afterwards minimize that for the opponent, or we maximize the difference between the current player's and its opponent's reward. The classification procedure works as shown in Algorithm 2. When going through the buckets in the specified order, we calculate the states that have to be inserted in each bucket by calculating the corresponding predecessors for the classifyable states using the pre-image function. These states are inserted into this bucket and the classifyable states are updated by deleting the newly classifed ones. After each classification step, the classified states as well as the unclassified ones are updated by adding and deleting the newly classified states, respectively.

To predict the runtime of the algorithm, we count the number of pre-images. As the strong ones can be converted to the weak ones we make no difference between them. Let d be the depth of the inverse game graph, which equals the number of traversals through the main algorithm's loop. During each iteration we calculate two strong pre-images (one for each player). This results in at most $2 \times d$ strong pre-image calculations. In the classification algorithm, we calculate one pre-image for each possible pair of rewards. Let $r = |\{(i, j) \mid (reward\,(0, i) \wedge reward\,(1, j)) \neq \bot\}|$ be the number of possible reward pairs. So, for classification we calculate r weak pre-images.

Algorithm 2. Classification of the classifyable states (*classifyStates*)

Input: *classifyable*, *player*, *bucket*
if *player* $= 0$ **then**
  foreach $i, j \in \{0, \ldots, 100\}$ **do**   // Go through the buckets in specific order.
    if *bucket* $(i, j) \neq \bot$ **then**
      newStates $\leftarrow$ *preImage* (*trans*, *bucket* (i, j)) $\wedge$ *classifyable*;
      bucket $(i, j) \leftarrow$ *bucket* $(i, j) \vee$ *newStates*;
      classifyable $\leftarrow$ *classifyable* $\wedge \neg$*newStates*;
      if *classifyable* $= \bot$ **then** *break*
  else // The same for the other player, with swapped bucket indices.
  return *bucket*;

In each iteration through the main algorithm's loop, we call the classification algorithm at most twice (once for each player). Thus, we get at most $2 \times d \times r$ weak pre-images and $2 \times d \times (1 + r)$ pre-images in total.

4.2 UCT with Symbolic Endgame Databases

The above algorithm is good for classifying all states of a game. But this total classification takes time – the calculation of the set of reachable states as well as the classification itself might take hours or days for complex games. If one is only interested in an estimate of the optimal rewards and has only a limited amount of time, the combination of the classification algorithm and UCT might be interesting. Thereto, we adapted the classification algorithm to create an endgame database and applied UCT to this.

To be able to create the endgame database as quickly as possible, it is best to omit the calculation of reachable states. This way, many states that are not reachable in forward direction will be created and classified, but the forward calculation drops out. In order to get better databases, one should calculate as many backward steps as possible, but if time is short, only few iterations should be performed.

In our UCT algorithm, we encode states as BDDs. Starting at the initial state, for each new node we first have to verify if it is in the endgame database (*classification*; cf. Algorithm 3). If it is, we get the state's value by subtracting the corresponding bucket's second index from the first to get the difference of the rewards. This value we set as the state's estimate, update its number of visits to 1, and return the value.

If the state is not in the database, we compute the applicable actions. Thereto, we build the conjunction of the current state with each action. Each applicable action is stored in a list along with the node. When reaching a node that was visited before, we verify if it has applicable actions. If it has not, it is a leaf node that is stored in the database and we only have to return its value. Otherwise, we further check if all actions have been applied at least once. If not, we randomly choose one of the unapplied actions and create its successor by calculating the image. If all actions have been used at least once, we take the action that maximizes the UCT value. In both cases we increment the number of applications of the action, update the estimate of this node, increment its number of visits, and return the value. For the calculation of the UCT value, we transform the estimated values into the interval $[0, 1]$ by adding 100 and dividing the result by 200, if it is the first player's move, or by -200, if it is the second player's.

Algorithm 3. UCT search function (*search*)

Input: *state, classification*
if *state.visits* $= 0$ **then**
  if $\exists x, y.\ (state.bdd \wedge classification\ (x, y) \neq \bot)$ **then**
    state.visits $\leftarrow 1$;
    state.estimate $\leftarrow x - y$;   // maximize the difference
    return *state.estimate*;
  foreach *action* $\in$ *trans* **do**
    if $(state.bdd \wedge action) \neq \bot$ **then** *state.actions* $\leftarrow$ *state.actions* $\cup$ *action*;
  if *state.actions* $= \emptyset$ **then return** *state.estimate*;
  if $|\{a\ |\ a \in state.actions \wedge a.visits = 0\}| > 0$ **then**
    action $\leftarrow$ *chooseActionRandomly* (*state.actions*);
    successor $\leftarrow$ *image* (*action, state.bdd*);
  else
    action $\leftarrow$ *findActionWithBestUCT* (*state.actions*);
    successor $\leftarrow$ *getSuccessor* (*state, action*);
  value $\leftarrow$ *search* (*classification, successor*);
  action.visits $\leftarrow$ *action.visits* $+ 1$;
  state.estimate $\leftarrow$ (*state.estimate* $\cdot$ *state.visits* $+$ *value*) / (*state.visits* $+ 1$);
  state.visits $\leftarrow$ *state.visits* $+ 1$;
  return *value*;

---

5   Experimental Results

We implemented the algorithms in Java using JavaBDD, which provides a native interface to Fabio Somenzi's CUDD package, and performed them on an AMD Opteron with 2.6 GHz and 16 GB RAM.

For *Clobber* [1], we set the reward for a victory depending on the number of moves taken. This way, we performed the classification for 3×4 and 4×5 boards that are initially filled with the pieces of both players alternating. While our older algorithm takes more than an hour for the 3×4 board, our improved version finishes after less than five seconds. The result is that the game is a victory for the second player with two pieces of each player remaining. On the 4×5 board, nearly 26.8 million states are reachable, while less than half a million BDD nodes suffice to represent them. The older algorithm cannot classify the game within 20 days, whereas the improved one is able to classify the reachable states within 2 hours and 15 minutes. This game is won for the first player with three own and two opponent pieces remaining.

With Clobber we also performed UCT with a timeout of one minute. Here, we compare the estimates for the initial state for different endgame database sizes (number of backward steps). Fig. 2 shows the difference of the estimates on the two players' rewards (higher values mean higher rewards for the first player). As can be seen, we get a solution much nearer to the optimal outcome of -40 after a shorter time with the larger databases. In fact, for databases that result from very few backward steps, after one minute the estimate for the initial state is far from the optimal one and thus might be misleading. Also, with the better databases the algorithm can perform more iterations.

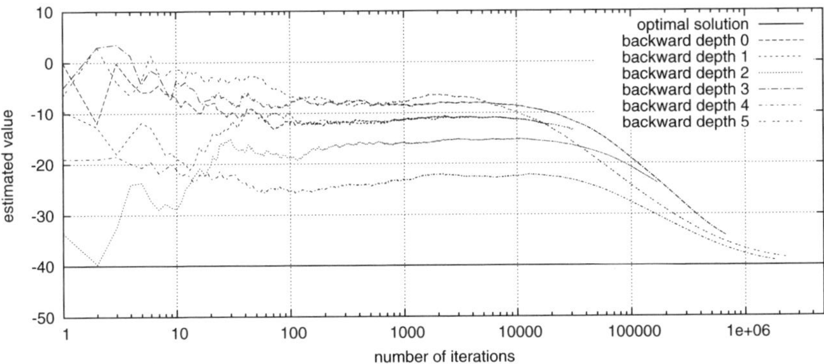

Fig. 2. Estimates for 3×4 Clobber with different endgame databases (averaged over 10 runs)

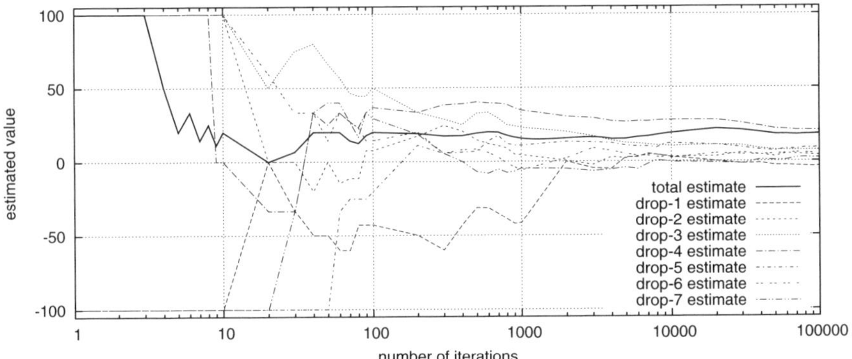

Fig. 3. The results for Connect Four

The game of *Crisscross* is a two-player adaptation of Chinese Checkers and was one of the games used in the qualification phase of the 2006 competition. Our algorithm can classify it completely in about twelve seconds. Nearly 150,000 states are reachable and the optimal outcome is $25/100$, i. e. both players can reach their goal positions but the second player is the first to reach it.

For *Connect Four* we performed UCT, as it takes too long (several hours) – and also too much RAM (more than 16 GB) – to be classified. Fig. 3 shows the results after about 2.5 hours. Nearly 100,000 iterations were performed with a final estimate of 18.3 at the initial state (again the value gives the difference of the estimate on the two players' rewards, higher being better for the first player). As the game is won for the first player, the correct value is 100. In Fig. 3, we can see that the estimate for the optimal move, dropping the first piece in the middle column (*drop-4*), is the highest (20.6 vs. 9.3 for the second best). Furthermore, the estimates for the moves resulting in a draw (*drop-3* and *drop-5*) are slightly higher than those for the losing moves. After no more than 300 iterations, which are performed in less than ten seconds, this already is the case. We also found out that in the end the optimal move has been analyzed about ten times as often as the other possible moves, thus it has the most reliable estimate.

6 Conclusion

In this paper, we have proposed a new symbolic algorithm for the classification of general non-simultaneous two-player games. We have shown that a linear number (in the depth of the game graph) of pre-images suffices to classify such a game. Furthermore, we have presented a symbolic variant of UCT that uses our algorithm beforehand to construct endgame databases.

We have presented results for several games. For Clobber we performed the classification algorithm as well as the UCT algorithm, while for Crisscross we performed only the classification. In Connect Four, we could not finish the classification and so used the classification calculated so far as an endgame database for the UCT algorithm.

With the classification, we are able to check how well players did in real matches. To do this, we can start with the initial state and perform the actions the players chose during the matches. For the successor state we can find the bucket it is in, which gives the optimal reward achievable from this state. Whenever the bucket is changed, the current player has performed a suboptimal move. This might help analyzing the quality of existing players.

References

1. Albert, M., Grossman, J.P., Nowakowski, R.J., Wolfe, D.: An introduction to clobber. INTE-GERS: The Electronic Journal of Combinatorial Number Theory 5(2) (2005)
2. Allis, V.: A knowledge-based approach of connect-four. Master's thesis, Vrije Universiteit Amsterdam (October 1988)
3. Bryant, R.E.: Graph-based algorithms for boolean function manipulation. IEEE Transactions on Computers 35(8), 677–691 (1986)
4. Clune, J.: Heuristic evaluation functions for general game playing. In: AAAI, pp. 1134–1139 (2007)
5. Edelkamp, S., Kissmann, P.: Symbolic exploration for general game playing in PDDL. In: ICAPS-Workshop on Planning in Games (2007)
6. Finnsson, H., Björnsson, Y.: Simulation-based approach to general game playing. In: AAAI, pp. 259–264 (2008)
7. Gelly, S., Wang, Y.: Exploration exploitation in Go: UCT for Monte-Carlo Go. In: NIPS Workshop on On-line Trading of Exploration and Exploitation (2006)
8. Genesereth, M.R., Fikes, R.E.: Knowledge interchange format, version 3.0 reference manual. Technical Report Logic-92-1, Stanford University (1992)
9. Genesereth, M.R., Love, N., Pell, B.: General game playing: Overview of the AAAI competition. AI Magazine 26(2), 62–72 (2005)
10. Kocsis, L., Szepesvári, C.: Bandit based Monte-Carlo planning. In: Fürnkranz, J., Scheffer, T., Spiliopoulou, M. (eds.) ECML 2006. LNCS (LNAI), vol. 4212, pp. 282–293. Springer, Heidelberg (2006)
11. Love, N.C., Hinrichs, T.L., Genesereth, M.R.: General game playing: Game description language specification. Technical Report LG-2006-01, Stanford Logic Group (April 2006)
12. McDermott, D.V.: The 1998 AI planning systems competition. AI Magazine 21(2), 35–55 (2000)
13. Schiffel, S., Thielscher, M.: Fluxplayer: A successful general game player. In: AAAI, pp. 1191–1196 (2007)

Partial Symbolic Pattern Databases for Optimal Sequential Planning*

Stefan Edelkamp and Peter Kissmann

Faculty of Computer Science
TU Dortmund, Germany

Abstract. This paper investigates symbolic heuristic search with BDDs for solving domain-independent action planning problems cost-optimally. By distributing the impact of operators that take part in several abstractions, multiple partial symbolic pattern databases are added for an admissible heuristic, even if the selected patterns are not disjoint. As a trade-off between symbolic bidirectional and heuristic search with BDDs on rather small pattern sets, partial symbolic pattern databases are applied.

1 Introduction

Optimal sequential planning for minimizing the sum of action costs is a natural search concept for many applications. For the space-efficient construction of planning heuristics, symbolic pattern databases [7] (a BDD-based compressed representation of explicit-state pattern databases) are promising. Their space requirements are often much smaller than explicit-state pattern databases. They correlate with the size of the pattern state space, but the request of a parameter that respects the RAM limit for all benchmark domains and problem instances makes it harder to apply symbolic pattern databases.

This paper combines symbolic pattern database heuristic search planning with perimeter search. Perimeters and pattern databases are very similar in their approach to speeding up search. Both techniques use backward search to construct a memory-based heuristic. The memory resources determine the abstraction level for pattern databases as they must completely fit into memory. Perimeters are built without any abstraction; the perimeter stops being expanded when a memory limit is reached. The good results for the hybrid of the two techniques align with recent findings in domain-dependent single-agent search: partial pattern databases [2] and perimeter pattern databases [11]. Besides general applicability, our approach covers different cost models; it applies to step-optimal propositional planning and planning with additive action costs. Moreover, the approach can operate with any pattern selection strategy. Instead of predicting the abstract search space, we set a time limit to terminate the pattern database construction.

The paper is structured as follows. First, we recall symbolic shortest-path and heuristic search. Then, we introduce planning pattern databases and address the addition of non-disjoint symbolic pattern database heuristics by an automated distribution of operator costs. Partial symbolic pattern databases, their symbolic construction, and their inclusion in domain-dependent heuristic search are discussed next. Finally, we provide some experiments and give concluding remarks.

---

* Thanks to DFG for support in the projects ED 74/3 and 74/2.

A. Dengel et al. (Eds.): KI 2008, LNAI 5243, pp. 193–200, 2008.

2 Symbolic Shortest Path Search

We assume a given planning problem $\mathcal{P} = (\mathcal{S}, \mathcal{A}, \mathcal{I}, \mathcal{G}, \mathcal{C})$ of finding a sequence of actions $a_1, \ldots, a_k \in \mathcal{A}$ from $\mathcal{I}$ to $\mathcal{G}$ with minimal $\sum_{i=1}^{k} \mathcal{C}(a_i)$. For symbolic search with BDDs we additionally assume a binary encoding of a planning problem.

Action are formalized as relations, representing sets of tuples of predecessor and successor states. This allows to compute the image as a conjunction of the state set (formula) and the transition relation (formula), existentially quantified over the set of predecessor state variables. This way, all states reached by applying one action to one state in the input set are determined. Iterating the process (starting with the representation of the initial state) yields a symbolic implementation of breadth-first search (BFS). Fortunately, by keeping sub-relations $Trans_a$ attached to each action $a \in \mathcal{A}$ it is not required to build a monolithic transition relation. The image of state set S then reads as $\bigvee_{a \in \mathcal{A}} (\exists x\, (Trans_a(x, x') \land S(x)))$.

For positive action costs, the first plan reported by the single-source shortest-paths search algorithm of Dijkstra [5] is optimal. For implicit graphs, we need two data structures, one to access nodes in the search frontier and one to detect duplicates.

Algorithm 1. Symbolic-Shortest-Path

Input: Problem $\mathcal{P} = (\mathcal{S}, \mathcal{A}, \mathcal{I}, \mathcal{G}, \mathcal{C})$ in symbolic form with $\mathcal{I}(x)$, $\mathcal{G}(x)$, and $Trans_a(x, x')$
Output: Optimal solution path

$Open[0](x) \leftarrow \mathcal{I}(x)$
for all $f = 0, \ldots, f_{\max}$
  for all $l = 1, 2, \ldots, g$ **do** $Open[g] \leftarrow Open[g] \setminus Open[g - l]$
  $Min(x) \leftarrow Open[f](x)$
  if $(Min(x) \land \mathcal{G}(x) \neq false)$ **return** $Construct(Min(x) \land \mathcal{G}(x))$
  for all $i = 1, \ldots, C$ **do**
    $Succ_i(x') \leftarrow \bigvee_{a \in \mathcal{A}, \mathcal{C}(a)=i}(\exists x(Min(x) \land Trans_a(x, x')))$
    $Succ_i(x) \leftarrow \exists x'(Succ_i(x') \land x = x')$
    $Open[f + i](x) \leftarrow Open[f + i](x) \lor Succ_i(x)$

If we assume that the largest action cost is bounded by some constant C, a symbolic shortest path search procedure is implemented in Algorithm 1. The algorithm works as follows. The BDD $Open[0]$ is initialized to the representation of the start state. Unless one goal state is reached, in one iteration we first choose the next f-value together with the BDD Min of all states in the priority queue having this value. Then for each $a \in A$ with $c(a) = i$ the transition relation $Trans_a(x, x')$ is applied to determine the BDD for the subset of all successor states that can be reached with cost i. In order to attach new f-values to this set, we add the result in bucket $f + i$.

An advanced implementation for the priority queue is a one-level bucket [4], namely an array of size $C + 1$, each of which entries stores a BDD for the elements. For large values of C, multi-layered bucket and radix-heap data structures are appropriate [1].

Let us briefly consider possible implementations for $Construct$. If all previous layers remain in main memory, sequential solution reconstruction is sufficient. If layers are eliminated as in frontier search [19] or breadth-first heuristic search [21], additional relay layers have to be maintained. The state closest to the start state in the relay layer is used for divide-and-conquer solution reconstruction. Alternatively, already expanded buckets are flushed to the disk [8].

The above algorithm traverses the search tree expansion of the problem graph. It is sound as it finds an optimal solution if one exists. It is complete and necessarily terminates if there is no solution. By employing backward search, therefore, it can be used to generate symbolic pattern databases. For the search, we apply a symbolic variant of A* [9,12,17,20], which, for consistent heuristics, can be deemed as a variant of the symbolic implementation of Dijkstra's algorithm.

3   Additive Symbolic Pattern Databases

Applying abstractions simplifies a problem: exact distances in these relaxed problems serve as lower bound estimates. Moreover, the combination of heuristics for different abstractions often leads to a better search guidance. Pattern databases [3] completely evaluate an abstract search space $\mathcal{P}' = (\mathcal{S}', \mathcal{A}', \mathcal{I}', \mathcal{G}', \mathcal{C}')$ prior to the concrete, base-level search in $\mathcal{P}$ in form of a dictionary containing the abstract goal distance from each state in $\mathcal{S}'$. Abstractions have to be cost-preserving, which implies that each concrete path maps to an abstract one with smaller or equal cost. The construction process is backwards and the size of a pattern database is the number of states it contains.

One natural concept for planning abstraction domains is to select a pattern variable support set $varset(R) \subseteq \mathcal{V}$ and project each planning state and each action to $varset(R)$. In domain-dependent planning, the abstraction functions are selected by the user. For domain-independent planning, the system has to infer the abstractions automatically [13,16,14].

Symbolic pattern databases [7] are pattern databases that are constructed symbolically for later use in symbolic or explicit-state heuristic search. They exploit that the representation of the actions is provided in form of a relation, which allows to perform backward search by computing the predecessors (the *PreImage*) of a given state set efficiently. Each state set with a distinct goal distance is represented as a BDD. Different to the posterior compression of the state set [10], the construction itself works on a compressed representation, allowing the generation of much larger databases.

Additive symbolic pattern databases allow to add abstract goal distances of different abstractions without affecting the admissibility of the heuristic estimate, i.e., the sum of the abstract goal distances is still a lower bound for the original space. Cost-preserving abstract problem tasks $\mathcal{P}_1, \ldots, \mathcal{P}_k$ wrt. state space homomorphisms $\phi_1, \ldots, \phi_k$ of a planning task $\mathcal{P}$ are trivially additive, if no original action contributes to two different abstractions. For disjoint patterns in single-agent challenges like the $(n^2 - 1)$-Puzzle, this assumption is immediate, as only one tile can move at a time. For general domains, however, the restriction limits pattern database design. Therefore, we introduce the weighted combination of different abstract tasks as follows. First, we determine the number of times an original action $a \in \mathcal{A}$ is valid in the abstractions. We define $valid(a) = |\{i \in \{1, \ldots, k\} \mid \phi_i(a) \neq noop\}|$, and $t = lcm_{a \in \mathcal{A}} valid(a)$, where lcm denotes the least common multiple, and $noop$ a void action. Then, we scale the original problem $\mathcal{P}$ to $\mathcal{P}^*$, by setting $\mathcal{C}^*(a) = t \cdot \mathcal{C}(a)$. All plans π now have costs $\mathcal{C}^*(\pi) = \sum_{a \in \pi} t \cdot Cost(a) = t \cdot \mathcal{C}(\pi)$, such that an optimal plan in $\mathcal{P}$ induces an optimal plan in $\mathcal{P}^*$ and vice versa. Next, we set all action costs $\mathcal{C}_i^*(a)$ in the abstract problem $\mathcal{P}_i^*$ to $\mathcal{C}_i^*(a) = t \cdot \mathcal{C}(a)/valid(a)$, if $\phi_i(a) \neq noop$, and to zero, otherwise. As t is the least common multiple, the cost values in the abstract domains remain integral.

The following result shows that the sum of the heuristic values according to the abstractions $\mathcal{P}_i$, $i \in \{1, \ldots, k\}$, is a lower bound.

Theorem 1. *The weighted combination of different abstract tasks $\mathcal{P}_1, \ldots, \mathcal{P}_k$ of $\mathcal{P}$ wrt. cost-preserving state space homomorphims $\phi_1, \ldots, \phi_k$ always results in an admissible heuristic for the scaled problem $\mathcal{P}^*$.*

Proof. Let $\pi = (a_1, \ldots, a_l)$ be any plan solving $\mathcal{P}^*$ and $\pi_i = (\phi_i(a_1), \ldots, \phi_i(a_l))$ be the corresponding solution in abstraction i, $i \in \{1, \ldots, k\}$. Some $\phi_i(a_j)$ are trivial and correspond to cost 0. The sum of costs of all π_i is equal to

$$
\sum_{i=1}^{k} \mathcal{C}_i^*(\pi_i) = \sum_{i=1}^{k}\sum_{j=1}^{l} \mathcal{C}_i^*(a_j) = \sum_{i=1}^{k}\sum_{j=1}^{l} t \cdot \mathcal{C}(a_j)[\phi_i(a) \neq noop]/valid(a_j)
$$

$$
= \sum_{j=1}^{l}\sum_{i=1}^{k} t \cdot \mathcal{C}(a_j)[\phi_i(a) \neq noop]/valid(a_j)
$$

$$
\leq \sum_{j=1}^{l} (t \cdot \mathcal{C}(a_j)/valid(a_j)) \sum_{i=1}^{k} [\phi_i(a_j) \neq noop]
$$

$$
= \sum_{j=1}^{l} (t \cdot \mathcal{C}(a_j)/valid(a_j))valid(a_j)
$$

$$
= \sum_{j=1}^{l} t \cdot \mathcal{C}(a_j) = \sum_{a \in \pi} t \cdot \mathcal{C}(a) = \mathcal{C}^*(\pi).
$$

As π is chosen arbitrarily, we conclude that $\min_\pi \sum_{i=1}^{k} \mathcal{C}_i^*(\pi_i) \leq \min_\pi \mathcal{C}^*(\pi)$.

Additive symbolic databases for non-disjoint patterns are novel, but similar techniques for explicit search have been introduced (though not empirically evaluated) [18].

4   Partial Symbolic Pattern Databases

Perimeter search [6] tries to reap the benefits of front-to-front-evaluations in bidirectional search, while avoiding the computational efforts involved in re-targeting the heuristics towards a continuously changing search frontier. It conducts a cost-bounded best-first search starting from the goal nodes; the nodes on the final search frontier, called the perimeter, are stored in a dictionary. Then, a forward search, starting from $\mathcal{I}$, employs front-to-front evaluation with respect to these nodes. Alternatively, in front-to-goal perimeter search all nodes outside the perimeter are assigned to the maximum of the minimum of the goal distances of all expanded nodes in the perimeter and an additionally available heuristic estimate. Although larger perimeters provide better heuristics, they take increasingly longer to compute. The memory requirements for storing the perimeter are considerable and, more crucially, the effort for multiple heuristic computations can become large.

In essence, a partial pattern database is a perimeter in the abstract space (with the interior stored). Any node in the original space has a heuristic estimate to the goal: if it is in the partial pattern database, the recorded goal distance value is returned; if not, the radius d of the perimeter is returned. This heuristic is both admissible and consistent. Building a partial pattern database starts from the abstract goal and stores heuristic values. When a memory or time limit is reached it terminates and the heuristic values are used for the forward search. The value d is the minimum cost of all abstract nodes outside the partial pattern database. On one extreme, a partial pattern database with no abstraction reverts to exactly a perimeter. On the other extreme, a partial pattern database with a coarse-grained abstraction will cover the entire space, and performs exactly like an ordinary pattern database. Bidirectional BFS is an interleaved partial pattern database search without abstraction.

An ordinary pattern database represents the entire space; every state visited during the forward search has a corresponding heuristic value in the database. The pattern database abstraction level is determined by the amount of available memory. Fine-grained abstraction levels are not possible, because the memory requirements increase exponentially with finer abstractions. Explicit-state partial pattern databases, however, generally do not cover the entire space.

A partial symbolic pattern database is the outcome of a partial symbolic backward shortest paths exploration in abstract space. It consists of a set of abstract states $\mathcal{S}'_{<d}$ and their goal distance, where $\mathcal{S}'_{<d}$ contains all states in $\mathcal{S}'$ with cost-to-goal less than d, and where d is a lower bound on the cost of any abstract state not contained in $\mathcal{S}'_{<d}$. The state sets are kept as BDDs $H[i]$ representing $\mathcal{S}'_{<i+1} \setminus \mathcal{S}'_{<i}$, $i \in \{0, ..., d-1\}$. The BDD $H[d]$ represents the state set $\mathcal{S}' \setminus \mathcal{S}'_{<d}$, such that a partial pattern database partitions the abstract search space. If $C_{\max}$ is the cost of the most expensive action, the construction works as shown in Algorithm 2.[1]

Multiple pattern databases can be *merged* by either taking the maximum, the sum, or the weighted combination of individual pattern database entries. Additionally we apply

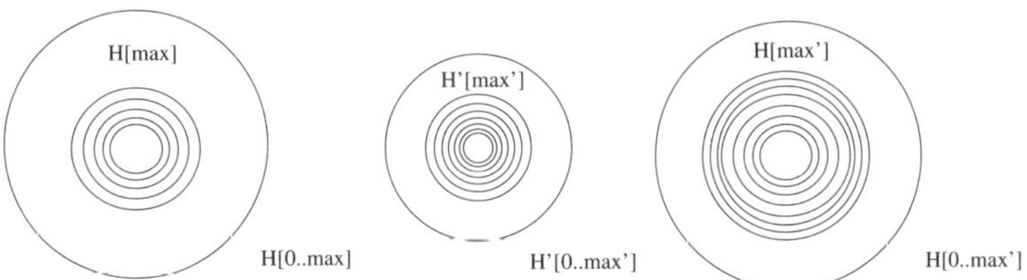

Fig. 1. Partial pattern database refinement: left 2 pattern databases are merged into the right one

---

[1] For large costs, pattern databases can become sparse, so that in practice we compress the BDD vector $H[0.. \max]$ to a list containing $(i, H[i])$ for every i with $H[i] \neq false$.

Algorithm 2. Construct-Partial-Symbolic-PDB

Input: Abstract planning problem $\mathcal{P}' = (\mathcal{S}', \mathcal{A}', \mathcal{I}', \mathcal{G}', \mathcal{C}')$
Output: Partial or full symbolic pattern database

$Reached \leftarrow H[0] \leftarrow \mathcal{G}'$
for all $d = 0, 1, 2, \ldots$ **do**
  $H[d] \leftarrow H[d] \wedge \neg Reached$
  if $(H[d..d + C_{\max} - 1] \equiv \textit{false})$ **then return** $(\textit{full}, H[0..d - 1])$
  if $(\textit{time-out})$ **then return** $(\textit{partial}, (H[0..d], \neg Reached))$
  $Reached \leftarrow Reached \vee H[d]$
  for all $i = 1, \ldots, C_{\max}$ **do**
    $Succ_i \leftarrow \bigvee_{a \in \mathcal{A}', C'(a)=i} PreImage(H[d], Trans'_a)$
    $H[d + i] \leftarrow H[d + i] \vee Succ_i$

---

iterative abstraction to refine the quality of a database. If pattern database construction is partial because of the time cut-off, an abstraction ϕ can be re-abstracted to ϕ' (e.g., by dropping another variable). The idea is illustrated in Fig. 1. Algorithm 3 shows how remaining states outside the previous perimeter are classified wrt. ϕ'. The process is iterated until all states are classified.

Algorithm 3. Refinement

Input: Planning problem $\mathcal{P} = (\mathcal{S}, \mathcal{A}, \mathcal{I}, \mathcal{G})$, multiple pattern database abstraction set Φ
Output: Symbolic pattern database set

for all $\phi \in \Phi$ **do**
  $(r, H[0..\max]) \leftarrow Construct\text{-}Partial\text{-}Symbolic\text{-}PDB(P_\phi)$
  while $(r = \textit{partial})$ **do**
    $\phi' \leftarrow Abstract(\phi); Rest \leftarrow \textit{false}$
    $(r, H'[0..\max']) \leftarrow Construct\text{-}Partial\text{-}Symbolic\text{-}PDB(P_{\phi'})$
    for all $i = 1, \ldots, \max -1$ **do** $Rest \leftarrow (H'[i] \wedge H[\max]) \vee Rest$
    for all $i = \max +1, \ldots, \max'$ **do** $H[i] \leftarrow H'[i] \wedge H[\max]$
    $H[\max] \leftarrow H[\max] \wedge (H'[\max] \vee Rest)$
    $\max \leftarrow \max'$
  $\mathcal{H} \leftarrow Merge(\mathcal{H}, H[0..\max])$
return $\mathcal{H}$

In the extension to weighted graphs, all successors of the set of states with minimum f-value are determined wrt. the current cost value g and action cost value i. To determine their h-values, the merged symbolic pattern database is scanned.[2]

---

[2] In our implementation, we avoid merging prior to the search and merge the results of the lookups. However, we observed that the advantages of such dynamic lookups are small.

5   Experiments

Table 1 compares symbolic search with partial symbolic pattern databases (PSPDB, construction time included) with symbolic bidirectional breadth-first search (SBBFS), PDB [14] and LFPA [15] on various competition problems on equivalent computational resources. For automated pattern selection (at least compared to [14]) we chose a rather simple policy. Abstractions are variable projections, combined via greedy bin-packing. PSPDB shows benefits to PDB, especially considering that automated pattern selection is less elaborated. Wrt. LFPA, the advantage is less impressive.

Table 1. Comparison between optimal planning approaches (on a 2.6GHz CPU, 2 GB RAM)

Problem	L^*	PSPDB	SBBFS	PDB	LFPA
log-10-0	45	137s	577s	497s	117s
log-10-1	42	174s	546s	405s	129s
log-11-0	48	68s	577s	377s	129s
log-11-1	60	65s	-	-	284s
log-12-0	42	47s	473s	545s	185s
log-12-1	68	861s	-	-	221s
sat-4	17	5s	2s	6s	11s
sat-5	15	41s	37s	110s	47s
sat-6	20	38s	27s	21s	634s
sat-7	21	567s	-	-	-
psr-48	37	10s	128s	787s	36s
psr-49	47	57s	-	-	67s

For TPP-9 (from IPC-5), a problem that according to [15] cannot be solved by any other heuristic search planner so far. It also can not be solved with bidirectional search. Two partial pattern databases were constructed. As some actions remain applicable in both abstractions, their costs are 1. The costs of all other actions in the abstract space and the costs of all actions in the original state space are multiplied with two. The exploration took 354 iterations to find a plan of 48 steps in about 2h. Between 32 and 64 million BDDs were used for the entire exploration, corresponding to about 1.5 GB.

6   Conclusion

We extended heuristic search planning to work with partial symbolic pattern databases, a recent compromise between bidirectional breadth-first search and a low number of variables in the pattern for a full database. Thereby, we obtained promising results for a selection of planning benchmarks. As a matter of fact, the results are highly selective, and given that both LFPA and PDB take different parameter settings in different domains, we are excited to directly compare the approaches in a competition.

Due to the scaling of action costs to preserve admissibility of the heuristic, we employed a bucket implementation of Dijkstra's single-source shortest paths algorithm for the cost-limited construction of each database. One possible future research avenue is to construct partial pattern databases on-the-fly; similar to bidirectional BFS, where the search continues in the direction that is likely to be the cheapest.

References

1. Ahuja, R.K., Mehlhorn, K., Orlin, J.B., Tarjan, R.E.: Faster algorithms for the shortest path problem. Journal of the ACM 37(2), 213–223 (1990)
2. Anderson, K., Holte, R., Schaeffer, J.: Partial pattern databases. In: Miguel, I., Ruml, W. (eds.) SARA 2007. LNCS (LNAI), vol. 4612, pp. 20–34. Springer, Heidelberg (2007)
3. Culberson, J.C., Schaeffer, J.: Pattern databases. Computational Intelligence 14(4), 318–334 (1998)
4. Dial, R.B.: Shortest-path forest with topological ordering. Communications of the ACM 12(11), 632–633 (1969)
5. Dijkstra, E.W.: A note on two problems in connection with graphs. Numerische Mathematik 1, 269–271 (1959)
6. Dillenburg, J.F., Nelson, P.C.: Perimeter search. Artificial Intelligence 65(1), 165–178 (1994)
7. Edelkamp, S.: Symbolic pattern databases in heuristic search planning. In: AIPS, pp. 274–293 (2002)
8. Edelkamp, S.: External symbolic heuristic search with pattern databases. In: ICAPS, pp. 51–60 (2005)
9. Edelkamp, S., Reffel, F.: OBDDs in heuristic search. In: Herzog, O. (ed.) KI 1998. LNCS, vol. 1504, pp. 81–92. Springer, Heidelberg (1998)
10. Felner, A., Korf, R.E., Meshulam, R., Holte, R.C.: Compressed pattern databases. Journal of Artificial Intelligence Research 30, 213–247 (2006)
11. Felner, A., Ofek, N.: Combining perimeter search and pattern database abstractions. In: Miguel, I., Ruml, W. (eds.) SARA 2007. LNCS (LNAI), vol. 4612, pp. 155–168. Springer, Heidelberg (2007)
12. Hansen, E.A., Zhou, R., Feng, Z.: Symbolic heuristic search using decision diagrams. In: Koenig, S., Holte, R.C. (eds.) SARA 2002. LNCS (LNAI), vol. 2371, pp. 83–98. Springer, Heidelberg (2002)
13. Haslum, P., Bonet, B., Geffner, H.: New admissible heuristics for domain-independent planning. In: AAAI, pp. 1163–1168 (2005)
14. Haslum, P., Botea, A., Helmert, M., Bonet, B., Koenig, S.: Domain-independent construction of pattern database heuristics for cost-optimal planning. In: AAAI, pp. 1007–1012 (2007)
15. Helmert, M., Haslum, P., Hoffmann, J.: Flexible abstraction heuristics for optimal sequential planning. In: ICAPS, pp. 176–183 (2007)
16. Hoffmann, J., Sabharwal, A., Domshlak, C.: Friends or foes? An AI planning perspective on abstraction and search. In: ICAPS, pp. 294–303 (2006)
17. Jensen, R.M., Bryant, R.E., Veloso, M.M.: SetA*: An efficient BDD-based heuristic search algorithm. In: AAAI, pp. 668–673 (2002)
18. Katz, M., Domshlak, C.: Combining perimeter search and pattern database abstractions. In: ICAPS-Workshop (2007)
19. Korf, R.E., Zhang, W., Thayer, I., Hohwald, H.: Frontier search. Journal of the ACM 52(5), 715–748 (2005)
20. Qian, K.: Formal Verification using heursitic search and abstraction techniques. PhD thesis, University of New South Wales (2006)
21. Zhou, R., Hansen, E.: Breadth-first heuristic search. In: ICAPS, pp. 92–100 (2004)

Optimal Scheduling with Resources for Application Execution in 3G Networks

Roman Englert

Deutsche Telekom Laboratories at Ben Gurion University
P.O.B. 653, Beer-Sheva 84150, Israel
`roman.englert@telekom.de`

Abstract. Today the number of data applications for cellular phones increases rapidly. Most of these data applications require a connection to the Service Delivery Platform (SDP) in the cellular network. As a consequence the booting of the cellular phone and the starting of mobile applications take several minutes. In this paper an approach is presented to optimise the start time of mobile applications. The idea is to partition the call set-up of mobile applications into modules and to apply them to scheduling. Main challenges are to take into account the required temporal and operational resources, and to generate the schedule in real-time. The applied scheduler is based on the Simplex method from Discrete Mathematics and delivers *en passant* the proof for the optimality. Experiments demonstrate the approach and the results are compared with former experiments based on a temporal planner from the field of Artificial Intelligence (AI).

1 Introduction

The third generation (3G) of mobile telecommunication, which is known in Europe as Universal Mobile Telecommunication Standard (UMTS), enables the simultaneous execution of data-intensive applications in mobile terminals. Quality of Service (QoS) ensures a robust and delay preventing execution of mobile data applications [1,2]. To start an application in a mobile terminal a call set-up with the radio bearer is required. This procedure takes between a couple of seconds for an interactive game like chess, and approximately 30 seconds for an access to the Service Delivery Platform (SDP) gateway. Often users start simultaneously several applications, and as a consequence, the waiting period takes minutes until their call set-ups are executed. Obviously, a high-speed 3G mobile telecommunications network without an attractive high-speed entry does not add value to the well-developed technology of UMTS. Therefore, it is the goal of this work to investigate how we can significantly improve the QoS for mobile data applications by reducing the time required for multiple call set-ups of these applications, as is typical for 3G telecommunication networks.

Scheduling enables the computation of a plan for an optimised execution of the call set-ups. The scheduler has to consider operational and temporal resources in order to ensure that required constraints like available bandwidth and

A. Dengel et al. (Eds.): KI 2008, LNAI 5243, pp. 201–208, 2008.

parameters for QoS are met. The developed approach is applicable to 3G mobile networks, for which UMTS is a wide-spread standard in Europe.

The UMTS network architecture is described and the execution of mobile-to-server and mobile-to-mobile applications are discussed. The execution of a mobile application requires a QoS that is made available by the radio bearer during the call set-up. Technically viewed the call set-up is a sequence of electro-magnetic signals. In order to apply a scheduler to the call set-up the sequence of signals has to be partitioned into in a fixed sequence of discrete modules. *Intelligent Software Agents* (ISA) [3,4] are a flexible technique to modularise the call set-up resulting in a discrete instead of a normal continuous call set-up.

As scheduling the Simplex method [5] from the field of Linear Programming is applied. A main feature of the Simplex method is that an optimal schedule is computed. However, the domain modelling of the beforehand optimisation task is a challenge, since resources with real numbers have to applied. This challenge is solved by introducing binary variables for fractioned resources. Experiments for the scheduling computation of several mobile applications are run. Subsequently, these experiments are compared to an implementation based on the Planning Domain Definition Language (PDDL$^+$) from the field of temporal planning within AI [6].

The alternative modelling [7,8] considers also constraints given by the operational and temporal resources. Current research considers QoS on a network level, more precisely, on a codec level for speech and data compression [9]. In contrast to this addresses the approach the semantic level, where applications are grouped according to QoS requirements like conversation parameters. The domain for temporal planning was also one of the six challenges at the International Planning Competition 2004 (IPC-4) [8]. In total 19 world-wide research groups tried to solve the planning challenges. Despite the on hand experiments with the Simplex method, only one planner with an optimality criterion for the plan generation, namely TP4 [10], was able to solve the challenges of the UMTS domain.

The paper is organised as follows: In Sect. 2 the structure of 3G networks and the call set-up of mobile applications are described. Then, the scheduling domain including the renewable and consumable resources is described (Sect. 3). The domain implementation is based on the Zuse Institute Modelling Program Language (ZIMPL) [11]. Sect. 4 contains a description of the Simplex method. Afterwards, in Sect. 5 experiments with ZIMPL are run and these results are compared to an earlier implemented domain [7] based on PDDL $^+$ from the field of AI. We conclude in Sect. 6.

2   3G Networks and Application Call Set-Up

The call set-up of a mobile application that requires a server in the back-end is based on a continuous signal from the client, through the cellular network to the back-end system in the Internet. In this section the partitioning of the continuous call set-up into discrete modules is described.

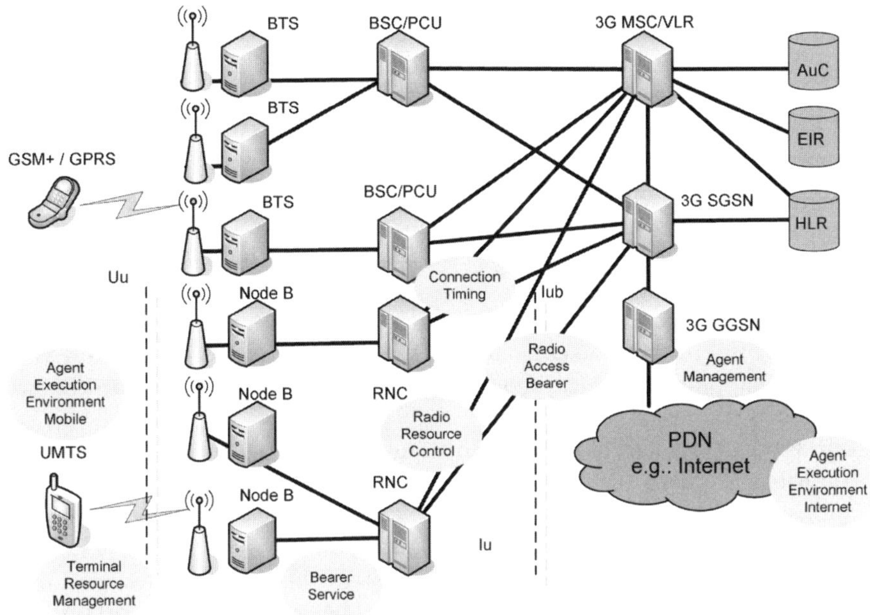

Fig. 1. UMTS terminal equipment, access, and core network domain adorned with the modules for the discrete call set-up (left to right)

The UMTS call set-up can be modularised adopting the concept of ISA, since agents are logical units and enable a discrete perspective of the continuous signalling process. The call set-up is partitioned into the following functional modules that have to be executed in sequential order [1,7] (Fig. 1):

Terminal Resource Management (TRM): The initial step is the initiation of an application on the mobile and the determination of the required resources for the execution. The resources of the mobile client (short: mobile), e.g. display and memory, are checked by the TRM and allocated, if available. Otherwise, the execution is aborted.

Connection Timing (CT): The wireless connection to the radio network is initiated via the dedicated control channel in the GSM part of the 3G network [1]. In case of success, the transmission of the message "Ready for service" is transferred via the node B to the mobile in order to ensure a proper CT.

Agent Management (AM): The information of the mobile, e.g. QoS and data handling capabilities, is sent to the application server in the Internet (cf. AEEI). The transmission can be done conveniently by a so-called service agent [12], that is controlled by the AM (Fig. 1). The task of a service agent is, that in case of failure, e.g. network resources are not sufficiently available,

the agent can negotiate with the mobile's agent about another QoS class or
different quality parameters.

Agent Execution Environment Mobile (AEEM): A service agent with
the required QoS class for the execution of the application and with parameters of the mobile application is sent from the mobile's AEEM to the
application server in the Internet (cf. AEEI).

Radio Resource Controller (RRC): The RRC provides the required QoS
through logical resources from the radio bearer's AND. A detailed description
is out of the scope of this work and can be found in Holma and Toskala's
UMTS book [1].

Radio Access Bearer (RAB): Subsequently, the bearer resources are sup-
plied on the physical level from the RAB in the CND. The call flow is set-
up by mapping the logical QoS parameters and the physical QoS resources
together.

Agent Execution Environment Internet (AEEI): The AEEI module establishes the data transfer from the core network to a PDN (e.g. Internet)
and sends a service agent (controlled by AM) to the application in the PDN
in order to ensure the QoS for the application.

Bearer Service (BS): Finally, the BS is established for the execution of the
mobile application with the required radio bearer resources and QoS. Messages are sent to the modules TRM and AEEI to start the execution of the
application.

These modules are applied to scheduling in order to optimise the starting time
for several mobile applications. In the following section the scheduling domain
including the temporal and operational resources is described.

3   Scheduling Domain Description

Scheduling has the goal to minimise the total execution time of k jobs on m
machines [13]. It is assumed, that the jobs (we also call them modules) are totally packed, i.e. there are no start or release times. In the existing scenario
we have to deal with an extension of scheduling, namely, with renewable and
consumable resources and with constraints ensuring the execution order of the
modules. The involvement of time into the scheduling process is a main challenge of scheduling tasks [14]. *Renewable resources* are not used-up by a module, e.g. processor time, and *consumable resources* vanish after the execution of
the module, e.g. accumulator energy of a mobile client [7]. Examples for the
resources are given in Table 1, where all resources are renewable, except the
energy is consumable for scheduling. Additionally, there are *intra-application*
constraints that ensures the execution of the modules in the given order (see
Sect. 2). Finally, there are *inter-application* constraints that prevent a concurrent execution of the resource allocating modules TRM, RAB, and RRC. Both types
of constraints are represented by the $BEFORE(A, B)$ operator, that defines the
execution of module A before module B. An example for an intra-application

Table 1. Consumable and renewable resources of the UMTS call set-up

mobile-cpu	used with x per cent per application
d-available	partition of the display, e.g. ticker and chess
energy	available accumulator energy
num-calls	mobile network load for a node B
max-no-pdp	max. no. of packet data protocols per mobile
max-no-apn	max. access point names (APN) per mobile

constraint is $TRM(app_1)\ BEFORE\ CT(app_1)$ and for an inter-application constraint is $TRM(app_1)\ BEFORE\ TRM(app_2)$. The former example ensures the execution of the modules in a sequential order, and the latter example guarantees the initiation order of the applications.

The domain has been modelled based on the Zuse Institute Mathematic Programming Language (ZIMPL) for the application to Simplex schedulers [11]. Since the Simplex method lacks in the representation of constraints that contain real-numbers, we partitioned resources into portions and introduced binary variables for each portion. As an example the display is partitioned into three portions of 20%, 40%, and 40% of the mobile's display. A news ticker may need only 20% of the display, whereas composing a SMS may need a little bit more of the display, e.g. 40%. The binary variables indicate whether a resource portion is used by a module or not. Additionally, binary variables were introduced for the constraints, in order to indicate whether a module can be executed or not. As an example, consider the terminal resource management TRM. As soon as this module is applied, a binary variable BV_{TRM} is set to false, indicating that no second TRM module can be executed in parallel. At the end of the module execution the variable is reseted to true.

A comprehensive description of the domain is given in [15]. In the following the Simplex method is briefly described before experiments are run.

4   Simplex Method

The Simplex method is based on Linear Programming problems, that consist of a collection of linear inequalities on a number of real variables and a fixed linear functional that is to be maximised (or minimised) [5,11]:

$$max\{a^T x = \sum_{i=1}^{n} a_j x_j : L(x) \leq c, x_1, \ldots, x_n \geq 0\} \tag{1}$$

where L is a $m \times n$ matrix, and a and c are vectors. The idea of the Simplex method is to represent a Linear Programming problem in a matrix form that is typically under-determined, since the number of variables exceed the number of equations. How does the Simplex method finds the optimum solution? The series of linear inequalities defines a polytope as a feasible region. The Simplex algorithm begins at a starting vertex and moves along the edges of the polytope

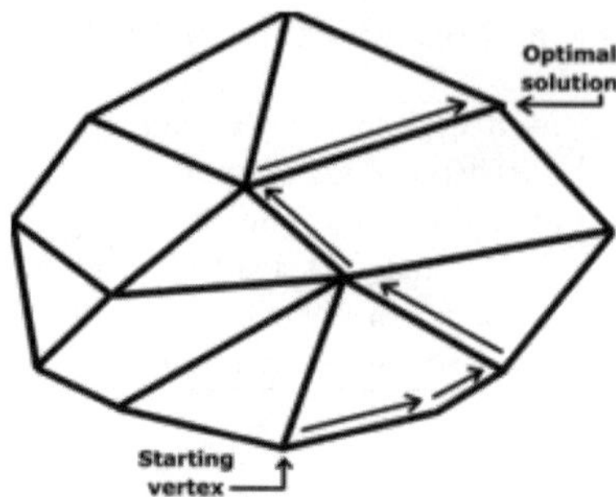

Fig. 2. Simplex method exploring the vertices of a polytope

until it reaches the vertex of the optimum solution. Or more precise: In geometric terms we are considering a convex polytope P, defined by intersecting a number of half-spaces in the n-dimensional Euclidean space. The objective is to maximise a linear functional L (Eq. 1). Consider the hyper-planes $H(c)$ defined by $L(x) \leq c$, with increasing c. If the problem is well-posed, then the goal is to find the largest value of c such that $H(c)$ intersects P. In this case it can be shown that the optimum value of c is attained, on the boundary of P [5]. Methods for finding this optimum point on P have a choice of improving a possible point by moving through the interior of P (so-called interior point methods), or starting and remaining on the boundary.

The Simplex method falls into the latter class of methods. The idea is to move along the facets of P in search of the optimum, from point to point (Fig. 2). The method specifies how this has to be done. It starts at some vertex of the polytope, and at every iteration an adjacent vertex with a gradient of the highest magnitude is chosen, such that the value of the objective function does not decrease. If no such vertex exists, a solution to the problem is found. In the worst case all vertices of the polytope (exponential number) have to be visited.

5   Experiments

The UMTS domain has been modelled based on ZIMPL as described in Sect. 3. Goal is to compute the makespan, i.e. optimum schedule, for 4, 6, 8, 10, and 12 applications. Table 2 shows the schedule generation times on a 3.2 GHz Pentium 4 (second row): The measured times are between 0.01 sec. for 4 applications up to 120 sec. for 12 applications.

A further measurement is the makespan optimisation. In the naive scenario all mobile applications are executed in sequential order. This scenario is compared with the makespan (Table 2): The makespan optimisation is between 53% for 4 applications and 68% for 12 applications.

The UMTS domain has been modelled earlier with the Planning Domain Definition Language (PDDL^{+}) [6] and applied to the International Planning Competition (IPC-4) from the field of AI [16]. 19 world-wide research groups

Table 2. Schedule generation times for the application of the Simplex method

#applications	4	6	8	10	12
schedule generation time (sec.)	00.1	0.8	2.7	10.5	120
makespan reduction	53%	65%	74%	72%	68%

tried to solve the domain for up to 20 applications. As a result only 6 planners from the field of AI were able to compute a schedule with temporal and operational resources [8]. Only one planner was able to compute a makespan with an optimum criterion: The admissible planner TP4 which is based on Iterative Deepening A* (IDA*) search [10]. The planner TP4 computed the makespan for up to 20 applications below 1 second (800 MHz Pentium 3).

The Simplex method has a poor performance, since for each applied resource an own parameter (column in a matrix) has to be defined. As a consequence, the matrix for the UMTS domain has a large number of columns. But, it performs for up to 8 applications nearly in real-time. IDA* as is used within TP4 applies a tableau to compute and store the possible successors of a node. The good results with TP4 are due to the property of the UMTS domain that only the minority of the actions (which represent call set-up modules) can be executed in parallel. Summarising, the Simplex makespan is in average 7% more compact compared to the computation with the planner TP4.

6   Conclusion

Scheduling domains for the set-up of mobile applications for UMTS have been described. The domains are based on the ZIMPL language for the Simplex method from the field of Discrete Mathematics, and on PDDL$^+$ for AI planners like TP4. The experimental results for the Simplex method provide an optimal schedule that is in average by 7% more compact than the one computed with the planner TP4. The planner TP4 has an average schedule generation time below 1 second for up to 20 applications. Compared to this provides the Simplex method for up to 8 applications a good schedule generation time that is nearly real-time.

In future we will try to optimise the Simplex domain, i.e. to reduce the number of binary variables that have to be used for the modelling. The number of columns in the ZIMPL matrix seems to be the bottleneck for the makespan generation time.

Acknowledgements

Grateful thanks to Thorsten Koch from the Konrad Zuse Institute Berlin for the intensive discussions we had and his sedulous effort to the modelling with ZIMPL.

References

1. Holma, H., Toskala, A. (eds.): WCDMA for UMTS. Wiley & Sons, U.K (2000)
2. Kotsis, G., Spaniol, O. (eds.): Euro-NGI 2004. LNCS, vol. 3427. Springer, Heidelberg (2005)
3. Hayzelden, A., Bigham, J. (eds.): Software Agents for Future Communication Systems. Springer, Heidelberg (1999)
4. Busuioc, M.: Software Agents for Future Telecommunication Systems. In: Hayzelden, A., Bigham, J. (eds.) Distributed Intelligent Agents - A Solution for the Management of Complex Telecommunications Services. Springer, Heidelberg (1999)
5. Maros, I.: Computational Techniques of the Simplex Method. Kluwer Academic Publishers, Dordrecht (2003)
6. McDermott, D.: The formal semantics of processes in PDDL. In: Proceedings of ICAPS Workshop on PDDL (2003)
7. Englert, R.: Planning to Optimize the UMTS Call Set-up for the Execution of Mobile Applications. Int. Journal of Applied Artificial Intelligence 19(2), 99–107 (2005)
8. Hoffmann, T., Edelkamp, S., Thiébaux, S., Englert, R., Liporace, F., Trüg, S.: Engineering Benchmarks for Planning: the Domains Used in the Deterministic Part of IPC4. Int. Journal of Artifical Intelligence Research (JAIR), 1–91 (2006)
9. Boche, H., Stanczak, S.: Optimal QoS Tradeoff and Power Control in CDMA Systems. IEEE Infocom. (2004)
10. Haslum, P., Geffner, H.: Heuristic planning with time and resources. In: Proc. IJCAI Workshop on Planning with Resources (2001)
11. Koch, T.: Rapid Mathematical Programming. PhD thesis, Technische Universität Berlin (2004)
12. Farjami, P., Görg, C., Bell, F.: Advanced service provisioning based on mobile agents. Computer Communications 23, 754–760 (2000); Special Issue: Mobile Software Agents for Telecommunication Applications
13. Ghallab, M., Milani, A. (eds.): New Directions in AI Planning. Series Frontiers in Artificial Intelligence and Applications, vol. 31. IOS Press, Amsterdam (1996)
14. Fox, M., Long, D.: PDDL2.1: An Extension of PDDL for Expressing Temporal Planning Domains. Journal of AI Research 20, 61–124 (2003)
15. Englert, R.: Execution scheduling for mobile applications in 3g networks, Habilitation thesis at the Faculty of Computer Science and Electrical Engineering, Technical University of Berlin (2008)
16. IPC: International planning competition (2004), http://ipc.icaps-conference.org

ESO: Evolutionary Self-organization in Smart-Appliances Ensembles

Stefan Goldmann and Ralf Salomon

Faculty of Computer Science and Electrical Engineering
University of Rostock, 18051 Rostock, Germany
{stefan.goldmann,ralf.salomon}@uni-rostock.de

Abstract. This paper presents an evolutionary algorithm applicable to the task of device adjustment in smart appliances ensembles. The algorithm requires very little environmental knowledge and is therefore complementary to the commonly applied rule based methods, such as ontologies. In contrast to traditional evolutionary algorithms, the new approach avoids any central processing scheme. Instead, the ensemble settings are distributed *physically* across all devices such that every parameter resides only in the device to which it belongs. This approach enables the correct handling of the dynamic nature of smart appliances ensembles, as will be shown in the course of the paper.

1   Introduction

The term "smart-appliances" refers to everyday-life devices, such as beamers, light bulbs, and the like, enhanced by some communication interfaces. Smart-appliances are considered an *ensemble* if they support the user in an autonomous and non-invasive way [1,8,10]. In a smart conference room, for example, the window blinds should automatically lower at the start of a presentation. If the room is large enough, a microphone for the speaking person as well as some speakers for the audience in the back rows should be switched on.

A common approach for solving this task is to utilize rule based methods, such as ontologies [3,4]. In this approach, the devices use rules to negotiate how to react to a given situation. Modelling such rule sets for all possible situations is a very complex and difficult task. For example, devices can join or leave an ensemble at any time without notification, e.g., due to users carrying devices along or due to device failures.

In order to reduce the required modelling, this paper proposes to employ an evolutionary algorithm to this task. However, the discussion presented in Section 3 suggests that traditional evolutionary algorithms are not suitable. They are based on a central processing scheme, which contradicts the dynamic nature of smart-appliances ensembles and may easily lead to inconsistencies between the actual physical ensemble and its representation on the central machine. Furthermore, each device knows best about the physical modality it controls and in what ranges this can happen. Therefore, Section 4 describes a new evolutionary algorithm, that

A. Dengel et al. (Eds.): KI 2008, LNAI 5243, pp. 209–216, 2008.

avoids any central processing. Instead, the ensemble setting is distributed *physically* across the devices, such that each parameter only resides in the device to which it belongs. First simulation results of a lighting scenario in a smart office, as presented in Section 6, indicate that the algorithm is capable of handling this dynamic task correctly. Section 7 concludes this paper with a short discussion and some future prospects.

2   Background: Evolutionary Algorithms

Evolutionary algorithms denote a class of numerical optimization methods for solving technical problems [2,7]. Herein, a problem is described as a set of parameters, also called genome, and an error criterion. The latter is also called fitness function and expresses the quality of a possible solution. An example could be the positioning of router stations in a large-area communication network such that the overall cable length becomes minimal. Here, the router locations are the parameters whereas the cable length is the fitness criterion.

The generic evolutionary processing scheme works as follows: (1) Start with some arbitrarily initialized genomes, also called parents. (2) During the mutation step, new genomes are generated from the parents. These are called children or offspring. For each child, a parent is chosen randomly and each of the child genes results from a Gaussian random modification of the associated parent gene. (3) The quality of each parent and child is evaluated using the fitness function. (4) Finally, the best out of the parents and children are selected for the next iteration, and the process continues with step 2.

This generic scheme has been successfully applied to a large number of different problems, such as machine learning, breast cancer detection, and VLSI design. They mainly differ in the number of parents and children, the selection scheme, and the variation operators applied. For a more detailed review, the interested reader is referred to the pertinent literature [2,5,7].

Please note that the remainder of this paper focuses on the $(1 + 1)$-evolution strategy. This notation indicates that the algorithm generates one child from one parent and that it selects the better one as the parent for the next iteration.

3   Problem Description

In smart-appliances ensembles, changes can occur at any time without notification. This does not only apply to devices joining or leaving the ensemble, but to any kind of change, such as varying user wishes or varying external influences like the sunshine.

In terms of evolutionary algorithms, this leads to the following properties of the task at hand: (1) The number of genes, i.e., devices, is not known in advance. (2) The relation of a genome to its fitness value is not necessarily persistent, since a change in the ensemble corresponds to a change of the fitness function. (3) Any genome assembled from the device settings and stored on some machine might be inconsistent with the actual ensemble.

This implies, that traditional evolutionary algorithms cannot be applied to this task. They are based on central processing of completely assembled genomes. Furthermore, they assume the number of genes and the fitness function to be fixed and known in advance; though these limitations might be resolved using the concept of variable-length genomes [6], as has been done in evolving the structure of neural networks [9], and by periodically revalidating the fitness of the genomes. Still, the bound to a central processing machine remains, which clearly contradicts the properties stated above.

4   The ESO Algorithm

Since the consideration of an assembled genome is not suitable for the application at hand, the evolutionary self-organizing algorithm, ESO for short, avoids any central processing, and instead distributes the genome *physically* across all devices such that each gene only resides in the device to which it belongs. Therefore, removing or adding devices automatically removes or adds the associated genes from the genome. With this conceptional modification in mind, the ESO algorithm works as follows:

1. All devices are split into two classes, sensors and actuators. Actuators are those devices that influence principal modalities, such as brightness, sound volume, and the like. Sensors are those devices that measure modalities. Each sensor is tagged with a target value. These values originate from a higher abstraction level, e.g., an intention module, and are beyond the scope of this paper. Each actuator stores its current gene, i.e., child gene, parent gene, as well as the parent fitness, which corresponds to the fitness of the so-far best ensemble setting found. This concept is the same for all actuators.
2. Each actuator determines its current gene from its parent gene according to the mutation operation mentioned in Section 2.
3. The sensors measure the changes in the physical modalities and broadcast their partial fitness values, e.g., the squared distance of target and measured value.
4. Each actuator uses the same formula to determine the overall fitness, e.g., the sum of the partial fitness values. The better of parent and current gene, along with the corresponding fitness, is selected as the new parent for the next iteration.
5. To cope with changes, the actuators periodically skip the Selection and subsequent mutation step. Instead, they swap their particular parent and current gene, and therefore revalidate the parent genome.
6. The process continues with step 2.

Please note, that a real-word implementation requires a suitable self-organizing synchronization mechanism to realize this scheme.

5   Methods

In order to validate the concepts of the proposed ESO algorithm, this paper is using a simulation. Such a simulation-based approach considers only those aspects, which are technically relevant for the algorithm. For the sake of simplicity, the simulation models the illumination of a certain number of desks by a certain number of light sources. The remainder of this section presents a description of the considered scenarios as well as the used parameter settings.

Configuration of the ESO algorithm: All experiments have been done with the (1+1)-ESO algorithm. This notation indicates that the algorithm generates one offspring from one parent and that it selects the better one as the parent for the next generation.

Sensors s_i**:** The ESO algorithm employs a user-specified number m of sensors s_i. In the validation study presented in Section 6, these sensors measure the illumination at various locations, e.g., the users' desks.

Fitness function f**:** All sensors calculate their particular fitness contribution $f_i = (s_i - s_i^t)^2$ as the square of the current sensor reading s_i and its target value s_i^t. By means of a global communication infrastructure, all sensors broadcast their fitness contributions f_i across the system such that every device can calculate the ensemble's total fitness as

$$f = \sum_{i=1}^{m} f_i = \sum_{i=1}^{m} (s_i - s_i^t)^2 \ . \tag{1}$$

Actuators a_i**:** The simulation employs n actuators a_i, which represent n light sources that are distributed in the environment. Every actuator employs its private mutation operator $a_i \leftarrow a_i + s * N(0,1)$, with s denoting a private step size. Without loss of generality, all actuator values a_i are bound to $0 \leq a_i$. Unless otherwise stated, the step size is set to $s = 0.1$ in all simulation scenarios. Please remember that these actuators are not explicitly known to the sensors, the fitness evaluation, or the system. Rather, the environment, i.e., the physics, autonomously mediate their modalities towards the sensors s_i.

Simulation setup: The simulation models the physics and the effects the actuators have on the sensors as follows:

$$s_j = g + \sum_{i=1}^{n} w_{ij} * a_i \ , \tag{2}$$

with g denoting a global illumination source, such as the sunshine, and w_{ij} denoting the influence of the i-th actuator onto the j-th sensor. The weights w_{ij} subsume all the relevant physical effects, such as the light sources' positions, their brightness, their illumination characteristics, etc.

In the scenarios described below, the weight configuration w_{ij} referres to the small office illustrated in Fig. 1. In this setup, all parameters are deliberately set to values such that the setup does not contain any symmetry, which might weaken the achieved results. Therefore, the sensor target values are (arbitrarily) set to $s_1^t = 0.7$ and $s_2^t = 1.2$.

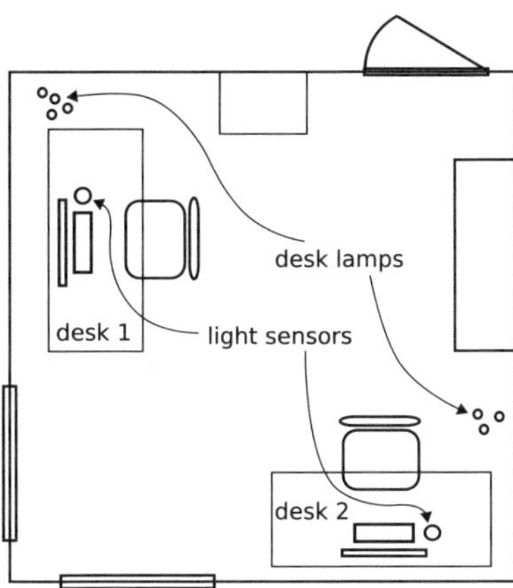

Fig. 1. The test scenario consists of some light sources as well as two desks with light sensors mounted close to the keyboards

Depending on the chosen scenario (please, see below), the weights, the sensor target values, and the number of actuators spontaneously change over time, i.e., $w_{ij}(t)$, $g(t)$, $s_i^t(t)$, and $n(t)$ are time dependent. For the purpose of readability, the time t is omitted.

Scenario 1, System Startup: At startup time, all actuator values are set to $a_i = 0$ and $g = 0$, and the sensor target values are set to $s_1^t = 0.7$ and $s_2^t = 1.2$. The ensemble is then responsible to power up the lamps to the desired level. This situation resembles an early winter morning, when the users enter the dark office.

Scenario 2, Scalability: This scenario increases the number of actuators from $n = 2$ to $n = 100$. The goal of this scenario is to test the scaling behavior of the ESO algorithm.

Scenario 3, System Dynamics: This scenario focuses on the ensemble behavior in more dynamic setups in which light sources might fail or join. It starts off with two light sources per desk, i.e., $n = 4$. In simulation step $t = 150$, a light source of each desk fails. Then, in simulation step $t = 300$, the sensor target value of desk 2 is increased from $s_2^t = 1.2$ to $s_2^t = 1.5$. This might model a situation in which a different person starts working and who might prefer a brighter desk. This situation implies that the illumination of the other desk remains unchanged.

Scenario 4, External Effects: This scenario starts off like the first one. But in simulation step $t = 150$, the external (sunshine) illumination is set to $g = 1$, and is reduced to $g = 0.5$ im simulation step $t = 300$. This scenario models the influence of external modalities, which are outside the control of the ensemble. A further challenge is that during time steps $150 \leq t \leq 300$, the ensemble cannot reach the specified target values, since the external illumination already exceeds target value s_1^t.

6   Results

The simulation results of the four experiment are summarized in Fig. 2 to Fig. 5. On the x-axis, these figures show the simulation time, and on the y-axis, they show the target sensor values s_i^t, their actual readings s_i, and the global ensemble fitness f (Eq. (1)). All data were obtained from 500 independent runs. From all these runs, the figures always present the values that was *worst* in the corresponding time step t.

All figures clearly indicate that the global system error decreases exponential, and that thus after some adaption time, the ESO algorithm arrives at the specified target values, i.e., $s_i \approx s_i^t$. This is not only observable in the simple Scenarios 1 and 2 but also in the more dynamic one (Fig. 4) in which the target values change over time. The only exception occurs in the fourth scenario (Fig. 5) between time steps 150 and 300. However, it has already been discussed above that the ESO algorithm cannot reach the target values, since the external illumination is brighter than the specification demands. In other words, the small deviation from the optimum is not due to the ESO algorithm but due to the physics; but even in this case, the ESO algorithm returns to the optimum shortly after the reduction of the external illumination in time step $t = 300$.

It might be quite interesting to take a look at the scaling behavior of the proposed self-organization algorithm. Normally, an increasing number of components slow down the system convergence speed. However, a comparison of Fig. 2 and Fig. 3 indicate that a larger ensemble (i.e., 50 light sources per desk) reaches the optimum faster than a smaller one (i.e, 2 light sources per desk). This effect is counter intuitive but probably due to a significantly increased number of actuator configurations that match the optimum sensor readings.

Rather than focusing on convergence speed, the application at hand focuses on a smooth adaption of its actuators. To this end, the step size s should be

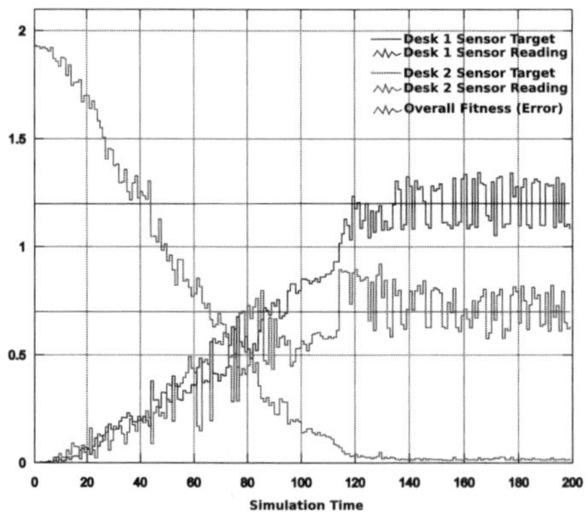
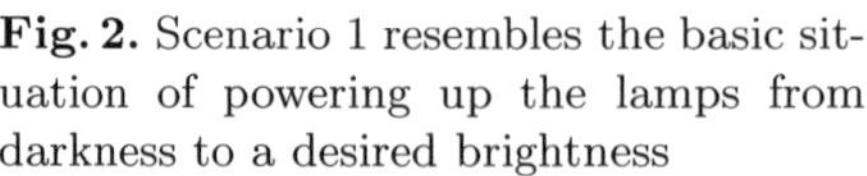
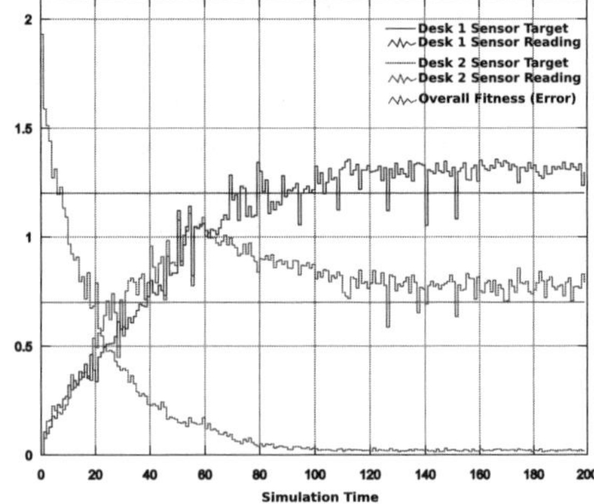

Fig. 2. Scenario 1 resembles the basic situation of powering up the lamps from darkness to a desired brightness

Fig. 3. Scenario 2 resembles Scenario 1, but with 100 instead of only 2 lamps

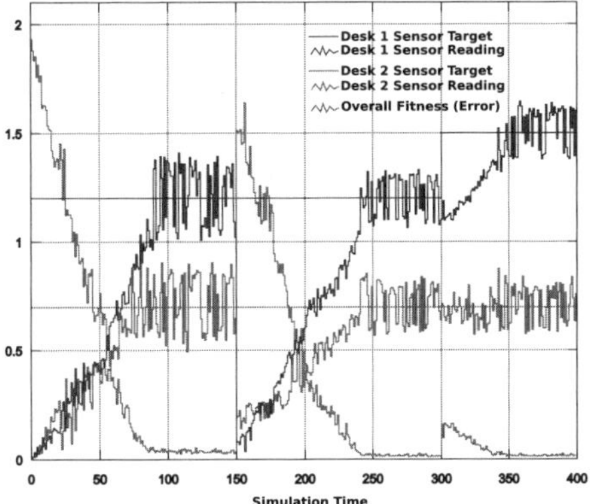

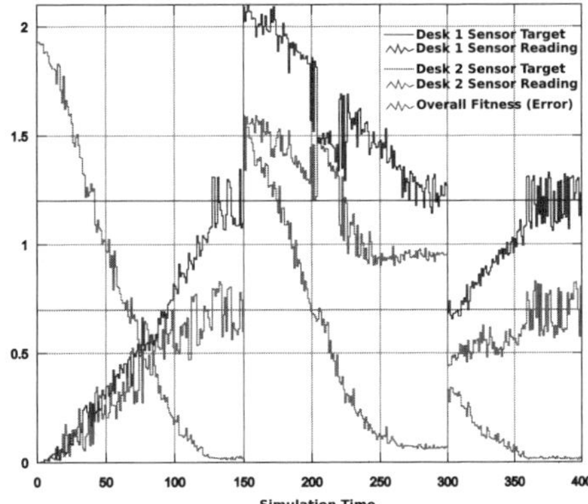

Fig. 4. Scenario 3 focuses on the ensemble behavior in more dynamic setups in which light sources might fail or join. At time $t = 150$ two of four lamps fail, at time $t = 300$ the second sensor target value is increased from $s_2^t = 1.2$ to $s_2^t = 1.5$.

Fig. 5. This scenario models the influence of external modalities, which are outside the control of the ensemble. Especially, between $t = 150$ and $t = 300$, the external influence already exceeds the target value s_1^t and therefore prevents the system to reach the specified goal.

set to rather small values; larger values would speed up the adaptation process, but would also induce significant fluctuations around the optimum, which might be rather annoying in real-world applications. Furthermore, a larger number of actuators, such as a total of 100 light sources used in scenario 2, requires a smaller step size, such as $s = 0.02$. Otherwise, the fluctuations would be way too large to be acceptable. Therefore, future versions of the ESO algorithm should employ a proper self-adaption scheme of the step size s.

7   Conclusions

This paper has proposed a distributed evolutionary algorithm, called ESO, for the self-organization of smart-appliances ensembles. A key feature of this algorithm is that it does not maintain assembled genomes in the traditional sense. Rather, the ESO algorithm physically distributes all gene values across all the devices, and evaluates only the resulting sensor modalities. Furthermore, the application of the mutation operators is done by the actuator rather than a central processing instance.

The presented simulation results indicate that the proposed method is suitable as the self-organization process for smart-appliances ensembles. In addition to the required basic adaptation capabilities, the ESO procedure scales well, and is also able to cope with the inherent system dynamics of those ensembles.

The experiments also show that the behavior of ESO algorithm depends on the chosen step size s of the mutation operators, privately employed in every actuator. Therefore, future research will be devoted to the development of an adequate self-adaption mechanism.

Ongoing research is developing hardware equipment that consists of remotely controllable light sources as well as remotely readable sensors. All these devices are equipped with a wireless communication module. Preliminary results indicate that in the real-world, the ESO algorithm also has to cope with varying time constants. For example, slight brightness changes are faster than high brightness changes. Furthermore, these timing constants also depends on the chosen lamp and potentially other system parameters.

Acknowledgements

The authors gratefully Ralf Joost and Ulf Ochsenfahrt for encouraging discussions and valuable comments on draft versions of this paper. This work was supported in part by the DFG graduate school 1424.

References

1. Aarts, E.: Ambient intelligence: A multimedia perspective. IEEE Multimedia 11(1), 12–19 (2004)
2. Bäck, T., Hammel, U., Schwefel, H.-P.: Evolutionary Computation: Comments on the History and Current State. IEEE Transactions on Evolutionary Computation 1(1), 3–17 (1997)
3. Bry, F., Hattori, T., Hiramatsu, K., Okadome, T., Wieser, C., Yamada, T.: Context Modeling in OWL for Smart Building Services. In: Brass, S., Goldberg, C. (eds.) Tagungsband zum 17. GI-Workshop über Grundlagen von Datenbanken, Wörlitz, Germany, Gesellschaft für Informatik, Institute of Computer Science, Martin-Luther-Univerity Halle-Wittenberg, pp. 38–42 (2005)
4. Chen, H., Perich, F., Finin, T., Joshi, A.: SOUPA: Standard Ontology for Ubiquitous and Pervasive Applications. In: International Conference on Mobile and Ubiquitous Systems: Networking and Services, Boston, MA (2004)
5. Fogel, D.B.: Evolutionary Computation: Toward a New Philosophy of Machine Learning Intelligence. IEEE Press, NJ (1995)
6. Lee, C.-Y., Antonsson, E.K.: Variable Length Genomes for Evolutionary Algorithms. In: Whitley, L.D., Goldberg, D.E., Cantú-Paz, E., Spector, L., Parmee, I.C., Beyer, H.-G. (eds.) Proceedings of the Genetic and Evolutionary Computation Conference (GECCO 2000), p. 806 (2000)
7. Rechenberg, I.: Evolutionsstrategie, Frommann-Holzboog, Stuttgart (1994)
8. Saha., D., Mukherjee, A.: Pervasive computing: A paradigm for the 21st century. IEEE Computer, 25–31
9. Schiffmann, W., Joost, M., Werner, R.: Application of Genetic Algorithms to the Construction of Topologies for Multilayer Perceptrons. In: Proceedings of Artificial Neural Networks and Genetic Algorithms, pp. 675–682 (1993)
10. Weiser, M.: Some computer science issues in ubiquitous computing. Communications of the ACM 26(7), 75–84 (1993)

On-Line Detection of Rule Violations
in Table Soccer

Armin Hornung and Dapeng Zhang

Albert-Ludwigs-Universität Freiburg, Department of Computer Science
Georges-Köhler-Allee 052, 79110 Freiburg, Germany
{hornunga,zhangd}@informatik.uni-freiburg.de

Abstract. In table soccer, humans can not always thoroughly observe
fast actions like rod spins and kicks. However, this is necessary in order
to detect rule violations for example for tournament play. We describe
an automatic system using sensors on a regular soccer table to detect
rule violations in realtime. Naive Bayes is used for kick classification, the
parameters are trained using supervised learning. In the on-line experiments, rule violations were detected at a higher rate than by the human
players. The implementation proved its usefulness by being used by humans in real games and sets a basis for future research using probability
models in table soccer.

1 Introduction

Table soccer – also known as *"Foosball"* – is a popular game, often played in pubs
or other social contexts. Two teams of up to two players compete at scoring goals
by controlling and kicking a ball on a game table with playing figures attached
to rods. Playing table soccer requires skill at controlling the rods and quick
reactions, because it is a very fast game. The ball can be kicked by rotating the
rods, and professional players can shoot the ball across the table within a few
hundred milliseconds.

Currently, the sport is also becoming more and more popular on a tournament
level. There are various local and national leagues, and internationally, the ITSF[1]
is organizing championships with a set of unified rules.

Because of the high speed of table soccer, a human is not always able to
thoroughly observe all actions in the game, such as the fast turning of a kicking
rod. However, this is necessary to make objective decisions about rule violations.

We present an approach to automatically detect rule violations on-line, using
high frequency sensor data to classify and evaluate game situations. This helps
humans playing table soccer, for example when practising for tournament games.

Most rules covering game mechanics depend on the detection of a playing
figure kicking the ball. Thus, we concentrate on this key element, and detect
violations of one of the most important rules: Rods may not be rotated by more
than 360 degrees before or after ball contact, the angles before and after ball

---

[1] International Table Soccer Federation, http://www.table-soccer.org/

A. Dengel et al. (Eds.): KI 2008, LNAI 5243, pp. 217–224, 2008.

contact are not added. This prevents players from spinning the rods in an uncontrolled manner when kicking. In this paper, the coherent rotation movement of a rod is referred to as a *spin*.

The sequential sensor data for the game rods and ball is segmented, and game situations are classified as kicks. Because the sensor data is noisy, a probabilistic model for kick detection is needed. We use supervised learning to train a naive Bayes classifier for continuous variables.

The paper is organized as follows. We will first describe related work in the domain of table soccer and probability models, while deriving the theoretical background in Section 2. The methods used for segmentation and classification are described in Section 3, followed by the experimental setup and results in Section 4. Finally, we will conclude our work in Section 5.

2   Related Work

2.1   Robotic Table Soccer

The first work on a robot playing table soccer was published by Weigel and Nebel [1] in 2002. A regular soccer table was equipped with control units to manipulate the rods. The state was perceived by a camera mounted over the table. Analyzing the game state and controlling the actuators with a regular PC, *KiRo* ("Kicker Robot") won most games against amateur players. In 2005, the first commercial product based on KiRo became available under the name "*Star-Kick*", posing a challenge even to advanced players [2].

The current implementations of KiRo and Star-Kick do not take the enemy figures into consideration, the world model includes just the positions of the ball and the controlled figures. Future improvements could be to play on a professional level using a more advanced ball control, avoiding the enemy figures and planning actions such as passes.

For research in this direction and the work presented in this paper, a regular game table was outfitted with sensors to record ball positions as well as rod positions and angles [3].

2.2   Probability Models

A common approach to model dependencies between random variables is a *Bayesian network* in the form of a directed graph [4]. Nodes represent random variables or classes, while directed edges denote dependencies. In a classification problem, the network is used to compute the posterior probabilities of all classes c_k given the observed values of the attributes: $P(c_k|x_1, \ldots, x_n)$. The class with the highest probability is then assigned to the observation. Using *Bayes' Theorem*, the posterior of a class c can be computed as

$$P(c|x_1, \ldots, x_n) = \frac{P(x_1, \ldots, x_n|c)P(c)}{P(x_1, \ldots, x_n)} \, , \tag{1}$$

which flips the conditioning to make the distribution easily learnable.

We assume that all classes are equally probable, all attributes independent given the class and that the prior is independent of the class. This greatly simplifies computation to:

$$P(c|x_1,\ldots,x_n) = \alpha \prod_{i=1}^{n} P(x_i|c) \ . \tag{2}$$

With α being a normalization constant, this is the naive Bayes classifier. Even though the "naive" assumptions seldom hold in reality, naive Bayes has been attested a good performance in many domains and can even be the optimal classifier with respect to the misclassification rate in some cases [5,6].

In this work, observed values are not discrete but continuous. One common approach is to discretize the attributes [7]. However, we believe that discretization will lead to system degradation by discarding information like the probability for a kick. Instead, we model the random variables to be Gaussian-distributed and use the joint probability of the distributions for each variable.

Common extensions to naive Bayes include *tree-augmented naive Bayesian networks* [8], where additional dependencies between the attributes can be modeled. The cost is a higher algorithmic complexity. To cope with the high sampling frequency to observe games on-line, an efficient implementation is needed in our case. Our experiments revealed that the classification performance of naive Bayes is good enough for this application.

With continuous and Gaussian-distributed attributes in (2), it follows that:

$$P(c|x_1,\ldots,x_n) = \alpha \prod_{i=1}^{n} \varphi_i(x_i) = \alpha \prod_{i=1}^{n} \frac{1}{\sqrt{2\pi\sigma_i^2}} e^{\frac{-(x_i-\mu_i)^2}{2\sigma_i^2}} \ . \tag{3}$$

Variance σ_i^2 and mean μ_i of each attribute i are determined with supervised learning for a known class label c. m attribute values $x_{i,1},\ldots x_{i,m}$ are sampled for each attribute i and the unbiased estimators for mean and variance are used on these values.

3   Segmentation and Classification

3.1   Sensor Input

To cope with the high speed of the game, the sensor data is read and recorded at roughly 250 Hz using a standard PC, distributing the data stream over network for further evaluation on other machines.

The ball is located with a *Sick LMS 400* laser measurement system, scanning through the gap between table surface and feet of the playing figures. Rod positions are measured with optical distance sensors and rod angles are observed with magnetic rotary encoders. Overall, there is almost no additional friction on the rods, enabling an unhindered game play.

To analyze the game state, one needs to segment the sequential sensor data and detect key events in it. The sensors introduce Gaussian-distributed noise on the measured signal.

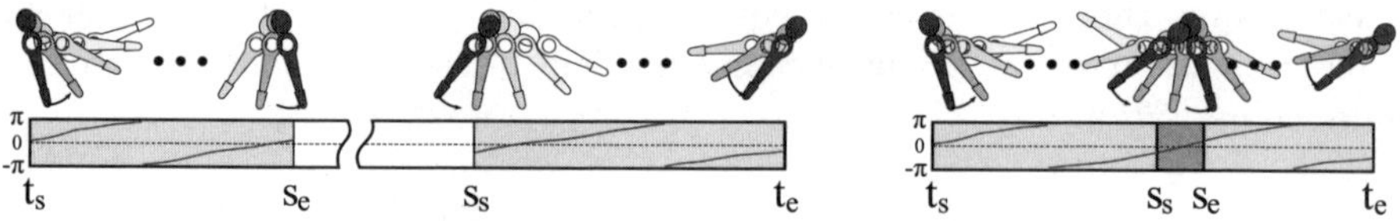

Fig. 1. The timestamps t_s, s_e, s_s and t_e defining one complete spin in a rotation covering an angle of more (left) and less than 4π (right)

Currently, three rods are observed by sensors, all in the same half of the table: blue attacker, red defender and red goalkeeper.

3.2 Rotation Segmentation

With respect to the rule violations described in Section 1, a rod spin is completely parametrized by four timestamps: t_s and t_e, the starting and end times of the whole spin, as well as s_e and s_s, the times when the first 2π part of the rotation ends and the time when the last 2π part starts (Fig. 1). These discrete timestamps need to be detected in the stream of angle measurements for each rod i.

To do so, the difference in between two successive rod angles α_i is observed as approximation of the derivative:

$$\Delta_i(t) = \alpha_i(t) - \alpha_i(t-1) \ . \tag{4}$$

When crossing from 2π to 0, the values are adjusted accordingly. As soon as $|\Delta_i(t)| \geq \varepsilon$ for some small threshold ε, the start of a spin movement is detected, and it lasts until $|\Delta_i(t)| < \varepsilon$. To reduce the influence of noise, the signal is smoothed by using the running average over a window of size three.

As soon as $\alpha_i(t_s)$ is passed for the second time, the time s_e is detected. s_s depends on $\alpha_i(t_e)$, the angle at which the spin stops, and is detected by using a circular array or ring buffer indexed by angle. For additional robustness against noise, monotonicity in between start and end of the spin is enforced when storing timestamps in the circular array. Also for noise robustness, all bins in between storing two successive timestamps need to be emptied.

3.3 Kick Detection

Naive Bayes classifiers are used to detect kicks, by using the relation of ball and active figure as input. The active figure is the one closest to the ball on the rod that is within range of the ball. The two directions of a kick are distinguished by classifying two cases, forward and backward.

Input for each classifier are the continuous attributes x_1, x_2, x_3, computed from coordinates of ball and active rod, and its angle. Figure 2 displays the coordinates relative to the playing figure, Fig. 3 displays the resulting Bayesian network. Note that x_3 is directly related to the vertical distance between ball and playing figure, because it is assumed that the ball always touches the table

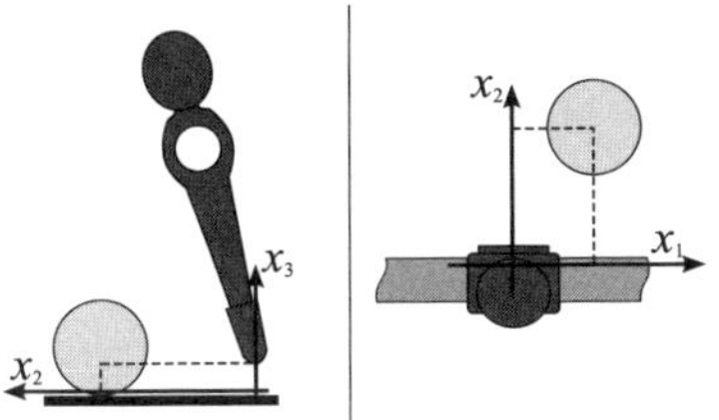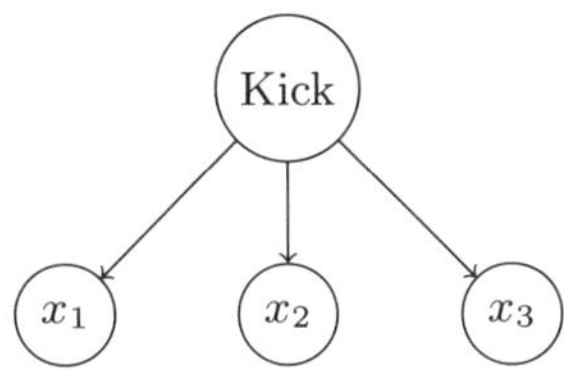

Fig. 2. Coordinates of ball relative to the active figure from the side (left) and top(right)

Fig. 3. Bayesian network of the kick classifier

when kicking. When it is not on the surface, e.g. resulting from a fast kick, it cannot be observed by the laser anyways.

The probability can now be computed according to (3) with $n = 3$ variables. In the implementation the log-likelihood is used instead, turning the products into sums. This is computationally more stable because very small probabilities are avoided, while it is equivalent in terms of classification. Furthermore, the constant α is ignored for classification because it is assumed to be identical for all classes.

To detect violations of the rod-spinning rule efficiently, the log-likelihood for kicks is constantly recorded. As soon as a spin covering an angle of more than 2π is detected, the peak of the kick likelihood in the intervals $[s_e, t_e]$ and $[t_s, s_s]$ (see 3.2) is compared to an experimental threshold. Depending on the direction of the spin, the forward or backward probability is used.

4   Experiments

In the experiments, we first evaluate the supervised learning performance of the kick classifier. Then, the parameters for the kick model for on-line detection are trained. Finally, we test the performance of violation detection in real table soccer games on-line.

4.1   Supervised Learning Performance

To evaluate the learning performance, 50 recorded kick actions were randomly partitioned into a test set of size 10 and a training set of size 40. In each single action k, the instant when the playing figure touches the ball is selected by hand, and the variable state $(x_{1,k}, x_{2,k}, x_{3,k})$ is extracted.

The parameters of the kick model μ_i, σ_i for $i \in \{1, 2, 3\}$ are then learned incrementally, using $k = 1, \ldots, 40$ samples of the training set as input. The performance on classifying the test set is evaluated for each step.

As performance measure V, the sum over the normalized probability of each test sample $x_{i,l}$ is used:

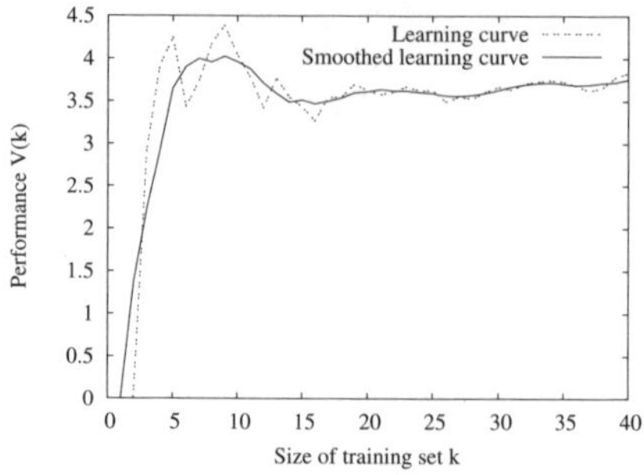

Fig. 4. Learning curve for 40 kick samples on a testing set of size 10. The smoothed curve displays the average over a running window of five values.

Fig. 5. Angles and first derivative for one spin, lasting about 0.3s. The detected times of t_s, s_e, s_s, t_e are marked.

$$V(k) = \sum_{l=1}^{10} \gamma_k \prod_{i=1}^{3} \varphi_{i,k}(x_{i,l}) \ . \tag{5}$$

The Gaussian distribution $\varphi_{i,k}$ is parametrized by mean and variance of variable x_i, using k samples of the training set. The normalization constant is determined by the maximum of the joint probability, reached at the mean:

$$\gamma_k = \prod_{i=1}^{3} \left(\varphi_{i,k}\left(\mu_{i,k}\right)\right)^{-1} \ . \tag{6}$$

This scales all Gaussian distributions in (5) to the range of $[0,1]$. Otherwise, their probabilities would not be comparable, because a more general model creates a more shallow distribution.

The resulting learning curve is shown in Fig. 4. A first peak is reached after using only six samples for training, and the performance of the learned model stays stable after using 16 samples for training. A training set significantly larger than 20 samples only leads to small improvements.

4.2 Parameter Learning

To classify kicks in running games, only the model for a forward kick is learned. The parameters for a back kick are obtained by negating μ_2.

To train the parameters of the kick model, the variable state (x_1, x_2, x_3) is sampled from several configurations with the ball in front of a playing figure, varying positions, angles and playing figures.

Table 1 displays the parameters trained through the sampling process, using 27 samples. As expected, the peak is centered in front of the playing figure. The rectangular footprint of the playing figure results in a larger variance in x_1 than in x_2.

Table 1. Parameters obtained from training by sampling kick positions

$i =$	1	2	3
Mean μ_i	-1.37	23.54	13.02
Variance σ_i^2	30.30	7.09	3.21

4.3 Spin Detection

Spins covering angles of more than 2π and the classifying timestamps t_s, s_e, s_s and t_e are successfully detected. Figure 5 exemplarily shows the sensor recording of one spin movement starting at $t_s = 31$ and ending at $t_e = 105$, with the detected timestamps marked. Note the noisy signal for $\alpha(t)$, and that the whole spin from the angle $\alpha(t_s) \approx 5.85$ to $\alpha(t_e) \approx 4.75$ (rad) lasts only 74 sensor ticks, which is about 0.3s.

4.4 Detecting Violations On-Line

To evaluate the detection rate in real games, test subjects not familiar with the system played several games on the table. The position of the players on the table (attacker or defender for the blue or red team) were exchanged regularly. Two players rated themselves as "good", two "average" and the remaining seven as "amateurs". In total, 9 games were recorded for about 1 hour and 45 minutes, silently logging the detected rule violations. Before the games, the rod spinning rule was explained to the players, and they were asked to evaluate whether they think they violated the rule after each shot. Afterwards, the recorded sensor logs were manually inspected in slow-motion for rule violations, and compared to the detections of the system and the players.

All in all, there were 42 rod spins of more than 360 degrees on the observed three rods, of which 19 were illegal kicks. The players themselves were aware of only two of them (10.52%), while the system detected 17 violations correctly (89.47%). Two violations were missed by the system, and one false positive detected. When a rod is spun and misses the ball closely instead of kicking it, there is a chance that a rod spin violation is detected. The noise on the rod and ball data might add up so that the ball is falsely located as being touched by the rotating figure, resulting in the false positive detection. False negatives occur for similar reasons.

The referee system runs efficiently on a standard PC with a 2.66 GHz Pentium 4 CPU and 1 GB RAM, running SuSE Linux 10.1. The application uses just a small fraction of the available CPU power. Most of the CPU power is used for a 3D display of the soccer table in the user interface, which shows the live representation of the soccer table and slow-motion replays of rule violations. The distributed implementation of the system allows the display to be easily outsourced to a dedicated machine, leaving more processing power for example to detect additional rules.

5   Conclusion

In this work, we presented an approach to automatically detect rule violations in table soccer games. A naive Bayes classifier is trained off-line to detect kicks, using the relation of ball and closest playing figure as input.

The classifier demonstrated a good performance in the on-line classification experiments, detecting 89.47% of all rule violations, while the human players only detected 10.52%. The efficient implementation of naive Bayes enables the system to run effortless on a standard PC, evaluating the high-frequency sensor data on-line during a running game. All this demonstrates the usefulness of our implementation. Additional robustness on the classification method can be achieved in future work by taking the ball movement into account for kick detection.

Finally, future research on table soccer can benefit from the classification method described here, such as learning by imitation or game analysis. The classifier could be used to detect various relations between ball and playing figures.

References

1. Weigel, T., Nebel, B.: Kiro – an autonomous table soccer player. In: RoboCup 2002, pp. 384–392. Springer, Heidelberg (2002)
2. Weigel, T., Nebel, B.: Kiro – a table soccer robot ready for the market. In: ICRA, Spain, pp. 4277–4282 (2005)
3. Zhang, D., Nebel, B.: Recording and segmenting table soccer games – initial results. In: ISSS 2007, pp. 193–195. Keio University, Tokyo (2007)
4. Pearl, J.: Bayesian networks: A model of self-activated memory for evidential reasoning. In: Proceedings of the 7th Conference of the Cognitive Science Society, August 1985, pp. 329–334. University of California, Irvine (1985)
5. Domingos, P., Pazzani, M.: On the optimality of the simple Bayesian classifier under zero-one loss. Machine Learning 29(2-3), 103–130 (1997)
6. Zhang, H.: The optimality of naive Bayes. In: Barr, V., Markov, Z. (eds.) FLAIRS Conference. AAAI Press, Menlo Park (2004)
7. Dougherty, J., Kohavi, R., Sahami, M.: Supervised and unsupervised discretization of continuous features. In: International Conference on Machine Learning, pp. 194–202 (1995)
8. Friedman, N., Geiger, D., Goldszmidt, M.: Bayesian network classifiers. Machine Learning 29(2-3), 131–163 (1997)

Extracting and Verifying Hyponymy Relations Based on Multiple Patterns and Features

Lei Liu[1], Sen Zhang[1], Lu Hong Diao[1], Shu Ying Yan[2], and Cun Gen Cao[2]

[1] College of Applied Sciences, Beijing University of Technology
{liuliu_leilei,zhangsen,diaoluhong}@bjut.edu.cn
[2] Institute of Computing Technology, Chinese Academy of Sciences
{cgcao,yanshuying}@ict.ac.cn

Abstract. Hyponymy relations play a crucial role in various natural language processing systems. Automatic acquisition and verification of hyponymy relations is a basic problem in knowledge acquisition from text. We present a method that acquires and verifies hyponymy relations based on multiple patterns and features. It initially obtains a set of removable patterns using Chinese lexico-syntactic patterns. Then the concepts of constituting hyponymy relation are acquired with removable patterns. Finally, hyponymy relations are verified with space structure features, semantic features and context features. Experimental results demonstrate good performance of the method.

Keywords: Hyponymy Relation, Relation Acquisition, Knowledge Acquisition, Hyponymy Verification.

1   Introduction

Automatic acquisition of concepts and semantic relations from text has received much attention. Especially, hyponymy relation acquisition is a more interesting and fundamental because hyponymy relations play a crucial role in various NLP (Natural Language Processing) systems, such as systems for information extraction, information retrieval, and dialog systems. Hyponymy relations are also important in accuracy verification of ontologies, knowledge bases and lexicons [1][2].

The types of input used for hyponymy relation acquisition are usually divided into three kinds: the structured text (e.g. database), the semi-structured text (e.g. dictionary), and free text (e.g. Web pages) [3][4]. Human knowledge is mainly presented in the format of free text at present, so processing free text has become a crucial yet challenging research problem.

In this paper, we present a method that acquires and verifies hyponymy relations based on multiple patterns and features from Chinese free text. Experimental results show that the method is adequate of extracting hyponymy relations from Chinese free text. The rest of the paper is organized as follows. Section 2 describes related work, section 3 shows few Chinese hyponymy patterns, section 4 elaborates the framework of this method, section 5 gives a performance evaluation, and finally section 6 concludes the paper.

A. Dengel et al. (Eds.): KI 2008, LNAI 5243, pp. 225–232, 2008.

2 Related Work

Given two concepts x and y, there is the hyponymy between x and y if the sentence "x is a (kind of) y" is acceptable. This relation is transitive and asymmetrical under the condition of the same sense[1]. We denote a hyponymy relation as hr(x, y). For example hr(dog, animal).

There are two main approaches for automatic/semi-automatic hyponymy acquisition. One is pattern-based (also called rule-based), and the other is statistics-based. The former uses the linguistics and natural language processing techniques (such as lexical and parsing analysis) to obtain hyponymy patterns, and then makes use of pattern matching to acquire hyponymy, and the latter is based on corpus and statistical language model, and uses clustering algorithm to acquire hyponymy. The pattern-based approach is dominant. The so-called patterns include special idiomatic expressions, lexical features, phrasing features, and semantic features of sentences. Patterns are acquired by using the linguistics and natural language processing techniques.

One of the first studies was done by Hearst [5][6]. Hearst proposed a method for retrieving concept relations from unannotated text (Grolier's Encyclopedia) by using predefined lexico-syntactic patterns, such as

$$...NP_1 \text{ is a } NP_2... \qquad ---hr(NP_1, NP_2)$$
$$...NP_1 \text{ such as } NP_2... \qquad ---hr(NP_2, NP_1)$$
$$...NP_1 \{, NP_2\}*\{,\} \text{ or other } NP_3 ... \quad ---hr (NP_1, NP_{3)}, hr (NP_2, NP_3)$$

Other researchers also developed other ways to obtain hyponymy. Most of these techniques are based on particular linguistic patterns.

Caraballo used a hierarchical clustering technique to build a hyponymy hierarchy of nouns like the hypernym-labeled noun hierarchy of WordNet from text [7]. Morin and Jacquemin produced partial hyponymy hierarchies guided by transitivity in the relation, but the method works on a domain-specific corpus [8]. Llorens and Astudillo presented a technique based on linguistic algorithms, to construct hierarchical taxonomies from free text. These hierarchies, as well as other relationships, are extracted from free text by identifying verbal structures with semantic meaning [9].

3 Chinese Lexico-Syntactic Patterns

In Chinese, one may find several hundreds of different hyponymy patterns based on different quantifiers and synonymous words, which is equivalent to the single hyponymy pattern in English. An example is as follows.

Pattern 1: <?C3><如|象|包括|分|包含|囊括|涵盖|有|指|即><?C1>{<或|及|以及|和|与|、 ><?C2>}*<等>

(Pattern: <?C3> such as <?C1>,<?C2>...)

Pattern 1 means "Pattern: <?C3> such as <?C1>,<?C2>...". Where the symbols "<"and ">" are a starting tag and an ending tag of an item. These items of the pattern are divided into constant item and variable item. Constant item is composed of one or more Chinese words or punctuations. Variable item is a non-null string variable. "<?C1>" is a variable item in the pattern, where the symbol "?" is an identifier. "|"

expresses logical "or". "如|象" means "such as"; "包括|分|包含|囊括|涵盖|有" means "include"; "指|即" means "namely"; "和|与" means "and"; "或" means "or"; "等" means "etc."; "及|以及" denotes "as well as"; Chinese dunhao "、" is a special kind of Chinese comma used to set off items in a series.

Chinese hyponymy patterns will be used to capture concrete sentences from Chinese free corpus. In this process, variables <?C> will be instantiated with words or phrases in a sentence, in which real concepts may be located. Let c and c' be the real concept in <?C>. If hr(c, c') is true, then we tag c by c_L, and c' by c_H, as shown below.

$\{\{\,$ 农 作 物 $\}c_H$ 主 要 $\}_{<?C5>}$／ 有 ／$\{\{\,$ 水 稻 $\}c_L\}_{<?C1>}$、 $\{\{\,$ 玉 米 $\}c_L\}_{<?C2>}$、 $\{\{\,$ 红 薯 $\}c_L\}_{<?C3>}$、 $\{\{$烟叶$\}c_L\}_{<?C4>}$／ 等／

(The farm crop mainly includes paddy rice, corn, sweet potato, tobacco leaves etc.)

We can acquire hr(水稻,农作物), hr(玉米,农作物), hr(红薯,农作物) and hr(烟叶, 农作物) from the above example.

The problem now is that after a sentence matches a hyponymy pattern, how can we identify the real concepts from the sentence and how can we verify that they satisfy the hyponymy? It is difficult to resolve those problems for several reasons [10]:

(1) As we know, Chinese is a language different from any western language. A Chinese sentence consists of a string of characters which do not have any space or delimiter in between; Chinese word order is strict; Chinese lacks morphological change, and has no the explicit variety tag of plural, possessive and part of speech.

(2) The structural degree of free text is very weak, and the expression is vivid and diverse. It is difficult to give a uniform structured representation (different from dictionary) or know the domain of knowledge (different from domain dependent text).

4   Extraction and Verification

Our method consists of three phases. Firstly, we pre-process the raw corpus. Raw corpus is gathered from Chinese free text, and is preprocessed in a few steps, including word segmentation, part of speech tagging, and splitting sentences according to periods. Then we acquire the processed corpus by matching Chinese hyponymy patterns. Processed corpus is divided into two groups: training corpus and testing corpus.

Secondly, to handle these problems in section 3, we present a semi-automatic algorithm for acquiring and analyzing non-concept components, and then convert them into removable patterns. Then we remove non-concept components using these removable patterns. After the outside layer removal to <?C>, we continue to analyze the structure of the remainder by lexical analysis (such as word segmentation, and part of speech tagging), and make use of these information as the proof to judge whether the concepts are correct.

Thirdly, we analyze the features of hyponymy and combine the space structure features, semantic features and context features of hyponymy together. Then, we match the candidate hyponymy with these features. If it attains a predefined certain threshold, we can conclude that it is a real hyponymy.

4.1 Acquisition of Concepts

Let s be a sentence that matches Chinese hyponymy patterns, p be a pattern. If p belongs to a part of non-concept components in s, then p is called removable pattern. The algorithm of acquisition of removable patterns is as follows.

Let P be a list of removal patterns initially empty

Input: Training corpus Cor_{train}, thresholds θ_1 and θ_2

Step1: Separate each of sentences in Cor_{train} according to punctuation marker. Let $S = \{s_1, s_2, \ldots, s_n\}$ be a list of separate string.

Step2: For each pair of strings $s_1, s_2 \in S$, compute common substrings S_{sub}, if there exists $S'_{sub} \subseteq S_{sub}$ satisfy conditions (i) the element amount of $S'_{sub} >= 2$; (ii) the sequence that common substrings appear is consistent, and can't cross; (iii) exist a common substring is a prefix or a suffix of s_1 or s_2, then add S'_{sub} to P.

Step3: Automatic filter P according to a set of rules (such as the length of common substring $<\theta_1$, the frequency of common substring $> \theta_2$)

Step4: Using the interactive manual, we attached additional rules. Otherwise, we carry on the merge and generalizations to the similar patterns in form.

Output: removable patterns P

A removable sentence pattern is defined as follows.

removable pattern{

Pattern: <?w1><如|就如|正如|恰如|正像|像|就像|正向|正像|如同><?w2><评价|所述|所说|一样|所言|所示|所讲|所见|那样|><, |。|. |? | ! |,|.|?|!| ; |;><?w3>

additional rules: notcontain(<?w2>, 。|. |? | ! |, |,|.|?|!| ; |;)

}

Additional rules for improving the result of pattern matching. In additional rules, notcontain(a, b) expresses that string   b is not a substring of string a, and <?w1> and <?w2> are pattern variables.

We use an algorithm that combine outside layer removal and inside layer gathering to acquire concepts of constituting hyponymy. The algorithm is as follows:

Let R be a list of candidate hyponymy relations initially empty

Input: Testing corpus Cor_{test}, removable patterns P

Step1: For each sentence $s \in Cor_{test}$, process <?C> according to Step2 – Step4.

Step2: Discover patterns that match with <?C> using P. If exist matching pattern, carry on the processing that remove non-concept components from <?C> outside layer to inside layer until it can't discover any removal word further.

Step3: According to the result of part of speech tagging, we gather noun phases and remove adjective fractions.

Step4: Acquire candidate c_L and candidate c_H that is no-tagged components of <?C>, and add them to R

Output: candidate hyponymy relations R

4.2 Verification of Hyponymy

We verify hyponymy using the features of hyponymy. Those features are changed into a group of heuristic rules. If a candidate hyponymy satisfies a certain threshold with matching those features, we say that it is a real hyponymy. The features of hyponymy are defined as follows:

Definition 1: The feature of hyponymy is a 3-tuple ISAF= {STF, SMF, CTF}, where

(1) **STF** (space structure feature): When a group of candidate hyponymy relations are correct or error, they often satisfy some space structure feature. In space structure analysis, we use the coordinate relation between concepts. The coordinate relations are acquired using a set of coordinate relation patterns including "、". Chinese dun-hao "、" is a special kind of Chinese comma used to set off items in a series. For example:

In a sentence of matching a coordinate pattern, if there exists concept c_1 and concept c_2 divided by "、", then c_1 and c_2 are coordinate, denoted as $cr(c_1, c_2)$. An example is as shown below.

农作物主要有{水稻}c_1、{玉米}c_2、{红薯}c_3、{烟叶} c_4等

(The farm crop mainly includes paddy rice, corn, sweet potato, tobacco leaves etc..)

cr(水稻, 玉米, 红薯, 烟叶) (cr(paddy rice, corn, sweet potato, tobacco leaves)) is acquired from the above example.

We can add some space structure features on the basis of above coordinate relations. A few important features are as follows:

Structure 1: (c_1, c_2), (c_2, c_3), (c_1, c_3). For example:
(番茄, 蔬菜), (番茄, 食品), (蔬菜, 食品)
((tomato, vegetable), (tomato, food), (vegetable, food))
Structure 2: (c_1, c_2), (c_2, c_3), (c_3, c_1). For example:
(游戏, 生活), (生活, 童话), (童话, 游戏)
((game, life), (life, fairy tale), (fairy tale, game))
Structure 3: (c_1, c), (c_2, c),..., (c_m, c), $cr(c_1, c_2, ..., c_m)$, $\exists c_i \in \{c_1, c_2, ..., c_m\}$, CoSuffix (c_i, c). For example:
c=马, cr(千里马, 骏马, 老骥, 白驹), CoSuffix(千里马, 马), CoSuffix(骏马, 马)
(c=horse, cr(swift horse, courser, nag, white horse), CoSuffix(swift horse, horse), CoSuffix(courser, horse))
Structure 4: (c_1, c), (c_2, c), ..., (c_m, c), $cr(c_1, c_2, ..., c_m)$, $\exists c_i \in \{c_1, c_2, ..., c_m\}$, lpf $(c_i, c) \geq 1$. For example:
c=职业, cr(警察, 医生, 歌手, 护士), lpf(歌手, 职业)=3
(c=profession, cr(policeman, doctor, singer, nurse), lpf(singer, profession)=3)
(2) **SMF** (semantic features): It is constructed by the assumption that c_L and c_H are semantically similar in $hr(c_L, c_H)$, and is subdivided into three features, i.e. SMF={WF, SF, AF}, where

WF(word-formation feature): A concept consists of one or several certain sequence Chinese characters. To some extent, Chinese characters can appear the semantic feature of concept. For each pair of candidate hyponymy (c_1, c_2), compute their common substrings. If substrings exist, the position (such as prefix and suffix), length and amount of substrings will provide the evidence for the existence of a hyponymy.

SF(synonymous word feature): Making use of dictionary of synonymous words[11] to compute the semantic similarity of candidate (c_1, c_2).

AF(attribute feature): The attributes of concept can be used to discriminate different concept. If two concepts have the same attributes, they should be semantic similar. We denote common attributes as $CoAttr(c_1,c_2)$. For example:

CoAttr(黄河, 河流) = { 上游 }
(CoAttr(yellow river, river) = { upriver })

(3) **CTF** (context feature): Here we verify hyponymy using contextual knowledge. The co-occurrence context features are subdivided into two features, CTF= {FF, DF}.

FF (frequency features): If candidate (c_1, c_2) appears frequently in a kind of hyponymy pattern or in various hyponymy patterns, the probability of $hr(c_1, c_2)$ is higher. The type number of pattern that can acquire (c_1, c_2) is denoted by $lpf(c_1,c_2)$. The total of number of pattern that can acquire (c_1, c_2) is denoted by $lef(c_1, c_2)$. For example:

DF (domain features): Our corpus comes from Web and includes some error knowledge. We can acquire many error hyponymy, such as (美丽, 罪) ((beauty, evil)). If candidate (c_1, c_2) appears in a certain scientific domain-specific context, (c_1, c_2) may be a true piece of scientific knowledge; otherwise it may be a pair of general concepts and may not have any value. The domain-specific context is discriminated with a domain dictionary [3]. Given a group of context $CT(c_1,c_2) = \{ct_1, ct_2, ..., ct_n\}$, where ct_i is the i context of (c_1,c_2), $fw(ct_i)$ is the number of domain word in ct_i, and $length(ct_i)$ is the byte length of ct_i. The classify formula is as follows.

$$\text{Classify}(c_1,c_2) = \frac{\sum_{i=1}^{n} fw(ct_i)}{\sum_{i=1}^{n} length(ct_i)} \times 1000 \tag{1}$$

Because the above features are all uncertainty knowledge, they must be converted into a set of creation type heuristic rules used in uncertainty reasoning. Here we use CF (certainty factors) that is the most common approach in rule-based expert system. The CF formula is as follows:

$$CF(CR, f) = \begin{cases} \dfrac{P(CR|f) - P(CR)}{1 - P(CR)}, & P(CR|f) \geq P(CR) \\[2ex] \dfrac{P(CR|f) - P(CR)}{P(CR)}, & P(CR|f) < P(CR) \end{cases} \tag{2}$$

Where CR is a set of candidate hyponymy, which has a precision P(CR). P(CR|f) is the precision of a subset of CR satisfying feature f. CF is a number in the range from -1 to 1. If there exists CF(CR, f)≥0, then we denote f as positive feature and CF(CR, f) denotes the support degree of feature f; if there exists CF(CR, f)<0, then we denote f as negative feature and CF(CR, f) denotes the no support degree of feature f.

5 Evaluation

We adopt three kinds of measures: R (Recall), P (Precision), and F (F-measure). They are typically used in information retrieval.

Let H be the total number of correct hyponymy relations in the test corpus. Let H_1 be the total number of hyponymy relations acquired. Let H_2 be the total number of correct hyponymy relations acquired. We can give the measure of evaluation metrics as follows:

(1) Recall is the ratio of H_2 to H, i.e. $R = H_2/H$
(2) Precision is the ratio of H_2 to H_1, i.e. $P = H_2/H_1$
(3) F-measure is the harmonic mean of precision and recall, i.e. $F = 2RP/(R+P)$

For verifying our method, we used two groups of raw corpus as raw corpus: the People's Daily corpus (1998-1999, 91503 papers) and Web corpus (10GB).

People's Daily corpus: Processed corpus contains about 190,000 sentences acquired by matching Chinese hyponymy patterns.

Web corpus (10GB): Processed corpus filtered web tags contains about 1200,000 sentences acquired by matching Chinese hyponymy patterns. We manually evaluated a 1% random sample of final result.

Then we divided it into two groups: training corpus (80%) and testing corpus (20%). Training corpus and Testing corpus (20%) is processed by above method. We verify hyponymy using a group of heuristic rules coming from the features of hyponymy. Acquired hyponymy satisfies a certain threshold (>0.8). The detailed result is shown in Table 1.

Table 1. The processed result of testing corpus

Corpus (matching sentences)	Hyponymy	Precision	Recall	F-measure
People's Daily corpus (38,458)	19,834	92.3%	78.2%	85.4%
Web corpus (245,652)	116,772	86.1%	75.4%	80.4%

From table1, we can see the final hyponymy relations have a precision of 92.3%. But its recall is only 78.2%. Some correct relations (21.8%) were filtered because they had a lower feature value. Certainly, the threshold can be regulated for the sake of the different demand. Web corpus has more free expression and more false knowledge than the People's Daily Corpus. So acquired hyponymy had a lower precision of 86.1%.

There are still some inaccurate relations in the result. There are mainly two reasons to cause those errors. First, the structure of a sentence is so complicated that removable patterns can't handle, and more syntactical information will be helpful for resolving this problem. Second, relations may represent a kind of metaphor, and more sophisticated verification methods are needed.

6 Conclusion and Future Work

In this paper we presented a method of hyponymy acquisition and verification based on multiple features. Experimental results demonstrate good performance of the method. However, it is necessary for us to solve some problems, such as the ambiguity of tag, the reasonable usage of semantic information, and the polysemy of concept word. In future, we will combine some methods (such as the morpheme analysis, web page tag, concept space etc.) to the further verification of hyponymy.

Acknowledgements. This work is supported by the Natural Science Foundation (grant nos. 60572125, 60773059). The Beijing University of Technology Science Foundation (grant nos. X0006014200803, 97006017200701).

References

1. Beeferman, D.: Lexical discovery with an enriched semantic network. In: Proceedings of the Workshop on Applications of WordNet in Natural Language Processing Systems, ACL/COLING, pp. 358–364 (1998)
2. Richardson, S.D., Dolan, W.B., Vanderwende, L.: Mindnet: acquiring and structuring semantic information from text. In: COLING-ACL 1998, pp. 1098–1102 (1998)
3. Cao, C., Shi, Q.: Acquiring Chinese Historical Knowledge from Encyclopedic Texts. In: Proceedings of the International Conference for Young Computer Scientists, pp. 1194–1198 (2001)
4. Shinzato, K., Torisawa, K.: Acquiring hyponymy relations from web documents. In: Proceedings of HLT-NAACL, pp. 73–80 (2004)
5. Marti, A.: Hearst: Automatic acquisition of hyponyms from large text corpora. In: Proceedings of the 14th International Conference on Computational Linguistics, Nantes, France, pp. 539–545 (1992)
6. Hearst, M.A.: Automated Discovery of WordNet Relations. In: Fellbaum, C. (ed.) WordNet: An Electronic Lexical Database and Some of its Applications, pp. 131–153. MIT Press, Cambridge (1998)
7. Sharon, A.: Caraballo, Automatic construction of a hypernym-labeled noun hierarchy from text. In: Proceedings of the 37th Annual Meeting of the Association for Computational Linguistics, pp. 120–126 (1999)
8. Morin, E., Jacquemin, C.: Projecting corpus-based semantic links on a thesaurus. In: Proceedings of the 37th Annual Meeting of the Association for Computational Linguistics, pp. 389–396 (1999)
9. Lloréns, J., Astudillo, H.: Automatic generation of hierarchical taxonomies from free text using linguistic algorithms. In: Advances in Object-Oriented Information Systems, OOIS 2002 Workshops, Montpellier, France, pp. 74–83 (2002)
10. Chun-xia, Z., Tian-yong, H.: The State of the Art and Difficulties in Automatic Chinese Word Segmentation. Journal of System Simulation 17(1), 138–143 (2005)
11. Mei, J.J., Zhu, Y.M., Gao, Y.Q., Yin, H.X.: Tongyici Cilin (Dictionary of Synonymous Words). Shanghai Cishu Publisher, China (1983)

News Annotations for Navigation by Semantic Similarity

Walter Kasper, Jörg Steffen, and Yajing Zhang[*]

DFKI GmbH
Stuhlsatzenhausweg 3
D-66123 Saarbrücken, Germany
{kasper,steffen,yajing.zhang}@dfki.de

Abstract. We present a system for browsing a news repository that is based on semantic similarity of documents. News documents get automatically annotated semantically using information extraction. Annotations are displayed to a user who can easily retrieve crosslingual semantically related documents by selecting interesting content items.

1   Introduction

In the MESH project (http://www.mesh-ip.eu) news from online sources get analyzed for automatic annotation of their content with respect to domain specific information about major reported events. News in English, German and Spanish from *Deutsche Welle* (http://www.dw-world.de) and *BBC* (http://news.bbc.co.uk) are used. The service enables professional users to search not only for news stories about specific events but also to get a comprehensive overview of the information available from different sources and supports intelligent navigation within related information and news about the same and similar events. The most basic information about such events that gets extracted is their time and location as well as involved persons and organizations.

The SENA system described here is based on a scenario like this: a professional user is searching a news archive or repository for information about some event. He already found some relevant document. Now he is interested to know whether there are other documents about the same event or similar events in the same region, etc. This is where SENA can help: when the user is viewing a document SENA also presents an overview of relevant content items of some semantic categories that the user might be interested in to find more information about. By selecting interesting content items the user can retrieve documents related and similar to the document he is viewing with respect to the selected items. As SENA is based on semantic annotations to the news documents and not just textual search, SENA supports crosslingual retrieval of documents in a straightforward manner for multilingual news repositories.

The rest of the paper is organized as follows. Section 2 gives an overview of the system environment and major components. SENA's concept of *semantic similarity* is discussed in section 3. Then the SENA user interface will be presented in section 4.

[*] This work was supported by the European Commission, under contract number FP6-027685 (MESH).

A. Dengel et al. (Eds.): KI 2008, LNAI 5243, pp. 233–240, 2008.

Section 5 discusses related attempts to news annotation and navigation. Finally, we will give an outlook towards further development of SENA.

2   System Environment

Figure 1 shows the overall architecture of the SENA system. Basically it consists of three layers: the *acquisition* layer where new documents get analyzed, the *storage* layer for storing the annotations and the *retrieval* layer that interacts with the user interface to retrieve documents from storage. The main components will be described in the next sections.

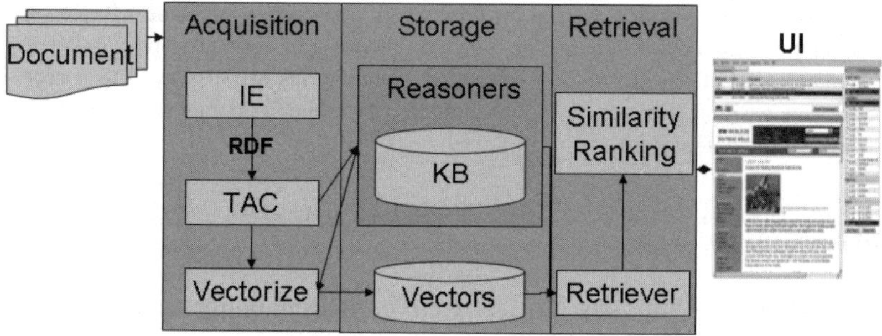

Fig. 1. SENA architecture

2.1   Knowledge Representation

Annotations about news documents are defined by OWL ontologies, the most important being a *domain ontology* derived from the IPTC subject code classification[1] and an ontology of news metadata that was derived from *NewsML*[2]. For multimedia documents a multimedia ontology based on the MPEG-7 standard is used.[3] Annotations are stored as instances of these ontologies in an RDF repository, linked to the document identifier or URL they belong to. Figure 2 shows parts of the domain ontology. The top-level class is *mesh_event*. Additionally, general properties of that class and some properties specific for the *earthquake* subclass are shown. The ontologies provide the conceptual core of the RDF knowledge base and the basis for reasoning processes. In addition a repository of external knowledge is used to support semantic reasoning. At present, this external repository mainly contains information about locations, especially containment relations among them. Its current size is about 1.2 million RDF triples. In SENA this allows to retrieve documents concerning e.g. Bavaria even if the document only mentions some place that is located in that region.

---

[1] http://www.iptc.org/NewsCodes/

[2] http://www.newsml.org

[3] We use the SmartWeb multimedia ontology: http://smartweb.dfki.de/ontology_en.html

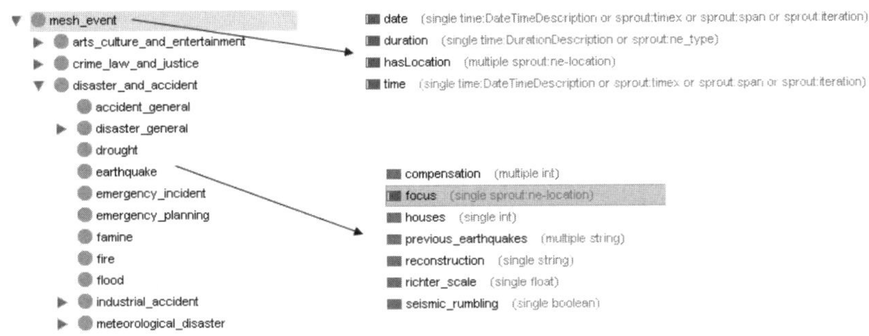

Fig. 2. Domain ontology

2.2   Information Extraction

The core annotation and information extraction engine is provided by the SProUT platform ([1]). SProUT combines finite state techniques with unification of typed feature structures (TFS). TFS provides a powerful device for representing and propagating information. Rules are expressed by regular expressions over input TFSs that get instantiated by the analysis. The uniform use of TFSs for input and output also allows for cascaded application of rule systems. For the MESH application the rule system is organized as

- A set of core named entity recognition (NER) grammars for date and time expressions, locations, persons and organizations.
- Domain specific annotation and extraction grammars based on the domain ontologies.
- An upper level merging component that merges partial annotations and information into larger chunks, usually at paragraph level. It uses unifiability as main compatibility test.

TFSs support well information extraction (IE) by allowing to model OWL domain ontologies: basically, the OWL classes become types, the properties are treated as features.[4] As a result the domain ontology is represented in TFS by complex types that provide templates that need to be filled by IE. This approach also makes it easy to map the results of IE to instances of the domain ontology for further semantic processing and storage in the RDF repository.

So, IE in SENA goes beyond semantic tagging of phrases to detect co-occurrence of entities but attaches the information to event types. The analysis would not just detect that an *earthquake* event and that a location like *Iran* is mentioned but if possible would link the location to the event by a *hasLocation* property.

The result of IE is an RDF representation such as this for *earthquake in Iran*:[5]

```
mesh:id1    rdf:type    mesh:earthquake ;
            mesh:hasLocation mesh:id2 .
```

---

[4] A similar idea is underlying the approach of [2].

[5] Actually UUIDs are used as unique instance identifiers.

```
mesh:id2    rdf:type   mesh:country ;
            mesh:name "Iran"  .
```

On average, IE generates about 300 (raw) RDF statements as annotations for a document.

The RDF model from IE is first passed to a temporal anchoring component (TAC; [3]) that resolves relative and underspecified temporal expressions, such as *on Monday* to specific dates and times to support cross-document comparison. Before updating the persistent knowledge base store (KB) with the data a set of reasoners is applied that are extensions of the JENA Generic Rule Reasoner and OWL reasoners (cf. http://jena.sourceforge.net). A *validator* validates the data against the ontologies and purges incompatible statements, e.g. temporary and low level linguistic information for TAC and update reasoning. An *identity reasoner* attempts to link especially NER instances to already known entities and merges the information sets.

3 Semantic Similarity of Documents

The most important information pieces for a document are contained in the KB as instances of the *mesh_event* class and its subclasses. The properties of an event specify its location, time and duration, etc. Given an unique document id (like its URL) a SPARQL query ([4]) is used to get all event instances for that document from the KB.

Each event instance can be interpreted as the root of a directed graph, with the properties as edges, instances and literals as nodes. We group all event instances of a document under a single artificial root node to get a single graph representing the events described in the document. We use these document graphs as a base for defining a similarity measure for documents. Numerous similarity measures for graphs exist. We use a similarity measure based on the maximal common subgraph of two graphs (see e.g. [5]). This similarity measure has also the advantage that it is a metric. So for two documents D_1 and D_2 represented by the two document graphs G_1 and G_2, we define the maximum common subgraph similarity (*mcs*) as:

$$sim_{mcs}(D_1, D_2) = \frac{|mcs(G_1, G_2)|}{max(|G_1|, |G_2|)} \tag{1}$$

This results in a *mcs* document similarity measure in an interval [0,1], 1 if G_1 and G_2 are isomorphic and 0 if G_1 and G_2 are completely distinct.

Our implementation of *mcs* document similarity is based on the SimPack Java library ([6]). Its implementation of the maximum common subgraph similarity is based on [7]. The SimPack accessor for Jena ontologies was extended to use instances of classes instead of just classes as we are interested in the instance similarity not in ontology similarity.

Besides the event annotations SENA creates additional annotations for information that cannot be related directly to the main events mentioned in the documents but can be of interest nevertheless such as persons or organizations mentioned. As we want to use all annotations to compare two documents, we define a second similarity measure that only considers these "unbound" instances.[6] We use a vector space model where the extracted instances are arranged in a vector where each instance corresponds to a dimension ([8]). The normalized frequency count of an instance c represents its weight in the vector:

$$weight(c) = \frac{|c|}{N_{type(c)}} \tag{2}$$

where $N_{type(c)}$ is the total number of occurrences of instances of that type.

The weight of instances from the headline of a document is boosted to accommodate the fact that the headline often holds the most important concepts of the document. Such instances are treated as if they appeared 5 times as often in the document. A similar mechanism is used for re-ranking user selected concepts on retrieval (cf. section 4).

For vector space modeling also a number of similarity measures exist. We use the cosine similarity. This measure quantifies the similarity between the two vectors as the cosine of the angle between the two vectors. The similarity of two documents D_1 and D_2 with the document vectors $\overrightarrow{D_1}$ and $\overrightarrow{D_2}$ is then defined as:

$$sim_{cos}(D_1, D_2) = \frac{\sum_{i=1}^{n} \overrightarrow{D_{1_i}} \times \overrightarrow{D_{2_i}}}{\sqrt{\sum_{i=1}^{n} \overrightarrow{D_{1_i}}^2} \times \sqrt{\sum_{i=1}^{n} \overrightarrow{D_{2_i}}^2}} \tag{3}$$

The resulting measure is also a value in the interval [0, 1].

Both similarity measures are combined to get a total similarity value for documents. For each similarity measure sim_i of a set of measures we define a weight w_i to control its influence on the total similarity value. The total similarity value is normalized to the interval [0,1]:

$$sim(D_1, D_2) = \sum_{i=1}^{n} \frac{w_i \times sim_i(D_1, D_2)}{\sum_{i=1}^{n} w_i} \tag{4}$$

In the current setup we use equal weights for w_{mcs} and w_{cos}.

In a similar way, additional similarity measure could get integrated into this framework, e.g. text based document similarity measures such as TFIDF.

4   Navigating Documents

SENA is realized as a web-service. So only a browser is needed to access and use the system. Figure 3 shows the main window of SENA's user interface, the *Document Selection* tab. A second tab (*New Document*) allows users to add new documents to the repository by entering their URL. The new document is analyzed immediately and added persistently to the repository including the extracted annotations. In the *Document Selection* view the new documents are immediately available on a par with the old documents. The *Document Selection* view is made up of three major areas:

---

[6] These instances, of course, have also properties with literals as values and could also be treated as graphs, but the identity reasoner makes sure that different occurrences of the same person, organization and location are represented by a single unique instance in the knowledge base. This is why we consider these instances as single nodes.

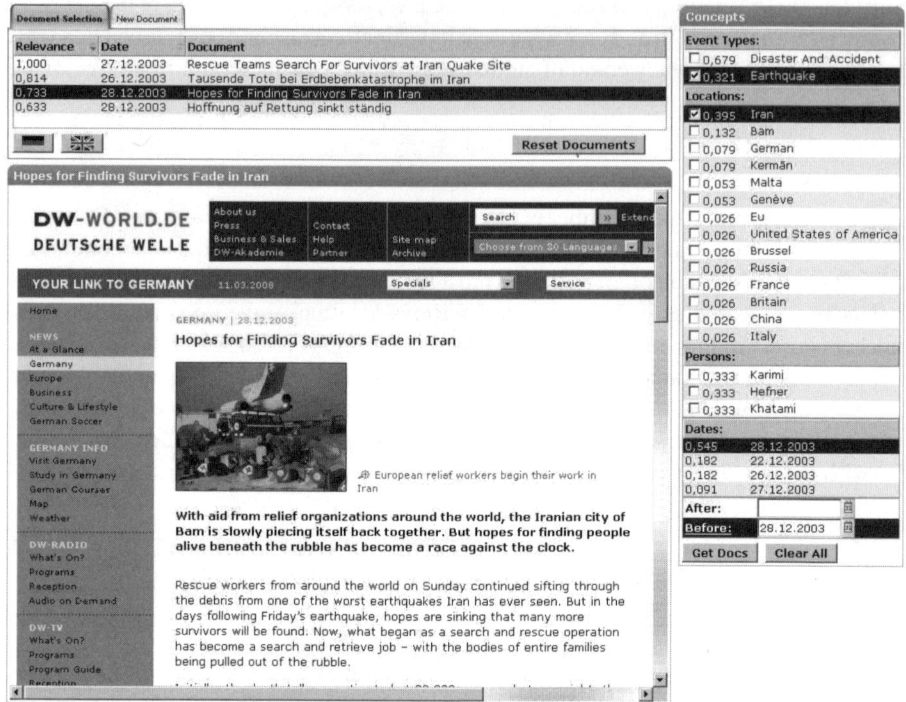

Fig. 3. SENA user interface for navigation

- The list of selected documents, the *Selection*
- The *Document* area displaying the actual document that also is the *start document* for navigation
- The *Navigator* area to the right.

The selection list displays the selected documents with their title. The similarity or relevance of the documents with respect to a start document the user had been viewing before creating the selection is indicated by a numerical value. This start document obviously always is the "most similar" selected document with a similarity value of 1. A second column shows the date, usually the publication or creation date of the document. The ordering of documents can be changed by clicking on the *Relevance* or *Date* header to sort according to that dimension. Sorting by date corresponds to a timeline view. The *language* buttons allow the user to exclude or include documents in that language.

Clicking into the document list displays the corresponding document in the *Document* area. For web documents, only the URL with the document's annotations are stored in the repository, not the document itself. So the document is retrieved directly by its URL from the web when requested. The selected document becomes the start document for further navigation.

The *Navigator* on the right side of the screen provides the main navigation device of SENA. Parts of the semantic annotations of the displayed document are shown there, according to semantic categories such as *event type*, *locations*, *persons* and *dates* mentioned. The date list contains not just the dates explicitly mentioned but also those indirectly referenced that have been resolved by TAC. Each item is accompanied by a relevance measure, based on the frequency, and a checkbox. The checkboxes allow the user to select a combination of interesting items that he would like to get further information about. Additionally, a time range for interesting documents and events can be specified, e.g. by using the date list that is indicative of related events. The *GetDocs* button then retrieves the documents according to the selections, computes their similarity to the viewed document (the *start document*) with respect to the selected items. The results of the retrieval then are displayed in the *Selection* area. From there the user can select other documents. This way the user easily can navigate within a large set of documents by going from one document to the next along a path defined by his selected concepts and the document similarity. Logically, the user selection corresponds to a Boolean *AND* query for the set of documents satisfying all the constraints.

In this first version of SENA, the content items are displayed as flat lists, giving an incomplete view on the extracted information, as it does not exhibit the semantic relations between the entities. For instance, that the earthquake event actually had been in Bam (and not in Malta, also mentioned in the text) is not obvious from that flat representation. Better ways of representing the semantic information, maintaining the simplicity of use, are under investigation.

5   Related Work

A large number of news services exist on the web that allow searching news archives. Not only newspapers, journals or TV stations operate web portals for searching their content, but there are also numerous cross-site news aggregators, e.g. from Google or Yahoo. Usually they are based on some text indexing, classification or clustering techniques that define textual document similarity measures though these usually are fixed and not dependent on users' choices. But there are also attempts to employ more semantic based techniques. Most of the approaches appear to use more shallow term based semantic tagging techniques, such as annotating documents with *WordNet* senses ([9]), e.g. [10]. Recent examples of approaches employing semantic web ontologies and being more similar to what we are investigating in SENA are the NEWS project ([11]) and especially the CALAIS project by Reuters ([12]) that provides a web based automatic annotation service for news that can be integrated into applications. But, besides also providing NER, their ontology framework seems to be more shallow than that of SENA, resembling more classification schemes rather than defining complex domain specific semantic objects that get instantiated by information extraction. In that, SENA's approach is close to the research done in the field of *Topic Detection and Tracking* (TDT; [13]).

Interesting navigation functionality beyond search is provided by the *Europe Media Monitor* with its *NewsExplorer* ([14,15]). But technically it seems to be closer to standard text based news services using clustering techniques rather than using a semantic web ontology framework. Similar to SENA it offers special navigation for some classes of named entities but does not allow selection of multiple items.

6 Conclusion and Outlook

We presented a system that provides users with easy to use navigation clues for retrieving related documents based on his choice of points of interest. In contrast to most other news services SENA's retrieval is based on semantic similarity with respect to domain specific ontologies that get instantiated by information extraction technologies and supports also crosslingual retrieval in a straightforward manner. First user evaluations with news professionals in the MESH project indicates high interest in navigation aids as provided by SENA.

SENA is at an early stage of development. For the future development, improvements to the user interface such as better visualization including the relations between entities from the underlying RDF graph are planned. Additional similarity measures such as text based measures that also support textual search will be included. Furthermore, SENA will be extended to multimedia news based on the semantic multimedia annotations of MESH.

References

1. Drozdzynski, W., Krieger, H.U., Piskorski, J., Schäfer, U., Xu, F.: Shallow processing with unification and typed feature structures — foundations and applications. Künstliche Intelligenz 1, 17–23 (2004)
2. Pérez, G., Amores, G., Manchón, P., Gonzáles, D.: Generating multilingual grammars from OWL ontologies. Research in Computing Science 18, 3–14 (2006)
3. Kasper, W., Zhang, Y.: Anchoring temporal expressions from news text (submitted, 2008)
4. Prudhommeaux, E., Seaborne, A.: SPARQL Query Language for RDF (2004), `http://www.w3.org/TR/2004/WD-rdf-sparql-query-20041012/`
5. Bunke, H., Shearer, K.: A graph distance metric based on the maximal common subgraph. Pattern Recognition Letters 19(3-4), 255–259 (1998)
6. Bernstein, A., Kaufmann, E., Kiefer, C., Bürki, C.: SimPack: A Generic Java Library for Similiarity Measures in Ontologies. Technical report, Department of Informatics, University of Zurich (2005)
7. Valiente, G.: Algorithms on Trees and Graphs. Springer, Heidelberg (2002)
8. Salton, G., Wong, A., Yang, C.: A vector space model for automatic indexing. Communications of the ACM, 613–620 (1975)
9. Fellbaum, C. (ed.): WordNet: An Electronic Lexical Database. MIT Press, Cambridge (1998)
10. Oleshchuk, V., Pedersen, A.: Ontology based semantic similarity comparison of documents. In: Proc. 14th International Workshop on Database and Expert Systems Applications, pp. 735–738 (2003)
11. Bernardi, A.: News engine web services (2004), `http://www.dfki.uni-kl.de/~berhardi/News`
12. OpenCalais: Calais (2008), `http://www.opencalais.com`
13. Allan, J. (ed.): Topic Detection and Tracking: Event-based Information Organization. Springer, Heidelberg (2002)
14. EMM: Europe news monitor: News brief (2007), `http://emm.jrc.it`
15. EMM: Europe media monitor: News explorer (2007), `http://press.jrc.it/NewsExplorer`

EANT+KALMAN: An Efficient Reinforcement Learning Method for Continuous State Partially Observable Domains

Yohannes Kassahun[1], Jose de Gea[1], Jan Hendrik Metzen[2], Mark Edgington[1], and Frank Kirchner[1,2]

[1] Robotics Group, University of Bremen
Robert-Hooke-Str. 5, D-28359, Bremen, Germany
[2] German Research Center for Artificial Intelligence (DFKI)
Robert-Hooke-Str. 5, D-28359, Bremen, Germany

Abstract. In this contribution we present an extension of a neuroevolutionary method called Evolutionary Acquisition of Neural Topologies (EANT) [11] that allows the evolution of solutions taking the form of a POMDP agent (Partially Observable Markov Decision Process) [8]. The solution we propose involves cascading a Kalman filter [10] (state estimator) and a feed-forward neural network. The extension (EANT+KALMAN) has been tested on the double pole balancing without velocity benchmark, achieving significantly better results than the to date published results of other algorithms.

1 Introduction

Neuroevolutionary methods have delivered promising results in recent years as methods of solving learning tasks, especially tasks that are stochastic, partially observable and/or noisy. Traditionally, neuroevolution methods solve partially-observable problems by using recurrent connections within the neural network, which provides a system with memory, enabling it to recover the missing state information. The main drawback to methods based on recurrent networks is that they require a significant amount of training time to find a solution.

In this work, the aim is at simplifying the topology of the evolved neural network, and as a consequence, reducing the time required to find a solution. The approach presented exploits the use of a Kalman filter as an input layer for the neural network to be evolved. This filter provides an estimation of missing (unobserved) state variables, and at the same time filters the noise that may be present in the measured input variables. In other words, the Kalman filter inherently provides the system with memory (as recurrent connections would) to estimate and thus recover the unobserved missing state variables. From the viewpoint of the neural network to be evolved (whose inputs are the outputs of the Kalman filter layer), the state information has been augmented and is noise-free. Clearly, this additional information provided to the system permits a simpler neural network solution, thus significantly reducing the time required to

A. Dengel et al. (Eds.): KI 2008, LNAI 5243, pp. 241–248, 2008.

find a solution. We have already shown that a Kalman filter can be used to accelerate neuroevolution methods [12]. The contribution of this paper, in contrast, is to use the ideas presented in [12] to extend a specific neuroevolutionary method (EANT) [11]. Unlike the results in [12], all results in this paper are obtained with the topologies of networks that were automatically determined by EANT.

This paper is organized as follows: first, a brief introduction to EANT will be given. After this, we provide a detailed description of the solution to be evolved and the $\alpha\beta$ Kalman filter used. Following this, experimental results will be presented, and important considerations related to the optimization of solutions will be discussed. Finally, we offer some conclusions and possible further research directions related to the presented approach.

2   Evolutionary Acquisition of Neural Topologies (EANT)

EANT [11] is an evolutionary learning method that evolves both the topology and weights of neural networks. The method starts with minimal structures, which are complexified along the evolution path. In this section, we will discuss the main components of EANT: (1) the genetic encoding that can be used to encode both the direct and indirect encodings of neural networks, (2) the genetic operators used in EANT, and (3) the exploration-exploitation of structures process that is used to evolve neural networks.

2.1   Common Genetic Encoding (CGE)

CGE [13] encodes a network using a *linear genome*. This genome consists of a string of genes, where each gene is either a *vertex gene* (representing a vertex in the network, also called a *neuron gene*), an *input gene* (representing an input in the network), or a *jumper gene* (representing an explicit connection between vertices). Each vertex gene has an associated unique identity number $id \in \mathbb{N}_0$, and to each input gene has an associated label, where input genes having the same label refer to the same input. The set of identity numbers and the set of labels are disjoint. A forward or a recurrent jumper gene stores the identity number of its source vertex gene.

2.2   Genetic Operators

Several genetic operators have been developed for CGE which operate on one or two linear genomes to produce another linear genome [13]. The three operators we used are *parametric mutation, structural mutation* and *structural crossover*. Parametric mutation changes only the values of the parameters included in the genes (e. g. the weights w_i); the order of the genes in the genome remains the same. The structural mutation operator inserts either a new randomly generated subgenome, a forward jumper, or a recurrent jumper after an arbitrarily chosen neuron gene. Inserted subgenomes consist of a new neuron gene (with a newly generated unique id) followed by an arbitrary number of input and jumper genes. This kind of structural mutation differs from the one used in NEAT [14] in that it allows whole subnetworks can be introduced at once without the need to add each node and edge of a subnetwork separately. This might help EANT find network topologies of sufficient complexity faster, though at the cost of possibly missing the simplest topologies. Initially, the weights of the newly added structures are set to 0, in order to prevent them from affecting the genome's overall performance. The third genetic operator is the structural crossover operator. This operator exploits the fact that structures which originate from the same ancestor structure have some parts in common. By aligning the common parts of two randomly selected structures, it is possible to generate a third structure that contains the common and disjoint parts of the two parent structures [14].

2.3 Exploitation and Exploration of Structures

The evolution of neural networks starts with the generation of an initial pool of genomes. The complexity of the initial genomes is determined by the domain expert. It then exploits the structures that are already have previously been generated by the system. By *exploitation*, we mean optimization of the weights of the structures. This is accomplished by an evolutionary process that occurs on a small time-scale, that uses parametric mutation as a search operator. During the exploitation phase, the weights of the structures are optimized using an advanced form of evolution strategy CMA-ES [1] (Covariance Matrix Adaptation - Evolution Strategy) [6]. *Exploration* of structures is done with the structural mutation and structural crossover operators. The structural selection operator is applied on a large time-scale, and selects the best structures (species) from which to form the next generation. Since sub-networks that are introduced are not removed, there is a gradual increase in the number of structures and their complexity along the evolution path. This allows the evolutionary process to search for a solution starting with neural networks having minimum structural complexity specified by a domain expert. The search stops when a neural network with the necessary optimal structure that solves a given task is obtained. The details of the exploitation and exploration of structures can be found in [11].

3 Description of the Solution to Be Evolved

The solution to be evolved is made up of a neural network and a predictor that can estimate the next state based on the *current partially-observable state* (which is possibly corrupted by noise). The predictor we use is composed of n Kalman filters ($\alpha\beta$ Kalman filters, see Section 4) $\{KF_1, KF_2, \ldots, KF_n\}$ one for each of the n sensory readings, as shown in Figure 1. The outputs of these Kalman filters are connected to a feed-forward neural network NN,

---

[1] We would like to thank Nikolaus Hansen and Andreas Ostermeier for making the MATLAB and C implementations of the CMA-ES algorithm publicly available on the web.

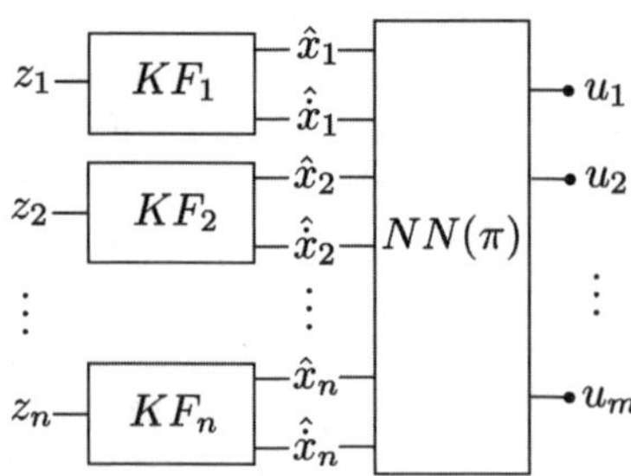

Fig. 1. The evolved network

whose outputs control the plant. A Kalman filter KF_i is used to estimate the sensor value $\hat{x}_i$ and the missing value $\hat{\dot{x}}_i$ from the measured (observed) value z_i, where $i \in [1, n]$ and n is the number of observable state variables. The quantity u_j, where $j \in [1, m]$, represents a control signal that is used to control a plant. The use of Kalman filters provides memory to the system and as a result enables the system to recover missing variables. Because of this, it is not necessary for the neural network to have a recurrent connection, and the use of a feed-forward neural network for the policy π to be learned is sufficient.

4   The $\alpha\beta$ Kalman Filter

A Kalman filter KF_i in Figure 1 is realized using an $\alpha\beta$ Kalman filter, which is a particular case of the general Kalman filter where the velocity is assumed to be constant. The filter is usually used in tracking applications.

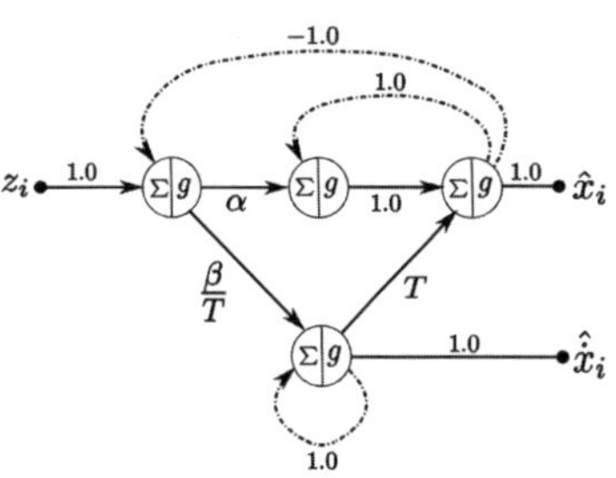

Fig. 2. An $\alpha\beta$ filter for a single input z_i. Note that the activation function g is an identity function and the recurrent connections have a unit delay.

The neural network equivalent of the filter is shown in Figure 2. As can be seen in the figure, all the weights of the filter have a magnitude of 1 except for α, β and T, where T is th sampling period. The optimal values for α and β are derived by Kalata [9] for assumed variance of both measurement and process noises (σ_v and σ_w) and are given by $\alpha = 1 - r^2$ and $\beta = 2(1 - r)^2$ respectively, where $r = \frac{4 + \gamma - \sqrt{8\gamma + \gamma^2}}{4}$ and $\gamma = \frac{T^2 \sigma_w}{\sigma_v}$. The term γ is referred to as a *tracking index*. Since the parameters α and β depend only on γ, an optimization algorithm *needs only to find a single parameter* γ per filter that results in the desired filter performance. An extension of the $\alpha\beta$ Kalman filter is the $\alpha\beta\gamma$ Kalman filter [4], which is based on a constant acceleration model and is better suited for the tracking of complex signals. Like the $\alpha\beta$ filter, an optimization algorithm needs only to find a single tracking index γ per filter to get the desired filter performance. This enables one filter to be easily exchanged with the other.

5   Evolving the Solution

There are two cases to consider while evolving the solution: noise-free partially observable domains and noisy partially observable domains.

In noise-free partially observable domains there are some state variables that are not observable. The purpose of the Kalman filter in this case is simply to estimate the missing variables. Therefore, we can set the tracking index γ_i of each Kalman filter KF_i to a higher value and optimize only the weights and topology of the feed-forward neural network. We found out empirically that a tracking index that results in $\alpha_i \geq 0.9$ performs well under noise-free scenario.

Partially observable domains which contain noise are the most general case, since virtually all real-world problems are noisy and partially observable. In this case the Kalman filter not only has to predict the missing state variables, but must filter the noise at the same time. We need, therefore, to optimize the tracking indices γ_i of each Kalman filter KF_i as well as the weights and topology of the feed-forward neural network. After optimization, a solution will be found that is robust against noise.

An attractive feature of using an $\alpha\beta$ or $\alpha\beta\gamma$ Kalman filter is that we do not need to optimize extra parameters in the case of noise-free partially observable domains and we need only to optimize n extra parameters in the case of noisy partially observable domains, where n is the number of observable variables of the state of the system.

6   Experiments and Results

6.1   The Double Pole Balancing Problem without Velocity

The inverted pendulum or the pole balancing system has one or more poles hinged to a wheeled cart on a finite length track. The movement of the cart and the poles are constrained within a 2-dimensional plane. The objective is to balance the poles indefinitely by applying a force to the cart at regular time intervals, such that the cart stays within the track boundaries. An attempt to balance the poles fails if either (1) the angle from vertical of any pole exceeds a certain threshold, or (2) the cart leaves the track boundaries. In the double pole balancing without velocity benchmark [2], the controller observes only x, θ_1 and θ_2 but *not* $\dot{x}$, $\dot{\theta}_1$ and $\dot{\theta}_2$. The fitness function used is the weighted sum of two

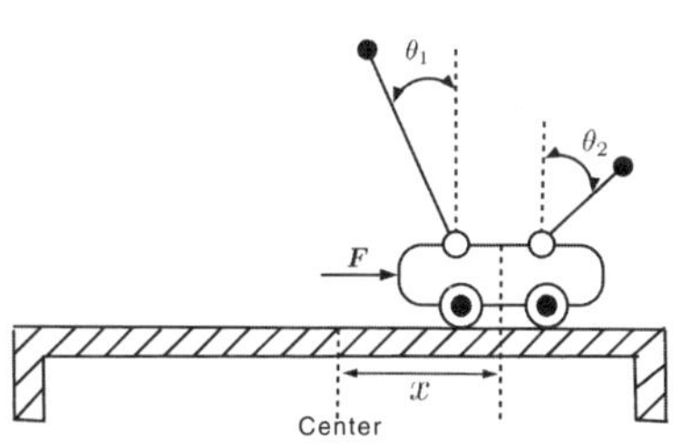

Fig. 3. The double pole balancing problem

---

[2] The implementation of the benchmark problem double pole balancing without velocity is based on the freely available source code by Faustino Gomez, Kenneth O. Stanley, and Risto Miikkulainen.

separate fitness measurements $f = 0.1f_1 + 0.9f_2$ taken over 1000 timesteps, where $f_1 = t/1000$ and $f_2 = \frac{0.75}{\sum_{i=t-100}^{t}(|x_i|+|\dot{x}_i|+|\dot{\theta}_{1,i}|+|\dot{\theta}_{i,2}|)}$ if $t \geq 100$, $f2 = 0$ if $t < 100$. The quantity t is the number of time steps the pole is balanced starting from a fixed initial position. In the initial position, all states are set to zero except $\theta_1 = 4.5\,\text{deg}$. The angle of the poles from the vertical must be in the range $[-36\,\text{deg}, 36\,\text{deg}]$. The evolution of the neural controllers is stopped when a champion of a generation passes two tests. First, it has to balance the poles for 10^5 timesteps starting from the 4.5 deg initialization. Second, it has to balance the poles for 1000 steps starting from at least 200 out of 625 different initial starting states. Each start state is chosen by giving each state variable $(x, \dot{x}, \theta_1, \dot{\theta}_1, \theta_2, \dot{\theta}_2)$ each of the values 0.05, 0.25, 0.5, 0.75, 0.95, 0, 0, scaled to the range of the input variable. The ranges of the input variables are ± 2.16 m for x, ± 1.35 m/s for $\dot{x}$, $\pm 3.6\,\text{deg}$ for θ_1, and $\pm 8.6\,\text{deg}$ for $\dot{\theta}_1$. The number of successful balances is a measure of the generalization performance of the best solution.

6.2   Noise-Free Scenario

The purpose of the noise-free experiment is to compare the results obtained by our approach with the current state-of-the-art results that are already published. Even if it is possible to set the tracking index of the Kalman filters and optimize only the weights and topology of networks as discussed in Section 5, for this experiment we let the tracking coefficients to be evolved along with the weights and topology of the network. Table 1 shows summary of results obtained for methods that evolve only the weights of neural networks, and both weights and topologies of neural networks. As can be seen in the table, EANT+KALMAN is at least 25 times faster than those methods evolving both the weights and topology of networks (*standard deviation = 320*). EANT+KALMAN also outperforms most of the methods that evolve the weights of the neural networks by at least a speed factor of 5. Obviously CMAES+KALMAN outperforms EANT+KALMAN since both use the same Kalman filter for recovering unobserved state variables and in the case of CMAES+KALMAN the topology of the neural network is manually predetermined. One further thing to note is that the generalization capability (*standard deviation = 60*) of EANT+KALMAN is significantly higher than all other methods tested on the task.

6.3   Noisy Scenario

The aim of this section is to test if our approach is robust in a noisy environment. We considered three different cases. In the first case, the sensors are left noiseless and a Gaussian noise with mean zero and variance of 0.5 is added to the actuator. In the second case, the actuator is left noiseless and a Gaussian noise with zero mean and variance of $10^{-2}S_{max}$ is added to each sensor, where S_{max} is 2.4 m for x and 0.62 rad for θ_1 and θ_2. Note that this noise level represents 6% of the maximum cart position and 12 % of the maximum angular positions of the poles. In the third case, we added the above described Gaussian noises to the sensors and actuators. In order to get a reliable estimate of

Table 1. Results for the double pole balancing task without velocities and noise-free scenario. Average of 50 simulations for EANT+KALMAN and CMAES+KALMAN. Generalization (referred to as **Gen.** in the table) is the average number of successful balances from 625 different initial positions. **#Eval.** stands for number of evaluations needed to solve the task.

Evolving only weights			Evolving both weights and topologies		
Method	#Eval.	Gen.	Method	#Eval.	Gene.
ESP [2]	169466	289	CE [5]	840000	300
CMA-ES [7]	6061	250	NEAT [14]	33184	286
CoSyNE [3]	3416	-	AGE [1]	25065	317
CMA-ES + KALMAN [12]	302	334	EANT [11]	15762	262
			EANT + KALMAN	653	468

Table 2. Performance of EANT+KALMAN under noisy environment. The results shown in the table are averages of 50 simulations. Note that every individual is evaluated five times before its fitness is determined. NA, NS and NA+NS stand for noisy actuator, noisy sensors, and noisy actuator and sensors, respectively. Numbers in brackets show standard deviation.

	NA	NS	NA+NS
Number of evalations needed	5335 (3530)	6123 (3480)	6470 (3400)
Generalization	484 (55)	436 (63)	485 (60)

the fitness value, every individual is evaluated in *five* episodes and its fitness value is calculated as the average of the fitness values obtained during those five episodes. The results in Table 2 show that EANT + KALMAN performs very well under noisy environment. In particular, it is interesting to see that under noisy environment, where an individual is evaluated five times before its fitness is assigned, EANT + KALMAN is two times faster as compared to the speed of the best neuroevolutionary method evolving the weights and topology of neural networks in noise-free environment.

7   Conclusions and Outlook

We have presented an extension of a neuroevolutionary method called Evolutionary Acquisition of Neural Topologies (EANT+KALMAN) that can be used to efficiently solve reinforcement learning tasks in continuous state partially observable domains. We have shown that EANT+KALMAN solves the double pole balancing without velocity problem efficiently. From the results obtained, one can conclude that this method can be applied to systems that exhibit non-linearity, partial observability and which have continuous states and actions. These properties manifest themselves in real systems such as robotic systems, aircraft control systems and many other.

In the future, we plan to test the algorithm on different benchmark problems and on real systems (for example, in learning a walking pattern for a legged robot), and assess its performance in comparison with other reinforcement learning algorithms.

References

1. Dürr, P., Mattiussi, C., Floreano, D.: Neuroevolution with analog genetic encoding. In: Proceedings of the 9th Conference on Parallel Problem Solving from Nature (PPSN IX), pp. 671–680 (2006)
2. Gomez, F.J., Miikkulainen, R.: Incremental evolution of complex general behavior. Adaptive Behavior 5, 317–342 (1997)
3. Gomez, F.J., Schmidhuber, J., Miikkulainen, R.: Efficient non-linear control through neuroevolution. In: Fürnkranz, J., Scheffer, T., Spiliopoulou, M. (eds.) ECML 2006. LNCS (LNAI), vol. 4212, pp. 654–662. Springer, Heidelberg (2006)
4. Gray, J.E., Murray, W.: A derivation of an analytic expression for the tracking index for alpha-beta-gamma filter. IEEE Transactions on Aerospace and Electronic Systems 29(3), 1064–1065 (1993)
5. Gruau, F.: Neural Network Synthesis Using Cellular Encoding and the Genetic Algorithm. PhD thesis, Ecole Normale Superieure de Lyon, Laboratoire de l'Informatique du Parallelisme, France (January 1994)
6. Hansen, N., Ostermeier, A.: Completely derandomized self-adaptation in evolution strategies. Evolutionary Computation 9(2), 159–195 (2001)
7. Igel, C.: Neuroevolution for reinforcement learning using evolution strategies. In: Sarker, R., Reynolds, R., Abbass, H., Tan, K.C., McKay, B., Essam, D., Gedeon, T. (eds.) Congress on Evolutionary Computation (CEC 2003), vol. 4, pp. 2588–2595. IEEE Computer Society Press, Los Alamitos (2003)
8. Kaelbling, L.P., Littman, M.L., Moore, A.P.: Reinforcement learning: A survey. Journal of Artificial Intelligence Research 4, 237–285 (1996)
9. Kalata, P.R.: Alpha-beta target tracking systems: A survey. In: American Control Conference, pp. 832–836 (1992)
10. Kalman, R.E.: A new approach to linear filtering and prediction problems. Transactions of the ASME-Journal of Basic Engineering, D35–45 (1960)
11. Kassahun, Y.: Towards a Unified Approach to Learning and Adaptation. PhD thesis, Technical Report 0602, Institute of Computer Science and Applied Mathematics, Christian-Albrechts University, Kiel, Germany (February 2006)
12. Kassahun, Y., de Gea, J., Edgington, M., Metzen, J., Kirchner, F.: Accelerating neuroevolutionary methods using a Kalman filter. In: Proceedings of the 10th Genetic and Evolutionary Computation Conference (GECCO 2008) (accepted, 2008)
13. Kassahun, Y., Edgington, M., Metzen, J.H., Sommer, G., Kirchner, F.: A common genetic encoding for both direct and indirect encodings of networks. In: Proceedings of the Genetic and Evolutionary Computation Conference (GECCO 2007), pp. 1029–1036 (2007)
14. Stanley, K.O.: Efficient Evolution of Neural Networks through Complexification. PhD thesis, Artificial Intelligence Laboratory. The University of Texas at Austin, Austin, USA (August 2004)

Where Temporal Description Logics Fail: Representing Temporally-Changing Relationships*

Hans-Ulrich Krieger

German Research Center for Artificial Intelligence (DFKI)
Language Technology Lab
Stuhlsatzenhausweg 3, D-66123 Saarbrücken, Germany
krieger@dfki.de

1   Introduction

Representing temporally-changing information becomes increasingly important for reasoning & query services defined on top of RDF and OWL, and for the Semantic Web/Web 2.0 in general. Extending binary OWL properties or RDF triples with a further temporal argument either lead to additional objects or to reification, as [1] have shown.

We argue that temporal description logics (e.g., [2]) are not able to represent temporally-changing information, since they are geared towards synchronic relationships, not diachronic ones. In this paper, we critically discuss several well-known approaches as presented in [1], add a new one, and finally reinterpret the OWL encoding of the 4D/perdurantist view. The reinterpretation has several advantages and requires no rewriting of an ontology that lacks a treatment of time. We also suggest that practical OWL reasoning should support means for temporal annotation of ABox relation instances.

The work discussed in this paper has been carried out in an EU-funded project called MUSING (http://www.musing.eu) and has been successfully applied to the PROTON upper-level ontology by equipping it with a concept of time and for representing temporally-changing relationships in MUSING.

2   Motivating Examples

The problem with *synchronic* relationships is that they all refer to only one (hidden) point/period in/of time. Here is an example:

Tony Blair was born on May 6, 1953.

Assuming a RDF-based representation, an information extraction (IE) system might compute the following set of triples:

* The research described in this paper has been partially financed by the European Integrated Project MUSING under contract number FP6-027097.

A. Dengel et al. (Eds.): KI 2008, LNAI 5243, pp. 249–257, 2008.

```
<tb, rdf:type, Person>
<tb, hasName, "Tony Blair">
<tb, dateOfBirth, "1953-05-06">
```

However, most relationships are *diachronic*, i.e., they embody the possibility to vary with time. Take, for instance, the following example:

Christopher Gent was Vodafone's chairman until July 2003. Later, Chris became the chairman of GlaxoSmithKline with effect from 1st January 2005.

Given this, an IE system might discover the following "tensed" facts:

```
[t₁, 2003-07-??]:  <cg, isChairman, vf>
[2005-01-01, t₂]:  <cg, isChairman, gsk>
```

Note that the temporal intervals in this example are open to the left and open to the right, resp. This is indicated by the two variable t_1 and t_2. Note further that the end point of the first interval is underspecified w.r.t. to the actual day of month (depicted by ??), assuming we measure time no finer than a day.

3   Applicability of Temporal Description Logics

Synchronic relations, like `dateOfBirth` in the above first example, "store" the temporal information (1953-05-06 in the example) *directly* as the range value of a relation instance. Since there is only *one* date of birth, this works perfectly well and uniquely capture the intended meaning. The deeper, albeit simple reason why this works comes from the fact that synchronic relations are simply functional *and* the temporal extent is directly stored as the range value of the synchronic relation.

Diachronic relationships, however, vary with time, i.e., in general, they are *no* longer functional. Representation frameworks such as OWL that are geared towards unary and binary relations can not directly be extended by a further (temporal) argument. By imposing a synchronic temporal representation scheme on a diachronic relationship, the association between the changing value and its temporal extent gets lost. Consider the following sentence:

Billy Bob Thornton was married with Pietra Dawn Cherniak from 1993 to 1997.

Now assume that the temporal extent of the marriage is stored in an additional relation, say `hasTime` (similar to `dateOfBirth`), we get the following:

```
<bbt, marriedWith, pdc>        <bbt, rdf:type, Person>
<bbt, hasTime, [1993,1997]>    <pdc, rdf:type, Person>
```

Note that this is different from the `dateOfBirth` example, since here the temporal extent of a relationship between two instances is encoded as the value of a second relation, viz., `hasTime`, even if `marriedWith` would be a functional relation. Let us consider another sentence that talks about B.B. Thornton:

Angelina Jolie was Thornton's spouse between 2000–2003.

This would give us further:

```
<bbt, marriedWith, aj>
<bbt, hasTime, [2000,2003]>
<bbt, rdf:type, Person>
```

This representation, however, mixes up the association between Thornton's spouses and the temporal extent of his marriages:

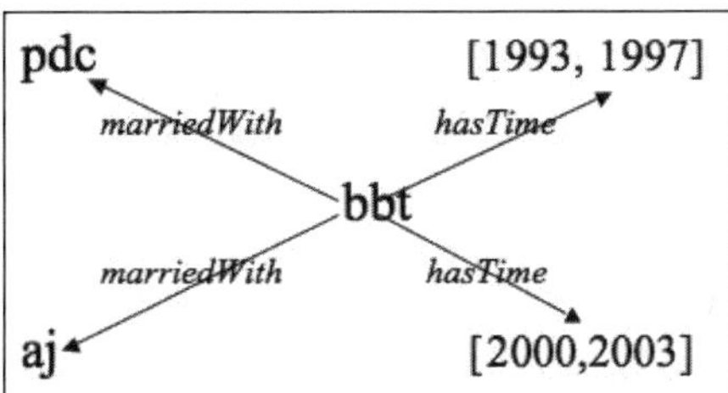

Assuming such a representation, we would have four possibilities of association, probably not what we want:

```
[1993, 1997]:  <bbt, marriedWith, pdc>
[1993, 1997]:  <bbt, marriedWith, aj>
[2000, 2003]:  <bbt, marriedWith, pdc>
[2000, 2003]:  <bbt, marriedWith, aj>
```

Now, temporal description logics, such as [2], are great when we aim at representing *synchronic* relations. Temporal description logics are essentially description logics extended with a concrete domain Temporal features are functional relations that can be employed to define feature path which in turn can be used to define temporal concepts through the use of constructors, such as <. These descriptive devices help to impose further constraints on concepts which ordinary description logics are not capable of, e.g.,

$$\text{Human} \sqsubseteq \exists(\text{hasMother dateOfBirth} < \text{dateOfBirth}) \sqcap$$
$$\exists(\text{hasFather dateOfBirth} < \text{dateOfBirth})$$

This implication says that children are born *after* their parents. Unfortunately, temporal description logics (like ordinary description logics) do not have any means to *directly* encode diachronic relationships.

4   Approaches to Diachronic Representation

There exist several well-known techniques of extending relations with additional arguments, or in our case, extending binary OWL properties with time. [1] present three of them—we add a fourth one and finally present a fifth way by reinterpreting the perdurantist or 4D view from [1]. We use a relational notation during the examples below.

4.1 Equip Relations with a Temporal Argument

This approach has been pursued in temporal databases and the logic programming community. Thus a binary relation, such as *hasCeo* between a company c and a person p becomes a ternary or quaternary relation with a third/fourth temporal argument t:

$$hasCeo(c, p) \longmapsto hasCeo(c, p, \underline{t}) \textbf{ or } hasCeo(c, p, \underline{t_1, t_2})$$

OWL and description logic in general only support unary (classes) and binary (properties) relations in order to guarantee decidability of the usual inference problems. Thus the above extended relation is not directly expressible in OWL.

Useful Extensions to OWL. Looking more closely at the extended relations, additional time arguments seem to behave more like *annotations* than real individual arguments that need to be strongly considered during OWL reasoning. Only *lightweight reasoning* is probably needed here. Assume that

$$ceoOf(\mathsf{schrempp, daimlerchrysler}, 1995, 2005)$$

is the case and that *hasCeo* has been defined as the *inverse* relation to *ceoOf*, using `owl:inverseOf`. Now it seems plausible to deduce that

$$hasCeo(\mathsf{daimlerchrysler, schrempp}, 1995, 2005)$$

also holds.

"Reasoning" over *symmetric* properties (`owl:SymmetricProperty`) is also easy, as we see from the next example. Given that `marriedWith` is a symmetric property, we infer from

$$married\,With(\mathsf{bbt, pdc}, 1993, 1997)$$

that

$$married\,With(\mathsf{pdc, bbt}, 1993, 1997)$$

must be true.

Let us finally focus on `owl:TransitiveProperty` and consider that *contains* is a *transitive* property. Assuming that

$$contains(\mathsf{dfki, room1.26}, s, t) \wedge contains(\mathsf{room1.26, chair42}, s', t')$$

is the case, the following inference is cheap, since it requires only a minimum/maximum computation in order to compute a temporal overlap:

$$contains(\mathsf{dfki, chair42}, max\{s, s'\}, min\{t, t'\})$$

Overall, only a few general expansion rules are needed to handle a temporal annotation. Unfortunately, such extensions to OWL are not available yet.

4.2 Apply a Meta-logical Predicate

McCarthy & Hayes' situation calculus, James Allen's interval logic, and the knowledge representation formalism KIF use the meta-logical predicate *holds* to encode a temporal change. Hence, our *hasCeo* relation becomes

$$\underline{holds}(hasCeo(c, p), \underline{t})$$

However, the property language in OWL (and again, description logic in general) is weak. Neither complex property definitions nor complex relation arguments are allowed in OWL. A kind of role composition would be needed here. Next generation OWL 1.1 only comes up with a very restricted form of role composition in role axioms to guarantee decidability. Unfortunately, general role composition has been shown to be undecidable, even for the relatively weak sublanguage $\mathcal{ALR}$ of KL-ONE [3].

4.3   Reify the Original Relation

Reification as used in RDF refers to the ability to treat a whole statement as a resource (and so as an argument of a RDF triple). Reifying a relation instance leads to the introduction of a new object and four additional new relationships. In addition, a new class needs to be introduced for each reified relation, plus accessors to the original arguments. Furthermore and very important, relation reification loses the original relation and thus needs rewriting of the original ontology. Coming back to our *hasCeo* example, we get something like this (*HasCeo* is the newly introduced class):

$$hasCeo(c, p, t) \longmapsto \exists hc \,.$$
$$type(hc, HasCeo) \wedge hasTime(hc, t) \wedge company(hc, c) \wedge person(hc, p)$$

4.4   Wrap Range Arguments

This approach wraps some of the original arguments in a new object in order to reduce the arity of a given relation instance. However, we do not wrap the original arguments (first and second), but instead the arguments in the range of the relation. The reason for this comes from the fact that the relation argument in first place often serves as an anchor during reasoning/querying. Coming back to our example, the diachronic ternary relation becomes

$$hasCeo(c, p, t) \longmapsto \exists et \,.$$
$$type(et, EntityTime) \wedge hasCeo(c, et) \wedge entity(et, p) \wedge hasTime(et, t)$$

The good thing with this apporach is that it does *not* lose the original relation name and only introduces one single container class (called *EntityTime*) that is used throughout every wrapped argument pair.

Both relation reification and argument wrapping can clearly be applied to relations with more than three arguments. As is the case for reification, wrapping, of course, introduces a new object, but needs less ontology rewriting when compared to reification—remember, we do not lose the original relation. [1] have shown that some forms of built-in OWL reasoning are no longer possible for relation reification. This fact also holds for the argument wrapping approach, but to a lesser extend.

4.5   Encode the 4D View in OWL

[1] have given a compact implementation of the 4D or perdurantist view in OWL. To do so, they define the notion of a *time slice* [4] that encodes the

time dimension of spacetime, occupied by an entity. Relations from the source ontology no longer connect the original entities, but instead connect time slices that belong to those entities. A time slice here is merely a container for storing time only. For an ontology that lacks time, such a representation requires a lot of rewriting:

$$\underline{hasCeo(c,p,t)} \longmapsto \exists ts_1, ts_2 \,.\, hasCeo(ts1, ts2) \,\wedge$$
$$type(ts_1, TimeSlice) \wedge hasTimeSlice(c, ts_1) \wedge hasTime(ts_1, t) \,\wedge$$
$$type(ts_2, TimeSlice) \wedge hasTimeSlice(p, ts_2) \wedge hasTime(ts_2, t)$$

4.6 Reinterpret the 4D View

What comes now is our reinterpretation of the perdurantist/4D view in that we reinterpret the original entries: *What has originally been an entity now becomes a time slice.* In the example above, c and p are no longer entities, but instead time slices that explain the behavior of an entity within a certain extent or point in time (e.g., that c is a time slice talking about a company and that p is a time slice representing a person). This reinterpretation does *not* need any ontology rewriting and makes it easy to equip arbitrary upper/domain ontologies with the concept of time. Coming back to our example, we have

$$hasCeo(c, p, \underline{t}) \longmapsto hasCeo(c, p) \wedge hasTime(c, t) \wedge hasTime(p, t)$$

Note that the time dimension is again encoded in the property `hasTime`, but now defined on the original entities. Now, in order *not* to lose the original association between the facts and their temporal extent (remember the `marriedWith` example from section 3), it is important to define `hasTime` as a functional property and to introduce a new time slice each time we are talking about a new temporal extension. In order to have access to the time slices of an entity, we define a new class `Perdurant` whose relational property `hasTimeSlice` exactly achieves this. Given all this and coming back to our `marriedWith` example, we now obtain the following *correct* representation:

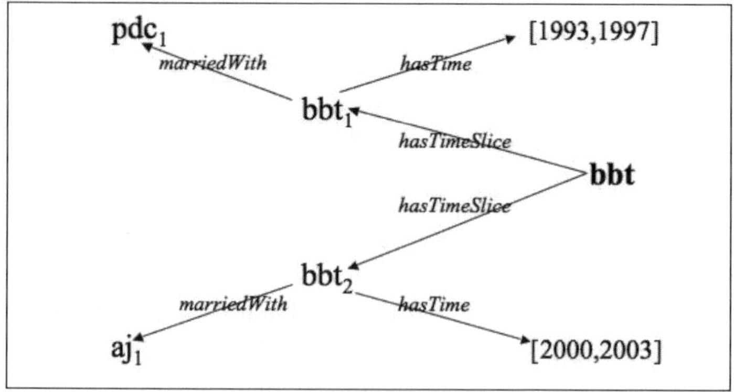

bbt_1, bbt_2, pdc_1, and aj_1 are still individuals of type `Person`, but now interpreted as *time slices*, i.e., they possess the property `hasTime` and describe the behavior of *perdurants* that act as persons, where bbt is such a perdurant, refering to

Billy Bob Thornton. In fact, this reinterpretation is easier than the original 4D formulation, viewed from the standpoint of complexity when looking at the domain and range of the above *hasCeo* property:

- **Welty et al. (2005)**
    - domain: $\exists hasCeo.\top \sqsubseteq \forall hasTimeSlice^-.Company$
    - range: $\top \sqsubseteq \forall hasCeo.(\forall hasTimeSlice^-.Person)$
- **4D reinterpretation**
    - domain: $\exists hasCeo.\top \sqsubseteq Company$
    - range: $\top \sqsubseteq \forall hasCeo.Person$

5   Advantages and Extensions

Let us summarize the advantages of the reinterpretation. **Firstly**, properties that do not change over time can be relocated from time slices (class: `TimeSlice`) of a perdurant to the perdurant itself (no duplication of information). Time-varying information instead is kept in a time slice. **Secondly**, if different properties of a perdurant change over the *same* period of time, we do not need to set up several time slices, but instead encode the changes in a *single* slice. **Thirdly**, the subtypes of `TimeSlice` (e.g., `Company`, `Person`) specify the *behavior* of a perdurant within a certain temporal interval (e.g., whether a perdurant *acts* as a company, a person). **Fourthly**, since `hasTimeSlice` is typed to `TimeSlice`, different slices of the same perdurant need *not* to be of the same type. For instance, the perdurant SRI might have a time slice for `Company` as well as a slice for `AcademicInstitution`, i.e., a perdurant/entity can act in different ways. Let us finally discus several extensions to the 4D reinterpretation.

5.1   Modalities

Representing modalities, such as *believe* can be achieved relatively easy. This is due to the fact that the properties that undergo a temporal change (those that are defined on `TimeSlice` and all its subclasses) again lead to time slices. In fact, a modality such as *believe* is similar—it is a property that connects special time slices. Consider the following example:

I believe, Jürgen Schrempp was the CEO of DC between 1995 and 2005.

The following RDF graph depicts the application of the modality *believe* to the propositional content that Schrempp was DC's CEO from 1995 until 2005. Time slice p_3 of perdurant i (I) is related to time slice p_2 via property `believe`:

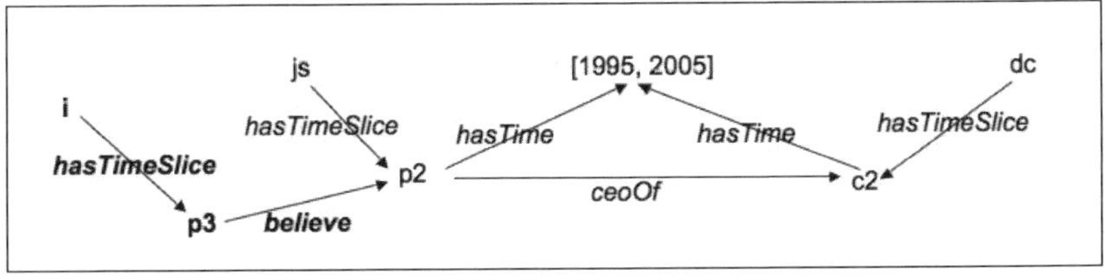

5.2   Allen Relations

[5] has defined 13 topological relations over temporal intervals. Given our treatment, we are able to define these relations now on time slices which we think is more reasonable, since a time slice usually encodes the propositional content of a piece of natural language text, anchored in time. Thus this representation can provide an explicit partially-ordered backbone over *temporally underspecified* facts. Consider the following example which focuses on instants, not intervals:

  (1) *In 1995, Schrempp was announced as CEO of DB.*
  (2) *Under his direction, DB and Chrysler became DC.*

This gives us the following constraint: 1995-??-??:(1) BEFORE ????-??-??:(2)

5.3   Temporally-Changing Objects

The approach also provides means to represent statements that deal with a temporal anchoring of a whole individual and its "overall" (perhaps political) change over time, as in the following sentence:[1]

The Tony Blair of 1997 is different to the one from 2004.

6   Summary and Outlook

We have presented a new framework for imposing time on ontologies which lack the possibility to represent temporally-changing relationships. The approach is based on a reinterpretation of the 4D OWL implementation by [1] and needs no ontology rewriting. We think that the 4D reinterpretation has numerous advantages as indicated in this paper. We have also shown that temporal description logics can only support synchronic relationships, not diachronic ones. However, by applying our reinterpretation to an ordinary ontology, temporal description logics can fruitfully be employed to impose further constraints on the temporal extent of a changing relationship.

Nevertheless, all the approaches presented here result in an overhead of further objects and relationships, due to the binary nature of OWL properties. In section 4.1, we have suggested that temporal information of ABox relation instances seems to behave more like an annotation than real further arguments of an original binary relation, since it requires only very lightweight reasoning capabilities (at most min/max computation between two arguments). We think that practical OWL reasoning should pay more attention to this area.

The apporach presented here has been applied to the PROTON ontology and several domain-specific ontologies and is used in MUSING's reasoning platform CROWL (Combining Rules and OWL), a hybrid reasoner that builds on Pellet, OWLIM, and Jena [6].

---

[1] Thanks to Hans Uszkoreit for providing me a similar example.

References

1. Welty, C., Fikes, R., Makarios, S.: A reusable ontology for fluents in OWL. Research Report RC23755 (W0510-142), IBM (2005)
2. Lutz, C.: Combining interval-based temporal reasoning with general TBoxes. Artificial Intelligence 152(2), 235–274 (2004)
3. Schmidt-Schauß, M.: Subsumption in KL-ONE is Undecidable. Technical Report SEKI SR-88-13, Universität Kaiserslautern, FB Informatik (1988)
4. Sider, T.: Four Dimensionalism. An Ontology of Persistence and Time. Oxford University Press, Oxford (2001)
5. Allen, J.F.: Maintaining knowledge about temporal intervals. Communications of the ACM 26(11), 832–843 (1983)
6. Krieger, H.U., Kiefer, B., Declerck, T.: A framework for temporal representation and reasoning in business intelligence applications. In: AAAI 2008 Spring Symposium on AI Meets Business Rules and Process Management, pp. 59–70 (2008)

Interpreting Motion Expressions in Route Instructions Using Two Projection-Based Spatial Models

Yohei Kurata and Hui Shi

SFB/TR 8 Spatial Cognition, Universität Bremen
Postfach 330 440, 28334 Bremen, Germany
{ykurata,shi}@informatik.uni-bremen.de

Abstract. This paper explores the applicability of two formal models of spatial relations, *Double Cross* and *RfDL$_{3\text{-}12}$*, to interpret some typical expressions that people use for describing a route. The relations in these two models allow the qualitative representation of the location and spatial extent of a landmark as seen from a route segment. We explore the correspondence between the relations in these two models and the motion expressions that refer to a point-like and a region-like landmark, respectively, which consist of the same set of direction-related expressions and specific sets of topology-related expressions. Through this exploration, we identify intrinsic ambiguities in the direction-related motion expressions that refer to a region-like landmark. Finally, we propose the generalization of our approach by using a spatial ontology, which potentially enables the mobile robots to interpret a large variety of expressions in human route instructions.

1 Introduction

A dialogue-based interface is a promising technology for the robots that work together with ordinary people. Especially, the mobile robots that navigate in human living spaces, such as the intelligent semi-autonomous wheelchair *Rolland* [1], should be equipped with an ability to communicate verbally with human users about their navigation plans. Formal models of qualitative spatial relations between a directed line segment (*DLine*) and another geometric object are helpful for developing such a technology [2], because those relations capture the prominent characteristics of route-landmark arrangements based on how people conceptualize the spatial context, and the existing studies report that many expressions in human route descriptions concern the actions associated with landmarks [3, 4]. Thus, this paper explores two formal models of such DLine-object relations, *Double Cross* [5, 6] and *RfDL$_{3\text{-}12}$* [7], to interpret some typical expressions that people use for describing motions in a flat human-scale space. DLine-point relations in Double Cross (in short, *DC relations*) allow us to represent the location of a point-like landmark as seen from a route segment. Similarly, DLine-region relations in RfDL$_{3\text{-}12}$ (in short, *RfDL$_{3\text{-}12}$ relations*) allow the representation of both the location and spatial extent of a region-like landmark as seen from a route segment. The presence of motion expressions that presume the landmark's spatial extent observed in [8-9], such as "*go into …*" and "*go to the end*"

A. Dengel et al. (Eds.): KI 2008, LNAI 5243, pp. 258–266, 2008.

of ...", motivated us to study RfDL$_{3\text{-}12}$ relations in addition to DC relations. This paper focuses on the motion expressions that concern the relation between a landmark and an entire route segment. Meanwhile, we do not discuss the interpretation of goal-oriented motion expressions, such as "*go to the left of ...*", which concern the spatial relations between a landmark and a potential goal. Such landmark-location relations are used not only for describing motions, but also for describing static scenes and, accordingly, they are already well studied (e.g., [12]).

The reminder of this paper is structured as follows: Section 2 reviews Double Cross and RfDL$_{3\text{-}12}$. Section 3 identifies the correspondence between DC relations and motion expressions. Then, Section 4 explores the correspondence between RfDL$_{3\text{-}12}$ relations and motion expressions, finding some issues in such associations. Finally, Section 5 concludes with a short discussion about the generalization of our approach.

2 Double Cross and RfDL$_{3\text{-}12}$

When an agent moves straightly in a space guided by a landmark, its movement pattern is represented by the spatial arrangement between a directed line segment (*DLine*) and another geometric object. The DLine represents the route segment on which the agent proceeds, while another object corresponds to the landmark. The landmark is represented by a point if its spatial extent is not significant; otherwise, the landmark is represented as a region (or possibly a line if its width is not significant).

Double Cross is a model of spatial relations between three points [5], which are also viewed as the relations between a straight DLine and a point [6]. Double Cross projects a ‡-shaped intrinsic frame of spatial reference [11] onto the space, by which twelve fields around the DLine and three fields on the DLine are defined (Fig. 1a). In this paper, these 15 fields are called *LF* (*left front*), *SF* (*straight front*), *RF* (*right front*), *LEx* (*left at exit*), *Ex* (*exit*), *REx* (*right at exit*), *LI* (*left of interior*), *I* (*interior*), *RI* (*right of interior*), *LEn* (*left at entry*), *En* (*entry*), *REn* (*right at entry*), *LB* (*left back*), *SB* (*straight back*), and *RB* (*right back*). Based on which field contains the point, Double Cross distinguishes 15 DLine-point relations.

RfDL (*Region-in-the-frame-of-Directed-Line*) is a series of projection-based models of spatial relations between a straight DLine and a simple region. Based on the use or non-use of *left-right*, *front-side-back*, and *entry-interior-exterior* distinctions with respect to the DLine, $2^3 = 8$ types of DLine-centric intrinsic frames are introduced. The spatial relation in each model is defined as the set of fields over which the region extends. The finest model, called *RfDL$_{3\text{-}12}$*, distinguishes 1772 DLine-region relations [7] based on the ‡-shaped frame that distinguishes 15 fields (Fig. 1a). These 15 fields are the same with those of Double Cross and, therefore, RfDL$_{3\text{-}12}$ has a strong correspondence with Double Cross.

Both DC relations and RfDL$_{3\text{-}12}$ relations are represented by icons (Fig. 1b). The icons have 3×5 blocks, which geometrically correspond to the 15 fields that each model considers. The icon of a DC relation has one marked block, which indicates the field that contains the point. The icon of an RfDL$_{3\text{-}12}$ relation has one or more marked block, which indicates the set of fields over which the region extends. Accordingly, in both models, relations are distinguished visually by the icons' marking patterns.

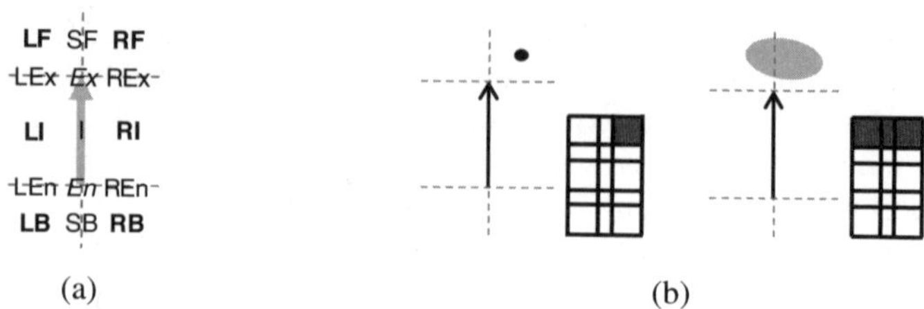

Fig. 1. (a) Fifteen fields that Double Cross and RfDL$_{3\text{-}12}$ consider. (b) Examples of a DC relation and an RfDL$_{3\text{-}12}$ relation represented by icons.

3 DC Relations and Motion Expressions

This section explores the correspondence between DC relations and some motion expressions that refer to a point-like landmark (e.g., a bus stop). Suppose that the route segment and the landmark are mapped to a DLine $\overrightarrow{ab}$ and a point p, respectively. Then, the movement pattern is mapped to a DC relation $\overrightarrow{ab}:p$. We assume that the distance between $\overrightarrow{ab}$ and p is small enough for people to recognize p as a landmark along $\overrightarrow{ab}$ whenever a DC relation holds between them.

"*Approach*" is a motion expression that refers to a movement during which the distance between the moving agent and the landmark decreases, but does not become zero. Thus, this expression corresponds to five DC relations where p is located at $\overrightarrow{ab}$'s *SF, LF, RF, LEx*, or *REx*. "*Approach*" is further distinguished into five sub-expressions: "*go toward*", "*go until … comes to the left*", "*go until … comes to the right*", "*approach … on the left front*", and "*approach … on the right front*". These five expressions are mapped uniquely to the five DC relations where p is located at $\overrightarrow{ab}$'s *SF, LEx, REx, LF*, and *RF*, respectively.

The motion expression opposite to "*approach*" is "*go away from*". This expression corresponds to five DC relations where p is located at $\overrightarrow{ab}$'s *SB, LB, RB, LEn*, or *REn*. Similarly, "*go away from*" can be distinguished into five sub-expressions, "*go straight away from*", "*go away from … on the left back*", "*go away from … on the right back*", "*go away from … on the left*", "*go away from … on the right*", which are mapped uniquely to the five DC relations where p is located at $\overrightarrow{ab}$'s *SB, LB, RB, LEn*, and *REn*, respectively.

"*Pass by*" is a motion expression that refers to the movement where the landmark is located ahead at the beginning and later comes behind the moving agent. Thus, this expression corresponds to two DC relations where p is located at $\overrightarrow{ab}$'s *LI* or *RI*. Naturally, "*pass by*" is distinguished into "*pass by … on the left*" and "*pass by … on the right*" depending on whether p is located at $\overrightarrow{ab}$'s *LI* or *RI*.

In this way, three expressions, "*approach*", "*go away from*", and "*pass by*", and their twelve sub-expressions are assigned distinctively to the twelve fields around $\overrightarrow{ab}$ (Fig. 2). All of these expressions presume the state where the point-like landmark is located around the route segment. If the landmark is located on the segment' start-point, interior, and end-point (i.e., if the movement is characterized by the DC relations where p is located at $\overrightarrow{ab}$'s *En*, *I*, and *Ex*), then *"depart from"*, *"pass"*, and *"arrive at"* are the typical expressions that fit with the movement pattern, respectively. These three expressions concern the topological characteristics of the route-landmark arrangement (i.e., how the route intersects with the landmark).

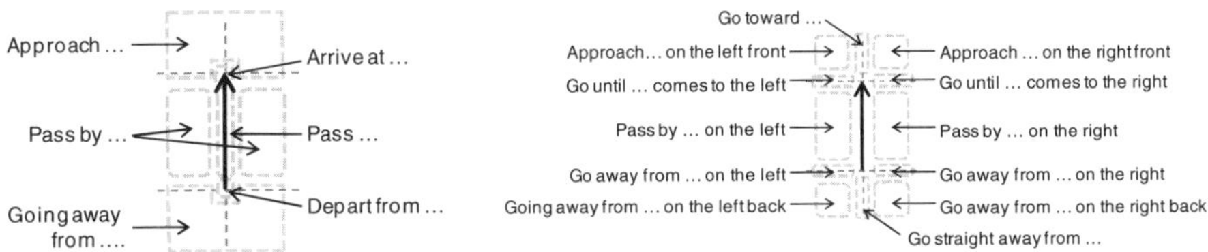

Fig. 2. Motion expressions assigned distinctively to the fields on/around a route segment

4 RfDL$_{3\text{-}12}$ Relations and Motion Expressions

Next, we consider the situation where the landmark is represented by a region. Suppose that the route and the landmark are mapped to a DLine $\overrightarrow{ab}$ and a region R, respectively. Then, the movement pattern is mapped to an RfDL$_{3\text{-}12}$ relation $\overrightarrow{ab}:R$.

For the situation where the landmark is located around the route segment, we can consider the same set of motion expressions as before (i.e., *"approach"*, *"go away from"*, *"pass by"*, and their sub-expressions). This time, however, the correspondence between these expressions and spatial relations is not clearly determined. For instance, let us think about *"approach"*. It is certain that *"approach"* fits with the movement pattern where the distance between the moving agent and every point on the landmark monotonically decreases (Fig. 3a). Moreover, it is clear that *"approach"* cannot fit with the movement pattern where the landmark has no point inside to which the distance from the moving agent decreases monotonically during the movement (Fig. 3f). Thus, we consider the following two conditions of *"approach"*:

- Strong condition: the movement pattern is mapped to $\overrightarrow{ab}:R$ where R extends at least one field among $\overrightarrow{ab}$'s *SF*, *LF*, *RF*, *LEx*, or *REx*, but no other field, and;

- Weak condition: the movement pattern is mapped to $\overrightarrow{ab}:R$ where R extends over at least one field among $\overrightarrow{ab}$'s *SF*, *LF*, *RF*, *LEx*, or *REx*, and neither *En*, *I*, nor *En*.

If a movement pattern satisfies the strong condition (e.g., Fig. 3a), this pattern always fits with the expression of *"approach"*. On the other hand, if a movement pattern does not satisfy the weak condition (e.g., Fig. 3f), this pattern never fits with this expression. Note the movement patterns in Figs. 3b-e satisfy the weak condition, but not the strong condition. The movement patterns in Figs. 3b and 3d may well be described as *"approach"*, but those in Figs. 3c and 3e probably not. This indicates the presence of borderline cases between Figs. 3b-c and between Figs. 3d-e, even though the movement patterns of each pair belong to the same RfDL$_{3\text{-}12}$ relation, respectively.

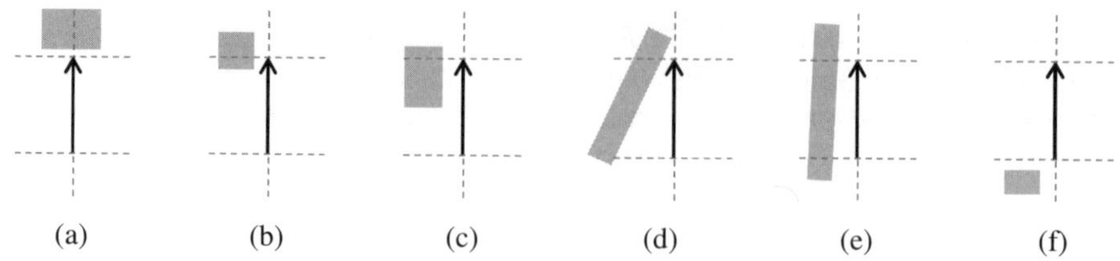

(a)   (b)   (c)   (d)   (e)   (f)

Fig. 3. Example of movement patterns that satisfy (a) both the strong and weak conditions of *"approach"*, (b-e) the weak condition only, and (f) neither

In order to evaluate how well *"approach"* fits with the movement pattern when it satisfies only the weak condition (e.g., Figs. 3b-e), we need further criteria; for instance, the relative length of period during which the nearest distance between the moving agent and the landmark decreases (compare Fig. 3b-c). In addition, the comparison of Fig. 3d-e indicates that it is better to consider a minimum threshold of the speed that the nearest distance decreases. The degree of fitness is then evaluated by the membership function of a fuzzy concept *"approach"* in Eqn. 1, where $\mathbf{a}$ and $\mathbf{b}$ are the location vector of the route segment's start-point a and end-points b, $\mathbf{x}$ is the location vector of the moving agent, $\frac{-\Delta \text{distance}(\mathbf{x},R)}{|\Delta \mathbf{x}|}$ is the relative speed that the nearest distance decreases, and α is its threshold ($\alpha > 0$).

$$\mu_{\text{approach}} = \frac{1}{|\mathbf{b}-\mathbf{a}|} \lim_{|\Delta \mathbf{x}| \to 0} \left(\sum_{\mathbf{x} \in [\mathbf{a},\mathbf{b}] \wedge \frac{-\Delta \text{distance}(\mathbf{x},R)}{|\Delta \mathbf{x}|} > \alpha} |\Delta \mathbf{x}| \right) \qquad (1)$$

As we considered in the previous section, *"approach"* is further distinguished into five sub-expressions. Their definitions, however, become slightly different when the landmark is represented by a region:

- *"go toward ..."* refers to the movement where the landmark stretches across the front extension of the route segment (Fig. 3a);
- *"go until ... comes to the left/right"* refers to the movement where the landmark appears the straight left/right of the moving agent when it comes to or *near* the end-point of the route segment (Fig. 3b); and
- *"approach ... on the left/right front"* refers to the movement where the landmark *mostly* extends over the left/right front of the moving agent when it arrives at the end-point of the route segment (Fig. 3b).

Table 1 summarizes the strong and weak conditions of *"approach"* and its sub-expressions using icons. These 'condition icons' follow the structure of the icons of $\text{RfDL}_{3\text{-}12}$ relations, where the icon's 15 blocks geometrically correspond to the 15 fields that $\text{RfDL}_{3\text{-}12}$ considers. Black, gray, and white blocks indicate the fields over which the region R must, may, and cannot extend, respectively. In addition to these conditions, it is possible to define membership functions that quantify how well each sub-expression fits with a given movement pattern, as we have defined such a function for "approach" (Eqn. 1). This is, however, beyond the scope of this paper.

"Go away from" is the motion expression that is opposite to *"approach"*. Consequently, the strong and weak conditions of this expression, as well as those of its sub-expressions, are derived from Table 1 simply by flipping the icons vertically.

Table 1. Conditions of the motion expression *"approach"* and its sub-expressions imposed on $RfDL_{3\text{-}12}$ relations, together with the number of relations that satisfy the conditions

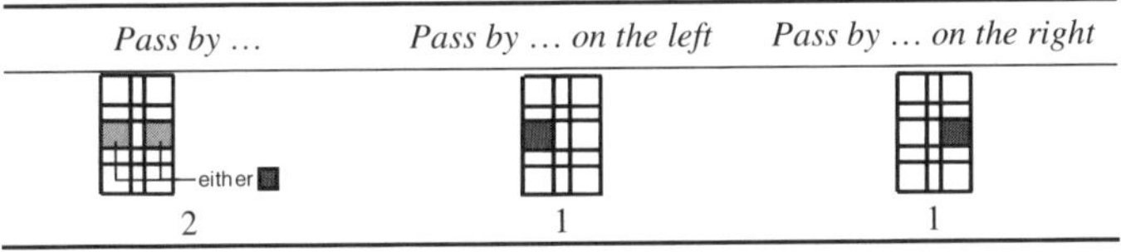

	Approach ...	*Go toward ...*	*Go until ... comes to the left*	*Approach ... on the left front*
Strong Condition	12	4	3	4
Weak Condition	102	46	7	67

The motion expression *"pass by"* that refers to a region-like landmark is essentially the same as *"pass by"* that refers to a point-like landmark. The condition of *"pass by"* (i.e., the landmark is located ahead at the beginning and later comes behind the moving agent) is satisfied if the R is entirely contained in $\overrightarrow{ab}$'s *LI* or *RI*, but not both. Table 2 shows the conditions of *"pass by"* and its two sub-expressions. As this table indicates, whether these expressions fit with a given movement pattern or not can be determined only from the $RfDL_{3\text{-}12}$ relation that characterize this movement pattern.

Table 2. Conditions of the motion expression *"pass by"* and its sub-expressions imposed on $RfDL_{3\text{-}12}$ relations, together with the number of relations that satisfy these conditions

Pass by ...	*Pass by ... on the left*	*Pass by ... on the right*
2	1	1

Next, we consider the situation where the landmark is located over the route segment. Such situations are covered by 1645 of the 1772 $RfDL_{3\text{-}12}$ relations. Some of these relations correspond to a unique motion expression. For instance, the $RfDL_{3\text{-}12}$ relations in Figs. 4a-b correspond to the expressions *"go across ..."* and *"pass the edge of ..."*, respectively. Meanwhile, the $RfDL_{3\text{-}12}$ relation in Fig. 4c corresponds to two different expressions: *"go into ..."* and *"go to the edge of ... and keep going"*. Like this example, the use of $RfDL_{3\text{-}12}$ relations for characterizing a movement pattern may leave certain ambiguity, because if a certain part of a DLine intersects with a region, the $RfDL_{3\text{-}12}$ relation tells the presence of this intersection, but does not specify whether this DLine's part intersects with the region's interior or boundary.

Alternatively, the model of topological DLine-region relations [2] serves as a nice framework for handling the motion expressions that refer to a region-like landmark located over a route segment, because for such a landmark people pay primal attention to whether the moving agent starts from, passes through, or ends at the landmark's inside, edge, or outside, which are the topological characteristics of route-landmark arrangements. Indeed, topological DLine-region relations capture such motion expressions as *"go into"*, *"go out of"*, *"go across"*, *"go within"*, *"leave the edge of"*, *"arrive at the edge of"*, *"pass the edge of"*, and so forth [2]. This, however, does not necessarily mean that $RfDL_{3\text{-}12}$ relations are useless for capturing such topology-related motion expressions. Actually, it is confirmed in [7] that:

- 76.2% of $RfDL_{3\text{-}12}$ relations are mapped to a unique topological relation; and
- 100% of $RfDL_{3\text{-}12}$ relations are mapped to a unique topological relation when the landmark is represented by a convex region (e.g., rectangles, circles).

Thus, many $RfDL_{3\text{-}12}$ relations can be associated with topology-related motion expressions by way of topological DLine-region relations without ambiguity, especially in the indoor environments where many landmarks (e.g., tables, sofas, beds, and closets) are represented by convex regions.

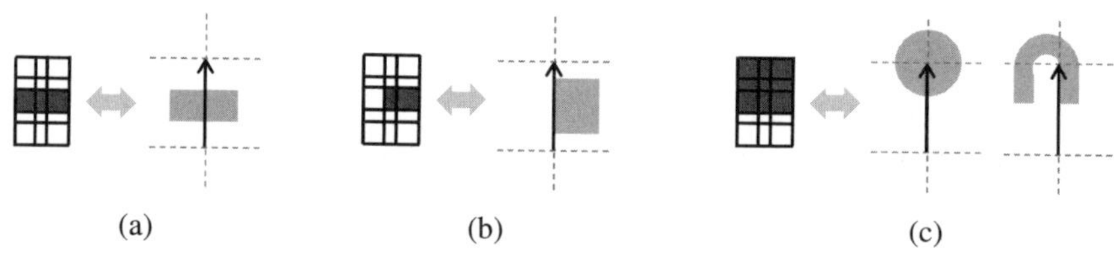

(a)   (b)   (c)

Fig. 4. Examples of $RfDL_{3\text{-}12}$ relations that capture topological characteristics of movement patterns (a-b) without and (c) with ambiguity

5 Conclusions and Future Work

Formal treatment of human route descriptions is essential for developing intelligent dialogue-based interface of the mobile robots that navigate in human living spaces. This paper demonstrated that DC relations and $RfDL_{3\text{-}12}$ relations, which allow the qualitative representation of landmarks' location and spatial extent as seen from a route segment, are useful for capturing some typical motion expressions in human route descriptions. We observed that the same set of motion expressions that concern the landmark's direction is applicable to both situations where the landmarks are represented by points or regions, although some expressions may have intrinsic ambiguities for the latter situation. For such ambiguous expressions, it is possible to evaluate how well each expression fits with the movement pattern as a membership value of a fuzzy concept. We also examined different sets of motion expressions that concern the topological characteristics of route-landmark arrangements.

The existing analyses of human route descriptions (e.g., [3, 9]) show the presence of several motion expressions that neither Double Cross nor $RfDL_{3\text{-}12}$ may cover, such as *"go to the left of …"* and *"go behind …"* These goal-oriented motion expressions presume landmark-centric frames of spatial reference, whereas both Double Cross and $RfDL_{3\text{-}12}$ consider path-centric frames. This indicates the necessity of additional spatial models for capturing more wide range of expressions in human route descriptions (e.g., [12]).

This paper examined some concrete motion expressions in English. This approach looks straightforward, but lacks generality. Alternatively, we are currently investigating the correspondence between DC/RfDL$_{3\text{-}12}$ relations and generalized concepts of motions in an upper-level spatial ontology, called the *Generalized Upper Model* [10]. For instance, the expressions *"go across ..."*, *"pass through ..."*, and *"gehen über ..."* correspond to the same concept in this ontology, called *Path-Representing-Internal*, if their slight nuance is neglected. If the set of RfDL$_{3\text{-}12}$ relations that corresponds to this ontological concept is clarified beforehand, mobile robots can determine the possible movement patterns that satisfy the user's such request as *"go across the parking lot"*, *"pass through the car park"*, and *"gehen Sie über den Parkplatz"*. This approach will definitely expand the applicability of our analysis in this paper as a foundation of dialogue-based interface of mobile robots.

Acknowledgments

This work is supported by DFG (Deutsche Forschungsgemeinschaft) through the Collaborative Research Center SFB/TR 8 Spatial Cognition – Subproject I3-SharC.

References

1. Lankenau, A., Röfer, T.: The Role of Shared Control in Service Robots - the Bremen Autonomous Wheelchair as an Example. In: Röfer, T., Lankenau, A., Moraz, R. (eds.) Service Robotics - Applications and Safety Issues in an Emerging Market, Workshop Notes, pp. 27–31 (2000)
2. Kurata, Y., Egenhofer, M.: The 9+-Intersection for Topological Relations between a Directed Line Segment and a Region. In: Gottfried, B. (ed.) 1st International Symposium for Behavioral Monitoring and Interpretation, pp. 62–76 (2007)
3. Denis, M.: The Description of Routes: A Cognitive Approach to the Production of Spatial Discourse. Current Psychology of Cognition 16, 409–458 (1997)
4. May, A., Ross, T., Bayer, S., Tarkiainen, M.: Pedestrian Navigation Aids: Information Requirements and Design Implications. Personal and Ubiquitous Computing 7, 331–338 (2003)
5. Freksa, C.: Using Orientation Information for Qualitative Spatial Reasoning. In: Frank, A., Formentini, U., Campari, I. (eds.) GIS 1992. LNCS, vol. 639, pp. 162–178. Springer, Heidelberg (1992)
6. Zimmermann, K., Freksa, C.: Qualitative Spatial Reasoning Using Orientation, Distance, and Path Knowledge. Applied Intelligence 6, 49–58 (1996)
7. Kurata, Y., Shi, H.: RfDL: Models for Capturing Directional and Topological Characteristics of Path-Landmark Arrangements (submitted)
8. Bugmann, G., Klein, E., Lauria, S., Kyriacou, T.: Corpus-Based Robotics: A Route Instruction Example. In: 8th Conference on Intelligent Autonomous Systems, pp. 96–103 (2004)

9. Shi, H., Tenbrink, T.: Telling Rolland Where to Go: HRI Dialogues on Route Navigation. In: Coventry, K., Tenbrink, T., Bateman, J. (eds.) Spatial Language and Dialogue. Oxford University Press, Oxford (to appear)
10. Bateman, J., Hois, J., Ross, R., Farrar, S.: The Generalized Upper Model 3.0: Documentation. Technical report, Collaborative Research Center for Spatial Cognition. University of Bremen, Bremen, Germany (2006)
11. Levinson, S.: Language and Space. Annual Review of Anthropology 25, 353–382 (1996)
12. Mukerjee, A., Gupta, K., Nautiyal, K., Singh, M., Mishra, N.: Conceputual Description of Visual Scenes from Linguistic Models. Image and Vision Computing 18, 173–187 (2000)

Repairing Decision-Theoretic Policies Using Goal-Oriented Planning

Christoph Mies[1], Alexander Ferrein[2], and Gerhard Lakemeyer[2]

[1] Fraunhofer-Institut für Intelligente Analyse-
und Informationssysteme IAIS
Schloss Birlinghoven, D-53754 St. Augustin
`christoph.mies@iais.fraunhofer.de`
[2] Knowledge-based Systems Group
RWTH Aachen University
Ahornstrasse 55, D-52056 Aachen
`{ferrein,gerhard}@kbsg.rwth-aachen.de`

Abstract. In this paper we address the problem of how decision-theoretic policies can be repaired. This work is motivated by observations made in robotic soccer where decision-theoretic policies become invalid due to small deviations during execution; and repairing might pay off compared to re-planning from scratch. Our policies are generated with READYLOG, a derivative of GOLOG based on the situation calculus, which combines programming and planning for agents in dynamic domains. When an invalid policy is detected, the world state is transformed into a PDDL description and a state-of-the-art PDDL planner is deployed to calculate the repair plan.

1 Introduction

Using decision-theoretic (DT) planning for the behavior specification of a mobile robot offers some flexibility over hard-coded behavior programs. The reason is that decision-theoretic planning follows a more declarative approach rather than exactly describing what the robot should do in a particular world situation. The programmer equips the robot with a specification of the domain, a specification of the actions and their effects, and the planning algorithm chooses the optimal actions according to a background optimization theory which assigns a reward to world states. With this reward, certain world states are preferred over others and goal-directed behavior emerges. The theory behind DT planning is the theory of Markov Decision Processes (e.g. [1]). The MDP model allows for stochastic actions, the solution of such an MDP leads to a behavior policy which optimizes the expected cumulated reward over the states which were traversed during planning. An interesting approach is the integration of DT planning techniques into other existing robot programming languages. One of these approaches is the language DTGOLOG proposed in [2], which marries DT planning with the well-known robot programming framework GOLOG [3,4]. They follow the idea to

A. Dengel et al. (Eds.): KI 2008, LNAI 5243, pp. 267–275, 2008.

combine decision-theoretic planning with explicit robot programming. The programmer has the control over the planning algorithm for example by restricting the state space for the search for a policy.

These techniques have been successfully deployed in dynamic real-time domains as well. In [5] it is shown how the GOLOG dialect READYLOG is used for formulating the behavior of soccer robots in the RoboCup [6] domain. In dynamic and uncertain domains like robotic soccer it turns out that policies become easily invalid at execution due to failing actions. The observation is that slight deviations in the execution can make the policy fail. Consider for example, when a ball should be intercepted and the robot does not have the ball in its gripper afterwards, or a move action where the robot deviated slightly from the target position. The result is, however, that the remainder policy cannot be executed anymore as preconditions for subsequent actions are violated. Nevertheless, often simple action sequences like $goto(x, y); turn(\theta); intercept\text{-}ball$ are the solution to the problem. In this paper we sketch our approach, how DT policies, which were computed with READYLOG, can be repaired. The rest of this paper is organized as follows. In Section 2 we briefly introduce the formalisms we use in our approach, namely the language READYLOG and the language PDDL. Section 3 addresses the execution monitoring for detecting when a policy has become inapplicable and outline the transformation of the world description between the situation calculus and PDDL. In Section 4 we show first results of the plan repair algorithm in simulated soccer as well as in the well-known Wumpus world. We conclude with Section 5, also discussing some related work there.

2 Planning Background

2.1 DT Planning in Readylog

READYLOG [5], a variant of GOLOG, is based on Reiter's variant of the situation calculus [4,7], a second-order language for reasoning about actions and their effects. Changes in the world are only due to actions so that a situation is completely described by the history of actions starting in some initial situation. Properties of the world are described by *fluents*, which are situation-dependent predicates and functions. For each fluent the user defines a successor state axiom specifying precisely which value the fluent takes on after performing an action. These, together with precondition axioms for each action, axioms for the initial situation, foundational and unique names axioms, form a so-called *basic action theory* [4]. READYLOG integrates several extensions made to GOLOG like loops, conditionals and recursive procedures, but also less standard constructs like the nondeterministic choice of actions as well as extensions exist for dealing with continuous change and concurrency allowing for exogenous and sensing actions and probabilistic projections into the future, or decision-theoretic planning employing Markov Decision Processes (MDPs), into one agent programming framework [5]. In this paper we focus on repairing DT policies. From an input program, which leaves several choices open, READYLOG computes an optimal policy (cf. also [2]). The policy is a tree branching over so-called *nature's choices.* These choice points represent the different outcomes of a stochastic action. *Agent choice points* in the plan skeleton, on the other hand, are optimized away when calculating the policy. An optimal policy π is then executed with READYLOG's run-time system. We refer to [8] for further details. Important for this work is to know that the planning process is done off-line, based on models of the world, while the execution of the policy naturally is on-line. Discrepancies between the model and the real execution might occur.

2.2 PDDL and SGPlan

The *Planning Domain Definition Language* (PDDL) is a family of formal standard languages to define planning domains, dating back to work by McDermott [9]. It is used as the description language for the bi-annual International Planning Competitions (IPC) and was since then further developed to meet the requirements of the planning competitions. Here, we basically use PDDL2.2. The most important language features of this PDDL-version are the following: (1) domain objects with types (2) basic actions and actions with conditional effects (3) fluents, which as opposed to READYLOG may only take numerical values (4) metrics which are used to measure the quality of a plan. A metric is defined in the problem definition and is a function over fluents. It can either be maximized or minimized. A PDDL *world state* is a collection of predicates that are *true* at certain time points. Using the closed world assumption all predicates not listed in a world state are assumed to be *false*. More details about PDDL can be found in [10]. For our system, we used the metric PDDL2.2 planner SGPLAN [11], which won the last two IPCs. The basic architecture of SGPLAN follows a hierarchical planning approach and decomposes the overall goal into several non-interfering sub-goals.

3 Policy Execution Monitoring and Policy Repair

3.1 Marking Possible Failures and Detecting Execution Flaws

As we pointed out, READYLOG distinguishes between an off-line mode for plan generation, and an on-line mode for executing the calculated policies and interact with the environment. Consequently, we can distinguish two classes of failures. One class contains failures which can be detected at planning time, failures of the other class can not. As we said earlier, READYLOG makes use of models during DT planning. These include (stochastic) action models together with their effects. The effect axiom (or successor state axioms, to be precise) of a primitive action in the situation calculus and with it in READYLOG describes how the world evolves from world situation to world situation due to actions. It is in general impossible to design these models in such a way that they are free of errors. It does happen that unforeseen action effects occur in reality. As this problem is inevitable, we thus need at least account for detecting these failures. We extend the previous concept of [8] of inserting special markers into the policy. Originally, for conditions φ occurring in loops or if-then statements a marker $\mathfrak{M}(\varphi, v)$ was inserted which stored the truth value of φ at planning time and allowed to re-evaluate it at execution time to compare the values. It could thus be detected, when a model assumption in conditions, test actions and loops became invalid. We add markers allowing to detect unforeseen action outcomes and failing action preconditions.

While the failure marker are a means to mark possible sources for a policy to become invalid at planning time, during execution we need to monitor if a policy is still executable by re-evaluating marker conditions and comparing the conditions with their reference values taken during plan generation. As described in [8] with some special transition in the implementation of READYLOG is it relatively straight-forward to check these conditions. For space reasons we cannot introduce it formally here. As, in general, it is hard to decide automatically in which cases it might be useful to repair or to cancel the execution of the policy, we lay it into the hand of the system designer to decide under which condition a policy shall be repaired or not. To this end, we introduce the construct $guardEx(\varphi, \pi)$ to attach condition φ to a policy π which must hold during its execution. A violation of the condition given by φ should initiate a re-planning in the current situation. Concurrent to the policy execution, a monitoring loop checks whether the condition attached to a policy hold or not. Additionally, after each execution of an action the monitoring loop simulates the remainder of the current policy in order to check if it is still executable. Therefore, all outcomes of each action are projected starting at the current world state. Thereby all precondition axioms and conditions in the policy are validated and tested in the projected world state where they are to be applied. If any of these checks fails the policy is corrupt and is subject to repairing or re-planning. Summarizing, there are two ways how a policy repair is initiated: (1) a failure marker occurs during policy execution; and (2) a policy failure in the remainder policy was detected by simulating it.

3.2   Repairing Policies

In the previous section we showed how we could determine whether a policy is corrupt, i.e. it cannot be executed until its end for some reasons, and shall be repaired. Now assume a policy π that has become invalid due to a violation of the condition φ in $guardEx(\varphi, \pi)$. To repair the policy we have to conduct the following steps:

1. *Calculate the desired world state:* The desired world state is the world state in which the remainder policy becomes executable again. This state serves as the goal description in step 2. With the formal specification of the actions in READYLOG it is possible to compute the preconditions for the remainder policy to become executable again.
2. *Translate state and goal description to PDDL.* Next, we translate the desired world state into the goal description of PDDL. Furthermore, in order to be able to perform goal-directed planning, we need an initial state description; we therefore translate the current state to a PDDL description and use it as the initial state. The state description has to be as small as possible as the run-time of PDDL-planners is, in general, directly related to the

number of ground actions. Therefore, we have to restrict our translation to the important and salient parts of our situation calculus world description.

- *Restrict to salient fluents.* The goal state should only refer to fluents that occur in the policy. Thus for the PDDL state translation, we can ignore all fluents that are not mentioned by the READYLOG policy.
- *Restrict fluent domains:* The idea is to translate only those values that are important to the remainder of the policy. A policy repair for a pass between player p_1 and player p_2 does not necessarily need anything to know about the positions of the other players. Therefore, we only translate p_1's and p_2's positions.

Note that in our current implementation, the translation between the situation calculus and PDDL has to be done by hand. For each pair of READYLOG and PDDL domain definitions, the domain axiomatizer has to provide a valid translation.

3. *Plan step.* The initial and goal state description are transmitted to an external planner via the file system, and the planning process is initiated. In our current implementation we make use of SGPlan. Note that, as we are using the abstract PDDL description, we can plug any other PDDL planner to our system easily.

4. *Re-translate the repair plan.* When a plan has been generated, we have to re-translate it to a READYLOG policy which can be executed then. After having calculated a PDDL repair plan with the external planner, it has to be executed before the remainder of the failed policy. Since PDDL is a deterministic language, no nondeterministic action can occur in the plan. Thus, each PDDL action can be translated to a sequence of READYLOG actions with specified outcome.

4   First Empirical Results

For proving the concept of the presented plan-repair scheme, we applied it to the toy domain WUMPUSWORLD, where an agent has to hunt a creature, the Wumpus, in a maze environment with pits and traps while searching for a pile of gold. The domain is modeled in a stochastic way, that is, the basic actions like move or grab_gold have stochastic effects, they can succeed or fail with certain probabilities. Moreover, exogenous events can occur. For example, the agent may lose the gold or move to a wrong position in the grid due to such events. In these cases, plan repair is invoked. The plans the agent performs are decision-theoretic policies with a plan horizon of three, meaning that the agent plans ahead the next three actions. In this paper we propose to connect READYLOG to an external PDDL planner. The first important question is whether or not plan repair pays off in the given application domain, as it was shown that, in general, plan repair is as complex as planning from scratch [12]. Therefore, we compared the run-times of planning from scratch (DT) each time a policy became invalid with an iterative deepening depth-first brute-force planner (BF) in READYLOG itself (this is very similar to the comparison made in [13]). Next, we compared this

Table 1. Results of WUMPUSWORLD

Setup	Run time		Plan. time		Repair time	
	# runs	av. [s]	# plans	av. [s]	# repairs	av. [s]
STANDARD_PDDL	91	19.53	740	1.23	722	0.99
STANDARD_BF	87	13.95	713	1.25	676	0.12
STANDARD_NR	93	17.26	1213	1.27	N/A	
SMALL_PDDL	87	14.62	713	1.21	759	0.33
SMALL_BF	89	13.57	728	1.26	617	0.11
SMALL_NR	95	16.58	1251	1.23	N/A	
HORIZON_PDDL	90	58.71	612	7.51	629	0.97
HORIZON_BF	89	53.13	606	8.21	704	0.12
HORIZON_NR	91	92.85	921	9.11	N/A	

to the PDDL plan repair scheme we propose in this paper to get information about the extra computational overhead to transfer the world states between READYLOG and the PDDL planner.

The results are shown in Tab. 1. We defined three evaluation setups STANDARD, SMALL and HORIZON. The scenarios where we applied the PDDL repair mechanism are suffixed with _PDDL, the ones with brute-force planning are suffixed with _BF and the ones without repairing with _NR (no repair). The first column represents the number of successful runs (out of 100 runs in total) and their average run-time. A run is successful when the agent arrives at the target cell with the gold in its hands. The second column contains the number of generated policies together with their average computation time; the number of calls of the repairing method as well as their average computation times are presented in the last column. All run-times are measured in real elapsed seconds. Note that there is no repairing data available if the plan repair was not performed (the _NR cases). Therefore, the number of generated plans for these scenarios roughly equals the number of repairs plus plans in the other cases. For the first row of the table, this means that each of the 91 successful runs took on average 19.53 seconds. In total, 740 policies have been generated, each of which took 1.23 second on average to be computed. 722 times the repair routine was called due to failing plans. On average, it took 0.99 seconds to establish the successful repair plan. Note that the number of repairs can exceed the number of plans. This is the case when a repaired plan fails again. The scenario STANDARD is the starting point for our evaluation. The DT planning depth is 3 in that setup. Thus, the lengths of the DT plans and the repair plans are roughly equal. The average computation time for PDDL repair (column 3) is slightly lower than the DT planning time (column 1). The average run-time of STANDARD_PDDL is higher than the one of STANDARD_DT. The advantage is taken by the overhead of the repair mechanism which is mostly determined by the policy execution simulation. The setup SMALL equals STANDARD except the smaller size of the WUMPUSWORLD. It contains 54 cells, 8 walls and 2 holes whereas STANDARD contains 180 cells, 16 walls and 2 holes. This leads to a significant faster PDDL repairing w.r.t. STANDARD_PDDL. The gap between repairing and DT re-planning is large enough to be faster than the pure DT planning approach SMALL_NR, which does not change in comparison to STANDARD_NR. The times of SMALL_BF are also very similar to the ones of STANDARD_BF. With the last setup, HORIZON, we wanted to check the influence of an increased plan horizon to the overall run-times. It has an increased plan horizon of 4 instead of 3 as in STANDARD and SMALL. This leads to a significantly higher DT planning time since our method has an exponential run-time w.r.t. the horizon. Therefore, HORIZON_PDDL is much faster than its pure DT planning companion. Again, the times of the brute-force approach are not influenced by this scenario variation. The times for plan repair for PDDL and BF equal those with horizon 3. As one can see from these results, the brute-force method is superior to PDDL in all WumpusWorld scenarios. The reason lies in the simple structure of the scenario. The computation time of the brute-force method depends on the number of ground actions. There exists only 9 ground actions (4 move actions, 4 shoot actions and grab_gold). Another issue is the very preliminary interface between Readylog and the PDDL planner. We make use of the file system to communicate the world states between PDDL and Readylog, the brute-force planner is integrated into Readylog. There is a connection between the simplicity of the action description and the run-time of the brute-force method. To check this, we applied the method to a more complex and realistic domain, the simulated soccer domain. Two teams of 11 agents play soccer against each other. The available actions are move, pass, dribble, score and intercept. Each action may take arguments. Thus, the number of ground actions (where the argument variables are substituted by values) is exponential in the size of the domain. This complexity enables the PDDL repairing to dominate the brute-force repairing due to SGPlan's superior search heuristics. Our scenario is as follows: an agent has to intercept the ball and to dribble to the opponent's goal. Thereby, the dribble path is planned, i.e. the field is divided into rectangular cells and the agent has to dribble from one cell to an adjacent one. We then artificially cause the policy execution to fail after some while in order to force policy repair. For a DT planing horizon of 2, i.e. the policy consists of one intercept and one dribble action, DT planning takes 0.027 and PDDL repairing 0.138 seconds on average. If the horizon is 3, the DT planning time increases to 0.302 seconds whereas the PDDL repairing time does not change. Thus, repairing with PDDL is faster than re-planning in this scenario. This confirms the evaluation of HORIZON. But in contrast to WumpusWorld, the BF repairing takes 0.305 seconds on average and is slower than the PDDL method. This gives more evidence to our observation that the performance of BF is related to the complexity of the action description. Our results show that plan repair in principle is useful in dynamic domains. We take these preliminary results as the starting point for future investigations.

5   Conclusions

In this paper we sketched our plan repair framework, where invalid DT policies generated by Readylog are repaired using an external PDDL planner. To this end, it must be detected when a policy becomes invalid, and how the world state must look like in order to continue the remainder policy. Then, we map our READYLOG world state to a PDDL description and calculate a repair plan with an external PDDL planner. In our current implementation we use SGPlan. The repair plan is then translated back into the READYLOG framework and executed before the remainder policy is continued. Although it was shown by Nebel and Koehler [12] that plan repair is at least as hard as planning from scratch, it turns out that plan repair works in practice under certain circumstances. In our case, the assumption is that only slight deviations in the execution make our policy invalid. Thus, only simple repair plans are needed to reach a state where the remainder policy can be executed again. A similar approach was proposed in [14], which integrate monitoring into a GOLOG-like language with a transition semantics similar to that of READYLOG. Their system also performs a monitoring step after each action execution and performs repair actions before the remainder of the failed policy in order to restore its executability. A similar idea, to combine GOLOG with an external planning system based on a PDDL description, was proposed in [13]. For the future work, we plan to combine their results with ours.

Acknowledgments

This work was supported by the German National Science Foundation (DFG) in the Priority Program 1125, *Cooperating Teams of Mobile Robots in Dynamic Environments*. We would like to thank the anonymous reviewers for their helpful comments.

References

1. Puterman, M.: Markov Decision Processes: Discrete Dynamic Programming. Wiley, New York (1994)
2. Boutilier, C., Reiter, R., Soutchanski, M., Thrun, S.: Decision-theoretic, high-level agent programming in the situation calculus. In: Proc. AAAI 2000, pp. 355–362. AAAI Press, Menlo Park (2000)
3. Levesque, H., Reiter, R., Lesperance, Y., Lin, F., Scherl, R.: GOLOG: A logic programming language for dynamic domains. Journal of Logic Programming 31(1-3), 59–83 (1997)
4. Reiter, R.: Knowledge in Action. MIT Press, Cambridge (2001)
5. Ferrein, A., Fritz, C., Lakemeyer, G.: Using golog for deliberation and team coordination in robotic soccer. KI 19(1), 24–30 (2005)
6. RoboCup: The Robocup Federation (2007), http://www.robocup.org
7. McCarthy, J.: Situations, Actions and Causal Laws. Technical report. Stanford University (1963)
8. Ferrein, A., Fritz, C., Lakemeyer, G.: On-line Decision-Theoretic Golog for Unpredictable Domains. In: Biundo, S., Frühwirth, T., Palm, G. (eds.) KI 2004. LNCS (LNAI), vol. 3238. Springer, Heidelberg (2004)
9. McDermott, D.: PDDL – The planning Domain Definition Language, Version 1.2. Technical Report TR-98-003/DCS TR-1165, Yale Center for Computational Vision and Control (1998)

10. Edelkamp, S., Hoffmann, J.: PDDL2.2: The Language for the Classical Part of the 4th International Planning Competition. Technical Report 195, Institute für Informatik (December 2003)
11. Hsu, W., Wah, B., Huang, R., Chen, Y.: New Features in SGPlan for Handling Soft Constraints and Goal Preferences in PDDL3.0. In: Proc. ICAPS 2006 (June 2006)
12. Nebel, B., Koehler, J.: Plan Reuse versus Plan Generation: A Theoretical and Empirical Analysis. Technical Report RR-93-33, DFKI GmbH (1993)
13. Claßen, J., Eyerich, P., Lakemeyer, G., Nebel, B.: Towards an integration of golog and planning. In: Proc. IJCAI 2007. AAAI Press, Menlo Park (2007)
14. Giacomo, G.D., Reiter, R., Soutchanski, M.: Execution Monitoring of High-Level Robot Programs. In: Proc. KR 1998, pp. 453–465 (1998)

A Recognition Interface for Bridging
the Semantic Desktop and the Physical World

Hiroshi Miyake[1], Koichi Kise[1], and Andreas Dengel[2]

[1] Dept. of Computer Science and Intelligent Systems,
Osaka Prefecture University, Japan
[2] German Research Center for Artificial Intelligence (DFKI), Germany

Abstract. The semantic desktop is an extension of the semantic web technologies to the field of desktops for the purpose of intelligent processing and share of data in the PC. Although most current technologies of the semantic desktop focus only on the virtual world in the PC, the field of user's activity is not limited to it; in a typical office environment, the user also interacts with various physical objects such as printed documents and office goods during the activity. The aim of this paper is to show some important concepts and a prototype of the interface necessary to extend the field of semantic desktop to the physical world. The key technologies for bridging the gap between virtual and physical worlds are fast and robust object recognition and document image retrieval.

Keywords: semantic desktop, interface, physical world, object recognition, document image retrieval.

1 Introduction

In order to share and leverage a huge amount of information stored on the Internet, the semantic web has been intensively investigated. The semantic web is a technology that enables automated collection, analysis and management of a huge amount of information with the help of semantic tags. As the available size of memory devices such as hard disk drives is getting larger, the amount of information stored on a personal computer also becomes far beyond the level of manual handling. In other words, problems to be solved by the semantic web technologies are now not only for the Internet but also for desktops of individual users. The semantic desktop, which is a concept similar to the semantic web but applied to individual desktops, helps us solve the above problems [5].

By combining the semantic web and the semantic desktop by, for example, making the latter social (i.e., social semantic desktop [5]), many problems on the "virtual world" could be solved. However we still have problems of the lack of "semantics" outside the virtual world. For example, we cannot get any semantic tags from objects in our physical world. The notion of "semantic reality" [6] captures potential needs for the connection of virtual and physical worlds based mainly on sensor networks in the physical world.

A. Dengel et al. (Eds.): KI 2008, LNAI 5243, pp. 276–283, 2008.

Our trial presented in this paper shares some awareness of the problems posed by the researchers of semantic reality. However our trial is different from it in the sense that our aim is to deal with the physical world around the desktop and/or the user, where the sensor networks are hardly of assistance. In such a situation, small or middle size objects play a central role within a limited area around the user. For example, we sometimes need to obtain relevant information from a part of a document we are looking at. With the help of semantic desktop technologies, it is fundamentally an easy task for an electronic document. However, this does not hold for a printed (physical) document. Other physical objects (posters, pictures, boxes, cans, calendars,...) would be keys to seek relevant information. In order to support the above functionality, it is necessary to have an easy interface for putting semantic tags to physical objects. As stated in the above examples, linking physical objects to the virtual world extends the usability of the semantic desktop.

One may think that the above function can be realized by using some existing technologies such as barcodes and RFID tags. Barcodes are however spoil the appearance of objects. For example, you cannot newly put a barcode to objects belonging to other persons without permission. RFIDs are almost invisible so that there should be no problem on this aspect. However, in exchange for this attractive property, we have to pay for tags and decoders which are too expensive for the use with almost priceless objects such as a sheet of paper.

This problem can be solved by introducing "image recognition technologies" to the interface. Information relevant to a physical object can be retrieved by just taking its picture. Semantic tags can also be added by taking a picture and sending it to the server with its tags. This makes objects "clickable" by pressing a shutter button of a mobile camera. In other words, image recognition technologies enable us to transform the appearance of physical objects to their IDs. Other functions including generation of links among multiple physical objects, and the use of physical objects as display of information will also be presented in this paper.

2   Related Technologies

Before going to the details of the proposed interface, let us discuss some related technologies. As technologies capable of "tagging" physical objects, barcodes and RFIDs are widely used. Barcodes play a central role in POS system. 2D barcodes such as QR codes [3] are commonly used in Japan to provide the user with richer information as compared to normal (1D) barcodes. For example, the user can access web pages by just recognizing QR codes. Designed 2D barcodes with colors are utilized to ease the deterioration of appearance as much as possible [1].

However, there is a serious drawback for linking physical objects by barcodes. It is generally difficult for barcodes to newly establish links from physical objects, because it is not always allowed to put user's barcodes to objects. In other words, the user can only follow the links that have already been prepared.

Although the RFID [4] has almost no problem about the deterioration of the appearance, the same problem on establishing new links also exists.

3   Interface Via Physical Objects

In this section, we would like to clarify necessary conditions and functions of the interface to physical objects, based upon the discussion in the previous section.

Let us start with necessary conditions for the interface. First, in order to have an advantage as compared to barcodes and RFIDs, relevant information of physical objects should be retrieved without putting additional materials to the objects. This enables us to add links freely to physical objects. Second, relevant information should be retrieved easily. In other words, usability of the interface should be as high as that of barcode readers; Users could accept to point a target object and press a button, but no more complicated steps.

Next consider the following example cases of the use of the interface in an office environment to know what functions are required for the interface:

1. (identity recognition) Retrieve electronic version of physical objects such as documents, line drawings and pictures from your computer;
2. (physical objects to information and services) Annotate physical objects with their relevant information, such as a good with a date of purchase, a document with a picture of its authors, a book with a review on the Internet, an instruction manual with a supplementary video, travel brochures with reservation;
3. (physical objects as icons) Physical objects associated with relevant information can be used as icons on the computer, such as a business card on a telephone to make a phone call;
4. (physical objects as devices for display) Physical objects can also be utilized as devices for displaying relevant information on them. For example, a part of a document can be used as a display of video with the help of augmented reality technologies.

So, what are fundamental functions necessary for realizing the above cases while satisfying the necessary conditions? When we use a physical object, we simply "recognize" them to know what it is and what we can do by using it. If this is also possible for the interface, the necessary conditions are naturally satisfied. Thus we have selected visual "recognition" as a way to implement the interface — the interface with a camera.

Using the recognition technologies, the above cases are realized as follows. The identify recognition is the recognition itself. Physical objects are related with electronic data through their recognition. Images of physical objects are recognized to find their IDs. Once IDs are available it is easy to obtain the predetermined relevant information or services. Information and services are associated with physical objects as the process of building models of recognition. An operation like drag-and-drop with a camera enables us to use physical objects as icons. With the help of augmented reality, it is possible for the interface to overlay relevant information on the object such as a news paper with a video like the world of Harry Potter.

In order to realize the function of the above "recognition", the following are required:

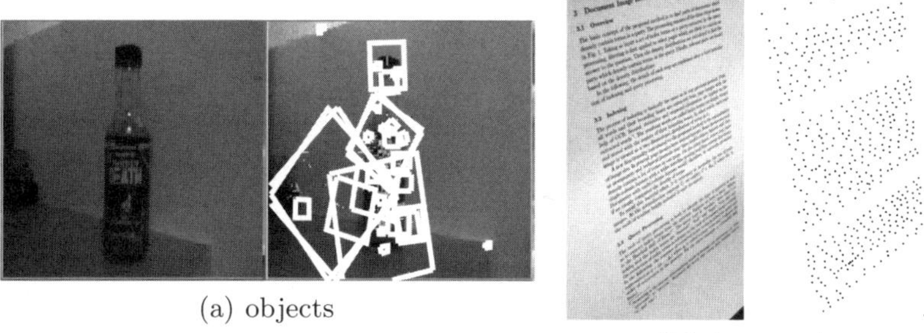

(a) objects

(b) documents

Fig. 1. Local descriptors

1. Fast and accurate recognition of object instances,
2. Recognition robust against viewing angle and illumination changes, occlusion and cluttered backgrounds
3. Recognition of multiple object instances in a single view,
4. Pose and position estimation

The first one is mandatory as a real-time interface with minimum user interactions. As stated in the second item, recognition should be robust enough in order to achieve high usability. The third one is to achieve high usability as well as to support the use of physical objects as icons. The fourth item is needed to display relevant information on physical objects naturally.

4   Recognition Methods

We have already proposed two methods of physical object recognition which provide us with fundamental recognition functionality for the interface. In this section, we briefly describe these two methods: a method of object recognition [8] and a method of document image retrieval [11].

4.1   Object Recognition

Recent progress of object recognition technologies enables us to realize a method that fulfill the above requirements: fast, accurate recognition with robustness. One of the key technologies is feature extraction methods from local salient regions of input images. Scale-invariant feature transform (SIFT) is a widely used local descriptor [9]. Our method employs a principal component analysis version of SIFT called PCA-SIFT [7].

From an image of QVGA size, from dozens to hundreds of feature vectors are typically extracted by PCA-SIFT. These feature vectors are highly discriminative: it is not difficult to find a feature vector extracted at the same part of the object stored in the database of more than a hundred million feature vectors. Figure 1 exemplifies local descriptors. In Fig. 1(a), local regions described by PCA-SIFT are illustrated.

The first step required to use the object recognition is to define object models by using local features. It is extremely simple: Store all feature vectors extracted from sample images with their categories. This is the reason why this technology is useful for the interface. All we have to do to establish a link from a new object is just to extract and store feature vectors.

The recognition process is as follows. Feature vectors are extracted from an unknown image. Then for each of them find its nearest neighbor vector from the object models by matching. This allows us to vote for an object from which the nearest neighbor is found. Finally objects which obtain votes larger than a predetermined threshold are the results of recognition.

The ability of our object recognition can be summarized as follows. We have tested the method through the recognition of planar objects such as pictures, posters and book covers. With the object models of up to 100,000 objects, the method is capable of achieving the recognition accuracy of 98% in 10 ms for matching and a few hundred ms for feature extraction.

4.2   Document Image Retrieval

Although the above mentioned method of object recognition is reliable, it is not applicable to all types of objects. In the office environment, documents are important objects to be recognized. However due to many similar local regions in documents, it is difficult to recognize documents using SIFT and its variants. This indicates the need of different feature extraction.

For this purpose we have developed a method of recognition for documents [11]. Since the term "document recognition" means character recognition in a document[1], we call our method "document image retrieval", which is the process of finding the electronic equivalent based on an image of a page as a key.

The feature extraction of our method is based on point representation of documents as shown in Fig. 1(b). Each point represents a centroid of a word region. We calculate a feature vector for each point based on local arrangements of nearby points.

The retrieval process is also fast and reliable. Using models of 10,000 pages, the method retrieves a corresponding electronic page with the accuracy of 98% in 100 ms including feature extraction. It has also been shown that the method is robust to occlusion, uneven lighting and non-linear deformation of page surfaces [2].

5   A Prototype of the Recognition Interface

We have developed a preliminary prototype of the recognition interface equipped with two recognition methods. The system is based on a simple client-server model. At the client side, an image is captured with a web cam. Then feature vectors are extracted from the captured image and sent to the server. At the server side, objects in the image are recognized by the voting process. Two

---

[1] For the case of documents, we can also apply a document recognition technology to annotate physical documents [10].

Fig. 2. A result of object recognition

recognition methods are individually and sequentially applied. All that have to be done by the user is just to point an object of interest and press the button of recognition.

Figure 2 shows major windows for the interface. The window 1 is for showing the captured image, the window 2 is for the recognition result and the window 3 represents the tag associated with the recognized object. Although a simple tag is shown in this figure, we can also provide any services if they are attached to the object. The user can construct an object model in a way similar to the recognition. First, the user point the object, press the store button and associate any information or services for the object.

Figure 3 illustrates the case of document image retrieval. As shown in this figure, the corresponding document page is correctly searched by the system even with an image captured from an angle.

Because the recognition process is based on the matching of local regions, we can obtain point correspondence as a side effect of recognition. Figure 4 indicates the point correspondence between a model and a captured image. It is also possible to obtain such correspondence for document images.

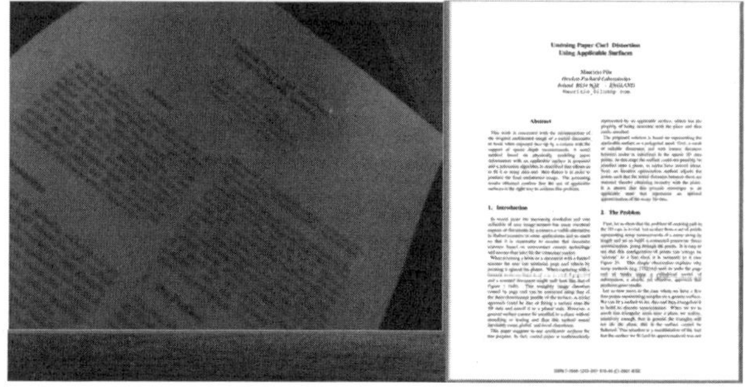

Fig. 3. A result of document image retrieval

Fig. 4. Point correspondence between a model and a captured image

The point correspondence enables us the following two important functions for the recognition interface. One is the case of recognizing multiple objects in a single image. In such a case, point correspondence indicates that there should be multiple objects in the image. If objects are planar, their regions in the captured image can be estimated based at least on the correspondence of 4 points. The prototype is capable of displaying such information on the window for the captured image shown in Fig. 5(a): Two objects are surrounded by two frames. It is allowed for the user to select one of them to execute an associated service.

Another important example is shown in Fig. 5(b): Relevant information is overlaid on the captured image. The user can view this information on a display of a portable device or through a head-mounted display. Note that it is not necessary to display the same information for all users; we can alter the information depending on the user. The information to be displayed can be extended from static to dynamic. By using a real-time capture-recognize cycle, a video can be played on the captured page image. Because the parameters of the geometric transformation of the paper surface are simultaneously calculated in real-time, the region for the video is also transformed as if it were on the paper.

Although all the functions shown above are still fundamental, we consider that it is not so difficult to combine these technologies to build a useful interface for the semantic desktop to extend their playing field to the real world.

(a) Multiple objects

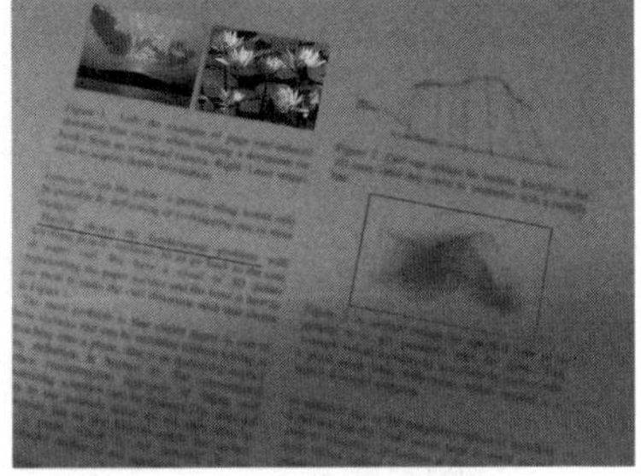

(b) Page for display

Fig. 5. Advanced functions of the recognition interface

6   Conclusion

In this paper, we have discussed demands and issues of the physical world interface that extends the field of semantic desktop to the real world. As an example we have proposed a prototype interface for tagging physical objects and retrieval of information using physical objects. In this interface, object recognition technologies for documents and other objects play the central role. Future work includes the improvement of the usability as well as the completion of the interface.

References

1. Color barcodes, `http://www.tradedot.com/2007/11/27/colorzip/`
2. Locally likely arrangement hashing, `http://www.m.cs.osakafu-u.ac.jp/LLAH/`
3. Qr code, `http://www.qrcode.com/`
4. Rfids, `http://en.wikipedia.org/wiki/RFID`
5. Dengel, A.: Knowledge technologies for the social semantic desktop. In: Proc. of the Second International Conference on Knowledge Science, Engineering and Management, pp. 2–9 (November 2007)
6. Hauswirth, M., Decker, S.: Semantic reality — connecting the real and the virtual world. In: Proc. SemGrail 2007 Workshop (June 2007)
7. Ke, Y., Sukthankar, R.: Pca-sift: A more distinctive representation for local image descriptors (2004)
8. Kise, K., Noguchi, K., Iwamura, M.: Simple representation and approximate search of feature vectors for large-scale object recognition. In: Proc. BMVC 2007, vol. 1, pp. 182–191 (September 2007)
9. Lowe, D.: Distinctive image features from scale-invariant keypoints. International Journal of Computer Vision 20, 91–110 (2003)
10. Maus, H., Dengel, A.: Semantic annotation of paper-based information. In: Proc. of the Second International Workshop on Camera-Based Document Analysis and Recognition, pp. 158–160 (September 2007)
11. Nakai, T., Kise, K., Iwamura, M.: Real-time document image retrieval with more time and memory efficient llah. In: Proc. of the Second International Workshop on Camera-Based Document Analysis and Recognition, pp. 21–28 (September 2007)

Learning by Observing: Case-Based Decision Making in Complex Strategy Games

Darko Obradovič and Armin Stahl

German Research Center for Artificial Intelligence (DFKI)
& Technical University of Kaiserslautern
{obradovic,stahl}@dfki.uni-kl.de

Abstract. There is a growing research interest in the design of competitive and adaptive Game AI for complex computer strategy games. In this paper, we present a novel approach for developing intelligent bots, which is based on the idea to observe successful human players and to learn from their individual decisions and strategies. These decisions are then reused by a bot in similar situations, resulting in a flexible and realistic strategic behaviour with low development and knowledge acquisition costs. Using Case-Based Reasoning (CBR) techniques, we implement this principle in the CYBORG system and achieve to outperform scripted opponents in a challenging multiplayer scenario.

1 Introduction

Computer games are a fast growing industry, in which realistic 3D graphics has been the main driving force for innovation and competitiveness in recent years. Since realism in graphics has come to stall at a high level, researchers predict that Artificial Intelligence is about to become the next decisive feature in future computer games [1]. We focus on complex computer strategy games that typically contain a lot of uncertainty and are played by more than two players. In contrast to classical strategy games like chess, gammon or poker, these games cannot be effectively described by Game Theory [2], as they are usually played with simultaneous hidden moves over a large number of turns. Thus they require abstractions and adapted or new approaches in order to become masterable by AI systems. Due to restrictions of resources and time, commercial Game AI developers often decide to use handcrafted rules (scripts). These however are static and, in consequence, uncompetitive against human opponents after a few repetitions. Hence researchers more and more investigate adaptive Game AI systems and predict that learning will be the "next big thing" [3].

Our idea is to use experiences of human strategy skills as the knowledge base for a game playing agent. The general knowledge, i.e. the situation description and the decision system mechanics, are defined by a domain expert, the situation-specific knowledge, that determines the strategy in the game, is elicited from the observation of successful players "in the wild". This results in an AI system as proposed in [4], which is very cheap to train and able to closely imitate any strategy demonstrated by a player. We use Case-Based Reasoning (CBR) [5] for

A. Dengel et al. (Eds.): KI 2008, LNAI 5243, pp. 284–291, 2008.

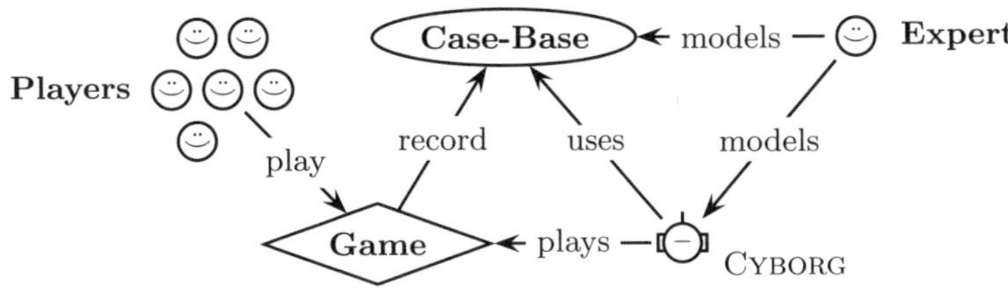

Fig. 1. System Functioning

our system, since the use of similarity enables an efficient situation space abstraction. The decision space is not abstracted, the system performs all low-level decisions based on the current game situation only. There are no explicit plans, goals or states. We argue that the current game situation, including all past and ongoing actions, is sufficient to compose a rational high-level strategy out of the low-level decisions, resulting in a very flexible system. Thus the challenge is to build an autonomously playing agent without any priorly induced strategies, which scales to games with more than two players. The functioning of our system, named CYBORG, is illustrated in Figure 1 and is implemented on the COSAIR platform[1].

The rest of the paper is organised as follows. In Section 2, we discuss related work and its influence on our idea. In Section 3, we explain in detail the conceptual and technical realisation of our idea, which is then empirically evaluated in Section 4. The paper ends with a conclusion in Section 5.

2   Related Work

Automatic Case Elicitation stores previously experienced knowledge into cases and uses these cases to come to decisions in new situations. The approach realises a trial-and-error method and requires only syntactical knowledge about the game rules in order to identify legal moves. Moves are tried by random and the resulting performance is recorded. It has been successfully applied in the domain of checkers [6], where the CHEBR system was able to defeat all other systems after a certain exploration phase. In chess however, the system did not perform well [7], because of the significantly larger situation space that needs to be explored. One reason certainly is the abdication of situation similarity, resulting in a one-to-one mapping of cases to situations. We thereby conclude that we need abstraction by similarity in order to deal with large situation spaces and that some form of knowledgeable guidance for learning in realistic scenarios is mandatory for successful learning.

A first complete and successful CBR system for strategy games is CAT [8], which performs *Case-Based Subplan Selection* in the real-time strategy (RTS) game "Wargus". It is based on previous work that abstracted the situation space to 20 key states, and the decision space to a number of handcrafted *subplans* for

---

[1] http://www.cosair.org/

each of these states. In a series of training games against a particular opponent, these subplans have been composed to player-specific counter-strategies by an evolutionary algorithm. CAT uses all this to realise an adaptive AI that can play against an initially unknown, random opponent from this set of scripted student bots. It therefor defines a simple case structure for game situations with a few attributes and starts exploring the use of subplans in specific situations. The performance is recorded and the cases are rated respectively, realising reinforcement learning. After the exploration phase, it manages to outperform all opponents. The most interesting thing for our work is the fact, that a relatively simple case specification covers the situation space very well. However, the system is limited to the existing subplans and to two-player games by design.

In *Case-Based Planning and Execution* [9], knowledge is acquired by observing human players in Wargus and by letting them annotate their individual decisions afterwards with goals that they were pursuing. In the game playing phase, the system selects behaviours from the case-base that are marked to accomplish the current goal of the AI. It thus dynamically composes and executes plans with promising results in a small two-player experiment with two annotated game traces. This approach also uses human gameplay to base AI decisions upon, but instead of an indirect approach via a planning engine, we target a direct game state to decision mapping. Furthermore this approach requires the player to annotate his complete game trace, what takes more time than the game itself, whereas we consciously avoid any additional effort beyond playing the game. Although having used the same game as CAT, the results are not comparable to each other because of different test scenarios.

3   Implementation

3.1   Game Platform

We implemented our approach for the COSAIR game platform, a multiplayer strategy game located in space, where each player controls a nation with several planets, fleets for defense and attacks and agents for espionage. Each of these objects has a number of characteristics and altogether this forms a strategic scenario far more complex than in any board game. The complete rules can be read in the game manual on the project website.

3.2   Situation Space

We decided to model the game situation by decomposing it into its relevant related game objects, which in turn are described by a domain expert with a number of attributes and weights. This is a very important step that decisively influences the resulting performance. Ideally, all aspects of a game situation that human players consider in their decisions should be included. The type of these attributes and the complexity for determining their values is not restricted, thus this allows to integrate AI techniques like terrain analysis or player profiling with this approach. As explained above, we are using an object oriented CBR system. The object relations correspond to the game objects, with all one-to-one relations consolidated into one object, e.g. a planet and its defending fleet.

Similarities of game situations are calculated on three levels, for attributes, objects and situations. Attribute similarity is calculated by type-specific standard similarity functions, e.g. Formula 1 shows the similarity function for set-type attributes a and $a*$, like the list of buildings on a planet for example. Object similarity is calculated by a weighted average over all attributes a_1 to a_n with their respective weights w_1 to w_n as described in Formula 2. Finally, game situation similarity is calculated as described in Formula 3, with involved objects o_1 to o_n. The n-root asserts a certain comparability of similarities between situations with a different number of objects involved, which is required in some decisions. The product asserts that situations with one very unsimilar object get low similarity values. For example, in case of an attack consideration, the involved objects are the own nation, the planet in question and its controlling nation.

$$sim_{attribute}(a, a^*) = \frac{|a \cap a^*|}{|a \cup a^*|} \tag{1}$$

$$sim_{object}(o, o^*) = \frac{\sum_{i=1}^{n} w_i * sim_{attribute}(a_i, a_i^*)}{\sum_{i=1}^{n} w_i} \tag{2}$$

$$sim_{situation}(s, s^*) = \prod_{i=1}^{n} \sqrt[n]{sim_{object}(o_i, o_i^*)} \tag{3}$$

For every decision, we have a specifically tailored representation of the relevant game situation with respect to the involved objects and their attribute weights. This decision-specific game situation is equivalent to a case in our system. Depending on this specification, the system retrieves all cases c_1 to c_n from the case-base, sorted downwards by their similarity to the current case c^*, i.e. case c_1 being the most similar one. In order to reduce retrieval costs, we apply a filter based on the turn number of the current game situation. Game situations in the middle of a game cannot be similar to game situations in early or late turns. This fact is also responsible for another effect. Just like in a game tree, situation space becomes broader over time. While ten game traces provide a dense situation space coverage in early turns, the coverage becomes thinner in later turns.

3.3 Decision System

As previously mentioned, CYBORG is designed to perform all low level decisions in the game without abstractions, thus we first take a closer look at the decision space in order to tailor the system as good as possible. Decision space is formed by eight different decisions that are listed in Table 1 and described in more detail in the bot programming manual on the platform website. In general we distinguish between the following types of decisions.

Decision point: There are specified points of time, where a decision has to be made, enforced by the game rules. This is typically the only decision type in traditional strategy board games.

Control variables: These are variables that are permanently in effect and can be adjusted each turn.

Optional decision: These are decisions that can be made at any point of time, given that certain preconditions are fulfilled, but never have to be made according to the game rules.

The last step is to specify the determination of the solution to be reused for a decision at issue. The solutions we want to reuse are regular attributes of the cases, e.g. in case we want to set the tax rate, we select a case while ignoring the tax rate attribute for the similarity calculation, and use the tax rate attribute as our solution value. This principle is applied to all decisions.

For decision points and control variables we use standard CBR methods. We either select the most similar case (MSC) c_1 to be reused, the confirmed most similar case (CMSC), i.e. the first solution that is suggested twice when searching from c_1 to c_n, or the average solution value of the k most similar cases (k-Avg), with k to be specified by the domain expert.

Optional decisions are more difficult to realise in a CBR system. We came up with two different methods for this. The first method is a two-step method (2-Step) decomposing the problem into a first step, in which we decide whether to perform an action or not, and a second step, only executed in case of a positive decision in step one, in which we choose the concrete action. Now, both these steps can be treated like decision points, and given that the first step is relatively cheap, it can be executed each turn. If the question whether to take an option or not cannot be decided upfront in a first step, we use a method named k-voting. This method selects all cases within c_1 to c_k that suggest an action. If this set is empty, no action is performed, otherwise we execute the action of the case with the highest similarity in this set. Both these methods have to be properly parameterised by a domain expert.

Table 1 summarises our model for the decisions of the game, which is the result of a care- and thoughtful application of domain knowledge.

Table 1. Overview of the CYBORG Decision Modules

Decision	Type	Method	Objects
Tax Rate	control variable	CMSC	1
Research Goal	decision point	CMSC	1
Agent Recruitment	optional decision	CMSC	1
Item Building	decision point	MSC	2
Production Rush	control variable	CMSC	2
Fleet Distribution	control variable	k-Avg	2
Agent Missions	optional decision	2-Step	1, 2-3
Attacks	optional decision	k-Voting	3

4 Evaluation

We evaluate our bot in the standard four-player test scenario of the platform, a game against three scripted bots authored by students, with series of 500 game simulations. We consider both performance measurements, absolute wins and the score from the relative ranking in the field of players. We use the game-traces of the top three participants of the available platform database, in which games of human players playing this scenario have been recorded. The data is summarised in Table 2, and the percentage of overall won games illustrates the ambitious difficulty level of this scenario. We constructed personalised case-bases from the won games only. This is our only filter against bad decisions. The value $t_{>=4}$ indicates the turn number up to which we have at least four game traces available. Experience shows that this is the minimum number needed for rational decision making and we thus define the turn limit by this value in order to obtain a fair evaluation of our bot.

Table 3 shows the results for the case-bases of the players A, B and C with wins and scores normalised to a series of 100 games. The score value indicates the share of the total points the player achieved at the end of the game and thus measures performance differences amongst the non-winning players, which would otherwise be ignored. We observe that all case-bases perform reasonably well and thus conclude that our principal approach is working. Obviously a larger case-base does not necessarily yield better results, as case-base C performs better than B does. There are strong indications that the complexity and the nature of the recorded strategies influence performance a lot, since player B preferred an aggressive and risky strategy, which has a lower tolerance for deviations, while player C used a more complex, but safe strategy instead.

We also let the observed players watch some game traces played using their personal case-base. All three players attested the system to follow their strategy surprisingly well, with the exception of fleet distributions, where the system misses some good opportunities for attacks in consequence.

In order to examine the influence of the case-base size on the performance, we used a reduced case-base from player A, called A^*, which contains every second game of the original record. Results for this case-base are also listed in Table 3 and indicate that there is a strong correlation.

Another aspect we evaluated is the similarity value of decisions during one game. We tracked the average decision similarity of a control variable, in a series

Table 2. Experiment Results by Players

Player	Games	Wins	in %	t_{avg}	$t_{>=4}$
A	35	32	91%	301	393
B	23	21	91%	266	361
C	17	17	100%	419	441
8 others	62	34	55%	406	

Table 3. Simulation Results per Case-Base

Case-Base	Rank	Bot	Wins	Score
A	1.	CYBORG	**50.6%**	**34.4**
	2.	eDevil	26.6%	24.1
	3.	Tombot	3.2%	24.0
	4.	BrainBug	19.6%	17.6
B	1.	eDevil	33.8%	**28.5**
	2.	CYBORG	27.8%	26.4
	3.	BrainBug	**35.2%**	23.6
	4.	Tombot	3.2%	21.6
C	1.	CYBORG	32.8%	**31.8**
	2.	BrainBug	**35.6%**	26.3
	3.	eDevil	30.2%	22.4
	4.	Tombot	1.4%	19.5
A^*	1.	CYBORG	31.4%	**27.7**
	2.	eDevil	**31.8%**	26.1
	3.	Tombot	8.4%	24.8
	4.	BrainBug	28.4%	21.4

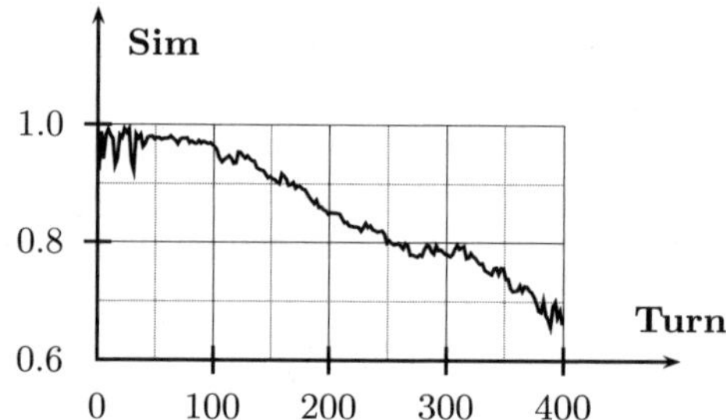

Fig. 2. Tax Rate Decision Similarity Over Time

of 50 games with case-base A. The result is plotted in Figure 2. In late turns, average similarity drops below 0.7, implying that we would need more cases in late turns to make better decisions. The similarity values for the other decisions show roughly the same gradient.

5 Conclusion

We have demonstrated the practicability of our suggested approach, which enables bots to play successfully in complex computer strategy games against multiple opponents at once. Our approach also opens up new possibilities in commercial games. Such a system is able to learn from a player during games and to assist her in a task she wishes to delegate. It can also function as a complete drop-in replacement in times of absence, which is a common problem in long-term web-based strategy games. It could also be used to create a database of unique bots with realistic behaviours which are then sharable over the Internet.

The evaluation reveiled that a better decision technique for the fleet distribution problem should further improve performance significantly. The second action to take should definitely be the enlargement of case-bases in later turns. Possible future research work comprises the questions, how to effectively combine multiple case-bases in order to improve performance and how to implement reinforcement learning during bot games, so that the system is able to learn the quality of cases automatically.

References

1. Laird, J.E., van Lent, M.: Human-level ai's killer application: Interactive computer games. AI Magazine 22, 15–25 (2001)
2. Carmichael, F.: A Guide to Game Theory. FT Prentice Hall, Englewood Cliffs (2004)
3. Rabin, S.: AI Game Programming Wisdom, vol. 2. Charles River Media (2003)
4. Thurau, C., Bauckhage, C., Sagerer, G.: Imitation learning at all levels of game-ai. In: Proceedings of the International Conference on Computer Games, Artificial Intelligence, Design and Education, pp. 402–408 (2004)
5. Aarmodt, A., Plaza, E.: Case-based reasoning: Foundational issues, methodological variations and system approaches. AI Communications 7(1), 29–59 (1994)
6. Powell, J.H., Hauff, B.M., Hastings, J.D.: Evaluating the effectiveness of exploration and accumulated experience in automatic case elicitation. In: Proceedings of the 6th International Conference on Case-Based Reasoning, pp. 397–407. Springer, Heidelberg (2005)
7. Kommuri, S.N., Powell, J.H., Hastings, J.D.: On the effectiveness of automatic case elicitation in a more complex domain. In: ICCBR 2005 Computer Gaming and Simulation Environments Workshop, pp. 185–192 (2005)
8. Aha, D.W., Molineaux, M., Ponsen, M.J.V.: Learning to win: Case-based plan selection in a real-time strategy game. In: Proceedings of the Sixth International Conference on Case-Based Reasoning, pp. 5–20. Springer, Heidelberg (2005)
9. Ontañón, S., Mishra, K., Sugandh, N., Ram, A.: Case-based planning and execution for real-time strategy games. In: Proceedings of the Seventh International Conference on Case-Based Reasoning, pp. 164–178. Springer, Heidelberg (2007)

Toward Alignment with a Virtual Human - Achieving Joint Attention

Nadine Pfeiffer-Leßmann and Ipke Wachsmuth

Artificial Intelligence Group, Faculty of Technology
Bielefeld University
33594 Bielefeld, Germany
{nlessman,ipke}@techfak.uni-bielefeld.de

Abstract. Alignment forms the basis of successful communication. It can be seen as the most efficient means for action coordination of co-operating agents, covering adaptation processes operating without an explicit exchange of information states. One critical condition of alignment consists of *joint attention*. We present work on equipping a virtual human with the capability of reaching joint attention with its human interlocutor. On the one hand, mechanisms to detect the human's focus of attention are employed. On the other hand, basic cognitive as well as intentional processes underlying the phenomenon of joint attention are incorporated in our agent's cognitive architecture. In this context, a dynamic working memory and a partner model accounting for theory of mind and intentionality are crucial constituents.

Keywords: joint attention, alignment, virtual humans, theory of mind.

1   Introduction

In order to build a believable virtual human, we must understand how to model its cognitive abilities to engage in natural, successful face-to-face communication. According to Pickering and Garrod [1] successful communication is based on efficient action coordination and adaptation mechanisms realizing a close connection between the interlocutors. These *alignment processes* are joined processes between the interactants allowing them to sufficiently reconstruct the meaning of the interaction. One central condition of these joint process consists of *joint attention*. Joint attention facilitates interaction processes and supports inferences about people's current and future activities, both overt and covert. It is a foundational skill in human social interaction and cognition and can be defined as simultaneously allocating attention to a target as a consequence of attending to each other's attentional states [2]. However, to distinguish *joint attention* from *joint perception*, Tomasello stresses the intentional aspect of joint attention by demanding that the interlocutors have to deliberatively focus on the same entity while being mutually aware of this [3].

We investigate *joint attention* in an interaction scenario with the virtual human *Max* [4]. The human interlocutor meets the embodied conversational agent

A. Dengel et al. (Eds.): KI 2008, LNAI 5243, pp. 292–299, 2008.

face-to-face in a CAVE-like virtual environment. This scenario allows for the inclusion of verbal as well as non-verbal communication channels (e.g. gaze and gestures) for both the human interlocutor as well as for the virtual agent.

After discussing the psychological background of joint attention in the next section, current research on implementing attentional behavior in virtual humans and robots will be presented. The essential interplay of intentionality and attention will be covered in the requirements section. Based on these insights, our own approach of modeling joint attention in the virtual human *Max* will be presented in (Sect 5).

2   Psychological Background

Attention has been characterized as consisting of an increased awareness with respect to internal as well as external aspects such as perceptions, conceptions, and behaviors. This awareness can be invoked by involuntary as well as deliberate processes [5]. Attention is therefore not a unitary process but a complex phenomenon. Attention can be defined as intentionally directed perception [3]. Its purpose lies in the allowance and maintenance of goal-directed behavior. Cohen et al. [6] follow [7] in assuming three attentional subsystems: an *anterior attentional system* concerned with cognitive control and action selection, a *posterior attentional system* associated with orienting and perceptual attention, and an *arousal system* covering alertness phenomena.

During interactions, human attention is modulated by the observation of gaze direction and by inferences derived from observations. Recent experiments suggest that interacting agents pay as much attention to each other's intentions as they do to each other's observable acts [8]. Hobson additionally underlines that agents need the capacity of joining one another by sharing an experience and registering an intersubjective linkage. To reach joint attention, the agents need to be aware of of the other's focus of attention as well as of the process of sharing attention itself [9]. Considering these deliberative aspects of joint attention, the attentional focus of *cognitive control* has to be taken into account. Cognitive control and attention can be seen as emergent properties of information representation in working memory [10]. In line with this research, Cowan claims that the focus of attention is controlled conjointly by voluntary processes (*central executive system*) and involuntary processes (*the attention orienting systems*) [11]. Oberauer adopts these ideas seeing working memory as a concentric structure with its parts being characterized by an increased state of accessibility for cognitive processes: (1) The activated part of long-term memory holds information over brief periods of time. (2) The region of direct access serves to keep a limited number of chunks available for ongoing cognitive processes. (3) The *focus of attention* itself holds at any time the one chunk being selected for the next cognitive operation to be applied upon [12].

One of the most comprehensive mod1els of joint attention comes from Baron-Cohen's work on autism [13]. He postulates a tiered model containing four modules including an intentionality detector, an eye-direction detector, a shared attention module, and a theory of mind module. Emphasis is put on the theory of mind as an endpoint and meta-representation as a process, but thereby some key relations, especially between attention and intentions are not described [14].

3   Related Work

In the area of virtual humans, researchers have mainly focused on modeling the perceptional attention focus as well as convincing gaze behavior [15] [16]. These computational models can be seen as prerequisites for joint attentional mechanisms. However, aspects of conjoining the attentional foci of the interlocutors are not covered.

A number of researchers in cognitive science and cognitive robotics use developmental insights as a basis for modeling joint attention. They show how a robot can acquire aspects of joint attention by supervised and unsupervised learning [2], [17], [18]. However, the aspect of intentionality and explicit representation of the other's mental state are not accounted for in these approaches. Work on a listener-robot explicitly addresses the issue of *joint attention* [19]. But in this robot, joint attention is modeled as an unconscious mechanism. The robot's behaviors are not subject to deliberate decisions but are implemented by if-then rules based on the redundancy of the interlocutor's attentional behavior and the communication mode. Breazeal et al. [20] work on a robot which is capable of rich social interactions and is provided with joint attention capabilities modeled as a collaborative process. The robot has an attention system which determines its attentional focus by calculating saliency values for all perceived objects. Additionally, the attention system is used to monitor and represent the human's focus of attention by attaching saliency tags to the respective objects. The robot's attention following and directing skills can be accompanied by conversational policies along with gestures and shifts of gaze accouting for repair, elaboration, and confirmation of the shared referential focus.

Hence, besides that the robotics community has recently demonstrated an increasing interest for modeling joint attention, most of the existing models focus only on partial and isolated elements of joint attention phenomena. They cover surface behaviors like simultaneous looking or simple coordinated behavior, but do not address the deeper, more cognitive aspects of joint attention [21].

4   Requirements

Following [21], we understand joint attention as an active bilateral process which involves attention alternation, and can only be fully understood by assuming that it is realized by intentional agents. To reach joint attention, the agents must be aware of the coordination mechanisms of understanding, monitoring and directing the intentions underlying the interlocutor's attentional behavior. To this end, the agent needs to be able to (r1) track the attentional behavior of other agents by gaze monitoring and has to (r2) derive the candidate objects the interlocutor may be focusing on from observation of the interlocutor's behavior and the situational context. Furthermore, the agent has to (r3) recognize, whether the attentional direction cues of the interlocutor are put out intentionally or not. This aspect can be covered by keeping a model of the interlocutor's mental state with respect to his focus of attention. The agent has to (r4) react instantly, as simultaneity plays a crucial role in joint attention. When in response to an attentional direction cue, the agent deliberately draws its focus of attention on the referred object, it should (r5) use an adequate overt behavior which can be observed by its interlocutor.

To manipulate the attentional behavior of its interlocutor, the agent should also be able to engage in proto-declarative pointing, the ability to point in order to comment or remark on the world to another person. This behavior can be applied when gaze alone does not suffice. Also verbal references can be constituted to draw the interlocutor's attention on a specific object.

5 Modeling Joint Attention in a Virtual Human

Reconciling the requirements and research on joint attention, we propose the following model of joint attention (see Fig. 1). In order to reach joint attention, three main aspects need to be considered. Firstly, the agent's *mental state* serves as the origin of the attention mechanisms allowing for intentionally guided behavior. To reach joint attention, information covering a *partner model* which accounts for the interlocutor's *focus of attention* is acquired. This attentional focus has to be inferred from the interlocutor's overt behavior. Additionally, the environmental context is taken into account in order to embrace situatedness. This is achieved by bottom-up *activation processes* marking relevant objects as salient in the current situation. These processes can be seen as the second main aspect of joint attention. Thirdly, the agent itself needs to display appropriate overt behaviors to accentuate its *focus of attention* and to manipulate the *interlocutor's mental state.*

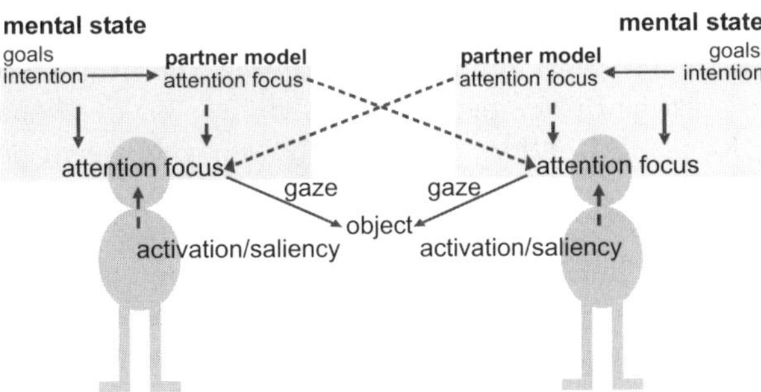

Fig. 1. Model of joint attention

5.1 Cognitive Architecture

Instead of modeling a separate module solely concerned with producing aligned gaze behavior of joint attention, we want to model the mechanisms behind the phenomenon of joint attention having their roots in the agent's cognition. We extend the *cognitive architecture* of our virtual agent [4]. In CASEC (Cognitive Architecture for a Situated Embodied Cooperator), ideas of basic cognitive mechanisms are integrated with explicit representations of mental states. The CASEC architecture adopts the BDI (*belief-desire-intention*) paradigm of rational agents [22] but additionally incorporates a dynamic memory model being inspired by work of Cowan and Oberauer (see Sect 2) to account for basic cognitive processes. Instead of static sets of beliefs in which entities have to be deliberately added and removed, the dynamic model employs automatic activation and decay processes. The activation values of the entities kept in working memory represent their saliency in the current context. They can be influenced by events, internal processing, and by the decision of the agent. In addition to *automatic processes* increasing the activation values whenever the object gets in the agent's gaze focus, *deliberative mechanisms* increase the activation values whenever the object is subject to internal processing. That is, in contrast to attention modeling in form of a stack on which objects are deliberatively pushed, we model the mechanisms behind attention so that an attention focus emerges out of the agent's goals, its behavior, and its interaction with the environment.

As attention is not seen as a unitary process, several attentional mechanisms are modeled in our architecture. Figure 2 outlines the agent's working memory. The rhombi represent relational chunks written into working memory functioning as explicit representations of the agent's *beliefs*. They descend from visual and auditive perception processes residing in the *phonological loop* and the *visual spatial memory*. Additionally, relevant chunks can be retrieved out of *long-term semantic memory* and are represented together with the agent's current conclusions. Activation impulses are sent by the agent's perception mechanisms. Whenever a new object is perceived or refocused, information about it covering its unique ID and its relevant attributes are written into working memory. The content of working memory is represented by use of relations which consist of a relation name e.g. *color, is_a, or inst_of* together with the respective attributes. Each of the relation entities has its own activation value. But not only the salience of the object's perceptual attributes makes them appear as attractors for the focus of attention. Additionally, when the agent perceives verbal input referring to an object, the respective activation values will be increased. Also when the agent notices that another agents focuses on a specific object, activation impulses are initiated. By this means, the content of working memory and thereby the candidate set of the agent's cognitive processes is influenced by its interlocutor providing the basis for *attention alignment processes*.

5.2 Attention Detection

As an indicator for the human interlocutor's focus of attention, we track the human's gaze as a basic manifestation of joint attention is gaze following. For the

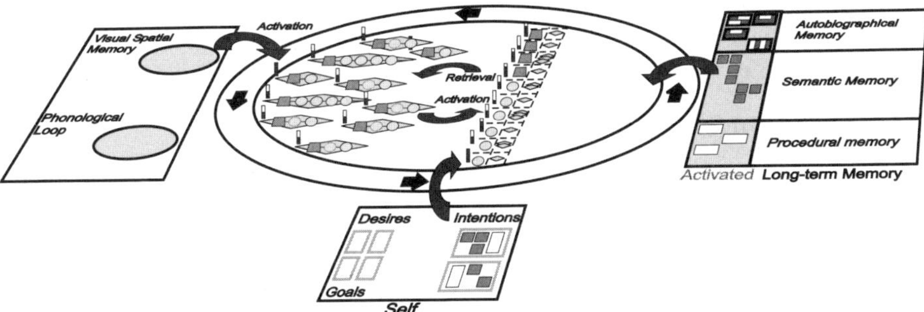

Fig. 2. Model of dynamic working memory

first requirement (r1) (see Sect 4) tracking the interlocutor's focus of attention, an eye-tracker is employed. Additionally, the human's head direction is tracked by an infrared camera tracking system providing the position of the eyes relative to the world coordinate system. To (r2) derive the objects lying in the in the human's focus of attention, a cone of 2.5 degrees is used. In addition to the boundary of the cone, activation values of the agent's working memory are incorporated in the calculation of the candidate set. By this means, we do not only take the humans's line of view into account but also the situational context. The determination of the selected object is done using histogram calculations. An object is detected as being in the human's focus of attention, when it has been focused at least for a sum of 400ms in a 600ms time frame. In addition to this heuristic, the use of other communication channels such as pointing gestures or verbal expressions are interpreted as *intentional direction cues*. If attentional direction cues of a certain intensity are detected, the resulting belief is written into working memory.

5.3 Partner Model

The incorporated partner model covers beliefs about the conversational state and the conversational role of the interactant. Additionally, the interlocutor's attentional focus is represented as beliefs in form of *"attention_focus $interlocutor $object"* being updated dynamically and thereby leading to new beliefs or increasing a belief's activation respectively. However, as only intentional focusing is reckoned as an invitation to *joint attention*, we follow [19] with respect of (r3) detecting the speaker's intention: The intensity of the speaker's overt behaviors are interpreted as indicators for the interlocutor's intention. The intensity is calculated by the communication channel used and the redundancy of the interlocutor's attentional direction cues. For the agent to ascribe the desire to reach joint attention to its interlocutor, we use the following heuristic: An object has be the focus of attention for several times with additional short glances addressing the agent inbetween (triadic relation). Otherwise the interlocutor may just be focusing on the object with respect to other than interactional aspects. When the activation value of an *attention-focus-belief* passes a threshold and the interlocutor has shown interactive glances, the agent believes *"achieve $interlocutor* *attention_focus self $object"* leading to an (iv) instantaneous activation of a *conclude*-plan. Thus the agent becomes aware of the interlocutor's intention and decides how to respond e.g. gazing at the same object to reach joint attention.

5.4   Attention Manipulation

While the attention detection mechanisms can be seen as prerequisites for engaging in joint attention, the agent also employs (r5) pro-active mechanisms to manipulate the interlocutor's focus of attention. Being an embodied conversational agent, the agent deploys different attentional direction cues e.g. intentional gaze, deictic gestures, and verbal expressions [4]. The agent's mental state is modeled using the BDI paradigm. Besides the explicitly represented beliefs, intentional states are represented as explicit goal states together with the means to achieve them. In case of the agent's goal to reach joint attention, the agent represents an explicit goal *"ACHIEVE attention_focus $interlocutor $object"* and pursues a plan in order to draw the interlocutor's attention toward the same object it is focusing on. When deciding which mechanism to use, the agent relies upon its plan library, usually adopting the plan with the least involved effort (e.g. gaze). If no suitable reaction of the interlocutor is perceived, more obvious attentional direction cues are applied. The plans are only carried out for as long as the goal state has not been achieved. As soon as the agent believes *"attention_focus $interlocutor $object"*, joint attention is achieved and the agent turns to the next goal of its agenda. Also, when the top-level goal which led to the instantiation of the *joint-attention-goal* is achieved or dropped, the *joint-attention-goal* is automatically abandoned. In case of not achieving joint attention, the failure part of the plan catches leading to a verbal expression making the agent's goal explicit.

6   Conclusion and Future Work

We presented work on equipping our virtual human Max with capabilities of joint attention. To this end, perception and detection mechanisms have been proposed and the processing in the agent's cognitive architecture has been presented. In future research, we want to evaluate how the model of joint attention is approved by naive human interactants. Additionally, we want to explore how parameters of joint attention mechanisms (e.g. timing, explicitness, intensity) can be tuned to adapt to the interactant's reactions applying for alignment processes.

Acknowledgments. This research is partially supported by the Deutsche Forschungsgemeinschaft (DFG) in the Collaborative Research Center SFB 673.

References

1. Pickering, M.J., Garrod, S.: Alignment as the basis for successful communication. Research on Language and Computation 4(2), 203–228 (2006)
2. Deak, G.O., Fasel, I., Movellan, J.: The emergence of shared attention: Using robots to test developmental theories. In: Proc. of the First Intl. Workshop on Epigenetic Robotics. Lund University Cognitive Studies, vol. 85, pp. 95–104 (2001)

3. Tomasello, M., Carpenter, M., Call, J., Behne, T., Moll, H.: Understanding and sharing intentions: The origins of cultural cognition. Behavioral and Brain Sciences 28, 675–691 (2005)
4. Leßmann, N., Kopp, S., Wachsmuth, I.: Situated interaction with a virtual human - perception, action, and cognition. In: Rickheit, G., Wachsmuth, I. (eds.) Situated Communication, pp. 287–323. Mouton de Gruyter, Berlin (2006)
5. Brinck, I.: The objects of attention. In: Proc. of ESPP 2003, Torino, pp. 1–4 (2003)
6. Cohen, J., Aston-Jones, G., Gilzenrat, M.: A systems-level perspective on attention and cognitive control. In: Posner, M. (ed.) Cognitive Neuroscience of Attention, pp. 71–90. Guilford Publications, New York (2004)
7. Posner, M., Petersen, S.: The attention system of the human brain. Annu. Rev. Neurosci. 13, 25–42 (1990)
8. Nuku, P., Bekkering, H.: Joint attention: Inferring what others perceive (and don't perceive). Conscious and Cognition 17(1), 339–349 (2008)
9. Hobson, R.P.: What puts the jointness into joint attention? In: Eilan, N., Hoerl, C., McCormack, T., Roessler, J. (eds.) Joint attention: communication and other minds, pp. 185–204. Oxford University Press, Oxford (2005)
10. Courtney, S.M.: Attention and cognitive control as emergent properties of information representation in working memory. Cognitive, Affective & Behavioral Neuroscience 4(4), 501–516 (2004)
11. Cowan, N.: An embedded-processes model of working memory. In: Miyake, A., Shah, P. (eds.) Models of Working Memory, Mechanisms of Active Maintenance and Executive Control, pp. 62–102. Cambridge University Press, Cambridge (1999)
12. Oberauer, K.: Access to information in working memory: Exploring the focus of attention. J. of Exp. Psych.: Learning, Memory, and Cognition 28, 411–421 (2002)
13. Baron-Cohen, S.: The eye-direction detector (edd) and the shared attention mechanism (sam): two cases for evolutionary psychology. In: Moore, C., Dunham, P. (eds.) Joint Attention: Its origins and role in development, pp. 41–61. L. Erlbaum, Mahwah (1994)
14. Tomasello, M.: Joint attention as social cognition. In: Moore, C., Dunham, P. (eds.) Joint Attention: Its origin and role in development, pp. 103–128. L. Erlbaum, Mahwah (1995)
15. Kim, Y., Hill, R.W., Traum, D.R.: Controlling the focus of perceptual attention in embodied conversational agents. In: Proceedings AAMAS, pp. 1097–1098 (2005)
16. Gu, E., Badler, N.I.: Visual attention and eye gaze during multiparty conversations with distractions. In: Gratch, J., Young, M., Aylett, R.S., Ballin, D., Olivier, P. (eds.) IVA 2006. LNCS (LNAI), vol. 4133, pp. 193–204. Springer, Heidelberg (2006)
17. Nagai, Y., Hosoda, K., Morita, A., Asada., M.: A constructive model for the development of joint attention. ConnectionScience 15(4), 211–229 (2003)
18. Doniec, M., Sun, G., Scassellati, B.: Active learning of joint attention. In: IEEE/RSJ International Conference on Humanoid Robotics (2006)
19. Ogasawara, Y., Okamoto, M., Nakano, Y., Nishida, T.: Establishing natural communication environment between a human and a listener robot. In: AISB Symposium on Conversational Informatics, pp. 42–51 (2005)
20. Breazeal, C., Brooks, A., Gray, J., Hoffman, G., Kidd, C., Lee, H., Lieberman, J., Lockerd, A., Chilongo, D.: Tutelage and collaboration for humanoid robots. International Journal of Humanoid Robots 1(2), 315–348 (2004)
21. Kaplan, F., Hafner, V.: The challenges of joint attention. Interaction Studies 7(2), 135–169 (2006)
22. Rao, A., Georgeff, M.: Modeling rational behavior within a BDI-architecture. In: Proc. Intl. Conf. on Principles of Knowledge Repr. and Planning, pp. 473–484 (1991)

Concerning Olga, the Beautiful Little Street Dancer: Adjectives as Higher-Order Polymorphic Functions

Walid S. Saba

American Institutes for Research
1000 Thomas Jefferson Street, NW, Washington, DC 20007 USA
wsaba@air.org

Abstract. In this paper we suggest a typed compositional semantics for nominal compounds of the form [*Adj Noun*] that models adjectives as higher-order polymorphic functions, and where types are assumed to represent concepts in an ontology that reflects our commonsense view of the world and the way we talk about it in ordinary language. In addition to [*Adj Noun*] compounds our proposal seems also to suggest a plausible explanation for well known adjective ordering restrictions.

1 Introduction and Overview

The sentence in (1) could be uttered by someone who believes that: (i) Olga is a dancer and a beautiful person; or (ii) Olga is beautiful as a dancer (i.e., Olga is a dancer and she dances beautifully).

$$\textit{Olga is a beautiful dancer.} \tag{1}$$

As suggested by Larson (1998), there are two possible routes to explain this ambiguity: one could assume that a noun such as 'dancer' is a simple one place predicate of type $\langle e,t \rangle$ and 'blame' this ambiguity on the adjective; alternatively, one could assume that the adjective is a simple one place predicate and blame the ambiguity on some sort of complexity in the structure of the head noun (Larson calls these alternatives *A*-analysis and *N*-analysis, respectively).

In an *A*-analysis, an approach predominantly advocated by Siegel (1976), adjectives are assumed to belong to two classes, termed predicative and attributive, where predicative adjectives (e.g. *red*, *small*, etc.) are taken to be simple functions from entities to truth-values and are extensional, and thus intersective: $\|Adj\ Noun\| = \|Adj\| \cap \|Noun\|$. Attributive adjectives (e.g., *former*, *rightful*, etc.), on the other hand, are functions from common noun denotations to common noun denotations – i.e., they are predicate modifiers of type $\langle\langle e,t\rangle,\langle e,t\rangle\rangle$ and are thus intensional and non-intersective (but are subsective: $\|Adj\ Noun\| \subseteq \|Noun\|$). On this view, the ambiguity in (1) is explained by posting two distinct lexemes (*beautiful*$_1$ and *beautiful*$_2$) for the adjective beautiful, one of which is an attributive while the other is a predicative adjective. In keeping with Montague's (1970) edict that similar syntactic categories must have the same semantic type, for this proposal to work, all adjectives are

A. Dengel et al. (Eds.): KI 2008, LNAI 5243, pp. 300–307, 2008.

initially assigned the type $\langle\langle e,t\rangle,\langle e,t\rangle\rangle$ where intersective adjectives are considered to be subtypes obtained by triggering an appropriate meaning postulate. For example, assuming the lexeme *beautiful*$_1$ is marked as +INTERSECTIVE, the meaning postulate $\exists P\forall Q\forall x[\text{BEAUTIFUL}(Q)(x) \leftrightarrow P(x) \wedge Q(x)]$ would then yield an intersective meaning when P is *beautiful*$_1$; and where a phrase such as 'a beautiful dancer' is interpreted as follows[1]:

$$\|a\ \textit{beautiful}_1\ \textit{dancer}\| \Rightarrow \lambda P[(\exists x)(\textbf{dancer}(x) \wedge \textbf{beautiful}(x) \wedge P(x))]$$
$$\|a\ \textit{beautiful}_2\ \textit{dancer}\| \Rightarrow \lambda P[(\exists x)(\textbf{beautiful}(^{\wedge}\textbf{dancer}(x)) \wedge P(x))]$$

While it does explain the ambiguity in (1), several reservations have been raised regarding this proposal. As Larson (1995; 1998) notes, however, this approach entails considerable duplication in the lexicon as this means that there are `doublets' for all adjectives that can be ambiguous between an intersective and a non-intersective meaning. Another objection, raised by McNally and Boleda (2004), is that in an *A*-analysis there are no obvious ways of determining the context in which a certain adjective can be considered intersective. For example, they suggest that the most natural reading of (2) is the one where beautiful is describing Olga's dancing, although it does not modify any noun and is thus wrongly considered intersective by modifying Olga.

$$\textit{Look at Olga dance. She is beautiful.} \qquad (2)$$

While valid in other contexts, in our opinion this observation does not necessarily hold in this specific example since the resolution of 'she' must ultimately consider all entities in the discourse, including, presumably, the dancing activity that would be introduced by a Davidsonian representation of 'Look at Olga dance' (this issue is discussed further below). A more promising alternative to the *A*-analysis of the ambiguity in (1) has been proposed by Larson (1995, 1998), who suggests that 'beautiful' in (1) is a simple intersective adjective of type e,t and that the source of the ambiguity is due to a complexity in the structure of the head noun. Specifically, Larson suggests that a deverbal noun such as dancer should have a Davidsonian representation such as $(\forall x)(\textbf{dancer}(x) \equiv_{df} (\exists e)(\textbf{dancing}(e) \wedge \textbf{agent}(e,x)))$; that is, any x is a dancer iff x is the agent of some dancing activity (Larson's notation is slightly different). In this analysis, the ambiguity in (1) is attributed to an ambiguity in what 'beautiful' is modifying, in that it could be said of Olga or her dancing activity. That is, (1) is to be interpreted as follows:

$$\|\textit{Olga is a beautiful dancer}\|$$
$$\Rightarrow (\exists e)(\textbf{dancing}(e) \wedge \textbf{agent}(e,\textit{olga}) \wedge (\textbf{beautiful}(e) \vee \textbf{beautiful}(\textit{olga})))$$

In our opinion, Larson's proposal is plausible on several grounds. First, in Larson's *N*-analysis there is no need for impromptu introduction of a considerable amount of lexical ambiguity. Second, and for reasons that are beyond the ambiguity of beautiful in (1), there is ample evidence that the structure of a deverbal noun such as `dancer'

---

[1] Note that as an alternative to meaning postulates that specialize intersective adjectives to $\langle e,t\rangle$, one can perform a type-lifting operation from $\langle e,t\rangle$ to $\langle\langle e,t\rangle,\langle e,t\rangle\rangle$ (see Partee, 2007).

must admit a reference to an abstract object, namely a dancing activity; as, for example, in the resolution of 'that' in (3).

Olga is an old dancer. She has been doing that for 30 years.   (3)

Furthermore, and in addition to a plausible explanation of the ambiguity in (1), Larson's proposal seems to provide a plausible explanation for why 'old' in (4a) seems to be ambiguous while the same is not true of 'elderly' in (4b): 'old' could be said of Olga or her teaching; while 'elderly' is not an adjective that is ordinarily said of objects that are of type activity:

a. *Olga is an old teacher*
b. *Olga is an elderly teacher*   (4)

With all its apparent appeal, however, Larson's proposal is still lacking. For one thing, and while it presupposes that some sort of type matching is what ultimately results in rejecting the subsective meaning of 'elderly' in (4b), the details of such processes are more involved than Larson's proposal suggests. For example, while it explains the ambiguity of 'beautiful' in (1), it is not quite clear how an *N*-Analysis can explain why 'beautiful' does not seem to admit a subsective meaning in (5).

Olga is a beautiful young street dancer   (5)

In fact, 'beautiful' in (5) seems to be modifying Olga for the same reason the sentence in (6a) seems to be more natural than that in (6b).

a. *Maria is a clever young girl*
b. *Maria is a young clever girl*   (6)

The sentences in (6) exemplify what is known in the literature as adjective ordering restrictions (AORs). However, despite numerous studies of AORs (e.g., see Wulff, 2003; Teodorescu, 2006), the slightly differing AORs that have been suggested in the literature have never been formally justified. What we hope to demonstrate below however is that the apparent ambiguity of some adjectives and adjective-ordering restrictions are both related to the nature of the ontological categories that these adjectives apply to in ordinary spoken language. While the general assumptions in Larson's (1995; 1998) *N*-Analysis seem to be valid, it will be demonstrated here that nominal modification seems to be more involved than has been suggested thus far. In particular, it seems that a proper semantics for nominal modification requires a much richer type system than currently employed in formal semantics. Before we proceed, therefore, in the next section we will briefly introduce a type system akin to that suggested Sommers (1963); a system that forms the foundation of a semantics grounded in an ontology that in reflects our commonsense view of reality.

2 Ontological Concepts as Types

We assume a Platonic universe that includes everything that can be spoken about in ordinary language, in a manner akin to that suggested by Hobbs (1985). However, in our formalism concepts belong to two distinct categories: (*i*) ontological concepts, such as animal, substance, entity, artifact, event, state, etc., which are assumed to exist in a subsumption hierarchy, and where the fact that an object of type human is ultimately an object of type entity is expressed as Human $\prec$ Entity ; and (*ii*) logical concepts, which are the properties (that can be said) of and relations (that may hold) between ontological concepts. Since adjectives are our immediate concern, consider the following illustrating the difference between ontological and logical concepts:

a. **dedicated**(x :: Human)   (7)
b. **clever**(x :: Animal)
c. **imminent**(x :: Event)
d. **old**(x :: Entity)
f. **beautiful**(x :: Entity)

These predicates are supposed to reflect the fact that, in ordinary spoken language, **dedicated** is a property that is ordinarily said of objects that must be of type Human (7a); that **clever** could be said of objects of type Animal (7b); **imminent** is a property that is said of objects that must be of type Event (7c); etc. In addition to logical and ontological concepts, there are also proper nouns, which are the names of objects; objects that could be of any type. A proper noun such as *sheba* is interpreted as $\|sheba\| \Rightarrow \lambda P[(\exists^1 x)(\mathbf{noo}(x :: \text{Thing}, 'sheba') \wedge P(x :: \text{t}))]$ where $\mathbf{noo}(x :: \text{Thing}, s)$ holds between some x (which could be any thing), and some s if (the label) s is the name of x, and t is presumably the type of objects that P applies to (to simplify notation we often write $\|sheba\| \Rightarrow \lambda P[(\exists^1 sheba :: \text{Thing})(P(sheba :: \text{t}))]$).

Consider now the following, where we have assumed that **thief**(x::Human), i.e., that **thief** is a property that is ordinarily said of objects that must be of type Human, and where $\mathbf{be}(x, y)$ is true when x and y are the same objects:

$$\|sheba\ is\ a\ thief\| \Rightarrow (\exists^1 sheba :: \text{Thing})(\exists x)(\mathbf{thief}(x :: \text{Human}) \wedge \mathbf{be}(sheba, x))] \qquad (8)$$

That is, there is a unique object named *sheba* (which could be any Thing) and some x such that x (which must be of type Human) is a **thief** and such that sheba is that x. Note now that sheba is associated with more than one type in a single scope, and this necessitates a type unification, where a type unification (s $\bullet$ t) between two types s and t, and where $Q \in \{\forall, \exists\}$ is defined (for now) as follows:

$$(Qx :: (s \bullet t))(P(x)) \equiv \begin{cases} (Qx :: s)(P(x)), & \text{if}(s \sqsubseteq t) \\ (Qx :: t)(P(x)), & \text{if}(t \sqsubseteq s) \\ (Qx :: \bot)(P(x)), & \text{otherwise} \end{cases} \qquad (9)$$

where $P(x :: \bot) = -$ and $(t \bullet \bot) = (\bot \bullet t) = -$. Since (Human $\prec$ Thing) , type unification required in (8) now proceeds as follows:

$\|sheba\ is\ a\ thief\|$
$\Rightarrow (\exists^1 sheba :: (\text{Human} \bullet \text{Thing}))(\exists x)(\mathbf{thief}(x) \wedge \mathbf{be}(sheba, x))]$
$\Rightarrow (\exists^1 sheba :: \text{Human})(\exists x)(\mathbf{thief}(x) \wedge \mathbf{be}(sheba, x))]$
Finally, and since $\mathbf{be}(sheba, x)$, we could replace x by *sheba* obtaining the following:

$\|sheba\ is\ a\ thief\|$

$\Rightarrow (\exists^1 sheba :: \mathsf{Human})(\exists x)(\mathbf{thief}(x) \wedge \mathbf{be}(sheba, x))]$

$\Rightarrow (\exists^1 sheba :: \mathsf{Human})(\exists sheba)(\mathbf{thief}(sheba) \wedge \mathbf{be}(sheba, sheba))]$

$\Rightarrow (\exists^1 sheba :: \mathsf{Human})(\mathbf{thief}(sheba) \wedge True)]$

$\Rightarrow (\exists^1 sheba :: \mathsf{Human})(\mathbf{thief}(sheba))]$

In the final analysis, therefore, 'Sheba is a thief' is interpreted as follows: there is a unique object named *sheba*, an object that must be of type Human, and such that *sheba* is a **thief**[2]. Finally, note the clear distinction between ontological concepts (such as Human), which Cocchiarella (2001) calls first-intension concepts, and logical (or second-intension) concepts, such as **thief**(x). That is, what ontologically exist are objects of type Human, not thieves, and **thief** is a mere property that we have come to use to talk of objects of type Human. Moreover, logical concepts such as **thief** are assumed to be defined by virtue of some logical expression, such as $(\forall x :: \mathsf{Human})(\mathbf{thief}(x) \equiv_{df} \phi)$ where the exact nature of ϕ might very well be suscep-

tible to temporal, cultural, and other contextual factors, depending on what, at a certain point in time, a certain community considers an **thief** to be. What is of particular interest to us here is that logical concepts such as **thief** (or **dancer**, **writer**, etc.), are defined by logical expressions that admit abstract objects such as activities, processes, states, etc., each of which could be the object of modification.

3 Types and Nominal Modifications

In this section we use the type system described above and the notion of type unification to properly formalize the intuitions behind Larson's proposal for nominal modification. Subsequently we show that our formalism explains the relationship between intersective/non-intersective adjectives and adjective-ordering restrictions.

3.1 Formalizing Larson's Proposal

First we begin by showing that the apparent ambiguity of 'beautiful' in (1) is due to the fact that beautiful applies to a generic type that subsumes many others. Consider the following, where we assume **beautiful**$(x::\mathsf{Entity})$; that is, **beautiful** is a property that can be said of any Entity:

$\|Olga\ is\ a\ beautiful\ dancer\|$

$\Rightarrow (\exists e :: \mathsf{Activity})(\exists olga :: \mathsf{Human})(\mathbf{dancing}(e) \wedge \mathbf{agent}(e, olga)$

---

[2] The removal of **be**$(sheba,x)$ essentially means that the copular `is' was in this case interpreted as the `is' of identity. This was due to the fact that in this case a subsumption relation exists between the types of the relevant objects. In other contexts, such as `Liz is aging', `Sheba is angry', etc., where it seems that we are dealing with the `is' of predication, removing **be**$(x,y)$ involves introducing some implicit relation between the different types that do not unify (human/process, human/state), essentially resulting in interpretations such as `Liz is-going-through-the-process-of aging', and `Sheba is-in-a-state-of anger'.

$$\wedge\,(\mathbf{beautiful}(e :: \mathsf{Entity}) \vee \mathbf{beautiful}(olga :: \mathsf{Entity})))$$

Note now that, in a single scope, e is considered to be an object of type Activity as well as an object of type Entity, while *Olga* is considered to be a Human and an Entity. This, as discussed above, requires a pair of type unifications:

$\|Olga\ is\ a\ beautiful\ dancer\|$
$\Rightarrow (\exists e :: \mathsf{Activity})(\exists olga :: \mathsf{Human})(\mathbf{dancing}(e) \wedge \mathbf{agent}(e, olga :: \mathsf{Human})$
$\qquad \wedge\,(\mathbf{beautiful}(e :: (\mathsf{Activity} \bullet \mathsf{Entity})) \vee \mathbf{beautiful}(olga :: (\mathsf{Human} \bullet \mathsf{Entity}))))$

Since all the types unifications succeed, the final interpretation is then the following:

$\|Olga\ is\ a\ beautiful\ dancer\|$
$\Rightarrow (\exists e :: \mathsf{Activity})(\exists olga :: \mathsf{Human})(\mathbf{dancing}(e) \wedge \mathbf{agent}(e, olga)$
$\qquad \wedge\,(\mathbf{beautiful}(e) \vee \mathbf{beautiful}(olga)))$

In the final analysis 'Olga is a beautiful dancer' is interpreted as: Olga is the agent of some dancing Activity, and either Olga or her **dancing** is beautiful (or, of course, both). However, consider now the following:

$\|Olga\ is\ an\ elderly\ teacher\|$
$\Rightarrow (\exists e :: \mathsf{Activity})(\exists olga :: \mathsf{Human})(\mathbf{teaching}(e) \wedge \mathbf{agent}(e, olga)$
$\qquad \wedge\,(\mathbf{elderly}(e :: (\mathsf{Activity} \bullet \mathsf{Human})) \vee \mathbf{elderly}(olga :: (\mathsf{Human} \bullet \mathsf{Human}))))$

Note now that e is considered to be an object of type Activity as well as an object of type Human. Since (Activity $\bullet$ Human) = the type unification in this case fails resulting in the following:

$\|Olga\ is\ an\ elderly\ teacher\|$
$\Rightarrow (\exists e :: \mathsf{Activity})(\exists olga :: \mathsf{Human})(\mathbf{teaching}(e) \wedge \mathbf{agent}(e, olga)$
$\qquad \wedge\,(\bot \vee \mathbf{elderly}(olga :: \mathsf{Human})))$
$\Rightarrow (\exists e :: \mathsf{Activity})(\exists olga :: \mathsf{Human})(\mathbf{teaching}(e) \wedge \mathbf{agent}(e, olga) \wedge \mathbf{elderly}(olga))$

3.2 Adjective Ordering Restrictions

Consider again the logical concepts given in (7). Note that beautiful can be said of objects of type entity, and thus it can be said of a cat, a person, a city, a movie, a dance, an island, etc. Therefore, beautiful can be thought of as a polymorphic function that applies to objects at several levels and where the semantics of this function depend on the type of the object, as illustrated in figure 1 below. Thus, and although beautiful applies to objects of type entity, in saying 'a beautiful car', for example, the meaning of beautiful that is accessed is that defined in the type physical (which could in principal be inherited from a supertype). Moreover, and as is well known in the theory of programming languages, one can always perform type casting upwards, but not downwards (e.g., one can always view a car as just an entity, but the converse is not true)[3]. Thus, and assuming **red**(x::Physical) and **beautiful**(x::Entity); that is, assuming that **red** can be said of Physical objects and **beautiful** can be said of any Entity, then, for example, the type casting that will be required in (11a) is valid, while that in (11b) is not.

---

[3] Technically, the reason we can always cast up is that we can always ignore additional information. Casting down, which entails adding information, is however undecidable.

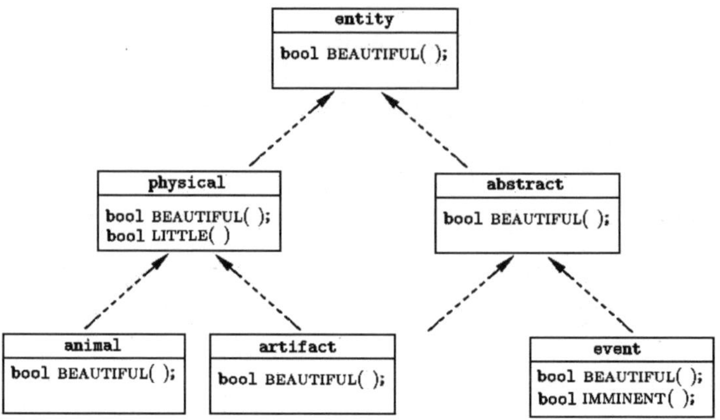

Fig. 1. Adjectives as higher-order functions

beautiful(**red**(x :: Physical) :: Entity)
red(**beautiful**(x :: Entity) :: Physical)
(11)

This, in fact, is precisely why 'Jon owns a beautiful red car' is more natural than `Jon owns a red beautiful car'. In general, a sequence $\mathbf{a}_1(\mathbf{a}_2(x::\mathsf{s})::\mathsf{t})$ is a valid sequence iff $(\mathsf{s} \prec \mathsf{t})$. Note that this is different from type unification, in that the unification does succeed in both cases in (11). However, before we perform type unification the direction of the type casting must be valid. The importance of this interaction will become apparent below.

3.3 How an Ambiguous Adjective Gets One Meaning

Let us explain the example in (5), where we argued that Larson's proposal cannot explain why 'beautiful', which is considered to be ambiguous in (1), does not admit a subsective meaning in (5).

$\lVert Olga\ is\ a\ beautiful\ young\ dancer \rVert$
(12)
$\Rightarrow (\exists e :: \mathsf{Activity})(\exists olga :: \mathsf{Human})(\mathbf{dancing}(e) \wedge \mathbf{agent}(e, olga :: \mathsf{Human})$
$\wedge (\mathbf{beautiful}(\mathbf{young}(e :: \mathsf{Activity}) :: \mathsf{Physical}) :: \mathsf{Entity})$
$\vee (\mathbf{beautiful}(\mathbf{young}(olga :: \mathsf{Human}) :: \mathsf{Physical}) :: \mathsf{Entity})))$

Note now that the casting required is valid in both cases. In other words, the order of adjectives is valid. This means that we can now perform the required type unifications. First note that since (Activity • Physical) =   , the term involving this type unification in (12) is reduced to $\bot$, and the term $(\bot \vee \beta)$ to β, hence:

$\lVert Olga\ is\ a\ beautiful\ young\ dancer \rVert$
(13)
$\Rightarrow (\exists e :: \mathsf{Activity})(\exists olga :: \mathsf{Human})(\mathbf{dancing}(e) \wedge \mathbf{agent}(e, olga :: \mathsf{Human})$
$\wedge \mathbf{beautiful}(\mathbf{young}(olga)))$

Note here that since **beautiful** was preceded by **young**, it could have not been applicable to an abstract object of type Activity, but was instead reduced to that defined at the level of Physical, and subsequently to that defined at the type Human. A valid question that comes to mind here is how then do we express the thought 'Olga is a young dancer and she dances beautifully'. The answer is that we usually make a statement such as this:

$$Olga\ is\ a\ young\ and\ beautiful\ dancer \qquad (14)$$

Note that in this case we are essentially overriding the sequential processing of the adjectives, and thus the adjective-ordering restrictions (or, the type-casting rules!) are no more applicable. That is, (13) is essentially equivalent to two sentences that are processed in parallel: $\|Olga\ is\ a\ young\ dancer\| \wedge \|Olga\ is\ a\ beautiful\ dancer\|$. Note now that 'beautiful' would again have an intersective and a subsective meaning, although 'young' will only apply to *Olga* due to type constraints.

4 Concluding Remarks

In this paper we have shown that nominal modification can be adequately treated in a semantics embedded in a strongly-typed ontology; an ontology that reflects our commonsense view of the world and the way we talk about it in ordinary language. While our concern in this paper was the semantics of [*Adj Noun*] nominals, our proposal seems to also provide an explanation for some well-known adjective-ordering restrictions.

References

1. Cocchiarella, N.B.: Logic and Ontology. Axiomathes 12, 117–150 (2001)
2. Gaskin, R.: Bradley's Regress, the Copula, and the Unity of the Proposition. The Philosophical Quarterly 45(179), 161–180 (1995)
3. Hobbs, J.: Ontological Promiscuity. In: Proc. of the 23rd Annual Meeting of the Association for Computational Linguistics, Chicago, Illinois, pp. 61–69 (1985)
4. Larson, R.: Olga is a Beautiful Dancer. The Winter Meetings of the Linguistic Society of America, New Orleans (1995)
5. Larson, R.: Events and Modification in Nominals. In: Strolovitch, D., Lawson, A. (eds.) Proceedings from Semantics and Linguistic Theory (SALT) VIII, pp. 145–168 (1998)
6. McNally, L., Boleda, G.: Relational Adjectives as Properties of Kinds. In: Bonami, O., Cabredo Hpfherr, P. (eds.) Empirical Issues in Formal Syntax and Semantics (2004)
7. Montague, R.: English as a Formal Language. In: Thomasson, R. (ed.) Formal Philosophy – Selected Papers of Richard Montague. Yale University Press, New Haven (1970)
8. Partee, B.: Compositionality & Coercion in Semantics – the Dynamics of Adjective Meanings. In: Bouma, G., et al. (eds.) Cognitive Foundations of Interpretation, pp. 145–161. Royal Netherlands Academy of Arts and Sciences, Amsterdam (2007)
9. Seigel, E.: Capturing the Adjective, Ph.D. thesis, University of Massachusetts (1976)
10. Sommers, F.: Types and Ontology. Philosophical Review lxxii, 327–363 (1963)
11. Teodorescu, A.: Adjective ordering restrictions revisited. In: Baumer, D., et al. (eds.) Proc. of the 25th West Coast Conference on Formal Linguistics, pp. 399–407 (2006)
12. Wulff, S.: A Multifactorial Analysis of Adjective Order in English. International Journal of Corpus Linguistics, 245-282 (2003)

FACT-Graph: Trend Visualization by Frequency and Co-occurrence*

Ryosuke Saga[1], Masahiro Terachi[2], Zhongqi Sheng[2,3], and Hiroshi Tsuji[2]

[1] Kanagawa Institute of Technology, School of Information Technology,
1030 Shimo-ogino, Atsugi, Kanagawa, 243-0292, Japan
[2] Osaka Prefecture University, Graduate School of Engineering,
1-1 Gakuencho, Nakaku, Sakai, 559-8531, Japan
[3] Northeastern University, School of Mechanical Engineering,
3-11, Wenhua Road, Shenyang, Liaoning, 110004, China
`saga@ic.kanagawa-it.ac.jp`,
`{terachi,sheng}@ mis.cs.osakafu-u.ac.jp`,
`tsuji@cs.osakafu-u.ac.jp`

Abstract. In order to visualize keyword trends embedded in documents, this paper proposes FACT-Graph (Frequency and Co-occurrence-based Trend Graph). First, we introduce four classes of keywords by TF (Term Frequency) and DF (Document Frequency). Then while some keywords stay in the same class between two periods, others stay in the difference classes. Paying attention to such class transition between periods, we make it a clue of trend analysis. Next, we identify relationship between keywords by their co-occurrence strength and their transition between two periods. Then, we propose FACT-Graph by combining class transition information and co-occurrence transition information. Finally, an application to newspaper article is also discussed.

Keywords: Frequency Analysis, Visualization, Text Mining, Trend Analysis Co-occurrence Analysis.

1 Introduction

The information society has been formed by the spread of computers, improvement of their performance, price reduction of hardware, and so on [1]. Then information might be miscellaneous. Because of such situation, there are demands of technical development of information compilation to find the useful information we need accurately by small cost. For example, in the field of text mining, abstraction technologies of the keyword [2] [3], algorithms for picking out trend information [4], and visualization methods [5] [6] are studied.

However, in the conventional static researches, it is not easy to grasp the changes in the trend. Also there is other problem. Some keywords that occur less frequently are paid little attention because those frequency fluctuations are small and they do not stand out even though they are actually important in describing articles.

---

* This research is partially supported by JST "Research for Promoting Technological Seeds 2007". Dataset is provided by NTCIR project.

A. Dengel et al. (Eds.): KI 2008, LNAI 5243, pp. 308–315, 2008.

First, this paper introduces the trend analytical method based on frequency fluctuation of keywords [2]. Then, we expand it to the analytical method which considers relations between keywords by strength of co-occurrence on them. Additionally, this paper proposes the visualization method which reflects the change of frequency and co-occurrence relation of keywords in time series and settles previous problems. We call the proposed visualization method and its output "FACT-Graph (Frequency and Co-occurrence-based Trend Graph)".

Joining in the NTCIR project[1], we drew up several FACT-Graph for newspaper article data. This paper indicates examples of FACT-Graph, and inspects about previous problems solving by FACT-Graph. Then, we examine about its utility and future works.

2 Motivation

Increase of computerized information and the fact that computer can accumulate of large quantity information promote building of large scale databases. And those databases are used for the decision-making of management strategy, and so on. Text mining methods are used to abstract characteristic information from such huge data, and to get knowledge only one document doesn't show by comparing multiple documents [7].

The area of text mining is wide-ranging, for example, journals of the academic meetings, newspaper articles, the complaints which reach an enterprise, obstacle report at factories, the bulletin board on the web, etc [8][9]. In these documents, handled theme and topics are changing every moment with passage of time, so it becomes important to grasp the vicissitude and change in trend [2]. Referring the data which relates to the elements such changes has generated also makes it possible to get new knowledge by investigating the background and causes around them.

TF-IDF algorithm [10] is one of the basic technologies in text mining. TF-IDF algorithm is one of the abstraction methods of keyword for indexing. This method calculates the contribution of each term and regards terms with high contribution as index words for document. The contribution of each term to each document is calculated by the product of the count of the term in document (TF: Term Frequency) and reciprocal (IDF: Inverted Document Frequency) of the count of the documents in which the term appears (DF: Document Frequency).

We can also predict the approximate tendency of trend by grasping a change of TF and DF in time series. However, it's impossible to know about what kind of topic exists in the document, or how it is related to each other between keywords only by comparing keywords and numerical value corresponding to each keyword in time series.

Consequentially, this paper proposes a trend graph called FACT-Graph, which visualizes what kinds of topics exist, and shows the change of trend in document data with date information. Then, we show that FACT-Graph is the effective supporting tool for discovering and analyzing trend information necessary to the user's decision-making and knowledge discovery from several application examples to real data.

---

[1] http://research.nii.ac.jp/ntcir/

3 FACT-Graph

FACT-Graph is the trend graph embedded co-occurrence graph and the information of keyword's class transition. The overview of the steps for outputting FACT-Graph is shown in Fig. 1. Firstly, it is necessary to make a morphological analysis for the text data in order to output FACT-Graph. A morpheme means smallest unit which makes the sense in sentence. Text data is divided into morphemes and the part of speech is also judged by using these tools. This step also abstracts the attribute of each document, such as date of issue, length of document, category of document, and so on. Then, the term database is built.

The user sets up the parameters such as analysis span, the filter of documents/terms, thresholds used in the analysis and so on. Then, the term database is divided into two databases, first half period and second half period, according to the analysis span. Each term's frequency is aggregated in respective databases, and keywords are abstracted from terms under the established conditions. These keywords go through procedures concerning the transition of keyword's classes and co-occurrence, which are introduced particularly after the next chapter. The output chart which reflects the respective processing results is our proposed FACT-Graph. The user can readjust the conditions as the need arises while watching an outputted FACT-Graph and it's possible to redraw FACT-Graph again and do a trial-and-error analysis.

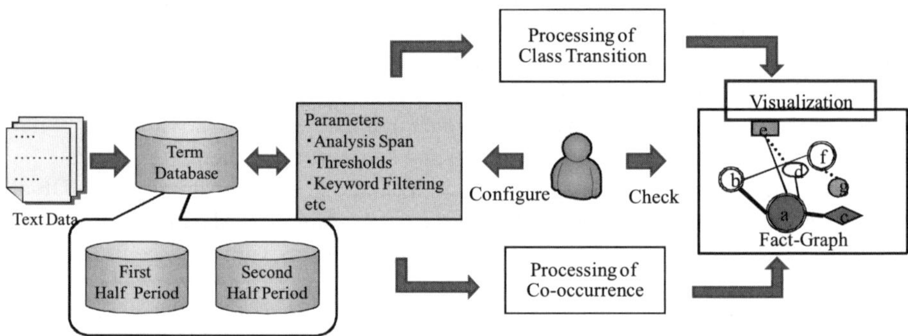

Fig. 1. Overview of Outputting FACT-Graph

3.1 Class Transition Analysis

Our proposed analytical method pays attention to the role of keyword in time series. This conception is based on the idea that keywords with characteristic frequency fluctuations would be important than stable keywords which always appear in documents.

At first, we decide which class indicated is assigned to each keyword in given analytical period by thresholds of TF and DF. This class is the clue to judge what kind of role the keyword plays roughly in the text. We can consider the keyword classified into class A is the keyword about main theme in target documents because its TF value and DF value are both high. Equally we can consider that the keyword classified into class B is the complementary keyword, class C is the intensive keyword, and class D is negligible keyword in analytical period.

Some keywords are classified into different classes in other period. We pay attention to such class transition between periods and make it a clue of trend analysis. We think there are some backgrounds about such keyword. And we give implications to the transition of class in time series at Table 1. For example, If the keyword classified into class C that appear frequently only in certain specific documents in the first analytical period, there is a possibility that the keyword relates to many characteristic keywords. If they appear in many more documents as time goes on, which move from class C to class A, or still occur as keywords belonging to class C as time series change, they probably include information important to the documents.

Moreover, we try to gather up class transition information, because it's too complex to catch changing of trends visually if we express each element in Table 1 individually. Then, we roughly divide the class transition information into 3 patterns, stable (white), on the raise (red), and on the decline (blue), as shown by colored cells in Table 1.

Table 1. Transition of Keyword Classes; class A (TF: High, DF: High), class B (TF: High, DF: Low), class C (TF: Low, DF: High), and class D (TF: Low, DF: Low)

		Second Half Period			
		Class A	**Class B**	**Class C**	**Class D**
First Half Period	**Class A**	Hot	Cooling	Bipolar	Fade
	Class B	Common	Universal	-	Fade
	Class C	Broaden	-	Locally Active	Fade
	Class D	New	Widely New	Locally New	Negligible

However, it's impossible to know what kind of topics exists in the document only by the class transition analysis. There is problem that we can't grasp the relation between the keywords because this analysis is based on word unit. As a result, when the user tries to refer to source documents of remarkable_keyword, it's necessary to search the all. If the attention keyword is appears over multiple topics, search cost becomes big. So it's important to make the analysis smoother by grasping which topic relates to the class transition. There is other problem that all the word with low frequency is not unimportant. In fact, some negligible keywords in Table 1 make the important sense in the specific topic. The purposes of this paper are to settle these problems by using co-occurrence information and to achieve the following 3 points.

- We analyze a macro trend by the topic unit.
- We analyze a topic which consists of multiple words efficiently.
- We find and analyze important keywords in spite of low frequency.

3.2 Co-occurrence Relationship Transition Analysis

In recent years, the keyword abstraction methods and visualization methods use the co-occurrence between the terms from the angle of chance discovery. Simpson coefficient is one of the indices that indicate the co-occurrence degree, which is expressed in the following formulation.

$$\text{Simpson}(a,b) = \text{count}(a \cap b) \, / \, \min(\text{count}(a), \text{count}(b)) \,. \tag{1}$$

Here, $\text{count}(a \cap b)$ is the number of documents in which term a and term b are both appeared, and $\text{count}(x)$ is the number of documents in which term x is appeared.

For outputting of co-occurrence graph, one of the visualization methods, we calculate the respective co-occurrence coefficients between the words of the target documents. Then, the edge is spread between the nodes with the co-occurrence efficient beyond the threshold. This paper also pays attention to transition of co-occurrence relation between the keywords. This transition is classified into the following 3 types.

(a) Co-occurrence relation continues in both analytical periods.
(b) Co-occurrence relation occurs in later analytical period.
(c) Co-occurrence relation goes off in later analytical period.

The relation of type (a) indicates that these words are very close, so we can consider they are essential elements of topics. On the other hand, relations of type (b) or type (c) indicate temporary topical changing.

3.3 Output of FACT-Graph

We propose the visualization method to describe information about keyword's class and co-occurrence relation as graph by expanding co-occurrence graph and call it FACT-Graph. FACT-Graph visualizes keyword's class by shape of node, class transition pattern by color of node, co-occurrence relation transition by type of link, and frequency (TF) by size of node, like Fig. 2.

4 Application to Newspaper Articles

In this case study, we analyze only the topics treated in the articles in The Mainichi Newspapers, Japanese popular paper, and text data is prepared by MuST Workshop [11] for dataset. We focus only on the nouns that compose topic's outline and use morphological analysis tool MeCab[2] equipped with the original dictionary maintained to connect the nouns which continues as a compound. Please note, however, that original nodes of FACT-Graph are Japanese, and in the following figures are translated into English.

We show the example of FACT-Graph which targets at front page in July 1998 (as first half period) and August 1998 (as second half period) in Fig. 3. It can be seen the cluster about political topics in Japan in the center of Fig. 3. The node *Obuchi* is a prime minister at that time. He was appointed to a prime minister at the end of July, so the fact will be expressed by major keyword, class A. On the other hand, former

---

[2] http://research.nii.ac.jp/ntcir/

Prime Minister Hashimoto and other candidates for premier (*Kajiyama* and *Koizumi* were defeated in *presidential election*) will be expressed fading keyword, Pattern 3. Also, there is the cluster about *curry-poisoning case* in the upper right of Fig. 3. This cluster shows that *arsenic compound* became the major keyword while *cyanide compound* became faded keyword. Actually, these characteristics correspond to the fact that occurred by interfusion of *arsenic compound* as a result of the reinvestigation (It had seemed the case caused by *cyanide compound*).

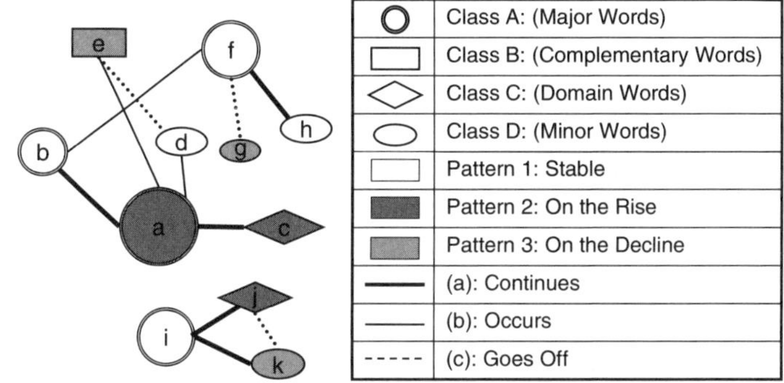

Fig. 2. Example of FACT-Graph

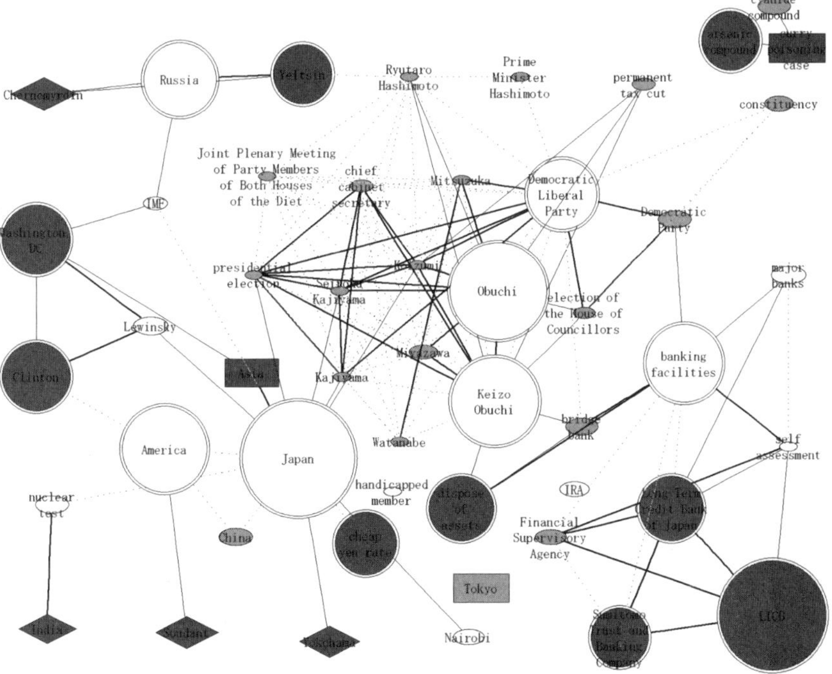

Fig. 3. Example of FACT-Graph for Dataset (Jul. 1998 - Aug. 1998, Front Page Keywords: 30, TF: 45, DF: 30, Simpson: 0.5)

Using the co-occurrence relationship makes it possible to analyze topic efficiently. For example, there is the spotlighted keyword *Yeltsin* in the upper side of Fig. 3. It is impossible to know why *Yeltsin* attracted much attention by traditional Class Transition Analysis. However, in FACT-Graph, analyzer can find that it related to *Chernomyrdin*. It also becomes easy to refer to sources by using multiple retrieval words. In fact, the topics that President *Yeltsin* appointed former Prime Minister *Chernomyrdin* to acting prime minister in August.

Furthermore, FACT-Graph supports analysis for keyword with low frequency. For instance, there is the keyword *Lewinsky* on the left side of Fig. 3. This keyword is regarded as negligible in the traditional Class Transition Analysis, because it is classified into Class D in both analytical periods. However, this keyword is worthy of attention. The reason is that this keyword continues the co-occurrence relationship with major keyword *Clinton* in both analytical periods. It indicates that the topic about *Lewinsky* and *Clinton* is reported in long term. In other words, *Lewinsky* is the necessary keyword in the topic about *Clinton* in this time in spite of its low frequency at the whole articles.

5   Relevant Research

KeyGraph [3] is one of the keyword visualization methods, which also uses co-occurrence relation. KeyGraph makes analysis for low frequency word possible by the original keyword abstraction algorithm using the degree of co-occurrence. However, FACT-Graph expresses co-occurrence relation and frequency together. Moreover, FACT-Graph describes these transitions in time series by overlooking way. (But, it's necessary to advance a research about original keyword abstraction algorithm which uses class transition information.)

ThemeRiver [5] and TrendGraph [6] are also keyword visualization methods. ThemeRiver expresses keyword's sequential trend by metaphor like river. But, ThemeRiver doesn't consider the relation between the keywords. We're designing mounting of the interface that enables FACT-Graph to change analytical periods flexibly. On the other hand, TrendGraph is the method focusing on the co-occurrence between 2 periods along with FACT-Graph. However, FACT-Graph can show the trend by multiple views such as links and keyword's shapes although TrendGraph expresses keyword's trend by links only.

6   Conclusion

This paper has proposed the visualization method named FACT-Graph to express frequency information and co-occurrence information of keywords as overview graph. Then, applying the proposed FACT-Graph to the dataset of newspaper articles, this paper exemplifies some example result and validates effectiveness in visualizing keyword trends embedded in volumes of text.

References

1. Inmon, W.H.: Building the Data Warehouse. Wiley Publishing Inc, Chichester (2005)
2. Terachi, M., Saga, R., Tsuji, H.: Trends Recognition in Journal Papers by Text Mining. In: IEEE International Conference on Systems, Man & Cybernetics (IEEE/SMC 2006), pp. 4784–4789 (2006)
3. Ohsawa, Y., Benson, N.E., Yachida, M.: KeyGraph: Automatic Indexing by Segmenting and Unifing Co-occurrence Graphs. IEICE D- I J82-D-I(2), 391–400 (1999)
4. Nanba, H., Okuda, N., Okumura, M.: Extraction and Visualization of Trend Information from Newspaper Articles and Blogs. In: Proceedings of the 6th NTCIR Workshop, pp. 243–248 (2007)
5. Havre, S., Hetzler, B., Nowell, L.: ThemeRiver(TM): In search of trends, patterns, relationships. IEEE Trans Visualization Computer Graphics 8, 9–20 (2002)
6. Feldman, R., Aumann, Y., Zilberstein, A., Ben-Yehuda, Y.: Trend graphs: Visualizing the evolution of concept relationships in large document collections. In: Żytkow, J.M. (ed.) PKDD 1998. LNCS, vol. 1510, pp. 38–46. Springer, Heidelberg (1998)
7. Salton, G.: Automatic Text Processing. Addison-Wesley Publishing Company, Reading (1989)
8. Kageyama, A., Tsuji, H.: Web-based Characteristics Analysis for Industrial Departments in Universities. World Multiconference on Systemics, Cybernetics and Informatics 10, 23–28 (2004)
9. Yamanishi, K., Li, H.: Mining Open Answers in Questionnaire Data. IEEE Intelligent Systems, 58-64 (2002)
10. Salton, G., Yang, C.S.: On the Specification of Term Values in Automatic Indexing. Journal of Documentation 29(4), 351–372 (1973)
11. Kato, T., Matsushita, M., Kando, N.: Information Compilation from Multi-modal Summarization for Trend Information. In: JSAI 2007 (The 21st Annual Conference of the Japanese Society for Artificial Intelligence), pp. 2H5-11 (2007)

Enhancing Animated Agents in an Instrumented Poker Game

Marc Schröder[1], Patrick Gebhard[1], Marcela Charfuelan[1], Christoph Endres[1], Michael Kipp[1], Sathish Pammi[1], Martin Rumpler[2], and Oytun Türk[1]

[1] DFKI, Saarbrücken and Berlin, Germany
`firstname.lastname@dfki.de`
[2] FH Trier, Umwelt-Campus Birkenfeld, Germany
`m.rumpler@umwelt-campus.de`

Abstract. In this paper we present an interactive poker game in which one human user plays against two animated agents using RFID-tagged poker cards. The game is used as a showcase to illustrate how current AI technologies can be used for providing new features to computer games. A powerful and easy-to-use multimodal dialog authoring tool is used for modeling game content and interaction. The poker characters rely on a sophisticated model of affect and a state-of-the art speech synthesizer. Through the combination of these methods, the characters show a consistent expressive behavior that enhances the naturalness of interaction in the game.

1 Motivation

An application area with growing relevance for AI technologies are computer games. Specifically, driving the behavior of non-player characters raises a whole range of challenges, such as interaction design, emotion modeling, figure animation, and speech synthesis [1].

Some research projects address these problems in combination, such as the interactive drama game Façade [2] or the Mission Rehearsal Exercise Project [3]. However, relevant research is also being carried out in a range of individual disciplines, including believable facial, gesture and body animation of virtual characters [4], modeling personality and emotion [5], expressive speech synthesis [6] and control mechanisms for story and interaction [7].

In the project IDEAS4Games, we investigate how modern AI technologies can help to improve the process of creating computer games with interactive expressive virtual characters. Based on the experience of a computer game company (RadonLabs, Berlin), we identified four main challenges in creating computer games: (i) fast creation of game demonstrators; (ii) localization of game content; (iii) easy interaction; and (iv) consistent quality. We address these challenges by combining, in an integrated system, flexible multimodal dialog and story modeling, sophisticated simulation of affect in real time, expressive synthesis, and rendering of 3D virtual characters.

We explain the concept of the integrated game demonstrator before providing some details about the individual component technologies.

A. Dengel et al. (Eds.): KI 2008, LNAI 5243, pp. 316–323, 2008.

2   Poker Demonstrator

In order to demonstrate that it is possible to meet the requirements stated above, we created a poker computer game and presented it at the CeBIT 2008 fair to a large number of people to collect initial feedback.

By using real poker cards with unique RFID tags, a user can play draw poker [8] against the two 3d virtual characters Sam and Max (see Fig. 1). Sam is a cartoon-like, friendly looking character, whereas Max is a mean, terminator-like robot character. Both are rendered by the open-source 3d visualization engine Horde3D [9]. The human user acts as the card dealer and also participates as a regular player.

As shown in Fig. 1, we use a poker table which shows three areas for poker cards: one for the user and one for each virtual character. These areas of the table are instrumented with RFID sensor hardware, so that the game logic can detect which card is actually placed at each specific position. A screen at the back of the poker table displays an interface which allows users to select their actions during the game, using a computer mouse. These include general actions, such as playing or quitting a game, and poker game actions: bet a certain amount of money, call, raise, or fold. This screen also shows the content relevant for the poker game: the face of the user's cards, the number of Sam's and Max's cards respectively, all bets, and the actual money pot. The two virtual characters are shown above this interaction screen on a second 42" monitor.

Fig. 1. Poker Demonstrator at the CeBIT 2008 fair

When a user approaches the poker table and initiates a game, Sam and Max explain the game setup and the general rules. In a next step the user has to deal the cards. During the game the animated agents react to events, notably when the user deals or exchanges their cards. Time is also considered – for example, Sam and Max start complaining if the user deals the cards too slowly, or they express their surprise about erratic bets.

Different poker algorithms are used to match Sam's and Max's individual character style. Sam, the human-like poker player, uses a rule based algorithm, whereas Max, the robot, relies on a brute-force algorithm that estimates a value for each of the 2.58 million possible combinations of five poker cards.

Based on game events, the affect of each character is computed in real-time and expressed through the character's speech and body. The richly modeled characters, as well as the easy interaction with the game using real poker cards, have stirred up a lot of interest in the CeBIT 2008 audience (see Fig. 1).

3   Enhancing Agents

In the following, we explain the key properties of the technologies used in the demonstrator.

3.1   Interaction and Story Authoring

The behavior of the two virtual players has been modeled with the authoring tool SceneMaker, which separates dialog content and narrative structure [10].

Dialog content is organized in *scenes* – pieces of contiguous dialog. Scenes are defined in a multimodal script that specifies the text to be spoken as well as the agents' verbal and nonverbal behavior. The utterances can be annotated with dialog act tags that influence the computation of affect (see below). In addition, system commands (e.g., for changing the camera position) can be specified. Scenes are created by an author with standard text processing software. The major challenge when using scripted dialog is variation. The characters must not repeat themselves because this would severely impact their believability. For this purpose we use blacklisting: once a scene is played, it is blocked for a certain period of time (e.g., five minutes), and variations of this scene are selected instead. For each scene, several variations can be provided that make up a *scene group*. In our poker scenario there are 335 scenes organized in 73 scene groups. The number of scenes in a scene group varies between two and eight scenes, depending on how much variation is needed.

The narrative structure – the order in which the individual scenes are played – is defined by the *sceneflow*. Technically, the sceneflow is modeled as a hypergraph that consists of nodes and edges (transitions). *Supernodes* contain subgraphs. Each node can be linked to one or more scenes and scene groups. Different branching strategies (e.g. logical and temporal conditions as well as randomization) can be used by specifying different edge types [10]. The sceneflow is modeled using the sceneflow editor, our graphical authoring tool that supports authors with drag'n'drop facilities to *draw* the sceneflow by creating nodes and edges.

At runtime the sceneflow graph is traversed by selecting nodes and edges based on the current game state and the actions of the three players. The scenes that are selected and executed during such a traversal control the multimodal behavior of the virtual agents. Transitions in the sceneflow are triggered either by the players' actions or as a result of context queries. In both cases sceneflow variables used for conditional branching are updated by an event handler.

Each time the user places or removes a card, an event is generated and sent to the poker event handler. The same happens when the user chooses an action (bet, call, raise, or fold) by pressing the respective button on the graphical user interface (see Fig. 1). The event handler receives these low-level events and updates the data model representing the game state as well as the graphical user interface. It also analyzes the situation and generates higher level events, e.g., that all cards have been removed from the table or that the user has changed the cards of a player in the drawing phase. At the end of this process, it updates the respective sceneflow variables, which may enable transitions in the sceneflow and trigger the next scenes.

Apart from scenes and scene groups the author can also attach commands to a node. These commands are executed by the sceneflow interpreter each time the node is visited. There are commands that modify the game state (e.g. selecting the next player after a scene has been played in which one of the players announces that he drops out) and commands that access the poker logic to suggest the next action (e.g. deciding which cards should be changed in the drawing phase and which action the virtual players should perform in the betting phase).

4   Affect Model

For the affect computation in real-time, we rely on ALMA, a computational model of affect [11]. It provides three affect types as they occur in human beings: (1) *emotions* reflect short-term affect that decays after a short period of time; (2) *moods* reflect medium-term affect, which is generally not related to a concrete event, action or object; and (3) *personality* reflects individual differences in mental characteristics and affective dispositions.

ALMA implements the cognitive model of emotions developed by Ortony, Clore, and Collins (OCC) [12] combined with the *BigFive* model of personality [13] and a simulation of mood based on the PAD model [14]. The three levels are interrelated: personality defines a default mood, and influences the intensities of different emotions; emotions as short term events influence the longer-term mood; and the mood, in turn, amplifies or dampens the intensities of emotional reactions to events.

ALMA enables the computation of 24 OCC emotions with *appraisal tags* [15] as input. Elicited emotions influence an individual's mood. The higher the intensity of an emotion, the higher the particular mood change. A unique feature is that the current mood also influences the intensity of emotions. This simulates, for example, the intensity increase of *joy* and the intensity decrease of *distress*, when a individual is in an *exuberant* mood. Mood is represented by a triple of the mood traits pleasure (P), arousal (A), and dominance (D). The mood's trait values define the mood class. If, for example, every trait value is positive (+P,+A,+D), the mood is *exuberant*.

The current mood and emotions, elicited during the game play, influence the virtual poker characters' affective behavior. *Breathing* is related to the mood's arousal and pleasure values. For positive values, a character shows fast distinct breathing. The breathing is slow and faint for negative values. *Speech quality* is related to the current mood or an elicited emotion. In *relaxed* mood, a character's speech quality is *neutral*. In *hostile* or *disdainful* mood or during negative emotions, it is *aggressive*, in *exuberant* mood or during positive emotions it is *cheerful*. In any other mood, the speech quality is *depressed*.

Initially, Sam's and Max's behavior is nearly identical, because they have the same default *relaxed* mood that is defined by their individual personality. However, Sam is slightly more *extroverted*, so his mood tends to become more *exuberant*. Max, on the other side, has a tendency to be *hostile*. This is caused by his negative *agreeableness* personality definition. This disposition drives Max towards negative mood (e.g. *hostile* or *disdainful*) and negative emotions faster than Sam.

In the poker game, events and user actions influence the selection of scenes and their execution. As a consequence the actions of Sam and Max as well as their appraisal of the situation changes. For example, if Sam loses the game because he has bad cards, a related scene will be selected in which he verbally complains about that. An *appraisal tag* [**BadEvent**] used in the scene lets Sam appraise the situation as a bad event during the scene execution. In the affect model, this will elicit the emotion *Distress*, and if such events occur often, they will lead to a *disdainful* or *hostile* mood. In general, appraisal tags can be seen as a comfortable method for dialog authors to define on an abstract level input for a affect computation module. For doing this, no conceptual knowledge about emotions or mood is needed.

The poker game scenario covers the simulation of all 24 OCC emotions. For example, prospect-based emotions such as *hope, satisfaction,* or *disappointment* can be triggered during the card change phases – Sam and Max sometimes utter their expectation of getting good new cards. Those utterances also contain the appraisal tag [**GoodLikelyFutureEvent**]. As a consequence, *hope* is elicited. After all cards have been dealt out, Sam's and Max's poker engine compares the current cards with the previous ones. Depending on the result either the [**EventConfirmed**] or the [**EventDisconfirmed**] appraisal tag is passed to the affect computation. In the first case (cards are better), *satisfaction* is elicited, otherwise *disappointment*. Appraisals leading to all other emotions are included in the scene logic in a similar way.

5   Expressive Synthetic Voices

Convincing speech generation is a necessary precondition for an animated agent to be believable. That is true especially for a system featuring emotional expressivity.

To address this issue, we have investigated the two currently most influential state-of-the-art speech synthesis technologies: unit selection synthesis and statistical-parametric synthesis. A suitable speech synthesizer should produce speech of *reliable quality* which at the same time shows a *natural expressivity*.

Unfortunately, in state-of-the-art speech synthesis technology, the two criteria are difficult to reconcile. Unit selection [16] can reach close-to-human naturalness in limited domains, but it suffers from unpredictable quality in unrestricted domains. Statistical-parametric synthesis [17] has a very stable synthesis quality, but typically sounds "muffled" because of the excessive smoothing involved in training the statistical models.

In IDEAS4Games, we investigated two methods for approximating the criteria formulated above. For our humanoid character, Sam, we created a custom unit selection voice with a "cool" speaking style, featuring high quality in the poker domain and some emotional expressivity; for our robot-like character, Max, we used a statistical-parametric voice, and applied audio effects to modify the sound to some extent in order to express emotions.

Sam's voice was carefully designed as a domain-oriented unit selection voice [18]. Domain-oriented voices sound highly natural within a given domain; they can also speak arbitrary text, but the quality outside the domain will be seriously reduced. We designed a recording script for the synthesis voice, consisting of a generic and a domain-specific part. We used a small set of 400 sentences selected from the German Wikipedia to cover the most important diphones in German [19]. In addition, about 200 sentences from the poker domain were recorded, i.e. sentences related to poker cards, dealing, betting, etc. The 600 sentences of the recording script were produced by a professional actor in a recording studio. The speaker was instructed to utter both kinds of sentences in the same "cool" tone of voice. In a similar way, the same speaker produced domain-oriented voice databases for a *cheerful*, an *aggressive* and a *depressed* voice. We built a separate unit selection voice for each of the four databases, using the open-source voice import toolkit of the MARY text-to-speech synthesis platform [20].

In our application, most of Sam's utterances are spoken with the neutral poker voice. As the voice database contains many suitable units from the domain-oriented part of the recordings, the poker sentences generally sound highly natural. Their expressivity corresponds to the "cool" speaking style realized by our actor. Selected utterances are realized using the *cheerful, aggressive*, and *depressed* voices. For example, when Sam loses a round of poker, he may utter a frustrated remark in the *depressed* voice; if the affect model predicts a positive mood, he may greet a new user in a *cheerful* voice, etc.

The voice of the robotic character, Max, uses statistical-parametric synthesis [21] which we incorporated into our MARY TTS platform. A voice is created by training the statistical models on a speech database. After training, the original data is not needed anymore; at runtime, speech is generated from the statistical models by means of a vocoder. HMM-based synthetic speech typically sounds muffled due to the averaging in the statistical models as well as the vocoding, but the quality is largely stable, independently of the text spoken.

Expressivity for Max' voice is performed using audio effects to modify the generated speech. At the level of the parametric input to the vocoder, we can modify the pitch level, pitch range and speaking rate. In addition, we apply audio signal processing algorithms that modify the generated speech signal, using linear predictive coding (LPC) techniques. In this framework, our code provides a vocal tract scaler, which can simulate a longer or shorter vocal tract; a whisper component, adding whisper to the voice; a robot effect, etc. Expressive speech is generated by combining various effects – for example, *aggressive* speaking style is simulated by a lengthened vocal tract, lowered pitch, increased pitch range, faster speech rate, and slightly whisperised speech.

6   Conclusion

We have presented an agent-based computer game employing RFID-tagged poker cards as a novel interaction paradigm. It is built around the integration of three components: a powerful and easy-to-use multimodal dialog authoring tool, a sophisticated model of affect, and a state-of-the-art speech synthesizer.

In our authoring approach, the separation of dialog content and narrative structure makes it possible to modify these two aspects independently and without expert knowledge.

In speech synthesis, we have illustrated the trade-off between reliable quality and naturalness, and have offered two partial solutions. We have created a very high quality unit selection voice for our character Sam, but the quality comes at the price of high effort and limited flexibility. The HMM-based voice used for our character Max is flexible so that it can speak arbitrary text reliably, but the naturalness is limited.

Our computational model of affect is integrated with the dialog script through the use of appraisal tags which can be authored without any deep knowledge about a model of emotion or mood. The continuous computation of short-term emotions and medium-term moods allows for a smooth blending of different aspects of affective behavior that help to increase the believability and the expressiveness of virtual characters.

Overall, we have shown how state-of-the-art research approaches can be combined in an interactive poker game to show new possibilities for the modeling of expressive virtual characters and for future computer game development.

Acknowledgements. The work reported here was supported by the EU ProFIT project IDEAS4Games (EFRE program) and by the DFG project PAVOQUE.

References

1. Gratch, J., Rickel, J., André, E., Cassell, J., Petajan, E., Badler, N.I.: Creating interactive virtual humans: Some assembly required. IEEE Intelligent Systems 17, 54–63 (2002)
2. Mateas, M., Stern, A.: Façade: An experiment in building a fully-realized interactive drama. In: Game Developers Conference, Game Design Track (2003)

3. Swartout, W., Gratch, J., Hill, R., Hovy, E., Marsella, S., Rickel, J., Traum, D.: Toward virtual humans. AI Magazine 27, 96–108 (2006)
4. Martin, J.-C., Niewiadomski, R., Devillers, L., Buisine, S., Pelachaud, C.: Multimodal complex emotions: Gesture expressivity and blended facial expressions. International Journal of Humanoid Robotics, Special Edition, Achieving Human-Like Qualities in Interactive Virtual and Physical Humanoids (2006)
5. de Rosis, F., Pelachaud, C., Poggi, I., Carofiglio, V., de Carolis, B.: From Greta's mind to her face: Modelling the dynamics of affective states in a conversational embodied agent. Int. Journal of Human Computer Studies 59, 81–118 (2003)
6. Schröder, M.: Emotional speech synthesis: A review. In: Proceedings of Eurospeech 2001, Aalborg, Denmark, vol. 1, pp. 561–564 (2001)
7. Prendinger, H., Saeyor, S., Ishizuk, M.: MPML and SCREAM: Scripting the bodies and minds of life-like characters. In: Life-like Characters – Tools, Affective Functions, and Applications, pp. 213–242. Springer, Heidelberg (2004)
8. Wikipedia: Draw Poker (2008), http://en.wikipedia.org/wiki/Draw_poker
9. Schulz, M.: Horde3D – Next-Generation Graphics Engine. Horde 3D Team (2006–2008), http://www.nextgen-engine.net/home.html
10. Gebhard, P., Kipp, M., Klesen, M., Rist, T.: Authoring scenes for adaptive, interactive performances. In: Proc. of the 2nd Int. Joint Conference on Autonomous Agents and Multi-Agent Systems, pp. 725–732. ACM, New York (2003)
11. Gebhard, P.: ALMA - a layered model of affect. In: Proc. of the 4th Int. Joint Conference on Autonomous Agents and Multiagent Systems, pp. 29–36. ACM, New York (2005)
12. Ortony, A., Clore, G.L., Collins, A.: The Cognitive Structure of Emotions. Cambridge University Press, Cambridge (1988)
13. McCrae, R., John, O.: An introduction to the five-factor model and its applications. Journal of Personality 60, 175–215 (1992)
14. Mehrabian, A.: Pleasure-arousal-dominance: A general framework for describing and measuring individual differences in temperament. Current Psychology: Developmental, Learning, Personality, Social 14, 261–292 (1996)
15. Gebhard, P., Kipp, K.H.: Are computer-generated emotions and moods plausible to humans? In: Gratch, J., Young, M., Aylett, R.S., Ballin, D., Olivier, P. (eds.) IVA 2006. LNCS (LNAI), vol. 4133, pp. 343–356. Springer, Heidelberg (2006)
16. Hunt, A., Black, A.W.: Unit selection in a concatenative speech synthesis system using a large speech database. In: Proceedings of ICASSP 1996, Atlanta, Georgia, vol. 1, pp. 373–376 (1996)
17. Yoshimura, T., Tokuda, K., Masuko, T., Kobayashi, T., Kitamura, T.: Simultaneous modeling of spectrum, pitch and duration in HMM-based speech synthesis. In: Proceedings of Eurospeech 1999, Budapest, Hungary (1999)
18. Schweitzer, A., Braunschweiler, N., Klankert, T., Möbius, B., Säuberlich, B.: Restricted unlimited domain synthesis. In: Proc. Eurospeech 2003, Geneva (2003)
19. Hunecke, A.: Optimal design of a speech database for unit selection synthesis. Diploma thesis, Universität des Saarlandes, Saarbrücken, Germany (2007)
20. Schröder, M., Hunecke, A.: Creating German unit selection voices for the MARY TTS platform from the BITS corpora. In: Proc. SSW6, Bonn, Germany (2007)
21. Zen, H., Nose, T., Yamagishi, J., Sako, S., Masuko, T., Black, A., Tokuda, K.: The HMM-based speech synthesis system version 2.0. In: Proc. of ISCA SSW6, Bonn, Germany (2007)

Shallow Models for Non-iterative Modal Logics

Lutz Schröder[1,*] and Dirk Pattinson[2,**]

[1] DFKI-Lab Bremen and Department of Computer Science, Universität Bremen
[2] Department of Computing, Imperial College London

Abstract. Modal logics see a wide variety of applications in artificial intelligence, e.g. in reasoning about knowledge, belief, uncertainty, agency, defaults, and relevance. From the perspective of applications, the attractivity of modal logics stems from a combination of expressive power and comparatively low computational complexity. Compared to the classical treatment of modal logics with relational semantics, the use of modal logics in AI has two characteristic traits: Firstly, a large and growing variety of logics is used, adapted to the concrete situation at hand, and secondly, these logics are often non-normal. Here, we present a shallow model construction that witnesses PSPACE bounds for a broad class of mostly non-normal modal logics. Our approach is uniform and generic: we present general criteria that uniformly apply to and are easily checked in large numbers of examples. Thus, we not only re-prove known complexity bounds for a wide variety of structurally different logics and obtain previously unknown PSPACE-bounds, e.g. for Elgesem's logic of agency, but also lay the foundations upon which the complexity of newly emerging logics can be determined.

Special purpose modal logics abound in applied logic, and in particular in artificial intelligence, where new logics emerge at a steady rate. They often combine expressiveness and decidability, and indeed many modal logics are decidable in PSPACE, i.e. not dramatically worse than propositional logic. While lower PSPACE bounds can typically be obtained directly from seminal results of Ladner [12] by embedding a PSPACE-hard logic such as K or KD, upper bounds are often non-trivial to establish. In this respect, non-normal logics have received much attention in recent research, which has lead e.g. to PSPACE upper bounds for graded modal logic [20] (correcting a previously published incorrect algorithm and refuting a previous EXPTIME hardness conjecture), Presburger modal logic [4], coalition logic [17], and various conditional logics [14].

The methods used to obtain these results can be broadly grouped into two classes. Syntactic approaches presuppose a complete tableaux or sequent system and establish that proof search can be performed efficiently. Semantics-driven approaches, on the other hand, directly construct shallow tree models. Both approaches are intimately connected in the case of normal modal logics with relational semantics: counter models can usually be derived directly from search trees [10]. However, the situation is quite different in the non-normal case, where the structure of models often goes far beyond mere graphs. We have previously shown [19] that the *syntactic* approach uniformly

* This work forms part of the DFG project *Generic Algorithms and Complexity Bounds in Coalgebraic Modal Logic* (SCHR 1118/5-1).
** Partially supported by EPSRC grant EP/F031173/1.

A. Dengel et al. (Eds.): KI 2008, LNAI 5243, pp. 324–331, 2008.

generalises to a large class of modal logics. Here, we present a generic *semantic* set of methods to establish uniform PSPACE bounds using a direct shallow model construction which in particular does not need to rely on an axiomatisation of the logic at hand. Apart from the fact that both methods use substantially different techniques, they apply to different classes of examples. Examples not easily amenable to the syntactic approach, because either no axiomatisation has been given or known axiomatisations are hard to harness, include probabilistic modal logic [6] and Presburger modal logic [4].

We emphasise that our methods go far beyond establishing the complexity of a particular logic: they employ the semantic framework of *coalgebraic modal logic* [16] to obtain results that are parametric in the underlying semantics of particular logics. In this paradigm, the role of models is played by coalgebras, which associate a structured collection of successor states to every state of the model: a coalgebra with state set C is a function $C \to TC$ where the parametrised datatype (technically: functor) T represents the structuring of successors, e.g. relational, probabilistic [6], game-oriented [17], or non-monotonically conditional [1]. Our approach is now best described as investigating *coherence conditions* between the syntax and the semantics, parametric in both, that guarantee the announced complexity bounds. While these methods have so far been limited to logics of *rank 1*, given by axioms whose modal nesting depth is uniformly equal to one, the present results apply to *non-iterative logics* [13], i.e. logics axiomatised without nested modalities (rank-1 logics additionally exclude top-level propositional variables). This increase in generality, achieved by working with *copointed functors* in the semantics, substantially extends the scope of the coalgebraic method, in particular where relevant to AI. E.g. all conditional logics covered in [14], Elgesem's logic of agency [5], and the graded version Tn of T [7] are non-iterative logics.

Our main technical tool is to cut model constructions for modal logics down to the level of *one-step logics* which semantically do not involve state transitions, and then amalgamate the corresponding *one-step models* into shallow models for the full modal logic. This requires the logic at hand to support a small model property for its one-step fragment, the *one-step polysize model property (OSPMP)*, which is much easier to establish than a shallow model property for the logic itself (e.g. to reprove Ladner's PSPACE upper bound for K, one just observes that to construct a set that intersects n given sets, one needs at most n elements). As a by-product of our construction, we obtain NP-bounds for bounded rank fragments, generalizing corresponding results for the logics K and T from [8]. We illustrate our method by concise new proofs of known PSPACE upper bounds for various conditional logics, probabilistic modal logic, and Presburger modal logic. As a new result, we prove e.g. that Elgesem's logic of agency is in PSPACE. Despite the emphasis we place on examples, we stress that the main intention of this work is to provide a *standard method* that goes beyond mere informal recipes, being based on formal theorems with easily verified and well-structured application conditions. A full version of this work is available as e-print arXiv:0802.0116.

1   Preliminaries: Coalgebraic Modal Logic

We give a self-contained introduction to the syntax and coalgebraic semantics of modal logics. A *(modal) similarity type* Λ is a set of modal operators with associated finite arity. The similarity type Λ determines two languages: firstly, the *modal logic* of Λ, whose set $\mathcal{F}(\Lambda)$ of Λ-*formulas* $\psi, \ldots$ is defined by the grammar

$$\psi ::= \bot \mid \psi_1 \wedge \psi_2 \mid \neg\psi \mid L(\psi_1, \ldots, \psi_n) \qquad (L \in \Lambda \ n\text{-ary}).$$

Propositional atoms are treated as nullary modalities and therefore do not explicitly appear in the syntax. Secondly, the signature Λ determines the *one-step logic* of Λ, whose formlas may be represented as *one-step pairs* (ϕ, ψ) where ϕ is a propositional formula over a set V of variables and ψ is a propositional combination of atoms $L(a_1, \ldots, a_n)$, with $a_1, \ldots, a_n \in V$ and $L \in \Lambda$ n-ary. Equivalently, one may use *one-step formulas*, i.e. propositional combinations of formulas $L(\phi_1, \ldots, \phi_n)$, where the ϕ_i are propositional formulas over V [18]; i.e. the modal logic of Λ is distinguished from the one-step logic in that it admits nested modalities. The *rank* $\mathrm{rank}(\phi)$ of $\phi \in \mathcal{F}(\Lambda)$ is the maximal nesting depth of modalities in ϕ. The *bounded-rank fragments* of $\mathcal{F}(\Lambda)$ are the sets $\mathcal{F}_n(\Lambda) = \{\phi \in \mathcal{F}(\Lambda) \mid \mathrm{rank}(\phi) \leq n\}$.

The semantics of both the one-step logic and the modal logic of Λ are parametrized coalgebraically by the choice of a *set functor*, i.e. an operation $T : \mathsf{Set} \to \mathsf{Set}$ taking sets to sets (w.l.o.g. preserving the subset relation) and maps $f : X \to Y$ to maps $Tf : TX \to TY$, preserving identities and composition. The standard setup of coalgebraic modal logic using all coalgebras for a set functor covers only *rank-1 logics*, i.e. logics axiomatised by one-step formulas [18] (a typical example is the K-axiom $\Box(a \to b) \to \Box a \to \Box b$). Here, we improve on this by considering the class of coalgebras for a given *copointed* set functor (in a slightly restricted sense), which enables us to cover the more general class of *non-iterative logics*, axiomatised by arbitrary formulas without nested modalities (such as the T-axiom $\Box a \to a$).

Definition 1. A *copointed functor* S with *signature functor* $S_0 : \mathsf{Set} \to \mathsf{Set}$ is a subfunctor of $S_0 \times Id$ (where $(S_0 \times Id)X = S_0 X \times X$). We say that S is *trivially copointed* if $S = S_0 \times Id$. An S-*coalgebra* $A = (X, \xi)$ consists of a set X of *states* and a *transition function* $\xi : X \to S_0 X$ such that $(\xi(x), x) \in SX$ for all x.

We view coalgebras as generalised transition systems: the transition function maps a state to a structured collection of successors, with the structure prescribed by the signature functor, which thus encapsulates the branching type of the transition systems employed. Copointed functors additionally impose local frame conditions that relate a state to the collection of its successors. We refer to elements of sets SX as *successor structures*. Generalising earlier work, *coalgebraic modal logic* [16] abstractly captures the interpretation of modal operators using predicate liftings:

Definition 2. An n-ary *predicate lifting* $(n \in \mathbb{N})$ for S_0 is a family $(\lambda_X : \mathcal{P}(X)^n \to \mathcal{P}(S_0 X))_{X \in \mathsf{Set}}$ of maps satisfying *naturality*, i.e. $\lambda_X(f^{-1}[A_1], \ldots, f^{-1}[A_n]) = (S_0 f)^{-1}[\lambda_Y(A_1, \ldots, A_n)]$ for all $f : X \to Y$, $A_1, \ldots, A_n \in \mathcal{P}(Y)$.

A coalgebraic semantics for Λ, i.e. a *coalgebraic modal logic* $\mathcal{L}$, consists of a copointed functor S with signature functor S_0 and an assignment of an n-ary predicate lifting $[\![L]\!]$ for S_0 to every n-ary modal operator $L \in \Lambda$. When S is trivially copointed, we will mention only S_0. *We fix $\mathcal{L}$, Λ, S, S_0 throughout.* The semantics of the modal language $\mathcal{F}(\Lambda)$ is defined inductively as a satisfaction relation $\models_C$ between states x of S-coalgebras $C = (X, \xi)$ and Λ-formulas. The clause for an n-ary modal operator L is

$$x \models_C L(\phi_1, \ldots, \phi_n) \iff \xi(x) \in [\![L]\!]_X([\![\phi_1]\!], \ldots, [\![\phi_n]\!])$$

where $[\![\phi]\!] = \{x \in X \mid x \models_C \phi\}$. Our interest is in the *satisfiability problem of* $\mathcal{L}$, which asks whether for a given formula ϕ, there exist an S-coalgebra C and a state x in C such that $x \models_C \phi$.

In contrast, the semantics of the one-step logic is defined over single successor structures, in particular does not involve a notion of state transition:

Definition 3. A *one-step model* (X, τ, t, x) over V consists of a set X, a $\mathcal{P}(X)$-valuation τ for V, $t \in S_0 X$, and $x \in X$ such that $(t, x) \in SX$. We omit the mention of x if S is trivially copointed. For a one-step pair (ϕ, ψ), τ induces interpretations $[\![\phi]\!]\tau \subseteq X$ and, using the given predicate liftings, $[\![\psi]\!]\tau \subseteq S_0 X$. We say that (X, τ, t, x) is a *one-step model of* (ϕ, ψ) if $[\![\phi]\!]\tau = X$ and $t \in [\![\psi]\!]\tau$.

Example 4.   1. *The modal logics K and T* [1] have a single unary modal operator $\Box$. The standard Kripke semantics of K is modelled coalgebraically over the powerset functor $\mathcal{P}$, whose coalgabras are just Kripke frames, by $[\![\Box]\!]_X(A) = \{B \in \mathcal{P}X \mid B \subseteq A\}$. The logic T (i.e. K extended with the non-iterative axiom $\Box a \to a$) is modelled by moving to the copointed functor R given by $RX = \{(A, x) \in \mathcal{P}X \times X \mid x \in A\}$. R-coalgebras are reflexive Kripke frames.

2. *Conditional logics* have a single binary infix modal operator $\Rightarrow$, read as a non-monotonic conditional (default, relevance, ...). The standard semantics of the conditional logic CK is modelled coalgebraically over the functor Cf given by $Cf(X) = \mathcal{P}(X) \to \mathcal{P}(X)$, with $\to$ denoting function space, by

$$[\![\Rightarrow]\!]_X(A, B) = \{f : \mathcal{P}(X) \to \mathcal{P}(X) \mid f(A) \subseteq B\}.$$

*Cf-coalgebras are *conditional frames* [1]. The conditional logic $CK + ID$ extends CK with the rank-1 axiom $a \Rightarrow a$. Its semantics is modelled by passing to the subfunctor Cf_{ID} of Cf defined by $Cf_{ID}(X) = \{f \in Cf(X) \mid \forall A \in \mathcal{P}(X). f(A) \subseteq A\}$. Similarly, the logic $CK + MP$ extends CK with the non-iterative axiom $(a \Rightarrow b) \to (a \to b)$. (This axiom is undesirable in default logics, but often considered in relevance logics.) Semantically, this amounts to passing to the copointed functor Cf_{MP} defined by

$$Cf_{MP}(X) = \{(f, x) \in Cf(X) \times X \mid \forall A \in \mathcal{P}(X). x \in A \Rightarrow x \in f(A)\}.$$

3. *Modal logics of quantitative uncertainty:* The similarity type of *likelihood* has polyadic modal operators expressing linear inequalities between likelihoods $l(\phi)$, variously interpreted as probabilities [6], upper probabilities, Dempster-Shafer degrees of belief, or Dubois-Prade degrees of possibility [9]. Extensions generalise likelihood to *expectations* [9] for linear combinations of formulas. These logics are captured coalgebraically by suitable distribution functors. E.g. the case of probabilities is modelled by the probability distribution functor D_ω, where $D_\omega X$ is the set of finitely supported probability distributions on X, with D_ω-coalgebras corresponding to Markov chains; and the case of upper probabilities is modelled by the sets-of-distributions functor $\mathcal{P} \circ D_\omega$. In the literature, one-step logics of quantitative uncertainty are often introduced independently and only later extended to full modal logics [6].

4. *Graded and Presburger modal logic:* Graded modal logic [7] has operators $\Diamond_k$ read 'in more than k successor states, it holds that ...'. Variants of these operators have found their way into modern description logics as *qualified number restrictions*. More generally, Presburger modal logic [4] has n-ary modal operators $\sum_{i=1}^{n} a_i\#(\_) \sim b$, where b and the a_i are integers and $\sim \in \{<,>,=\} \cup \{\equiv_k|\ k \in \mathbb{N}\}$, with $\equiv_k$ read as equality modulo k. The original Kripke semantics is equivalent to a coalgebraic semantics over the finite multiset functor $\mathcal{B}$, which maps a set X to the set of maps $B : X \to \mathbb{N}$ with finite support, understood as multisets containing $x \in X$ with multiplicity $B(x)$. $\mathcal{B}$-coalgebras are graphs with $\mathbb{N}$-weighted edges, over which the given modal operators are interpreted as suggested by the notation, adding up multiplicities [3].

5. *Agency:* Logics of agency, concerned with agents bringing about states of affairs, play a role in planning and task assignment in multi-agent systems [2,11]. A standard approach due to Elgesem [5], intended as an abstraction of pure agency to be used as a building block in more complex logics, has modalities E and C, read 'the agent brings about' and 'the agent is capable of realising', respectively. Their semantics is defined over a certain restricted class of conditional frames $(X, f : X \to (\mathcal{P}(X) \to \mathcal{P}(X)))$ (see above) by $x \models E\phi$ iff $x \in f(x)(\llbracket\phi\rrbracket)$ and $x \models C\phi$ iff $f(x)(\llbracket\phi\rrbracket) \neq \emptyset$. Most of the information in such models is disregarded: one only needs to know whether $f(x)(A)$ is non-empty and contains x. We may thus equivalently use the following coalgebraic semantics: put $3 = \{\bot, *, \top\}$, ordered $\bot < * < \top$ (to code the cases $f(x)(A) = \emptyset$, $x \notin f(x)(A) \neq \emptyset$, and $x \in f(x)(A)$), and take as signature functor the *3-valued neighbourhood functor* N_3 given by $N_3(X) = (\mathcal{P}(X) \to 3)$. The restrictions on conditional frames imposed by Elgesem translate into using the copointed functor $\mathcal{A}$ over N_3 where $(f, x) \in \mathcal{A}(X)$ iff for all $A, B \subseteq X$, $f(\emptyset) = \bot$, $f(X) = \bot$ ('the agent cannot bring about logical truths'), $f(A) \wedge f(B) \leq f(A\cap B)$, and $f(A) = \top \Rightarrow x \in A$ ('what the agent brings about is actually the case'). The operators E, C are interpreted by $\llbracket E \rrbracket_X A = \{f \in N_3(X) \mid f(A) = \top\}$ and $\llbracket C \rrbracket_X A = \{f \in N_3(X) \mid f(A) \neq \bot\}$. Notably, the logic is non-monotone, i.e. Ea does not imply $E(a \vee b)$, which makes typical approaches to proving PSPACE bounds hard to apply.

2   A Generic Shallow Model Construction

We now turn to the announced construction of polynomially branching shallow models for modal logics whose one-step logic has a small model property; this construction leads to a PSPACE decision procedure. For finite X, we assume given a representation of elements $(t, x) \in SX$ as strings of size $\mathrm{size}(t, x)$ over some finite alphabet, where crucially do not require that all elements of SX are representable.

Definition 5. The logic $\mathcal{L}$ has the *one-step polysize model property (OSPMP)* if there exist polynomials p and q such that, whenever a one-step pair (ϕ, ψ) over V has a one-step model (X, τ, t, x), then it has a one-step model (Y, κ, s, y) such that $|Y| \leq p(|\psi|)$, (s, y) is representable with $\mathrm{size}(s, y) \leq q(|\psi|)$, and $y \in \kappa(a)$ iff $x \in \tau(a)$ for all $a \in V$.

It is crucial that the polynomial bound depends only on ψ, as the proof of Thm. 7 below uses one-step pairs (ϕ, ψ) with exponential-size ϕ. In terms of one-step formulas, this amounts to discounting the potentially exponential-sized inner propositional layer.

Definition 6. A *supporting Kripke frame* of an S-coalgebra (X, ξ) is a Kripke frame (X, R) such that for each $x \in X$, $\xi(x) \in S_0\{y \mid xRy\} \subseteq S_0X$.

Theorem 7 (Shallow model property). *The OSPMP implies the* polynomially branching shallow model property: *There exist polynomials p, q such that every satisfiable Λ-formula ψ is satisfiable in an S-coalgebra (X, ξ) which has a supporting Kripke frame (X, R) such that removing all loops xRx from (X, R) yields a tree of depth at most* $\mathrm{rank}(\psi)$ *and branching degree at most $p(|\psi|)$, and $(\xi(x), x) \in S\{y \mid xRy\}$ is representable with* $\mathrm{size}(\xi(x), x) \leq q(|\psi|)$.

This theorem leads to a nondeterministic decision procedure for satisfiability that recursively traverses shallow models in a depth-first fashion, guessing at each level a polynomial-size successor structure. Checking whether such a structure satisfies the local requirements imposed by the input formula is encapsulated as follows:

Definition 8. The *one-step model checking problem* of $\mathcal{L}$ is to check, given a string s, a finite set X, $A_1, \ldots, A_n \subseteq X$, and $L \in \Lambda$ n-ary, whether s represents some $(t, x) \in SX$ and whether $t \in [\![L]\!]_X(A_1, \ldots, A_n)$.

Theorem 9. *Let $\mathcal{L}$ have the OSPMP.*

1. If the one-step model checking problem of $\mathcal{L}$ is in PSPACE, then the satisfiability problem of $\mathcal{L}$ is in PSPACE.

2. If the one-step model checking problem of $\mathcal{L}$ is in P, then the restriction of the satisfiability problem of $\mathcal{L}$ to the bounded-rank fragment $\mathcal{F}_n(\Lambda)$ is in NP for every $n \in \mathbb{N}$.

Theorem 9.2, which generalises known results for K and T [8], follows from the fact that Thm. 7 implies a polynomial-size model property for bounded-rank fragments.

In cases where the OSPMP fails, one can often use a relaxed criterion, the *one-step pointwise polysize model property (OSPPMP)*, provided that S_0 is *pointwise κ-bounded*, i.e. $|S_0X| \leq \kappa^{|X|}$, for some cardinal κ; we then assume $S_0X \subseteq \kappa^X$. E.g. the functors $\mathcal{P}$ and D_ω, but not Cf and $\mathcal{P} \circ D_\omega$, are pointwise bounded. Assuming a (partial) representation of elements of κ, we put $\mathrm{maxsize}(t) = \max_{x \in X} \mathrm{size}\, t(x)$ for $t \in S_0X \subseteq \kappa^X$. The OSPPMP essentially requires the existence of one-step models (X, τ, t, x) such that $\mathrm{maxsize}(t)$ is polynomially bounded. Then we have

Theorem 10. *If $\mathcal{L}$ has the OSPPMP and one-step model checking of (X, τ, t, x) is decidable on a non-deterministic Turing machine with input tape that uses space polynomial in* $\mathrm{maxsize}(t)$ *and accesses each input symbol at most once, then the satisfiability problem of $\mathcal{F}(\Lambda)$ is in PSPACE.*

The crucial point in the proof is that if input symbols are read at most once, they can be guessed without having to be stored.

Example 11. In all example applications, tractability of one-step model checking is straightforward, so that we concentrate on small one-step models.

1. *Modal logics K and T:* To verify the OSPMP for K, let (X, τ, A) be a one-step model of a one-step pair (ϕ, ψ) over V; w.l.o.g. ψ is a conjunctive clause over atoms

$\square a$, where $a \in V$. For $\neg \square a$ in ψ, there exists $x_a \in A$ such that $x_a \notin \tau(a)$. Taking Y to be the set of these x_a, we obtain a polynomial-size one-step model (Y, τ_Y, Y) of (ϕ, ψ), where $\tau_Y(a) = \tau(a) \cap Y$ for all a. The construction for T is the same, except that the point x of the original one-step model (X, τ, A, x) is retained in Y. By Thm. 9, this reproves Ladner's PSPACE upper bounds for K and T [12], as well as Halpern's NP upper bounds for bounded-rank fragments [8].

2. *Conditional logic:* To avoid exponential blowup, we represent elements of $Cf(X) = \mathcal{P}(X) \to \mathcal{P}(X)$ as *partial* maps, extended to total maps using default value $\emptyset$. The OSPMP for CK is proved as follows: given a one-step model (X, τ, f) of a one-step pair (ϕ, ψ), where ψ is w.l.o.g. a conjunctive clause, retain values of f only at the sets $\tau(a_i)$ and cut down to a polynomial-size set Y containing elements y_{ij} of the symmetric difference of $\tau(a_i)$ and $\tau(a_j)$ whenever $\tau(a_i) \neq \tau(a_j)$ and elements $z_i \in f(\tau(a_i)) \setminus \tau(b_i)$ whenever ψ contains $\neg(a_i \Rightarrow b_i)$. The proofs for $CK + ID$ and $CK + MP$ are the same, up to changing the default value for $f(A)$ in the representation of $(f, x) \in Cf_{MP}(X)$ to $A \cap \{x\}$. Thus, we reprove that CK, $CK + ID$, and $CK + MP$ are in PSPACE [14] (hence PSPACE-complete, as they contain known PSPACE-hard sublogics) and obtain a new NP upper bound for their bounded-rank fragments.

3. *Modal logics of quantitative uncertainty:* Polynomial size model properties for one-step logics and complexity estimates for one-step model checking have been proved for various logics of quantitative uncertainty [6,9]. The polynomial bounds are stated in the cited work as depending on the size of an entire one-step formula ψ; however, inspection of the given proofs shows that the bounds are in fact independent of the inner propositional layer, and hence actually establish the OSPMP, with ensuing (tight) upper complexity bounds as in Thm 9. A proof of the PSPACE upper bound for the modal logic of probability is sketched in [6]; the NP upper bound for bounded-rank fragments is new. In the remaining cases, also the PSPACE upper bounds are new, if only because just the one-step versions of these logics appear in the literature so far.

4. *Elgesem's logic of agency:* For X finite, we let a *partial* map $f_0 : \mathcal{P}(X) \rightharpoonup 3$ represent the element $f \in N_3(X) = \mathcal{P}(X) \to 3$ that maps $B \subseteq X$ to the maximum of $\bigwedge_{i=1}^{n} f_0(A_i)$, taken over all sets $A_1, \ldots, A_n \subseteq X$ such that $\bigcap A_i = B$ and $f_0(A_i)$ is defined for all i. In the proof of the OSPMP, a one-step model $(X, \tau, f : \mathcal{P}(X) \to 3, x)$ of a one-step pair over V is reduced to polynomial size by restricting f to the $\tau(a)$, $a \in V$, and cutting down to a set Y containing: the point x; an element $y_{ab} \in \tau(a) \setminus \tau(b)$ whenever $\tau(a) \not\subseteq \tau(b)$; an element $z_a \in \bigcap \{\tau(b) \mid b \in V, \tau(a) \subseteq \tau(b), f(\tau(b)) > f(\tau(a))\} \setminus \tau(a)$ for each $a \in V$; and an element $w_0 \in \bigcap \{\tau(b) \mid f(\tau(b)) > \bot\}$, where w_0 and the z_a exist by the definition of $\mathcal{A}$. Thus, *the modal logic of agency is in PSPACE*, and its bounded-rank fragments are in NP. Both results (and even decidability) seem to be new. We conjecture that the PSPACE upper bound is tight.

5. *Presburger Modal Logic:* The functor $\mathcal{B}$ is pointwise ω-bounded. It follows easily from estimates on solution sizes of integer linear equalities [15] that Presburger modal logic has the OSPPMP, and hence is in PSPACE [4]. One easily incorporates non-iterative frame conditions such as reflexivity (modelled by the copointed functor $SX = \{(B, x) \in \mathcal{B}X \times X \mid B(x) > 0\}$) or e.g. the condition that at least half of all transitions from a given state are loops (modelled by the copointed functor $SX = \{(B, x) \in \mathcal{B}X \times X \mid B(x) \geq B(X - \{x\}\})$. In particular, this implies that

graded modal logic over reflexive frames (i.e. the logic Tn of [7]) *is in PSPACE,* to our knowledge a new result. This extends straightforwardly to show that the concept satisfiability problem in description logics with role hierarchies, reflexive roles, and qualified number restrictions is in PSPACE over the empty TBox.

Acknowledgements. The authors wish to thank Joseph Halpern for useful discussions. Erwin R. Catesbeiana has contributed his opinion on consistent formulas.

References

1. Chellas, B.: Modal Logic. Cambridge Univ. Press, Cambridge (1980)
2. Cholvy, L., Garion, C., Saurel, C.: Ability in a multi-agent context: A model in the situation calculus. In: Toni, F., Torroni, P. (eds.) CLIMA 2005. LNCS (LNAI), vol. 3900, pp. 23–36. Springer, Heidelberg (2006)
3. D'Agostino, G., Visser, A.: Finality regained: A coalgebraic study of Scott-sets and multisets. Arch. Math. Logic 41, 267–298 (2002)
4. Demri, S., Lugiez, D.: Presburger modal logic is only PSPACE -complete. In: Furbach, U., Shankar, N. (eds.) IJCAR 2006. LNCS (LNAI), vol. 4130, pp. 541–556. Springer, Heidelberg (2006)
5. Elgesem, D.: The modal logic of agency. Nordic J. Philos. Logic 2, 1–46 (1997)
6. Fagin, R., Halpern, J.Y.: Reasoning about knowledge and probability. J. ACM 41, 340–367 (1994)
7. Fine, K.: In so many possible worlds. Notre Dame J. Formal Logic 13, 516–520 (1972)
8. Halpern, J.: The effect of bounding the number of primitive propositions and the depth of nesting on the complexity of modal logic. Artificial Intelligence 75, 361–372 (1995)
9. Halpern, J., Pucella, R.: Reasoning about expectation. In: Uncertainty in Artificial Intelligence, UAI 2002, pp. 207–215. Morgan Kaufmann, San Francisco (2002)
10. Halpern, J.Y., Moses, Y.O.: A guide to completeness and complexity for modal logics of knowledge and belief. Artificial Intelligence 54, 319–379 (1992)
11. Jones, A., Parent, X.: Conventional signalling acts and conversation. In: Dignum, F.P.M. (ed.) ACL 2003. LNCS (LNAI), vol. 2922, pp. 1–17. Springer, Heidelberg (2004)
12. Ladner, R.: The computational complexity of provability in systems of modal propositional logic. SIAM J. Comput. 6, 467–480 (1977)
13. Lewis, D.: Intensional logics without iterative axioms. J. Philos. Logic 3, 457–466 (1975)
14. Olivetti, N., Pozzato, G.L., Schwind, C.: A sequent calculus and a theorem prover for standard conditional logics. ACM Trans. Comput. Logic 8 (2007)
15. Papadimitriou, C.: On the complexity of integer programming. J. ACM 28, 765–768 (1981)
16. Pattinson, D.: Coalgebraic modal logic: Soundness, completeness and decidability of local consequence. Theoret. Comput. Sci. 309, 177–193 (2003)
17. Pauly, M.: A modal logic for coalitional power in games. J. Log. Comput. 12, 149–166 (2002)
18. Schröder, L.: A finite model construction for coalgebraic modal logic. J. Log. Algebr. Prog. 73, 97–110 (2007)
19. Schröder, L., Pattinson, D.: PSPACE reasoning for rank-1 modal logics. In: Logic in Computer Science, LICS 2006, pp. 231–240. IEEE, Los Alamitos (2006)
20. Tobies, S.: PSPACE reasoning for graded modal logics. J. Log. Comput. 11, 85–106 (2001)

Homography Based State Estimation
for Aerial Robots

Jakob Schwendner

DFKI Robotics Lab
Robert-Hooke-Str. 5, D-28359, Bremen, Germany

Abstract. Low-cost Unmanned Aerial Vehicles have large potential for applications in the civil sector. Cheap inertial sensors alone can not provide the degree of accuracy required for control and navigation of the UAVs. Additional sensors, and sensor fusion techniques are needed to reduce the state estimation error. In the context of this work, the possibility of using visual information to augment the inertial data is investigated. The projected image of the ground plane is processed by a homography constrained optical flow description and used as a measurement update for an EKF. A simulation environment is used, which models the UAV dynamics, the rate-gyro and accelerometer errors as well as a low resolution vision system.

1   Introduction

Over the recent years Unmanned Aerial Vehicles (UAV) have received an increased academic, commercial and military interest. A 2004 report by NASA [4] projects one third of US Airforce aircraft to be unmanned by 2010. UAVs current main application is in the military sector, and most of the current development effort is geared towards it. Decreasing costs in Sensor and Embedded System technology however make UAV technology an interesting prospect for civil use.

In order to be applicable to civil applications, the technology at use must be safe, reliable and also cost effective. An increased level of autonomy can help with this. One of the main challenges of operating an autonomous aerial robot is the sensing and situation awareness part. It combines multi-sensory input into an internal representation of the current system state and puts it into relation to known world facts and mission objectives.

Low-cost inertial sensors can provide the basis for a state estimation system. Due to the nature of the integration process however they are prone to extensive drift. Often GPS systems provide absolute references to compensate the drift. A different approach which is also followed by this work is to use visual means to aid the state estimation process. A method is described which constraints the optical flow using a homography transformation of the perceived ground plane. Further, an Extended Kalman Filter is developed where the state transition is based on a generic rigid body model using the inertial sensors as system inputs, and the homography matrix is used to perform a measurement update.

The validity of the approach is tested using a simulation environment. A scenario is presented where the homography based update is compared to inertial state estimation alone.

A. Dengel et al. (Eds.): KI 2008, LNAI 5243, pp. 332–339, 2008.
© Springer-Verlag Berlin Heidelberg 2008

2   Related Work

A number of ways have been investigated to augment inertial navigation with visual sensors to minimise localisation drift. An important concept in this context is the *motion field*, which is the projection of 3-D points and their velocity vectors onto a 2-D plane [9]. Egomotion, the motion of the observer in space, can be partly recovered due to the nature of the motion field [2]. If the speed of the observer is known, the motion field can also be used to calculate distance information, which is very helpful for obstacle avoidance [10]. *Optical flow* methods can under certain conditions be used to recover the motion field information. The optical flow is specified as the vector field that maps the brightness of points in one image to the brightness values in the next image in the sequence. The field is under-specified and suffers from what is known as the aperture problem. Various algorithms can be used to calculate an approximation of the motion field based on an image sequence. Because of the under-specified nature of the problem, these methods are recursive in nature and computationally expensive. A good overview of the different algorithms and their performances is given by Galvin et al. [5].

A number of different approaches have been used to apply optical flow measurements to UAV navigation and control. Barrows and Neely [1] have proposed a hardware centred approach in which an array of photoreceptors is used to measure the optical flow, thus making it possible to be used in micro aerial vehicles (MAV) with a weight requirement of less than a gram per sensor. A similar approach is used by Stocker [14], who has successfully demonstrated a solution for a 2D optical flow sensor in silicon.

A simulated environment is used by Muratet et al. [11] to test the application of optical flow in navigating urban canyons. The simulation assumes the presence of a sensor array for inertial, GPS and altimeter data. This data is combined with optical flow information in a traditional control scheme using PID controllers in order to avoid obstacles.

The previously shown examples for using optical flow information for controlling UAVs were mostly based for collision detection. However, when assuming a camera motion relative to a plane, it is possible to express the optical flow as a combination of orientation, linear velocity, rotational velocity and distance to the plane [12]. The optical flow field can thus be expressed as a 3x3 homography matrix, relating the projective coordinates from one frame to the next.

A homography based approach for landing is used by Bosse et al. [2], who propose an algorithm to fuse the information with existing state information with the help of a Kalman filter. The algorithm uses image sequences for which it calculates the time derivative of the projection matrix, which is equal to the optical flow. The global smoothness constraint of [8] is replaced by the quadratic function of image location, relating the optical flow at a certain image point to the derivative of the projection matrix. The system state values are extracted using eigenvector decomposition. The system's current estimate of altitude is used to solve the scale ambiguities.

3   Method

The system model is defined by its state transition function $f(x)$ and the measurement function $h(x)$. The environment is modeled as an inertial right-handed reference frame $\mathcal{I} = \{E_1, E_2, E_3\}$, with the positive base $e_2 \in E_2$ pointing upwards. A ground plane is assumed to be at position $0 \in \mathcal{I}$ with a normal of $(0, 1, 0) \in \mathcal{I}$. The gravity vector $-ge_2$ is assumed constant with $g = 9.81 m/s^2$, friction and air resistance are not part of the model.

Rigid Body Dynamics. The position of the centre of mass of the flyer is given by $\xi \in \mathcal{I}$. A second frame of reference, the body frame, is given by $\mathcal{B} = \{E_1^a, E_2^a, E_3^a\}$ its origin is the flyers centre of mass and fixed to the flyers orientation. The orientation of the body frame $\mathcal{B}$ is identified by the unity quaternion q. The systems dynamics are described using Newton's equations of motion. [7]

$$\dot{\xi} = v \tag{1}$$

$$m\dot{v} = F - mge_2 \tag{2}$$

$$\dot{q} = \frac{1}{2} \begin{bmatrix} 0 \\ \omega \end{bmatrix} q \tag{3}$$

$$I\dot{\omega} = -\omega \times I\omega + \tau \tag{4}$$

where $v \in \mathcal{I}$ is the linear velocity and $F \in \mathcal{I}$ the external forces acting on the flyer not including gravity. Furthermore $\omega \in \mathcal{B}$ describes the angular velocity and $\tau \in \mathcal{B}$ the torque in the body reference frame. I is the inertia tensor in the body fixed frame $\mathcal{B}$ and m the overall mass of the flyer. We consider F and τ to be unknown noise inputs for now.

Image Sensor. A perspective projection with known image plane distance is assumed for the optical system of the sensor. The image sensors output is then $\tilde{G} = G_{\xi,q} + n_G$, with $G_{\xi,q}, \tilde{G} \in \mathbb{R}_{0:1}^{w \times h}$ being a arrays of image brightness values in the range $\mathbb{R}_{0:1} = \{x \in \mathbb{R} | 0 \leq x \leq 1\}$, noise term n_G and (w, h) the image resolution. With the assumption of a constant ground plane texture, the noise free image $G_{\xi,q}$ is uniquely identified by the sensor position ξ and the orientation q.

To be able to relate consecutive frames, the image brightness constraint states that for any given point $x = \begin{bmatrix} x_1 & x_2 \end{bmatrix}^T$ and an image brightness $I(x, t)$ at time t,

$$I(x, t) = I(x + \delta x, t + \delta t). \tag{5}$$

With the ground plane as the only scene object assumed, Bosse et al. [2] give a method to relate image sequences to the camera state. A perspective image coordinate m is defined as,

$$x = \psi(m) = \frac{f}{m_2} \begin{pmatrix} m_1 \\ m_3 \end{pmatrix}. \tag{6}$$

where the focal length f can be assumed unity. We can also define the inverse of ψ by setting the second component of a three vector to unity.

$$m = \psi^{-1}(x) = \begin{pmatrix} x_1 \\ 1 \\ x_2 \end{pmatrix} \tag{7}$$

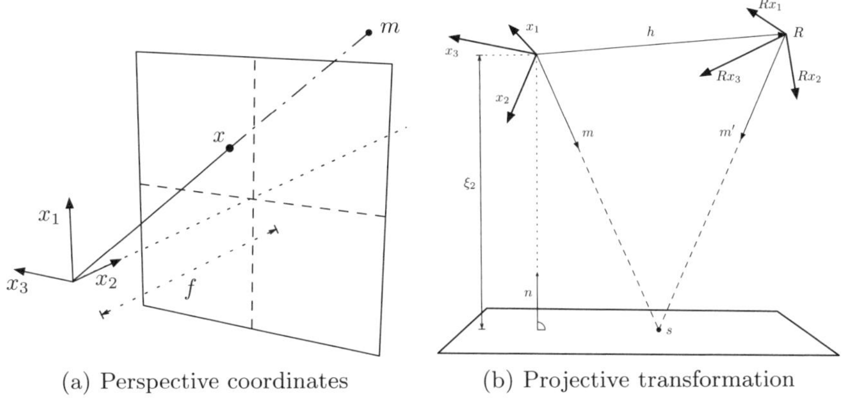

(a) Perspective coordinates   (b) Projective transformation

Fig. 1. Point on plane related to different camera states

If we now assume a camera with position ξ and an orientation q which is exerted to a translation of h and a rotation of R the projective transform of a perspective coordinate can be given as

$$\mathcal{A} = \frac{1}{k}(I + ph^T)R, \quad k = \sqrt[3]{1 + p^T h} \tag{8}$$

where p is the ground plane surface normal expressed in the reference frame given by R_q, divided by the distance of the camera to the plane ξ_2.

$$p = \frac{R_q^T e_2}{\xi_2} \tag{9}$$

Because perspective coordinates can be scaled arbitrarily, the k factor makes sure that $\det(\mathcal{A}) = 1$ is normalised.

The transform $\mathcal{A}$ allows now to relate the image point x to the projected point x' by using (6) and (7). The concept is illustrated in Figure 1.

$$x' = \psi(\mathcal{A}^T \psi^{-1}(x)) \tag{10}$$

The problem is to find the transformation matrix $\mathcal{A}$ which fits best for the brightness constraint criterion (5). Given a set of image points X we can formulate such a constraint easily using (10) and (5).

$$\min_{\mathcal{A} \in \mathbb{R}^{3 \times 3}} \sum_{x \in X} \left(I(\psi(\mathcal{A}^T \psi^{-1}(x)), t + \delta t) - I(x, t) \right)^2 \tag{11}$$

Equation (11) has the geometric interpretation of projecting the image at $t + \delta t$ using the transformation and aligning it with the image at t. Because the problem is an optimisation of a least square formulation, the requirements for applying the Levenberg-Marquardt method are fulfilled.

The image sensor provides an array of brightness values. However, the optimisation criterion (11) requires a spatially continuous function $I(x)$. This is achieved by using a bicubic interpolation method [13].

Sensor Fusion. The state estimation approach is partly based on [6] and [15]. An Extended Kalman Filter with a directly formulated state is used, where the state vector x consists of

$$x = (\xi, v, q, \omega)^T \tag{12}$$

The state vector carries the angular velocity ω, which is necessary to form the measurement estimate of the perspective transform for the vision system (8).

$$\dot{\xi} = v \tag{13}$$

$$\dot{v} = R_q a + g e_2 \tag{14}$$

$$\dot{q} = -\frac{1}{2}\Omega(\omega)q \tag{15}$$

The derivative of q which for the system description (3) is given as a quaternion multiplication is replaced with the matrix equivalent $\Omega(\omega)$ defined for the body reference frame [3] as

$$\Omega(\omega) = \begin{pmatrix} 0 & \omega^T \\ -\omega & -[\omega]_\times \end{pmatrix} = \begin{pmatrix} 0 & \omega_1 & \omega_2 & \omega_3 \\ -\omega_1 & 0 & \omega_3 & -\omega_2 \\ -\omega_2 & -\omega_3 & 0 & \omega_1 \\ -\omega_3 & \omega_2 & -\omega_1 & 0 \end{pmatrix} \tag{16}$$

This system description is discretised using a first-order Euler update.

$$p_{k+1} = p_k + v\Delta t \tag{17}$$

$$v_{k+1} = v_k + (R_q a + g e_2)\Delta t \tag{18}$$

$$q_{k+1} = \left(I(\cos(s) + \eta\Delta t\gamma) - \frac{\sin(s)}{2s}\Omega(\omega)\Delta t \right) q_k, \quad \gamma = 1 - |q_k|^2 \tag{19}$$

Where γ a Lagrange multiplier term [6], which ensures that q always converges to unity, with $\eta\Delta t < 1$ determining the speed of convergence.

The descriptions for the discrete dynamics (18) both still contain the measurement noise vectors n_a and n_ω, which can easily be extracted.

$$v_{k+1} = v_k + (R_q\tilde{a} + g e_2)\Delta t + p_a \tag{20}$$

$$q_{k+1} = \left(I(\cos(s) + \eta\Delta t\gamma) - \frac{\sin(s)}{2s}\Omega(\tilde{\omega})\Delta t \right) q_k + p_\omega \tag{21}$$

with

$$p_a = R_q n_a \Delta t, \quad p_\omega = \frac{-\sin(s)}{2s}\Omega(n_\omega \Delta t)q_k \tag{22}$$

The noise vector n_a and n_ω are given in the continuous time domain. However, for the discrete filter a discretised form of the noise vectors is required. We specify $\bar{n}_a$ and $\bar{n}_\omega$ to be the discrete versions of n_a and n_ω. The covariances of $\bar{n}_a$ and $\bar{n}_\omega$ can actually be measured and are given by

$$\bar{n}_a \sim N(0, P_a), \quad \bar{n}_\omega \sim N(0, P_\omega) \tag{23}$$

For the filtering process however, we need the covariance of the direct noise terms p_a and p_ω.

$$Q_v = \text{cov}(p_a) = R_q P_a R_q^T \Delta t^2 \tag{24}$$

The case is not as straightforward for p_ω. However, it is possible to find a solution for the equation $p_k = \Gamma(q)\omega$ by defining $\Gamma(q)$ as

$$\Gamma(q) = \begin{pmatrix} q_2 & q_3 & q_4 \\ -q_1 & q_4 & -q_3 \\ -q_4 & -q_1 & q_2 \\ q_3 & -q_2 & -q_1 \end{pmatrix} \tag{25}$$

so that we can give the covariance of p_ω as

$$Q_q = \text{cov}(p_\omega) = \Gamma(q_{sk})P_\omega \Gamma(q_{sk})^T, \quad q_{sk} = \frac{-\sin(s)\Delta t}{2s} q_k \tag{26}$$

The covariance of the angular velocity is of course equal to that measurement covariance of the gyro sensors, $Q_\omega = P_\omega$. Further, the cross covariance of Q_q and Q_ω is given by

$$Q_{q,\omega} = \Gamma(q_{sk})P_\omega \tag{27}$$

so that we can give the complete process noise covariance matrix Q as

$$Q_k = \begin{pmatrix} 0 & 0 & 0 & 0 \\ 0 & Q_v & 0 & 0 \\ 0 & 0 & Q_q & Q_{q,\omega} \\ 0 & 0 & Q_{q,\omega}^T & Q_\omega \end{pmatrix} \tag{28}$$

Recalling that the measurement for the filter is the projection matrix given in (8), we need to formulate $h(x)$ to map to the projection matrix $\mathcal{A}$ based on the current state vector x.

$$\mathcal{A}_s = \left(I + \frac{R_{q_k}^T e_2 (v_k \Delta t)^T}{\xi_2} \right) e^{[\omega_k] \times \Delta t} \tag{29}$$

The matrix exponential can be expressed in the closed form as

$$e^{[\omega_k] \times \Delta t} = I + \frac{\sin(s)}{s}[\omega \Delta t]_\times + \frac{1-\cos(s)}{s^2}[\omega \Delta t]_\times^2, \quad s = |\omega \Delta t| \tag{30}$$

the measurement function is then given by the matrix in (29) normalised to unity.

$$h(x) = \frac{1}{\sqrt[3]{\det(\mathcal{A}_s)}} \mathcal{A}_s \tag{31}$$

The covariance of the measurement noise is specified by R_k. This matrix is directly calculated from measurements with known ground truths and not analytically derived.

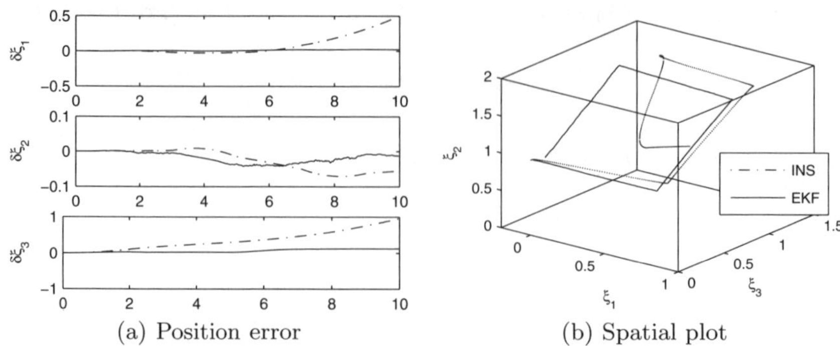

(a) Position error   (b) Spatial plot

Fig. 2. Evaluation scenario position comparison

4   Experiments

To verify the results of the EKF implementation a simulated environment based on the rigid body and image sensor description given in the previous section is used. The accelerometer and gyro noise has been modelled to be similar in properties to that of low-cost sensor components. A raytracer is used to simulate the image sensor output, and an additional pixel noise term is added. The ground texture wich is used is a grayscale random pattern.

Figure 2 shows a plot of the position errors for an evaluated example scenario, where the simulated UAV was made to perform a rectangular movement on a tilted plane. The start and end position are the same.

The overall error is significantly smaller for both plots in $\delta\xi_2$ than for the other two components. This is related to the fact that the e_2 component of the accelerometer has a lower noise part to reflect the same situation in an already implemented real system. For all three components of the position vector, the aided EKF offers a better performance than the inertial sensors alone. Figure 2 (b) gives a direct comparison of the position and allows a spatial interpretation of the results. For a total distance of about $5m$ the INS (Inertial Navigation System) solution has an error of more than $1m$, where the aided estimation is around $5cm$. A similar trend is visible in other evaluated scenarios.

5   Conclusion

The results of the simulation are very promising, as they show that the combination of the visual information and the inertial sensors can for particular situations perform better by a factor of 20 when compared to simple integration of low-cost inertial sensors. Previous works using similar methods did not use the projection matrix directly as a measurement. They extracted the state information from the matrix and relied on third sensors to solve the ambiguity in the velocity and height data. The proposed approach is able to handle this situation without the need for additional reference information and is stable even if the state estimate is inaccurate. The proposed method uses the constant brightness constraint method for relating consecutive images. Another possible solution is to perform feature extraction and relate the features of consecutive images. For this approach to work, the image requires enough visual information so that features can accurately be identified. A combination of the two approaches could potentially solve a wider class of problems. The sensor fusion method applied in the context of this work uses the Extended Kalman Filter, which has known problems with highly non-linear systems. An alternative approach which might have advantages in this area is the class of Unscented Kalman Filters.

References

1. Barrows, G., Neely, C.: Mixed-mode vlsi optic flow sensors for in-flight control of a micro air vehicle. In: SPIE 45th Annual Meeting (2000)
2. Bosse, M., Karl, W., Castanon, D., DeBitetto, P.: A vision augmented navigation system. In: Intelligent Transportation System, Boston, MA, USA, pp. 1028–1033 (1997)
3. Coutsiasy, E.A., Romeroz, L.: The quaternions with an application to rigid body dynamics
4. Cox, T.H., Nagy, C.J., Skoog, M.A., Somers, I.A.: Civil UAV capability assessment. Technical report, NASA (December 2004)
5. Galvin, B., McCane, B., Novins, K., Mason, D., Mills, S.: Recovering motion fields: An evaluation of eight optical flow algorithms. In: Proceedings of the British Machine Vision Conference 1998 (1998)
6. Gavrilets, V.: Autonomous aerobatic maneuvering of miniature helicopters. PhD thesis, Massachusetts Institute of Technology (2003)
7. Hamel, T., Mahony, R., Lozano, R., Ostrowski, J.-N.: Dynamic modelling and configuration stabilization for an x4-flyer. In: 15th IFAC World Congress 2002 (2002)
8. Horn, B.K., Schunck, B.G.: Determining optical flow. Artificial Intelligence 17(1-3), 185–203 (1981)
9. Jähne, B., Haußecker, H.: Computer Vision and Applications. Academic Press, London (2000)
10. Low, T., Wyeth, G.: Obstacle detection using optical flow. In: Proceedings of the 2005 Australasian Conference on Robotics & Automation (2005)
11. Muratet, L., Doncieux, S., Briere, Y., Meyer, J.-A.: A contribution to vision-based autonomous helicopter flight in urban environments. Robotics and Autonomous Systems 50(4), 195–209 (2005)
12. Negahdaripour, S., Horn, B.K.P.: Direct passive navigation: Analytical solution for planes. IEEE Robotics and Automation. Proceedings. 3, 1157–1163 (1986)
13. Press, W.H., Teukolsky, S.A., Vetterling, W.T., Flannery, B.P.: Numerical Recipes in C: The Art of Scientific Computing. Cambridge University Press, New York (1992)
14. Stocker, A.A.: Analog integrated 2-d optical flow sensor. Analog Integr. Circuits Signal Process 46(2), 121–138 (2006)
15. van der Merwe, R., Wan, E.A., Julier, S.: Sigma-point kalman filters nonlinear estimation and sensor fusion - applications in integrated navigation. In: AIAA Guidance Navigation and Controls Conference (March 2004)

A Symbolic Pattern Classifier for Interval Data Based on Binary Probit Analysis

Renata M.C.R. de Souza[1], Francisco José A. Cysneiros[2], Diego C.F. Queiroz[1], and Roberta A. de A. Fagundes[1]

[1]Centro de Informática, [2]Departamento de Estatística - Centro de Ciências Exatas, Universidade Federal de Pernambuco
Av. Prof. Luiz Freire, s/n, Cidade Universitaria CEP: 50740-540, Recife, Brasil
`rmcrs@cin.ufpe.br, cysneiros@de.ufpe.br, dcfq@cin.ufpe.br, raaf@cin.ufpe.br`

Abstract. This paper introduces a classifier for interval symbolic data based on the probit regression model. Each example of the learning set is described by a feature vector, for which each feature value is an interval. Two versions of this classifier are considered. First fits the classic probit regression model conjointly on the lower and upper bounds of the interval values assumed by the variables in the learning set. Second fits the classic probit model separately on the lower and upper bounds of the intervals. The prediction of the class for new examples is accomplished from the computation of the posterior probabilities of the classes. To show the usefulness of this method, examples with synthetic symbolic data sets with overlapping classes are considered. The assessment of the proposed classification method is based on the estimation of the average behaviour of the error rate in the framework of the Monte Carlo method.

1   Introduction

Symbolic data analysis (SDA) [2] aims to provide suitable methods (clustering, factorial techniques, decision tree, etc.) for managing aggregated data described through multi-valued variables, where there are sets of categories, intervals, or weight (probability) distributions in the cells of the data table (for more details about SDA, see www.jsda.unina2.it). A symbolic variable is defined according to its type of domain. For example, for an object, an interval variable takes an interval of $\Re$ (the set of real numbers). A symbolic modal takes, for a object, a non-negative measure (a frequency or a probability distribution or a system of weights). If this measure is specified in terms of a *histogram*, the modal variable is called *histogram variable*.

In classical data analysis, the items to be grouped are usually represented as a vector of quantitative or qualitative measurements where each column represents a variable. However, this model is too restrictive to represent complex data, which may, for instance, comprehend variability and/or uncertainty. We may have 'native' interval data, when describing ranges of variable values - for example, daily stock prices. Interval variables also allow to deal with imprecise data, coming from repeated measures or confidence interval estimation.

A. Dengel et al. (Eds.): KI 2008, LNAI 5243, pp. 340–347, 2008.

Suppose there are 2 classes labelled $0, 1$. Let $\Im = \{(\mathbf{x}_i, y_i)\}$ $(i = 1, \ldots, N)$ be a symbolic learning data set. Each item i is described by a vector of p symbolic variables $\mathbf{x}_i = (X_1, \ldots, X_p)$ and a discrete quantitative variable $y_i = Y$ that takes values in discrete set $G = \{0, 1\}$. A symbolic variable X_j $(j = 1, \ldots, p)$ is an interval variable when, given an item i of $\Im$, $X_j(i) = x_{ij} = [a_{ij}, b_{ij}] \subseteq \mathcal{A}_j$ where $\mathcal{A}_j = [a, b]$ is an interval.

Concerning supervised classification tools, SDA has extended several methods to handle interval data. D'Oliveira, De Carvalho and Souza [3] presented a region oriented approach in which each region is defined by the convex hull of the objects belonging to a class. Mali and Mitra [5] extended the fuzzy radial basis function (FRBF) network to work in the domain of symbolic data. Appice, D'Amato, Esposito and Malerba [1] introduced a lazy-learning approach (labeled Symbolic Objects Nearest Neighbor SO-SNN) that extends a traditional distance weighted k-Nearest Neighbor classification algorithm to interval and modal data. Silva and Brito [7] proposed three approaches to the multivariate analysis of interval data, focusing on linear discriminant analysis.

The main contribution of this paper is to introduce a symbolic classifier for interval data based on a statistical model. Recent advances in statistical modelling has been related to generalized linear models [6] providing more flexible model-based tools for data analysis. A special class of generalized linear models are the model with categorical discrete responses. The binary probit modelling arises from the desire to model the posterior probabilities of the class level given its observation via linear functions in the predictor variables [4].

This paper addresses the use of the probit regression model for interval data. Here, two versions of the posterior probability are considered. In the first one, the posterior probability has a single component and is based on the lower and upper bounds of the intervals conjointly. In the second one, it has two-components and is based on the lower and upper bounds of the intervals separately . New objects are assigned to the class with the highest posterior probability.

The structure of the paper is as follows: Section 2 introduces the classifier for interval data based on the binary probit statistical model. Section 3 discusses the experimental evaluation considering two synthetic interval data sets. The performance of this classifier is measured on the prediction accuracy. This measurement is assessed in the framework of a Monte Carlo experience with 100 replications of each data set. Section 5 concludes the paper.

2 A Binary Classifier for Symbolic Interval Data

In this section, a symbolic classifier for interval data based on the probit model is presented. Two main steps are involved in the construction of this classifier. In the learning step, a linear function to model the posterior probabilities of the classes of a training data set is constructed. Two versions to model the posterior probabilities are introduced. In this first one, the class posterior probability of each class has only a single component, whereas it has two components that are combined in the second version. In the allocation step, new examples of a test data set are affected to a class according to the estimated posterior probability.

2.1 Learning Step

This step aims to construct a fitting linear model $\hat{f}(\mathbf{x})$ that models the posterior probability $E(Y/\mathbf{x}) = \pi = Pr(g = 1/\mathbf{x})$ associated to the class 1 where g is the predictor that takes values in discrete set $G = \{0, 1\}$ and the posterior probability associated to the class 0 is $Pr(g = 0/\mathbf{x}) = 1 - \pi$. The new method fits a probit regression model on the values assumed by the interval symbolic variables in the learning set. These symbolic variables are independents.

One-Component Posteriori Probability. The analyze is performed using lower and upper bounds of the intervals conjointly and a classic binary probit model is fitted.

Let $\mathbf{x} = ([a_1, b_1], \ldots, [a_p, b_p])$ be a vector of intervals, $\boldsymbol{\beta} = (\beta_0, \beta_1, \ldots, \beta_{2p})$ be a vector of unknown parameters of dimension $(2p+1)$ and $\mathbf{z} = (1, a_1, b_1, \ldots, a_j, b_j, \ldots, a_p, b_p)$ be a input vector with $(2p + 1)$ components.

Given the vector $\mathbf{z}$, the one-component probit model is performed using lower and upper bounds of the intervals conjointly in the following form:

$$E(Y/\mathbf{x}) = \pi = Pr(g = 1/\mathbf{x}) = \Phi([\boldsymbol{\beta}]^T \mathbf{z})$$

$$Pr(g = 0/\mathbf{x}) = 1 - \Phi([\boldsymbol{\beta}]^T \mathbf{z})$$

where $\Phi([\boldsymbol{\beta}]^T \mathbf{z})$ is the standard normal distribution function.

The equation (1) can be written equivalently as

$$\Phi^{-1}(\pi) = \beta_0 + \beta_1 a_1 + \beta_2 b_1 + \ldots + \beta_{(2p-1)} a_p + \beta_{(2p)} b_p$$

where ϕ^{-1} is the inverse of the standard normal distribution function.

The model with one-component posteriori probability is specified in terms of a classic probit transformation in which $\mathbf{z}$ is the vector of input with $(2p+1)$ components. The vector of coefficients $\boldsymbol{\beta}$ in this binary model is estimated by the maximum likelihood method [4].

Two-Component Posteriori Probability. The analyze is performed using lower and upper bounds of the intervals separately. Two binary probit models are fitted.

Let $\boldsymbol{\beta}^L = (\beta_0^L, \beta_1^L, \ldots, \beta_p^L)$ and $\boldsymbol{\beta}^U = (\beta_0^U, \beta_1^U, \ldots, \beta_p^U)$ two vectors of unknown parameters both of dimension $(p + 1)$.

Consider $\mathbf{x}^L = (a_1, \ldots, a_j, \ldots, a_p)$ and $\mathbf{x}^U = (b_1, \ldots, b_j, \ldots, b_p)$. Let $\mathbf{z}^L = (1, \mathbf{x}^L)$ and $\mathbf{z}^U = (1, \mathbf{x}^U)$ two input vectors each one with $(p + 1)$ components related to the lower and upper bounds of the interval variables, respectively.

Given the vectors $\mathbf{z}^L$ and $\mathbf{z}^U$. The two-component probit model is performed by the binary components π^L and π^U related to the lower and upper bounds of the symbolic interval variables, respectively. These components are combined to define a probit model that is given by:

$$E(Y/\mathbf{x}) = \pi = Pr(g = 1/\mathbf{x}) = \frac{(\pi^L) + \pi^U)}{2} \tag{1}$$

$$Pr(g = 0/\mathbf{x}) = 1 - \pi = 1 - \left[\frac{(\pi^L) + \pi^U)}{2}\right] \tag{2}$$

where π^L is the classic probit model for lower bounds defined as:

$$E(Y/\mathbf{x}^L) = \pi^L = \Phi([\boldsymbol{\beta}^L]^T \mathbf{z}^L) \tag{3}$$

and π^U is the classic probit model for upper bounds defined as:

$$E(Y/\mathbf{x}^U) = \pi^U = \Phi([\boldsymbol{\beta}^U]^T \mathbf{z}^U) \tag{4}$$

The equations (6) and (7) can be written, respectively, equivalently as

$$\Phi^{-1}(\pi^L) = \beta_0^L + \beta_1^L x_1^L + \ldots, \beta_p^L x_p^L \tag{5}$$

$$\Phi^{-1}(\pi^U) = \beta_0^U + \beta_1^U x_1^U + \ldots, \beta_p^U x_p^U \tag{6}$$

where ϕ^{-1} is the inverse of the standard normal distribution function.

The model with two-component posteriori probability is specified in terms of one classic probit transformation in which $\mathbf{z}_L$ is the vector of input with $(p + 1)$ components and one probit transformation in which $\mathbf{z}_U$ is the vector of input with $(p + 1)$ components. The vectors of coefficients $\boldsymbol{\beta}^L$ and $\boldsymbol{\beta}^U$ in this binary model are estimated by the maximum likelihood method [4].

2.2 Allocation Step

Let ω be a new item, which is candidate to be assigned to one of the classes $\{0, 1\}$, and its corresponding symbolic description: $\mathbf{x}_\omega = (X_1(\omega), \ldots, X_p(\omega))$

The *classification rule* is defined as follow: ω is affected to the class 1 if

$$\pi = Pr(g = 1/\mathbf{x}_\omega) \geq 0.5) \tag{7}$$

otherwise ω is affected to the class 0.

3 Experimental Evaluation

To show the usefulness of this method, experiments with synthetic symbolic interval data sets of different degrees of overlapping classes are considered. The evaluation was performed in the framework of a Monte Carlo experience: 100 replications are considered for each interval data set.

In this work, two synthetic interval data sets are considered. They are initially generated from two standard quantitative data set in $\Re^2$. These data sets have 250 points scattered among two classes of unequal sizes: one class with ellipse shape and size 150 and one class with spherical shape of size 100. Each class in these quantitative data sets were drawn according to a bi-variate normal distribution with non-correlated components and vector $\boldsymbol{\mu}$ and covariance matrix $\boldsymbol{\Sigma}$ represented by:

$$\boldsymbol{\mu} = \begin{bmatrix} \mu_1 \\ \mu_2 \end{bmatrix} \text{ and } \boldsymbol{\Sigma} = \begin{bmatrix} \sigma_1^2 & \rho_{12}\sigma_1\sigma_2 \\ \rho_{12}\sigma_1\sigma_2 & \sigma_2^2 \end{bmatrix}$$

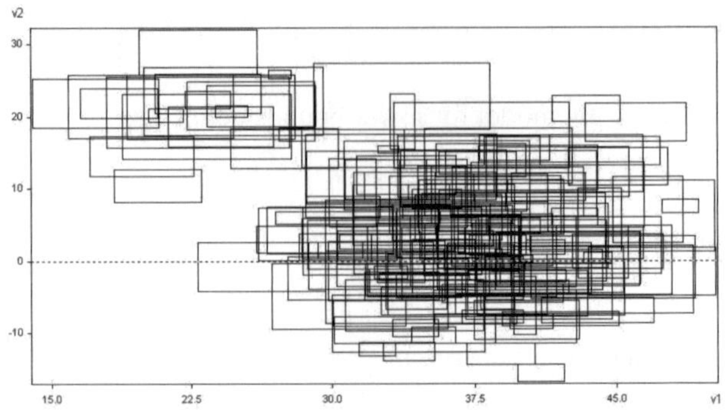

Fig. 1. Symbolic interval data set 1

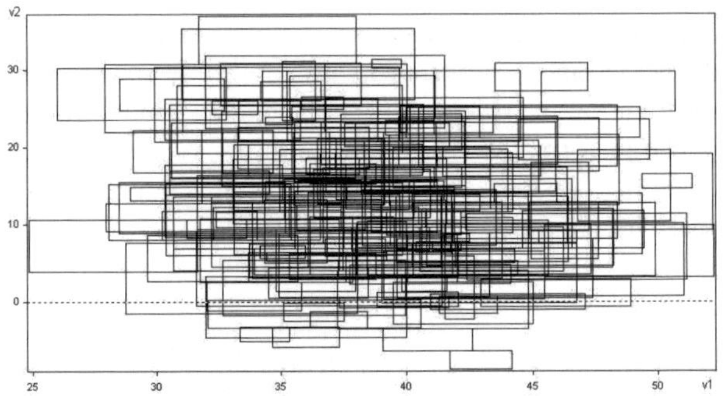

Fig. 2. Symbolic interval data set 2

Standard data set 1 was drawn according to the following parameters:

a) Class 2: $\mu_1 = 38$, $\mu_2 = 5$, $\sigma_1^2 = 16$, $\sigma_2^2 = 49$ and $\rho_{12} = 0.0$;
b) Class 3: $\mu_1 = 23$, $\mu_2 = 20$, $\sigma_1^2 = 9$, $\sigma_2^2 = 9$ and $\rho_{12} = 0.0$;

Standard data set 2 was drawn according to the following parameters:

a) Class 2: $\mu_1 = 40$, $\mu_2 = 12$, $\sigma_1^2 = 16$, $\sigma_2^2 = 49$ and $\rho_{12} = 0.0$;
b) Class 3: $\mu_1 = 35$, $\mu_2 = 25$, $\sigma_1^2 = 9$, $\sigma_2^2 = 9$ and $\rho_{12} = 0.0$;

Each data point (z_1, z_2) of each one of this synthetic quantitative data set is a seed of a vector of intervals (rectangle) defined as:

$$(v_1 = [z_1 - \gamma_1/2, z_1 + \gamma_1/2], v_2 = [z_2 - \gamma_2/2, z_2 + \gamma_2/2])$$

These parameters γ_1, γ_2 are randomly selected from the same predefined interval. The intervals considered in this paper are: $[1, 10], [1, 20], [1, 30], [1, 40]$ and $[1, 50]$. Figures 1 and 2 illustrate, respectively, these synthetic interval data sets ranging from easy to moderate cases of overlapping classes with parameters γ_1 and γ_2 randomly selected from the interval $[1, 10]$.

Table 1. The average (%) and the standard deviation of the error rate for synthetic interval data set 1

Predefined intervals for γ_1 and γ_2	One-component probability		Two-component probability	
	Average	St. Deviation	Average	St. Deviation
$[1, 10]$	1.33	0.0184	0.08	0.0149
$[1, 20]$	1.06	0.0139	1.00	0.0141
$[1, 30]$	1.10	0.0148	1.04	0.0164
$[1, 40]$	1.14	0.0168	0.07	0.0122
$[1, 50]$	1.13	0.0161	0.09	0.0143

To evaluate the performance of this classifier test and learning sets are randomly selected from each synthetic interval data set. The learning set corresponds to 75% of the original data set and the test data set corresponds to 25%. The performance is measured by the error rate of classification that is computed for each class of the test set. The estimated error rate of classification corresponds to the average of the error rates found between the 100 replications of the test set.

Table 1 and 2 show the values of the average and standard deviation of the error rate for one-component and two-component probability and synthetic interval data sets 1 and 2, respectively. From the results, it can observed that, in all situations of the interval data set 1, the average error rate for the two-component posteriori probability approach is highly lesser than the average error rate for the one-component posteriori probability approach. However, for interval data set 2, the average error rate for the one-component approach is highly lesser than the two-component approach in the most of the situations of this data set.

Table 2. The average (%) and the standard deviation of the error rate for synthetic interval data set 2

Predefined intervals for γ_1 and γ_2	One-component probability		Two-component probability	
	Average	St. Deviation	Average	St. Deviation
$[1, 10]$	7.29	0.0210	7.40	0.0320
$[1, 20]$	7.28	0.0271	6.95	0.0333
$[1, 30]$	6.92	0.0235	7.32	0.0350
$[1, 40]$	7.17	0.0239	7.13	0.0328
$[1, 50]$	7.39	0.0283	7.53	0.0319

The comparison between the one-component and two-component approaches is achieved by a independent Student's t-test at a significance level of 5%. Table 3 shows the suitable (null and alternative) hypothesis and the observed values of the test statistics following a Student's t distribution with 198 degrees of freedom. In this table, μ_1 and μ_2 are, respectively, the average of the error rate for the one-component and two-component approaches.

Table 3 shows that in all the simulations of the synthetic interval data set 1, the two-component approach is superior to the one-component approach and in 60% of the simulations of the synthetic interval data set 2, the one-component approach is superior to the two-component approach.

Table 3. Statistics of Student's t-tests comparing the methods

Predefined intervals for γ_1 and γ_2	$H_0 : \mu_1 = \mu_2$ $H_1 : \mu_2 < \mu_1$			
	Interval data set 1	Decision	Interval data set 2	Decision
$\gamma \in [1, 10]$	-527.95	Reject H_0	28.73	Not Reject H_0
$\gamma \in [1, 20]$	-30.30	Reject H_0	-76.86	Reject H_0
$\gamma \in [1, 30]$	-27.16	Reject H_0	94.88	Not Reject H_0
$\gamma \in [1, 40]$	-515.35	Reject H_0	-9.85	Reject H_0
$\gamma \in [1, 50]$	-482.96	Reject H_0	32.83	Not Reject H_0

4   Conclusions

In this paper, a binary probit regression model for interval data is presented. Each example of the learning set is described by a feature vector, for which each feature value is an interval. In this model, a linear function in the predictor variables to model the posterior probabilities of the classes is constructed. Two versions of the posterior probability are considered. In the first one, the posterior probability has a single component and is based on the lower and upper bounds of the intervals conjointly. In the second one, it has two-components and is based on the lower and upper bounds of the intervals separately. The new examples are assigned to the class with the highest posterior probability.

Experiments with synthetic interval symbolic data sets ranging from easy to moderate cases of overlapping classes are considered. The accuracy of the results furnished by the one-component and two component approaches are assessed by the error rate and compared based on t-Student's test for independent samples at a significance level of 5%. Concerning these interval symbolic data sets, the error rate is calculated in the framework of the Monte Carlo experience with 100 replications. For both types of data configurations with overlapping classes the approaches based on the lower and upper bounds of the intervals separately and conjointly presented good performances in terms of the error rate.

Acknowledgments. The authors would like to thank CNPq (Brazilian Agency) for its financial support.

References

1. Appice, A., D'Amato, C., Esposito, F., Malerba, D.: Classification of symbolic objects: A lazy learning approach. Intelligent Data Analysis. Special issue on Symbolic and Spatial Data Analysis: Mining Complex Data Structures (accepted for publication, 2005)

2. Bock, H.H., Diday, E.: Analysis of Symbolic Data: Exploratory Methods for Extracting Statistical Information from Complex Data. Springer, Heidelberg (2000)
3. D'Oliveira, S., De Carvalho, F.A.T., Souza, R.M.C.R.: Classification of sar images through a convex hull region oriented approach. In: Pal, N.R., Kasabov, N., Mudi, R.K., Pal, S., Parui, S.K. (eds.) ICONIP 2004. LNCS, vol. 3316, pp. 769–774. Springer, Heidelberg (2004)
4. Fahrmeir, L., Tutz, G.: Multivariate Statistical Modelling Based on Generalized Linear Models, 2nd edn. Springer, Heidelberg (2001)
5. Mali, K., Mitra, S.: Symbolic classification, clustering and fuzzy radial basis function network. Fyzzy sets and systems 152, 553–564 (2005)
6. Nelder, J.A., Wedderburn, R.W.M.: Gneralized Linear Models. Journal of the Royal Satistical Society A 135, 370–384 (1972)
7. Silva, A.P.D., Brito, P.: Linear Discriminant Analysis for Interval Data. Computational Statistics 21, 289–308 (2006)

Object Configuration Reconstruction from Incomplete Binary Object Relation Descriptions

H. Joe Steinhauer

Department of Computer Science, Linköpings Universitet, Sweden
`joest@ida.liu.se`

Abstract. We present a process for reconstructing object configurations described by a set of spatial constraints of the form (A northeast B) into a two-dimensional grid. The reconstruction process is cognitively easy for a person to fulfill and guides the user to avoid typical mistakes. For underspecified object configuration descriptions we suggest a strategy to handle coarse object relationships by representing a coarse object in a way that all disjunctive basic relationships that the coarse relationship consists of are represented within one reconstruction.

Keywords: Qualitative Reasoning, Spatial Reasoning, Diagrammatic Reasoning.

1   Motivation

Suppose you are planning an outdoor trip visiting your friend that left earlier and by now has established a tent camp 'somewhere' in the mountains. Your friend, a little confused as usual, has forgotten to bring a map and also missed to carefully memorize the route he took. Therefore, he can not provide directions to the camp. You talk on the cell phone and he describes the place so that you can compare his description with a map in order to find his whereabouts. However, you do not have the map accessible while you are talking so you try to establish an external diagrammatic representation for later comparison with a geographic map.

This paper firstly presents a general strategy how complete specified object configurations can be reconstructed into a diagrammatic representation. Secondly, it suggests a strategy how underspecified object configuration descriptions can be reconstructed using coarse relationships where specific relationships are not available.

2   Introduction

Object configurations can be described using binary object position relations. At least eight position classes, for instance: north, northeast, east, southeast, south, southwest, west, as used in the reference frame presented in figure 1, need to be distinguished in order to provide enough global information for a map-like diagrammatic reconstruction. The frame of reference used is rectangular, which

A. Dengel et al. (Eds.): KI 2008, LNAI 5243, pp. 348–355, 2008.

north west	north	north east
east	neutral	west
south east	south	south west

Fig. 1. Rectangular frame of reference with nine position classes

according to [1] is often preferred by people. Furthermore, it is projection-based with neutral zone [2] and has been proven sufficient in several tasks using cardinal directions [3] or intrinsic orientations [4],[5],[6].

Even though our focus lies on a physical, rather than on a mental reconstruction as for instance in [7] and [8], mental model theory (MMT) [9], [10] helps to understand typical human preferences considering object configuration reconstruction processes in general. It suggests that people use a symmetrical mental spatial array where they insert the objects. Experiments in [11], [12], and [13] show that people even for indeterminate problems construct one mental model only, the so called preferred mental model (PMM) that is consistent with all given premises. Ragni [14] introduces the first free fit (fff) theory that shows that objects are inserted within the first free array cell, that fulfills the relation given in the premise and that people avoid to move objects that have been inserted previously.

A reconstruction strategy should be cognitively easy and should allow to reconstruct a configuration of many objects quickly with the least amount of backtracking or rearranging of once placed objects. Furthermore it has to guide the user to use provided information in an efficient order and to avoid that underspecified configurations lead to 'wrong' reconstructions.

3   Reconstruction

The object configuration is considered an allocation of objects on a two-dimensional grid. The grid is defined similar to the grid used by Ragni [15]. However, Ragni's representation algebra, AC, has been developed to model the rule based approach of human reasoning. The problem of reconstructing an object configuration has therefore been transformed into a problem of solving inequalities.

In our approach the reconstruction is done incrementally directly into the grid (reconstruction area). The first object can be placed anywhere within this area. Following objects' positions are calculated by successively intersecting all qualitative regions to reference objects whose relationships are provided in the description. Figure 2 shows the successive reconstruction for the object configuration description (ocd): {(2 northeast 1), (3 east 1), (3 southeast 2), (4 northeast 1), (4 northeast 2), (4 northeast 3)}. Object 1 is placed into the reconstruction

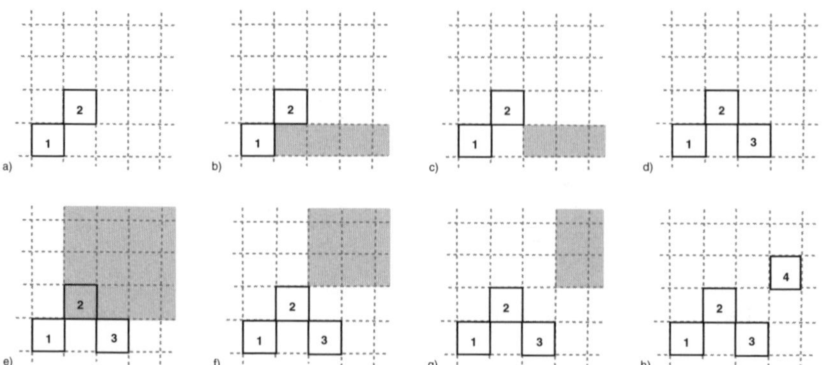

Fig. 2. Objects' positions are calculated by successively intersecting all qualitative regions to reference objects

as first reference object. Object 2 is placed northeast of it. As the first free fit theory [14] shows, people tend to place the object as close to the reference object as possible. Therefore, we suggest the distance default [16] to insert an object in the resulting intersection at a position that is closest to the rest of the reconstruction. The potential area for object 3, east of object 1 is coloured grey in figure 2b). This area is intersected with the area southeast of object 2. The remaining intersection is the grey area in figure 2c). Object 3 is inserted in figure 2d). The first potential area for object 2 is northeast of object 1 (figure 2e)), it is intersected with northeast of object 2 (figure 2f)) and further intersected with northeast of object 3 (figure 2g)). Following the distance default, object 4 is placed in the southwest corner of the remaining intersection shown in figure 2h).

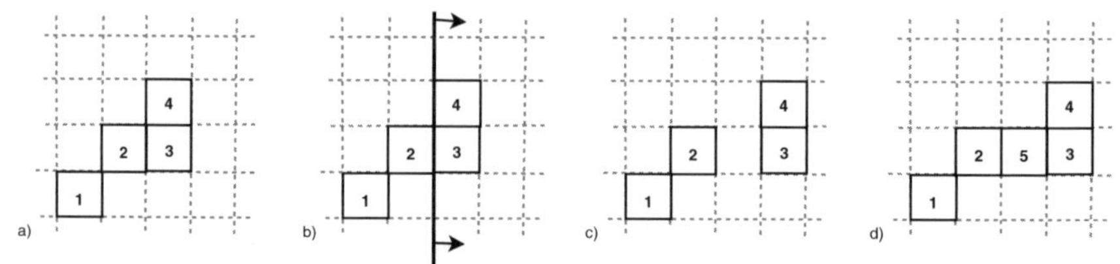

Fig. 3. Horizontal split

When a new object is inserted in between already placed objects it might occur that the intersected region is to small to hold another object. In this case new space at the desired position is made by splitting the reconstruction as described in [16] and [17] by a straight line in x- or y-dimension, and moving one half of it one step away from the rest. Figure 3 illustrates the process for the horizontal split. Object 5 to has be inserted into the intersection of northeast 1, east 2, west 3, and southwest 4.

A split like this will not change any considered qualitative relationship of the reconstructed configuration. The intuitive proof for the horizontal spit is as follows: At the beginning, the reconstruction correctly contains all objects' relationships. Objects will only be moved in increasing x-dimension and will therefore never cross any reference frame line that separates regions horizontally. For every object that is unmoved and situated west, northwest, or southwest of the moved object, the moved object is northeast, east, or southeast of it. These regions are infinite to the east and the object will never leave them by moving eastwards. All other objects are moved in the same way as the first-hand object and therefore its relationships to these objects do not change. The vertical split to make space in the vertical dimension works accordingly [17].

4   Coarse Object Positions

So far it was assumed that the object configuration was fully specified by the ocd. We now consider the case of underspecified ocds. It is our aim to still be able to reconstruct the object configuration with all relationships correctly represented within the reconstruction; even those relationships that are not provided in the ocd.

The questions that arise are of the kind: Given that object 3 is northeast of object 1 and we already know that object 2 is northwest of object 1. What is the relationship of object 3 to object 2? If this relationship is not provided it could be either northeast, east or southeast. In other words object 3 will be somewhere EAST of object 2. We call the relationship EAST a coarse relationship. Other coarse relationships are: WEST (northwest ∨ west ∨ southwest), NORTH (northeast ∨ north ∨ northeast), SOUTH (southwest ∨ south ∨ southwest), NEUTRAL-H (west ∨ neutral ∨ east), NEUTRAL-V (north ∨ neutral ∨ south), and OPEN (the disjunction of all nine basic relations). A coarse relationship can be looked up in a composition table.

Objects described by coarse relationships are called coarse objects. When a coarse object is inserted into the reconstruction, enough space has to be allocated so that every possible disjunctive relation that is included in the coarse relation is possible. In figure 4a) object 3 is a coarse object. Its relationship to object 1 is northeast, and its relationship to object 2 is EAST. Therefore, three cells are allocated for object 3 one northeast of 2, one east of 2 and one southeast of 2. Of course, all three cells must have the relationship northeast of 1.

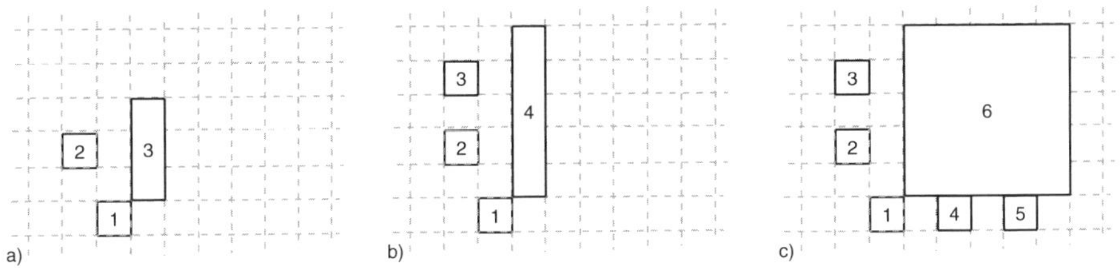

Fig. 4. Ambiguous object positions

When the object has more than one coarse relationship, the area allocated for it might extend accordingly. In figure 4b) object 4 is northeast of object 1, EAST of object 2, and EAST of object 3. The area allocated for object 4 must contain all possible basic relationships that are possible in this combination of coarse relationships. This area ranges from a cell northeast of 3 to a cell southeast of object 2. At the same time the relationship southeast of 3 and northeast of 2 must be possible, which only is the case if there is one cell space between object 2 and object 3. If this is not the case before object 4 is inserted, the configuration has to be split vertically to obtain this space.

When an object's position is described by coarse relationships in both dimensions, the object's area expands in both. In figure 4c) object 6 is northeast of 1, EAST of 2, EAST of 3, NORTH of 4, and NORTH of 5. In order to be EAST of object 2 and 3 five cells in vertical dimension are allocated. In order to be at the same time NORTH of 4 and 5, five cells in horizontal dimension are needed. Thus object 6 is represented with the 5 x 5 cell area. This way, all possible relationship combinations of object 6 to the objects 2, 3, 4, and 5 are represented by the area.

As far as possible, coarse objects are treated the same way as single cell objects. They are moved as a whole by horizontal or vertical splits. In case a split intersects a coarse object, the parts of the object are moved apart and the cells appearing in between become part of the object. When a coarse object is reference object in a provided relationship, it depends on the relationship if the target object will also become a coarse object. When its relationship is northwest, northeast, southwest, or southeast, the reference object's kind has no influence on the target object's size, but if the relation is north, east, south, or west the target object's size has to be adjusted accordingly to the reference object's size.

5   The Reconstruction Process

The reconstruction process described above can be summarized by the following algorithm:

```
Algorithm 1. Object Configuration Reconstruction
------------------------------------------------------------------
  1 The first object can be inserted anywhere in the grid.
  2 The second object is inserted according to the distance default
    and its relationship to the first object.
  3 Following objects are inserted by intersecting the regions
    described by their relationships to all objects already
    inserted in the reconstruction. The intersection is called the
    target region.
  4 If the intersection is to small, it is broadened by using a
    horizontal or vertical split.
  5 If all relationships to objects already in the reconstruction
    have been provided, the object is inserted into the
    reconstruction as a single cell object.
```

```
 6 For all objects in the reconstruction, that are north, east,
   south, or west of the target region and whose relationship
   to the target object is not provided, the relationship is
   calculated by reasoning.
 7 If the calculated relationship is not coarse, it has no
   influence on the objects position.
 8 For a coarse relationship enough space to fulfill all
   included disjunctive relationships is allocated for the
   object's area.
 9 If the object area does not fit into the target region, the
   target region is extended by vertical and horizontal splits.
10 All objects that are north, south, east or west of the object
   area and so far did not have a direct influence on the
   position of the object area that have not been considered
   so far, are moved by horizontal and vertical splits out of
   that regions.
11 When all necessary objects have been considered the
   resulting object area is treated as one object.
-----------------------------------------------------------------
```

The example in figure 5 demonstrates how space for coarse objects is allocated. Suppose the object configuration has been described by (2 northwest 1), (3 southeast 1), (4 east 1), (5 northeast 1), (6 northeast 5), (7 east 5), and (7 southeast 6). To place object 4 reasoning provides the additional relation (4 SOUTH 2). To fulfill this relation object 1 and 3 have to be moved one step eastwards figure 5b). The further relationships (5 EAST 2) and (5 NORTH 3) are needed to cover object 5's relationships to its influencing objects 2 and 3. Object 1, 3, and 4 need to be moved one step southwards and object 3 one step eastwards in order to fulfill all basic relationships contained in the combined relationships (figure 5d)). Object 7's position is dependent on object 5's therefore object 7 becomes a coarse object with the same measures as object 5 in north-south dimension (figure 5e)). Assuming that later on the new information (5 north 3), (5 northeast 2), and (4 southwest 2) becomes available. Object 5's new position becomes the intersection of object 5 with the regions north of object 3 and northeast of object 2. Object 7's area decreases accordingly due to its relation to object 5 and object 4's decreases to one cell southwest of object 2. The process is illustrated in figure 5 f)-i).

6   Outlook

For completely described object configurations, the process described above works perfectly. For underspecified ocds, two further situations not mentioned above need to be dealt with in order to achieve a complectly trouble free reconstruction strategy.

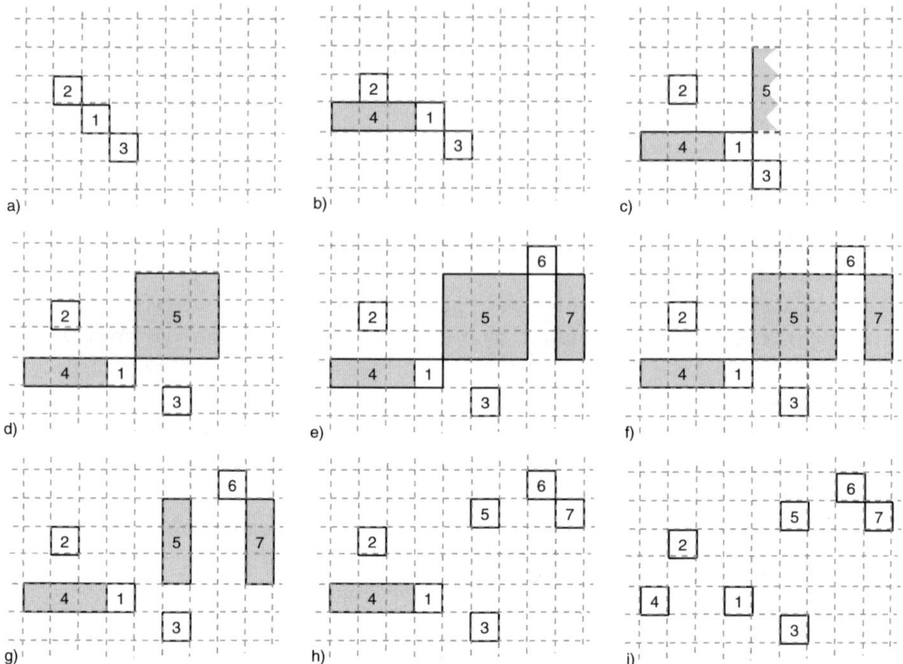

Fig. 5. Reconstruction example

1. When several target objects have unknown relationships to the same reference object, the coarse target objects might overlap.
2. In certain cases only a part of a coarse object influences another object's position.

In both cases it is necessary to resize objects that have already been placed into the reconstruction. For a system it is no problem to solve these situations, but for a person, overlapping object positions and resizing objects increases the cognitive load. Therefore, our ongoing research aims at finding a convenient strategy to handle these cases or to avoid them right from the beginning.

References

1. Zimmermann, K., Freksa, C.: Qualitative spatial reasoning using orientation, distance and path knowledge. Applied Intelligence 6, 49–58 (1996)
2. Frank, A.U.: Qualitative spatial reasoning with cardinal directions. In: Kaindl, H. (ed.) Seventh Austrian Conference on Artificial Intelligence, Informatik Fachberichte, Wien, Austria, September 1991, pp. 157–167 (1991)
3. Goyal, R.K., Egenhofer, M.J.: Similarity of cardinal directions. In: Jensen, C.S. (ed.) SSTD 2001. LNCS, vol. 2121, pp. 36–55. Springer, Heidelberg (2001)
4. Fernyhough, J., Cohn, A.G., Hogg, D.C.: Constructing qualitative event models automatically from video input. Image and Vision Computing 18, 81–103 (2000)

5. Ligozat, G.: Reasoning about cardinal directions. Journal of Visual Languages and Computing 9(1), 23–44 (1998)
6. Mukerjee, A., Joe, G.: A qualitative model for space. In: Proceedings of the AAAI, Boston, pp. 721–727 (1990)
7. Boeddinghaus, J., Ragni, M., Knauff, M., Nebel, B.: Simulating spatial reasoning using act-r. In: Proceedings of the ICCM 2006, pp. 62–67. Lawrence Erlbaum Associates, Mahwah (2006)
8. Ragni, M., Knauff, M., Nebel, B.: A computational model for spatial reasoning with mental models. In: Bara, B., Barsalou, B., Bucciarelli, M. (eds.) Proceedings of the 27th Annual Cognitive Science Conference (CogSci 2005), pp. 1064–1070. Erlbaum, Mahwah (2005)
9. Johnson-Laird, P.N., Byrne, R.M.J.: Deduction. Erlbaum, Hove (1991)
10. Johnson-Laird, P.N.: Mental models and deduction. In: Trends in Cognitive Science, vol. 5, Elsevier Science Ltd, Amsterdam (2001)
11. Rauh, R., Hagen, C., Knauff, M., Kuss, T., Schlieder, C., Strube, G.: Preferred and alternative mental models in spatial reasoning. In: Spatial Reasoning & Computation, vol. 5, pp. 239–269 (2005)
12. Jahn, G., Johnson-Laird, P.N., Knauff, M.: Reasoning about consistency with spatial mental models: Hidden and obvious indeterminacy in spatial descriptions. In: Freksa, C., Knauff, M., KriegBrückner, B., Nebel, B., Barkowsky, T. (eds.) Spatial Cognition IV - Reasoning, Action, Interaction, Frauenchiemsee, Germany, pp. 165–180. Springer, Berlin (2004)
13. Knauff, M., Rauh, R., Schlieder, C.: Preferred mental models in qualitative spatial reasoning: A cognitive assessment of allens's calculus. In: Proceedings of the Seventeenth Annual Conference of the Cognitive Science Society, pp. 200–2005. Lawrence Erlbaum Associates, Mahwah (1995)
14. Ragni, M., Fangmeier, T., Webber, L., Knauff, M.: Complexity in spatial reasoning. In: Proceedings of the 28th Annual Cognitive Science Conference (CogSci 2006), Lawrence Erlbaum Associates, Mahwah (2006)
15. Ragni, M.: An arragement calculus, its complexity, and algorithmic properties. In: Günter, A., Kruse, R., Neumann, B. (eds.) KI 2003. LNCS (LNAI), vol. 2821, pp. 580–590. Springer, Heidelberg (2003)
16. Steinhauer, H.J.: Qualitative reconstruction and update of an object constellation. In: Güsgen, H.W., Ligozat, G., Renz, J., Rodriguez, R.V. (eds.) ECAI 2006 Workshop on Spatial and Temporal Reasoning, Riva del Garda, Italy, 28 August - 1 September 2006, pp. 11–19 (2006)
17. Steinhauer, H.J.: Object configuration reconstruction for descriptions using relative and intrinsic reference frames. In: Proceedings of the 18th European Conference on Artificial Intellicence ECAI 2008, Patras, Greece. IOS Press, Amsterdam (to appear, July 2008)

Visual-Based Emotion Detection for Natural Man-Machine Interaction

Samuel Strupp, Norbert Schmitz, and Karsten Berns

Faculty of Computer Science, Robotics Research Lab,
University of Kaiserslautern, D-67653 Kaiserslautern, Germany

Abstract. The demand for humanoid robots as service robots for everyday life has increased during the last years. The processing power of the hardware and the development of complex software applications allows the realization of "natural" human-robot interaction. One of the important topics of natural interaction is the detection of emotions which enables the robot to react appropriately to the emotional state of the communication partner. Humanoid robots designed for natural interaction require a short response time and a reliable detection. In this paper we introduce a emotion detection system realized with a combination of a haar-cascade classifier and a contrast filter. The detected feature points are then used to estimate the emotional state using the so called action units. Final experiments with the humanoid robot ROMAN show the performance of the proposed approach.

1   Introduction

Humanoid robots are - as it can be seen in many movies - of great interest. Either as entertainment robots, enduring workers or for the care of elderly people. All these possible scenarios have one thing in common: the robot is an accepted member of society and therefore it must behave in a way the human can accept. To reach this goal it is necessary that a robot can interact and communicate in a natural way similar to the interaction process between humans.

The communication between humans is not limited to speech. Moreover it is a complex combination of speech, gestures, mimics and body pose. Robots interacting with people therefore have to be able to use these communication skills. This paper focuses on the aspect of emotion detection using face images which influences the behavior of a robot according to the situation of the human interaction partner.

The detection of emotions is often divided into two steps. The first step is capturing information of the current expression of the face. This includes the detection of features like eyes, eyebrows or mouth. Different approaches for feature detection like feature point tracking ([13], [16]), optic flow fields ([12], [1], [11]), neural networks ([4], [6], [7]), deformable contours ([9], [3], [15]) or difference imaging ([1], [6], [7]) have been presented in many publications.

Beside the feature detection algorithms several approaches use specific properties of the human face. Kawato for example [10] uses the blinking of the eyes

A. Dengel et al. (Eds.): KI 2008, LNAI 5243, pp. 356–363, 2008.
© Springer-Verlag Berlin Heidelberg 2008

to detect them. Depending on the frame rate and resolution of the camera the detection can fail. Another approach of Soderstrom uses the nostrils to detect the region of the mouth [14]. To recognize the nostrils in a picture you must have a viewer perspective under the tip of the nose which is generally not realizable for a humanoid robot. The detection of features is a well know research topic therefore this paper focuses on the emotion interpretation and integration into the humanoid robot.

The second step in the detection process is to interpret the feature information and associate them to an emotion. Canzler [2] proposes a system which interprets gestures including the detection of facial expressions. The disadvantage of his system is the slow processing and the robustness of the system. The system by Wimmer and Fischer [8] for an emotional sales agent uses a classifier based on the optical flow within the face to interpret an emotion. The disadvantage is that an emotion can only be interpreted if the facial expression of the emotion starts from a neutral expression. The initialization must be done manually when the person looks neutral which leads to an unnatural behavior for humanoid robots.

In the following a concept for feature detection and emotion interpretation is introduced. Final experiments performed with the humanoid robot ROMAN will show the performance and integration of the system.

2   Feature Detection

To interpret emotions, it is necessary to have simple information about the current state of the face. So the first step is to find exact positions of some feature points. Therefore detectable and useful feature points for the interpretation of emotions have to be selected. In coherency to the emotion interpretation in section 3, 10 significant features points have been defined: 2 points for each eyebrow, the pupils, the mouth corners as well as the upper and the lower lip. Fig. 1 shows the location of these points as red crosses. The selection of these feature points is based on the Facial Action Coding System (FACS) [5].

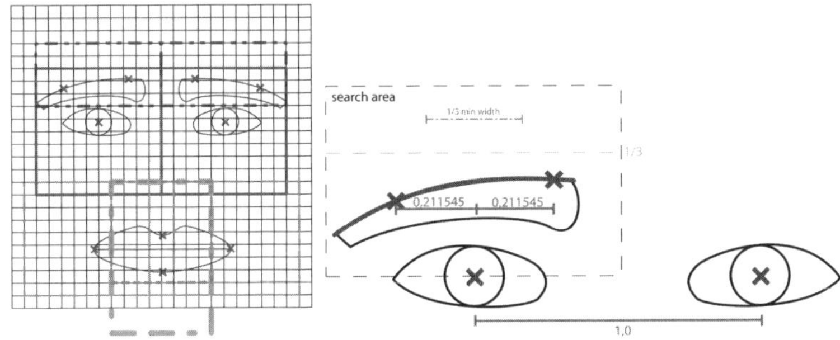

Fig. 1. (left) Feature points (red crosses) and their search regions. Blue: eyebrows; Red: pupils; Green: mouth corners; Cyan: lower lip; Magenta: upper lip. (right) Sketch of the search for the right eyebrow.

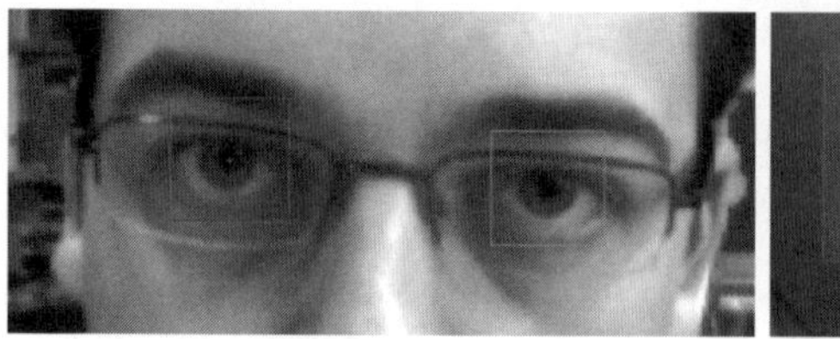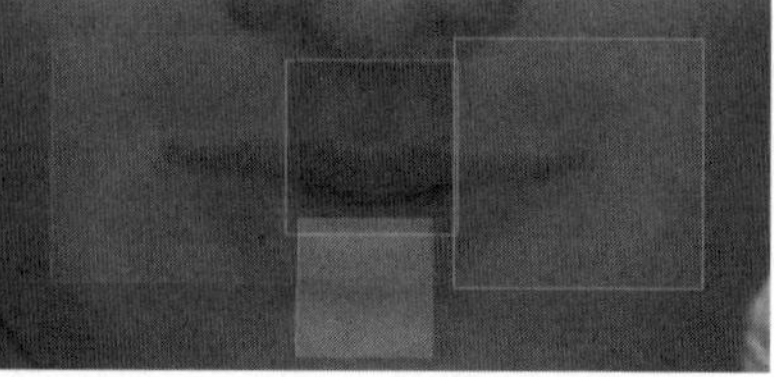

Fig. 2. (left) Pupil classifier region for the right and left pupil. The center of the region is located in the center of the pupil. (right) Mouth classifier regions. [green] right and left mouth corner regions [blue] upper lip region [yellow] lower lip region.

The goal of the feature detection is to localize these 10 points in the face region. Our emotion interpreter and the feature detection assume that all faces are frontal to the camera. Although this is an strong limitation it can be assumed that a person is facing a robot during a conversation. The face region itself can be found with the frontal face classifier of OpenCV[1]. For each feature point a search region within the face region has been defined to reduce the search space. In Fig. 1 the search regions are drafted. Two different techniques are used to find the feature points in this region. The eyebrows are extracted using an adaptive contrast filter which uses the min and max gray values in the eyebrow region to detect the borders. This filtering results in an exact representation of the eyebrow as a polygon. Using these polygons two points on the eyebrows are localized at a certain distance from the pupils. Fig. 1 shows a sketch of the eyebrow search regions. The other feature points are detected with self-trained classifiers based on the technique of Viola and Jones [17] implemented in OpenCV. The trained classifiers are shown in Fig. 2.

The detection of all these mentioned feature points is realized in a detection pipeline. This pipeline has the following steps:

1. Detection of the face region with the frontal-face classifier of OpenCV.
2. Localization of the pupils in their search areas with the pupil classifier. The pupil classifier provides very few false positives. Therefore the first detection is taken and the rest discarded.
3. The eyebrow detection is performed using the adaptive contrast filter.
4. The mouth corners are classified using a heuristic which searches the minimal vertical distance from the center point of the search area.
5. Detection of the lower lip point. If more than one area is classified the lowest match with minimal distance to the vertical axis in the center of the search area is chosen.
6. Finally the upper lip point is searched. If multiple detections occur the point with minimal distance to the vertical axis is chosen.

With the help of the detected points the emotion interpretation is performed as described in the following section. The detection rates of the specified regions

---

[1] See http://www.intel.com/technology/computing/opencv/index.htm for details.

in out test data set are: left pupil 86.2%, right pupil 87.2%, inner left brow 74.4%, outer left brow 63.6%, inner right brow 78.5%, outer right brow 63.6%, left mouth corner 83.6%, right mouth corner 80.0%, upper lip 59.0% and lower lip 49.7%.

3   Emotion Interpretation

To interpret an emotion with the detected feature points the "Facial Action Coding System" (FACS) [5] is used. This system defines so called "Action Units" for every muscle in the face which influences the facial expression. The general idea is to conclude from the positions of the feature points to the activity of some action units. The goal is to interpret the emotion of a facial expression activities of the action units. All detectable action units corresponding to the feature points described in 2 are listed in Tab. 1.

To interpret the activity of an action unit a neutral and a maximum distance from the eye line has been defined. The eye line is the straight line from the right pupil center to the left pupil center. If the distance between the feature point and the eye line is less than or equal to the neutral distance, the activity of the corresponding action unit is 0. If the distance is greater than or equal to the maximum distance the activity is 1. If the distance lies between the neutral and maximum distance then the activity value is linearly interpolated. For the calculation of the activities only vertical or horizontal distances from the eye line or the normal of the eye line which intersects the middle between the two pupils are used. In Fig. 3 the feature points with their corresponding action units and activation directions are illustrated.

The distance to the main axis is a simple point to straight line distance. To be independent from the size of the face we build the ratio of the distance between feature point and eye line and the distance between the pupils. So the distance between the pupils is 1 on all faces. Since only frontal faces are considered this unit definition is usable. With this method we measure the neutral and maximum distances from different faces for all action units. We simply choose the average values for each action unit distance. In Tab. 2 you can see the results. With this data it is possible to associate a point position to an activity value of an action unit.

Based on the activities of the action units the displayed emotion has to be estimated. The emotional state is divided into the 6 basic emotions anger, disgust, fear, happiness, sadness and surprise. If none of these emotions is expressed the emotion should be classified as neutral. Each of the presented emotions has an activity between 0 and 1. Starting with the original emotion formulas from

Table 1. Detectable action units with feature points from section 2

AU	Description	AU	Description	AU	Description
1	Inner Brow Raiser	10	Upper Lip Raiser	20	Lip Stretcher
2	Outer Brow Raiser	12	Lip Corner Puller	24	Lip Presser
4	Brow Lowerer	15	Lip Corner Depressor	26	Jaw Drop

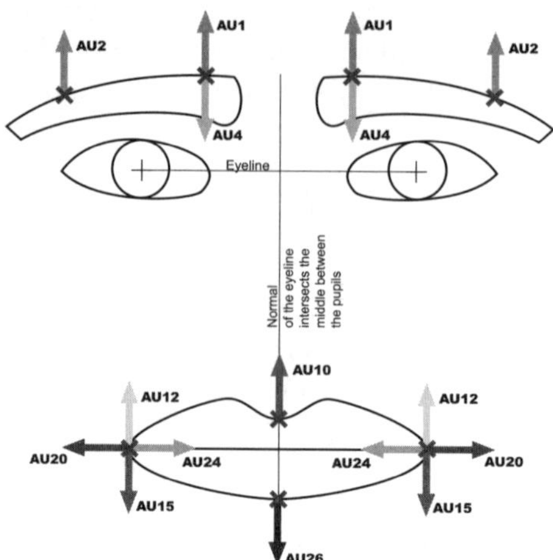

Fig. 3. Detectable action units. The arrow shows the direction a feature point must move to to activate this action unit.

Table 2. Neutral and maximum distances for each action unit extracted from example images of different persons. The reference distance with the value 1 is between the pupils. The action units with * use the vertical distance. If the maximum distance is smaller than the neutral distance then the action unit gets active if the feature point gets closer to the reference line.

AU	neutral	maximum	AU	neutral	maximum	AU	neutral	maximum
1	0,268	0,443	10	0,982	0,875	20*	0,396	0,513
2	0,284	0,446	12	1,056	0,902	24*	0,396	0,276
4	0,268	0,148	15	1,056	1,111	26	1,223	1,714

Ekman shown in Tab. 3 an improvement step has been performed. Using example images of faces showing a specific emotion various combinations and weights have manually been optimized to gain better results. The resulting formulas are also presented in Tab. 3. Using these activity values an emotion has been defined as active when the activity is higher than a heuristically determined threshold $t_{emo} = 0.26$, which has been extracted from the test data set. Additionally an emotion is defined as being dominant if the emotion has the highest activity of all 6 emotions.

To test the emotion interpretation and the emotion detection two tests on the same data set have been performed. The data set has multiple examples of different faces for each of the six basic emotions and the neutral expression. In the first test the feature points have been annotated manually. This test allows a testing of the emotion interpreter without feature detection errors. In the second test we use the feature detection described in section 2 to get the positions of the feature points. Fig. 4 shows the detection rates of the first and second test.

Table 3. Original and adapted action unit combinations to express the six basic emotions. L and R indicate the action units on the left and right side of the face. The emotion sadness for example is said to be active when action unit 1,4 and 15 are active. The adapted version only uses the action units 4 and 15 an normalizes the sum to the interval $[0..1]$.

Emotion	Original Definition	Adapted Definition
Fear	$1 + 2 + 4 + 5 + 20 + 25$	$(1L + 1R + 2L + 2R + 20L + 20R)/6$
Surprise	$1 + 2 + 5 + 26$	$(1L + 1R + 2L + 2R + 26 + 26)/6$
Anger	$4 + 5 + 7 + 24$	$(4L + 4R + 24L + 24R)/4$
Sadness	$1 + 4 + 15$	$(4L + 4R + 15L + 15R)/4$
Disgust	$4 + 9 + 10 + 17$	$(4L + 4R + 10)/3$
Happiness	$6 + 12 + 25$	$(12L + 12R)/2$

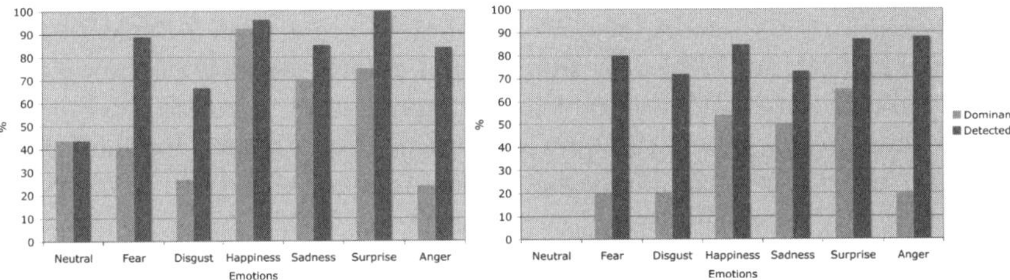

Fig. 4. (left) Emotion detection rates of our emotion interpreter. The feature points are marked manually (right) Emotion detection rates of our emotion detection. The feature points are detected automatically.

As expected the detection rate decreases with the automatic detection of the feature points. The missing neutral facial expression is a result of the feature detection process. Minor localization errors of a single point have a great influence on the activity of an action unit which leads to an unwanted activity of an emotion. These tests show that the emotions happiness, surprise and sadness can be detected correctly in over 70% of the cases. In about 50 % of the cases they are even dominant.

3.1 Integration

For interaction purposes the robot (Fig. 5) must have a description of the communication partner. Therefore it is necessary to generate a description for each person in the environment. The description includes all information which are necessary to react properly. These information include body pose, gaze direction, emotional state and many more. The object descriptions generated in the perception system are then transferred to the motives and the habits of interaction. Further details concerning the control architecture will be given in future publications.

The control of the robot itself uses these models of the communication partner to adapt and modify the behavior of the robot. The emotional state of a

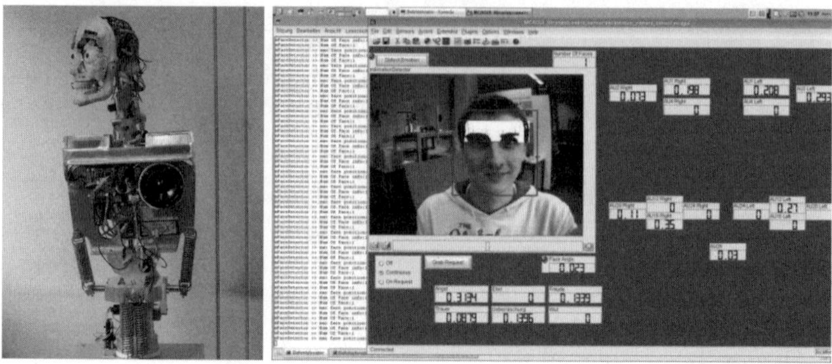

Fig. 5. (left) The humanoid robot ROMAN without clothes and silicone skin (right) Screen shot of the graphical user interface for the emotion detection

communication partner influences the internal emotional state of the robot. All motions and reactions are based on this emotional state. A robot in a emotional state of sadness reacts slower than in the neutral state. The development of these models and the correct reaction of the robot are a current research topic which will be addressed in near future.

4   Conclusions and Future Work

This paper presents a camera-based system for the detection of emotions of a human interaction partner. This system is divided into two main parts the feature detection and the emotion interpretation. In the first part important features in the face are localized with the help of a haar cascade classifier and a contrast filter. Based on the exact position of these features a probability for each of the six basic emotions fear, happiness, sadness, disgust, anger and surprise can be assigned. Final test experiments with reference images show correct detection rates of about 60% for the emotions happiness, sadness and surprise.

For the future it is necessary to detect the feature points with a better accuracy to improve the results of the emotion interpreter. Another aspect is to detect more feature points to describe a larger set of action units. One possible candidate is the nose wrinkler which is influenced by action unit 9. The detection of this action unit will improve the results for the emotion disgust.

References

1. Bartlett, M.: Face Image Analysis by Unsupervised Learning and Redundancy Reduction. PhD thesis, University of California, San Diego (1998)
2. Canzler, U.: Automatische erfassung und analyse der menschlichen mimik. In: Bildverarbeitung fr Medizin 2001. Algorithmen, Systeme, Anwendungen (2001)
3. Chiou, G.I., Hwang, J.-N.: Lipreading by using snakes, principal component analysis, and hidden markov models to recognize color motion video. unknown (1996)

4. Dailey, M.N., Cottrell, G.W.: Pca = gabor for expression recognition. Technical Report CS1999-0629, University of California, San Diego, October 26 (1999)
5. Ekman, P., Friesen, W.: Facial Action Coding System. Consulting psychologist Press, Inc. (1978)
6. Fasel, B., Luettin, J.: Recognition of asymmetric facial action unit activities and intensities. In: Proceedings of International Conference on Pattern Recognition (ICPR 2000), Barcelona, Spain (2000)
7. Fellenz, W., Taylor, J., Tsapatsoulis, N., Kollias, S.: Comparing Template-based, Feature-based and Supervised Classification of Facial Expressions from Static Images. In: Computational Intelligence and Applications. World Scientific and Engineering Society Press, Singapore (1999)
8. Fischer, S., Dring, S., Wimmer, M., Krummheuer, A.: Experiences with an emotional sales agent. In: André, E., Dybkjær, L., Minker, W., Heisterkamp, P. (eds.) ADS 2004. LNCS (LNAI), vol. 3068. Springer, Heidelberg (2004)
9. Kass, M., Witkin, A., Terzopoulos, D.: Snakes:active contour models. International Journal of Computer Vision, 321–331 (1988)
10. Kawato, S., Tetsutani, N.: Detection and tracking of eyes for gaze-camera control. In: VI (2002)
11. James Lien, J.-J.: Automatic recognition of facial expressions using hidden markov models and estimation of expression intensity. Technical Report CMU-R1-TR-31, Robotics Institute, Carnegie Mellon University, Pittsburgh, PA (April 1998)
12. Mase, K., Pentland, A.: Recognition of facial expressions from optical flow. IEICE Transactions (Special Issue on Computer Vision and its Applications) (1991)
13. Rosenblum, M., Yacoob, Y., Davis, L.S.: Human expression recognition from motion using a radial basis function network architecture. IEEE transactions on neural networks 7, 1121–1138 (1996)
14. Soderstrom, U., Li, H.: Customizing lip video into animation for wireless emotional communication. Technical Report DML-TR-2004:06, Department of Applied Physics and Electronics, Umea University, Sweden (2004)
15. Terzopoulos, D., Waters, K.: Analysis and synthesis of facial image sequences using physical and anatomical models. Pattern Analysis and Machine Intelligence PAMI 15(6), 579–596 (1993)
16. Tian, Y.-L., Kanade, T., Cohn, J.: Recognizing action units for facial expression analysis. IEEE Transactions on Pattern Analysis and Machine Intelligence 23(2), 97–115 (2001)
17. Viola, P., Jones, M.: Robust real-time object detection. In: Second International Workshop on Statistical and Computational Theories of Vision, Vancouver, Canada, July 13 (2001)

Simplest Scenario for Mutual Nested Modeling in Human-Machine-Interaction

Rustam Tagiew

Institute for Computer Science, Technical University Bergakademie Freiberg, Germany
tagiew@informatik.tu-freiberg.de

Abstract. The research aim of this paper is to represent everyday-life patterns of thought like "Because I know, what you think I think ..." by a process on a machine, which is involved in an interaction with humans, who also think in such patterns. It analyzes from the perspective of AI the relevant results of such fields as epistemic logic, game theory and psychological investigations of human behavior in strategic interactions. As a result of this interdisciplinary research, a very simple basic interaction scenario and a computer algorithm for managing machine "thoughts" and behavior are designed.

1 Introduction

In this paper, *modeling* denotes the representation of an interacting agent's mind in one's own mind. In some situations it is useful to know a lot about other agents and because of this, advanced agents are expected to model each other. Consider, for example, two agents. If the first agent models the second agent, then he would know, what the second one believes and desires. This implicates that if the second agent is an advanced one, the first has to model how the second models him and so on. They must have nested models. This point was investigated by psychologists, logicians and game theorists.

Theory of Mind (ToM) is a notion from psychology that means a model of another humans mind. Psychologists also use the notion *order* to define the quantity of steps in nested models [10].

Epistemic logic, as a formalism to represent multiagent knowledge, deals also with *ToM reasoning* [5]. Unfortunately, some conditions necessary for unextended epistemic logic like *logical omniscience* and *unbounded computing resources* mismatch real-world agents.

The *game theory* regards in contrast the concept of *rationality* in strategic interactions, called *games* [12]. A rational individual or player is expected to maximize his payoff. The goal in the classical game theory is often to find an *equilibrium*, a solution of a game for rational players with the assumption of common knowledge of rationality of all players. This solution is a combination of involved players strategies, which matches all rational players regardless of which kind of intelligence they are. Interacting rational players have to compute the equilibrium of a situation for hypothetical agents without modeling minds of real opponent's or partners. As example can be mentioned the experiment of [3], where a world-class poker player could not beat significantly a program that used an approximated game theoretical solution without typical for poker

A. Dengel et al. (Eds.): KI 2008, LNAI 5243, pp. 364–371, 2008.

opponent modeling techniques. Some interdisciplinary works like [7] consider nested observing of a situation from the points of view of different players as *ToM reasoning*, which is called in the game theory *iterated analysis* or *backward propagation*. Nevertheless, the classical game theory can not be used solely to model observed human behavior [6, p. 22]. The investigation of human behavior in games, *behavioral game theory*, yields some principles of human ToM reasoning [14].

From the perspective of AI, the motivation for the research goal of this paper can be considered from both possible definitions of AI, where there are *rational* or *human-like systems*. In the rational case, if a machine is designed to interact with humans and is expected to have a good performance, then it must, among other duties, read their minds. It needs to have algorithms for choice of an action with maximal payoff considering mind reading. But one can also define a task, in which the machine is expected to model other humans humanly and behave without strict payoff maximization.

The goal of this paper is to design a very simple basic interaction scenario and an algorithm for an artificial agent, where mutual nested modeling between human and machine take place. A straight utilization of knowledge about human behavior gathered in appropriate literature is the way, how we think to achieve this goal.

2 Involved Fields and Related Works in AI

Due to the fact that an adequate discussion of the results from the involved fields is very extensive, this chapter contains only the most relevant approaches:

1. Humans can accredit desire to a machine as it is shown in the experiment of [11]. In their experiment computer synthesized speech could be understood much better by humans if it was rendered from a predictably acting robot as from a laptop. But it is not clear how similar is a ToM of a machine to a ToM of a human.
2. Humans make with a certain probability irrational decisions, also called errors. So, if a human interacts with other individuals, she or he can make nested assumption of irrationality like $\Box_i(\neg Rational(j))$ or $\Box_j\Box_i(\neg Rational(j))$ or generally $\Box_{\_}^k(\neg Rational(i))$ ($\Box_i(X)$ denotes "i knows X"). The common knowledge of rationality of all involved individuals is not guaranteed in an interaction with humans. Beauty contest game, mentioned in [6, p. 210], is a good example to show that.
3. Computation of optimal strategies in large games like e.g. chess can not be done either by humans nor by machines. Real players need to use suboptimal heuristics.
4. Computation and using of mixed strategies in non-primitive games is problematical for human players, because they can not calculate the exact probabilities [6, p. 121] and can not produce a really unpredictable random sequence [9].
5. The last three points make it clear that human behavior in strategic interactions diverges from equilibria predicted by the game theory. It is of crucial relevance for natural and also artificial agents, who interacts with humans, to model other agents.

[4] have a similar research goal. They designed a multi-agent system for verbalisation of reasoning in a simplified variant of *wise men puzzle* (similar to *muddy children puzzle*). The result was a system that produced verbalized reasoning steps very similar to such of an expert in this puzzle, who was involved in the study. Muddy children puzzle is one of

the introductory examples for epistemic logic and can also be modelled as a game [16]. Investigation of this specific game showed that most people do not reason with nested models deeper than a few steps. People distrust simply in the rationality of other people or others trust in their own rationality and so on (assumption 2). That's why people can not properly conclude the state of the world from others behavior. It was shown that the feature of verbalisation is of crucial importance to prove the achievement of the research goal.

3   Scenario Design

The first question is, wether a mental model of a human could ever be formalized. Formalization of human mental models means that reasoning patterns of a concrete human can be represented mathematically. The first goal is to design a mechanism, which makes humans interested in thinking and behaving in a structured way. The second goal is avoiding of misunderstanding of a situation by interacting humans. Successfully merchandised parlor games satisfy this pattern, and are also expected to provide enough entertainment. A repeated null-sum game with two players, a machine and a human player, can be taken as basic interaction scenario. It must have following features:

1. Stimulation of mutual nested modeling.
2. Absence of risk, because of complex human risk estimation [8].
3. Middle level of complexity, because of avoiding frustration and boredom.

As we saw in the related work, humans would diverge in such games from optimal equilibria. The standard solution for this problem is to use *reinforcement learning*. But this does not satisfy the goal, because we have then only one step of nested modeling.

Another conceivable situation is that the artificial player diverges also from equilibria, but in such way that the human opponent can easy understand his strategy and exploit it. But if the artificial player notices that, he starts to exploit the exploiting strategy of the human player. In this moment, the artificial player has a nested model of the human player, who models the artificial player. Following table shows the upper left corner of the payoff matrix of the fictitious game describing this design:

Human\Machine	0-order	2-order	4-order
intuitive	a ; ≤ 0 $\downarrow$	c ; ≈ 0	c ; ≈ 0
1-order	b ; 1 $\rightarrow$	a ; -1 $\downarrow$	c ; ≈ 0
3-order	c ; ≈ 0	b ; 1 $\rightarrow$	a ; -1 $\downarrow$

$$a > b > c$$

The upper right corner of every cell contains column player's (machine's) payoff and the opposite corner contains the payoff of the row player (human) as the convention of the game theory is. Payoff of human player ranges between win (1) and lose (-1). At first the machine plays a deterministic, evident and exploitable strategy and the human player is playing his own intuitive strategy. Its strategy is deterministic, because a deterministic strategy is easy to be predicted and exploited. The human player is interested only in winning the game and would recognize the strategy of the machine. The machine is interested in building nested models of the human player. The machine has to recognize the strategy of the human player and calculate its fictitious playoffs (a, b or c). Arrows on the table show the drift of strategies of both players. If the human models the 0-order strategy of the machine, he has a 1-order ToM of the machine and so on.

A very good candidate for the required game is the card game *"Pico 2"* (invented by Doris & Frank). This game is inspired by the Knobeln Contest of [13] and is similar to *Silverman's Game* [1]. It is a null-sum two players symmetric extensive game of imperfect information. A round of this game consists of two phases, which are exactly equal besides of reversed hands. There are only 11 cards with numbers 4-13 and 16. Every number is weighted with points ($4 \rightarrow 1, 5\text{-}7 \rightarrow 2, 8\text{-}13 \rightarrow 3, 16 \rightarrow 4$). At the beginning of a phase each player gets 5 cards and the remaining card is kept open on the board and because of this every player can deduce the cards the other player has. In every turn, every player chooses secretly one card and puts it covered on the board. Then they open the cards and the winning card will be discarded. The card with higher number wins, unless the number is more than twice as high as the lower number. The owner of the winning card gets its points. The other card returns to its owners hand. A phase continues till one of the players have only one card. The player with the biggest sum of gathered points from both phases wins.

In "Pico 2", there are 2772 variants of card distributions and only 318 of them have the *subgame-perfect equilibrium* (SPE) in pure strategies. To play optimally "Pico 2", one must calculate mixed strategies with foresight and product a sequence of random choice, what is really hard to be done by humans. Fig.1(left) shows an example of *mixed-strategy equilibrium* (MSE) for the first turn at the beginning of a game.

The first point in algorithm design is choosing an evident and exploitable default strategy for the artificial player. Foresight is really hard in this game, even for subjects who played it a long time. The solution is to choose a *myopic* deterministic strategy. Myopic means that only the payoff (points) of one turn is considered. Fig.1(right) is a myopic payoff matrix for one turn, where rows and columns are assigned with the cards of the human player and machine respectively. In empirical investigations of one-shot

Hand 1	Nr.	4	5	6	7	11
	P	.067	.362	.287	.055	.288
Versus						
Hand 2	Nr.	8	10	12	13	16
	P	.18	.19	.283	.093	.254

H. \ M.	5	6	10	12	
4	-2	-2	1	1	
7	2	2	-3	-3	
8	3	3	-3	-3	D=MBR(C)
11	-2	3	3	-3	
13	-2	-2	3	3	B=MBR(A)
	C=MBR(B)		E=MBR(D)	A=MSM	

Fig. 1. Examples for MSE and myopic payoff matrix

3×3 matrix games [14], 24% of humans use the *sum maximalisation heuristic*, 49% think that the opponent uses sum maximalisation heuristic and only 27% use Nash equilibrium. Sum maximalisation is an exploitable and psychologically proven human pattern of thought in matrix games. The required deterministic default strategy is defined in following way, called further deterministic myopic sum maximalisation (DMSM): 'Behold the myopic payoff matrix from your point of view and choose the smallest card number of alternatives with highest minimum payoff, which are chosen from alternatives with maximum sum'. This strategy contains *risk aversion* ('highest minimum') for additional choice restriction, which is also a psychologically proven pattern [8]. The cells of this table are filled with points of the discarded cards from the point of view of the row player. The payoffs of the column player are the payoffs of the row player multiplied with -1. The card 12 is chosen by DMSM.

The evaluation of the $2 * 2772$ games of DMSM against SPE shows in Pico 2 that this strategy is only about $16,5\%$ worse ($\frac{\text{difference of wins}}{\text{played games}} * 100\%$), whereas the random strategy is about 67% worse. As the exploiting strategy is suggested myopic best response (MBR). Best response on sum maximalisation is the strategy, which was used by 49% of the subjects in the experiment of [14]. This strategy suggests multiple cards, which all can achieve best payoff on the chosen card of the machine beholding the myopic payoff matrix. If the human player is a Pico 2 master and has a choice between using SPE and MBR, he would choose the last, because it will cause a winning probability of nearby 100%. The exploiting strategy of the machine on MBR of the opponent is called deterministic MBR (DMBR) and is an application of DMSM on lines (rows or columns) chosen by own MBR. DMBR is just a predictable MBR. Fig.1(right) shows several applications of defined response strategies. The card 12 has the highest sum of payoffs for the machine and it is chosen by DMSM. The card 13 is the best response of the human player on the card 12. Then, if the machine uses MBR, it would chose 5 or 6 as best response on 13. But 5 has a higher sum and is predestined to be chosen by DMBR.

The definition of the algorithm for the artificial player is depicted as a state transition diagram on the following diagram:

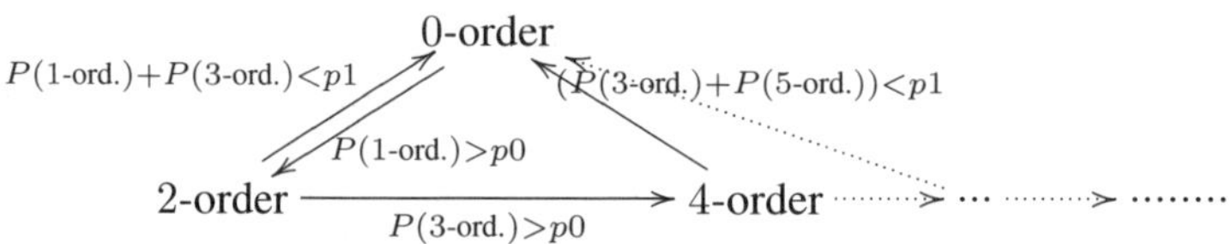

In the default state, annotated as 0-order, the artificial player plays DMSM and is only ready to recognize the intuitive strategy and the exploiting strategy, annotated as 1-order. The intuitive strategy is modelled as a linear combination of myopic sum maximalisation (MSM) and random choice. As a variant, SPE can be added to this linear combination analog to the nash equilibrium in the experiment of [14]. In each other state (depicted as dots), annotated as X-order, the artificial player uses X-order strategy and can recognize the intuitive, $(X - 1)$-order and $(X + 1)$-order strategies.

The rules of transitions on this diagram are designed using the principle: 'If human player exploits you, exploit him and if he plays the intuitive strategy, play the default strategy'. There is an arrow from each state to the default state, if the probability of opponent playing intuitive strategy is too high and an arrow from each state to the next higher state, if the opponent exploits.

For the computation of probabilities for each strategy, Bayes' theorem is used with equal a priori probabilities for all strategies:

$$P(\text{Str}_a \mid \text{Data}) = \frac{P(\text{Data}|\text{Str}_a)}{\sum_{k=1}^{h} P(\text{Data}|\text{Str}_k)} \ , \text{Str} - \text{strategy} \ , h - \text{quantity of strategies}$$

Data denotes a list of observations with fixed length, on which the probabilities $P(\text{Data} \mid Str_a)$ are calculated:

$$P(\text{Data} \mid Str_a) = \prod_{j=1}^{g} p_{aj} \ , g - \text{memory size}$$
$$p_{aj} - \text{Probability of an observation in the example j to be produced by strategy a}$$

The probabilities of the observations to be produced by different strategies are defined under the consideration of human inaccurateness. If a human player chooses a strategy, he does not play it exactly - sometimes he makes errors. The definition of the required probabilities is derived from the *quantal response equilibrium* (QRE) [6, p. 33]

$$P(s_i) = \frac{e^{\lambda \sum_{s_{-i}} P(s_{-i})u_i(s_i,s_{-i})}}{\sum_{s_k} e^{\lambda \sum_{s_{-i}} P(s_{-i})u_i(s_k,s_{-i})}}$$

assuming a 5% (human error rate) probability of choosing one of the wrong cards and 95% probability of choosing one of the correct cards. The utilities u_i for wrong and correct are 0 and 1 (draw and win) and the opponent card probability $P(s_{-i})$ is set to 100% for the deterministically chosen card. The derived solution is the following formula:

$$p_{aj} = P(s_i|R(Str_a)) = \begin{vmatrix} \dfrac{19}{n + m * 19}, if & s_i \in R(Str_a) \\ \dfrac{1}{n + m * 19}, if \ \neg(s_i \in R(Str_a)) \end{vmatrix}$$
$$m - \text{Quantity of cards selectable by a} \ , n - \text{Quantity of cards unselectable by a}$$
$$R(Str_a) - \text{The cards selectable by a} \ , s_i - \text{The card chosen in observation j}$$

Card 13 on Fig.1(right) as response to card 12 can be chosen by the human player with probability of about $82,6\%$. For the random strategy, all probabilities are equal.

One important detail in the algorithm design is the exclusion of observation examples, which can not be used to calculate in reasonable way opponent's strategy, if $n = 0$, dominant columns exist or the combination of the chosen row and the chosen column is a Nash equilibrium. An evaluation of 27720 games shows that there are in middle ≈ 6.3 useful observation examples in a match [15].

The memory size is set after some test runs against computer simulated human strategies in the default state to 9 and in other states to 11. These numbers are a compromise between proper recognition and ability for quick change of behavior. The constants $p0$ and $p1$ from transition diagram are set in same way to 0.96 and 0.0005. At the beginning and after every transition, the memory of the artificial player is empty. The algorithm does not make any transition before its memory is full.

4   Implementation

The implementation uses a state of the art client-server architecture with a user friendly GUI. It can be downloaded from author's homepage. It has a functionality of managing computer-computer, human-computer and human-human network matches and also an output of machine reasoning key points. The algorithm from the previous section is integrated in this environment and can be tested against other player types (SPE, Random and so on). Network gaming can be used for gathering data of human behavior.

As is described in the previous section, the algorithm has a memory of recent observations. This memory can be printed as a matrix in the output of the program:

```
[0.2     0.2     ...  0.2    0.2    0.2  ]
[0.043   0.024   ...  0.024  0.043  0.043]
[0.826   0.826   ...  0.826  0.043  0.043]
[0.826   0.826   ...  0.826  0.043  0.826]
```

The lines of this matrix while using X-Order strategy are associated with strategies of human player in following order: Random, (X+1)-Order, MSM and (X-1)-Order. The strategies of human player can be calculated on this matrix and also printed on the output:

```
> You play with probability 0.3 % Random.
> You play with probability 0.0 % 3-order.
> You play with probability 0.04 % MSM.
> You play with probability 99.66 % 1-order.
```

The ToM reasoning of the machine is also printed on the output in following way:

```
> The card proposed by DMSM is: 13.
> Your respond: 5
> So, I prefer card 10.
```

5   Conlusion

While designing the scenario and the algorithm we made a lot of assumptions. These assumptions are based on previous work and still need to be proven. We can specify here a preliminary list of points of criticism for further research:

- How understandable is the initial strategy of the machine?
- Is the design of strategy recognition adequate?
- What, if the human player is playing intuitively the best respond strategy without really understanding it?
- How easily is the performing of the best response strategy, if the human player understands the initial strategy?
- Does the estimation of the human error rate suit the reality?

In an econometical study subjects earn money if they play well and win respectively. It makes gathering lots of data for solid results expensive. The other possibility for gathering data is tracking of self-motivated users of the WWW. Another detail is the distinguishing between human behavior showed in the interaction with another humans or with artificial agents. Humans could also be misinformed about what or who they are interacting with, if the interaction takes place over a medium like internet. In human-machine interaction can be considered a combination with an embodied agent, who provides some emotional feedback [2]. This emotional feedback is correlated to actual state and distribution of probabilities similar to a human without perfect pokerface.

This paper showed a way, how to design an interaction scenario for mutual nested modelling between human and machine. It provides a brief but comprehensive overview of terms, involved fields and state of art. It introduced a running program for the designed scenario and considerations for an empirical study.

References

1. Heuer, A.G., Leopold-Wildburger, U.: Silverman's Game. Springer, Berlin (1995)
2. Becker, C., Prendinger, H., Ishizuka, M., Wachsmuth, I.: Evaluating affective feedback of the 3d agent max in a competitive cards game. In: Tao, J., Tan, T., Picard, R.W. (eds.) ACII 2005. LNCS, vol. 3784, pp. 466–473. Springer, Heidelberg (2005)
3. Billings, D., Burch, N., Davidson, A., Holte, R., Schaeffer, J., Schauenberg, T., Szafron, D.: Approximating game-theoretic optimal strategies for full-scale poker. In: IJCAI 2003 (2003)
4. Brazier, F.M.T., Treur, J.: Compositional modelling of reflective agents. Int. J. Hum.-Comput. Stud. 50(5), 407–431 (1999)
5. Fagin, R., Halpern, J., Moses, Y., Vardi, M.: Reasoning about Knowledge. The MIT Press, Cambridge (1995)
6. Camerer, F.C.: Behavioral Game Theory. Princeton University Press, New Jersey (2003)
7. Hedden, T., Zhang, J.: What do you think I think you think?: Strategic reasoning in matrix games. Cognition 85(1), 1–36 (2002)
8. Kahneman, D., Slovic, P., Tversky, A.: Judgement under uncertainty - Heuristics and biases. Cambridge University Press, Cambridge (1981)
9. Kareev, Y.: Not that bad after all: Generation of random sequences. Journal of Experimental Psychology: Human Perception and Performance 18(4), 1189–1194 (1992)
10. Mol, L., Verbrugge, R., Hendriks, P.: Learning to reason about other people's minds. In: Hall, L., Heylen, D. (eds.) Proceedings of the Joint Symposium on Virtual Social Agents. The Society for the Study of Artificial Intelligence and the Simulation of Behaviour (AISB), pp. 191–198 (2005)
11. Ono, T., Imai, M.: Reading a robot's mind: A model of utterance understanding based on the theory of mind mechanism. In: Proceedings of the Seventeenth National Conference on Artificial Intelligence and Twelfth Conference on Innovative Applications of Artificial Intelligence, pp. 142–148. AAAI Press / The MIT Press (2000)
12. Osborne, M., Rubinstein, A.: A Course in Game Theory. MIT Press, Cambridge (1994)
13. Prechelt, L.: A motivating example problem for teaching adaptive systems design. SIGCSE Bull. 26(4), 25–34 (1994)
14. Stahl, D.O., Wilson, P.W.: Experimental evidence on players' models of other players. Journal of Economic Behavior and Organisation 25, 309–327 (1994)
15. Tagiew, R.: Gegenseitiges geschachteltes Modellieren in der Interaktion mit einem künstlichen Agenten. Diplomarbeit, Universität Bielefeld (2007)
16. Weber, R.: Behavior and learning in the dirty faces game. Experimental Economics 4(3), 229–242 (2001)

Bayesian Network for Future Home Energy Consumption

Atsushi Takahashi[1], Shingo Aoki[1], Hiroshi Tsuji[1], and Shuki Inoue[2]

[1] Graduate School of Engineering, Osaka Prefecture University,
1-1 Gakuentyo, Nakaku, Sakai, Osaka, Japan, 599-8531
`takahashi@mis.cs.osakafu-u.ac.jp,`
`{aoki,tsuji}@cs.osakafu-u.ac.jp`
[2] Department of Energy Use R&D Center The Kansai Electric Power Co., inc.,
11-20, Nakoji 3-cheme, Amagasaki, Hyogo, Japan, 661-0974
`inoue.shuuki@d4.kepco.co.jp`

Abstract. In order to measure purchase preference on future electronic equipments, this paper proposes a method using Bayesian Network that considers causal relationships for questionnaire data. The proposed method is based on three layers network where the causal relationship is sequentially chained by personal attributes, senses of value and future services. Then, there is a problem that the probability value sensitivity becomes fewness when a model becomes complicated even if the evidence for same nodes is assigned. To overcome this problem and to perform probabilistic inference, this paper proposes stepwise assignment for a variety of evidence. As a result, the purchase preference such as "persons who make much of safety prefer products for an inventory management in their house" is found.

Keywords: Bayesian Network, Causal Relationships, Sensitivity Analysis, Future Home Energy Consumption, Preference Mining.

1 Introduction

The idea of "Smart House" which aims at getting a more comfortable environmental space was suggested twenty years ago [1]. Since then, a variety of future service images were described as a context of smart house and the basic technology for realizing them has been developed as seeds of the product providers. However, there are varieties of people who are different from each other in senses of value. Therefore, it is necessary to find out the consumers needs as well as the technical seeds.

To measure consumers' purchase preference, marketing survey with a questionnaire beforehand is often carried out. There are a lot of analytical methods for questionnaire data [2] [3], for example, the cross-tabulation or Bayesian Network (BN) [4] [5]. The cross-tabulation is a simple way to find out the relations among data on small sizes of axes, and BN is powerful to find out the sensitivity of causal relation. However, the cross-tabulation can not work when there are large sizes of axes for the data, and BN might indicate only small change for the probability in the case that the model has many layers when the evidence is assigned.

A. Dengel et al. (Eds.): KI 2008, LNAI 5243, pp. 372–379, 2008.

This paper discusses how to use BN for measuring purchase preference under the multi-layers model. To exaggerate sensitivity of the causal relation, this paper proposes stepwise assignment for a variety of evidence. This preference structure consists of three layers: personal attribute, sense of value and future home service. First we will describe the problem. Next we will show the proposed technique. Then, we will describe the simulation results for the collected samples of 1,030 consumers. Finally, the concluding remarks will be discussed.

2 Problem Description

2.1 Future Home Energy Consumption Assessment

The questionnaire consists of three classifications: personal attribute, senses of value and future home services. The responses are collected by internet search provided by a marketing research company.

We suppose that there are six kinds of senses of value (Comfort, Security, Health, Ecology, Convenience and Housework), and six future home services (Inventory Control Service etc.). At first, the responders are asked their personal data (Gender, Occupation, Age, etc.). The items and the examples of Personal Attribute are shown in Table 1. Next they are asked priority between senses of value by fifteen paired ($_6C_2$) comparison questions to place in the order. Then they are asked about future home services. An example of service visualization for inventory control service is shown in Fig. 1. Each service has at least four electronic equipments with future technology, and total number of products is forty-nine.

Table 1. Overview of Personal Attributes

Item	Example							
	1	2	3	4	5	6	7	8
Gender	Male	Female						
Occupation	Agriculture, Student	Self-Employed, Disemployment	Operation Work	Desk Job	Executive Officer	Specialist Personnel	Homemaker	
Age	20's	30's	40's	50's	Over 60			
Area	Kanto	Tyubu	Kinki					
Family Constitution	Living Alone	Couple No Children	Couple, Child (Living separetely)	Couple, Single Child (Under 20)	Couple, Single Child (Over 20)	Couple, Married Child	More than 3 Generations	Others
Relation	Householder	Spouse	Child	Others				
House Type	City Single House	Suburb Single House	Country Single House	City Apartment	Suburb Apartment	Country Apartment		
Energy	Gasification, Electrification (Private Power Generation)	Gasification, Electrification	All Electrification (Private Power Generation)	All Electrification	Others			

Taking care of the structure of questionnaire, there might be causation among their three classifications. This is because the preference on the home equipment shall be strongly dependent on people's sense of value based on their demography. For example, while some people make a point on convenience for purchasing electronic equipments, others do on health for them. Then we need to choose the suitable analysis technique.

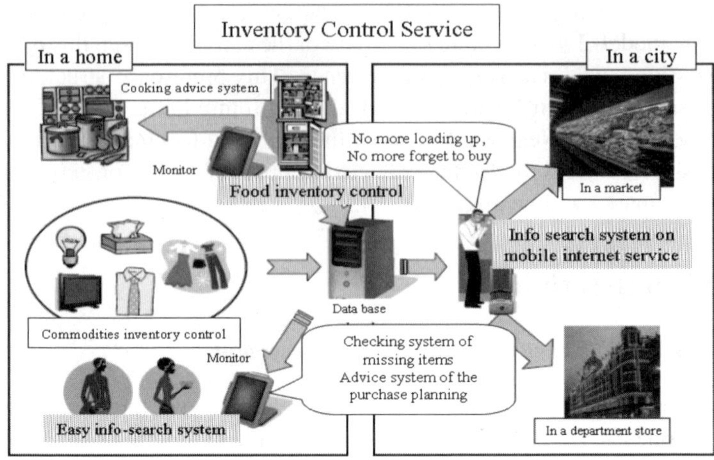

Fig. 1. An example of future home service

2.2 Problem for Appling Bayesian Network

Bayesian Network (BN) is a graphical model which represents quantitative dependencies among the nodes which consist to set of random variables [4] [5]. Nodes are connected by direct links. The intuitive meaning of the link from node X to node Y is that X has a direct influence on Y. Another way to say, there is a causal relationship between X and Y.

Because each connected pair has a conditional probability table that quantifies the effects, BN can compute the posterior probability distribution of the set of query variables, when the evidence variables are assigned. It is called probabilistic inference which allows understanding which nodes of the model have the greatest impact.

To construct a BN model, a rough causal relationship is assumed beforehand. This paper assumes the causal relationship sequentially chained by (I) personal attribute, (II) sense of value and (III) future home service. In this view, we construct the same structure of BN by service. The constructed model for inventory control service is presented in Fig. 2. The structure of personal attributes in this model is semi-automatically constructed by commercial software called BayoNet [6].

Using this model, it is possible to measure consumers' preference as assigning the evidence to the one of the bottom nodes. However, the deviations between prior and posterior probabilities are too small to find the preference. This is because that this model has five layers as shown in Fig. 2, and probabilities are calculated by nearer node. These cause the deviations of the probability getting fewer in the remote layer on the model. Therefore, it becomes difficult to get the knowledge following causal relationships. To clarify the deviations, this paper proposes a stepwise assignment. The proposed technique is taken up in next chapter.

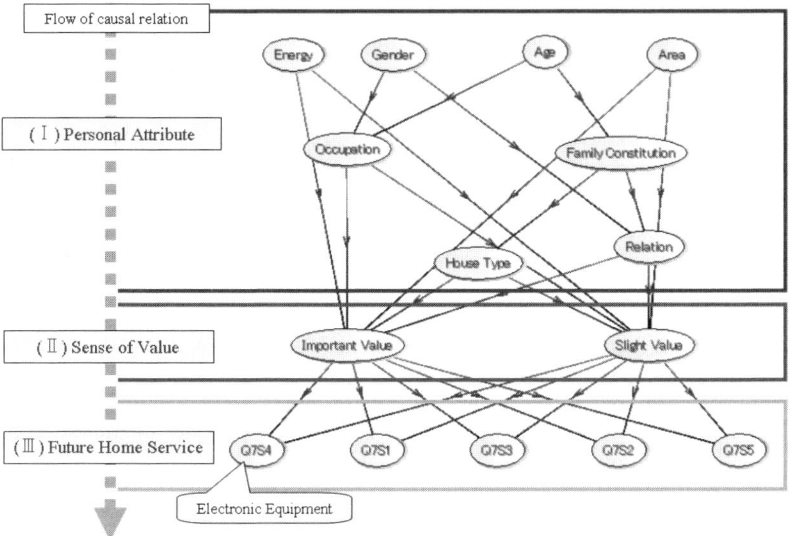

Fig. 2. Constructed BN Model

3   Stepwise Assignment

In order to improve precision of the probabilistic inference, a learning method based on additional data has been made [7]. This method does not change graph structure but updates CPT (Conditional Probability Table). The conception diagram is illustrated in Fig. 3.

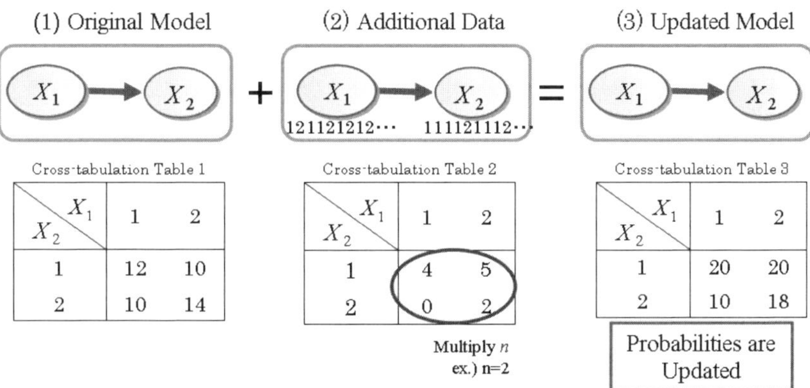

Cross-tabulation Table 1

X_2 \ X_1	1	2
1	12	10
2	10	14

Cross-tabulation Table 2

X_2 \ X_1	1	2
1	4	5
2	0	2

Cross-tabulation Table 3

X_2 \ X_1	1	2
1	20	20
2	10	18

Fig. 3. CPT Learning Method with Additional Data

The original model of CPT is introduced as prior knowledge, and the observed data is introduced as additional data. Fig. 3 shows the case where the additional data confirm about X_1 to X_2, for example, four times $X_2=1$ is observed when $X_1=1$. Hereby the cross-tabulation list for the original model is updated by the additional data. On this occasion, the additional data are often multiplied by "n" as the meaning how this value far makes much, so this magnification "n" rises when additional data are regarded as important. Finally, the new cross-tabulation is normalized in order to be new CPT.

To sum up this, the original model (1) is combined with the additional data (2) and we get the updated model (3). This is an outline of the method. Compared with this conventional technique, the proposed technique is additional data (2) as artificial additional data from the deviations of probability (2)'. One explanation for the parameter which probability is increased by probabilistic inference can be said that the number of the samples gets increased. Therefore, this technique pays attention to the deviations that increase, and find an increment ratio. Then the updated data is considered as new evidence. The procedure is as follows:

<u>Step1)</u> Assigning evidence for target future products, the reasoning is executed.
<u>Step2)</u> Assigning the following evidence for the parent nodes of the target in step1, the reasoning is repeated.

$$x_i = \frac{y_i + n\beta_i}{m + n} \tag{1}$$

x_i : Evidence for parameter i

y_i : Number of sample for parameter i

m : Total number of sample

n : Number of added sample

α : $\alpha = \sum \max(\Delta_i, 0)$

β_i : Rate of change $\left(\beta_i = \Delta_i / \alpha\right)$

Δ_i : Deviation between prior and posterior probability

4   Experimentation

4.1 Analysis Result

We have collected 1,030 samples where there are the same sample size on genders and ages. To measure consumers' purchase preference, as mentioned before, the evidence is assigned to one of the bottom nodes and parameter "Want". (the value of evidence as 1.0) Then BN makes an inference on conditional probabilities. Experimentation is carried out the following procedures by the proposed technique. In addition, the numerical result is based on software "BayoNet" [6].

(a)   Assigning the evidence "Want" for the bottom node,
(b)   Assigning the evidence for the parent nodes "Important Value" and "Slight Value", followed by the formula (1),
(c)   Assigning the evidence for the nodes "Occupation", "Area", "Relation", "House Type" and "Energy" followed by the formula (1).

where m=1,030, n=100 in our experimentation. Note that the first procedure (a) is the conventional technique, while the others are the proposed technique.

One of the analysis results is illustrated in Table 2. This result shows purchase preference on node "Q7S1" for a refrigerator which manages the inside foods status automatically. Note the difference between prior distribution and posterior probabilities which show the rate of persons who want to purchase the special refrigerator.

Table 2. Analysis Result of Probabilistic Inference

Node	Parameter	Prior Probability	Posterior Probability "Who wants the refrigerator?"		
			(a)	(b)	(c)
Gender	Male	0.5000	0.4996	0.5007	0.5043
	Female	0.5000	0.5004	0.4993	0.4957
Occupatin	Agriculture, Student	0.1227	0.1224	0.1195	0.1121
	Desk Job	0.2373	0.2386	0.2481	0.3033
Age	20's	0.2000	0.1999	0.1995	0.1905
	30's	0.2000	0.2002	0.2014	0.2218
Area	Kanto	0.6128	0.6142	0.6213	0.5361
	Tyubu	0.1317	0.1312	0.1269	0.1577
Family Constitution	Couple, Single Child (Under 20)	0.2308	0.2312	0.2331	0.2641
	Couple, Single Child (Over 20)	0.2140	0.2139	0.2128	0.1874
Relation	Householder	0.4885	0.4885	0.4938	0.5126
	Child	0.1482	0.1476	0.1436	0.1351
House Type	City, Single House	0.0946	0.0940	0.0913	0.0861
	Suburb, Single House	0.3498	0.3513	0.3556	0.3744
Energy	Gas, Electricity (Private Power Generation)	0.0367	0.0365	0.0343	0.0340
	Gas, Electricity	0.8889	0.8897	0.8962	0.8980
Important Value	Ecology	0.1445	0.1311	0.1320	0.1436
	Housework	0.1434	0.1551	0.1740	0.1414
Slight Value	Comfort	0.1444	0.1537	0.1500	0.1439
	Health	0.1272	0.0940	0.1160	0.1249
Q7S1	Want	0.5292	1.0000	0.5357	0.5294

Q7S1: A refrigerator which manages the inside foods status automatically.
(a): the conventional technique (b) and (c): the proposed technique.

4.2 Findings and Considerations

When the evidence is assigned to the node "Q7S1", the probabilities of the parameters in the nodes ("Important Value" and "Slight Value") change greatly. This is because these nodes are direct parents. Especially, the probability of person who makes much of the housework increases. That is to say, this equipment tends to be preferred by such persons. On the other hand, the deviations of probabilities are fewness in the layer of Personal Attribute. This is typical phenomena as mentioned before. Based on that, their nodes are located remotely in their layers.

The results (a), (b) and (c) make it clear that this equipment is preferred by the persons in 30's whose sense of value is on Housework, and who works as a desk job and has a single child (under 20). In order to reduce a burden of the housework, this refrigerator is preferred by such a person. The conventional technique is able to find out only the relations between sense of value and purchasing preference. However, from these results by the proposed technique, we can recognize the relation along the causal relation assumed beforehand. In addition, a probability value of "Female" increases by the results of (a) as that of "Male" does by the results of (c). This result makes it clear that this product is preferred regardless of gender.

4.3 Validation

In this section, we consider the validity to the results by the proposed technique. From the results of section 4.2, we notice eight parameters which have the greatest deviations to plus in each node. In order to confirm that these parameters make increase the probability of "Want", we take one as evidence from eight parameters, and assign the value of evidence as 1.0. Likewise, next we take all of these parameters, and assign each variable of them as 1.0. If the probability of "Want" increases, then we can confirm that the result of proposed technique can find out the parameter which the conventional technique can not. Table 3 summarizes the results of probabilistic inference. "Prior" or "Posterior" means the probability of parameter "Want".

Table 3. Results of Validation

Parameters Used for Assigning Evidence	Demand Degree "Want"		
	Prior	Posterior	Deviations
Male	0.5292	0.5298	0.0006
Desk Job	0.5292	0.5320	0.0028
30's	0.5292	0.5297	0.0005
Kanto	0.5292	0.5304	0.0012
Couple, Single Child (Under 20)	0.5292	0.5301	0.0009
Householder	0.5292	0.5300	0.0008
Suburb, Single House	0.5292	0.5313	0.0021
Gas, Electricity	0.5292	0.5296	0.0004
"All"	0.5292	0.5456	0.0164

(*"All" indicates the state that eight parameters are assigned at the same time).

Table 3 confirms that the probability of parameter "Want" increases in the all cases. Especially, the case of parameter "All" has the greatest deviations (0.5292 => 0.5456). This result indicates that the proposed technique find out the parameters correctly.

In addition, about the parameter "All" which have the greatest deviations, we consider the validity from a statistical criterion. We hypothesize that the distribution (=21/31) of "Want" (21samples) concerning the case "All" (31samples) is equal to the distribution (=557/1,030) of that (557samples) concerning the original data (1,030samples). As the result of t-test, t is equal to -0.885. This value rejects the hypothesis in one-sided 0.190.

5 Conclusion

This paper has described Bayesian Networks for constructing causal relationship model on future home energy consumption. To clarify the deviations, we proposed the method on stepwise assigning evidences. These results have led to the conclusion that we could find out the relations among three classification of questionnaire. The future work includes watching the differences by changing the value of parameters of formula (1).

Acknowledgment

The authors would like to express sincere thanks to Mr. Eiji Mimura and Mr. Kazunari Asari for their useful advises and encouragement. The special thanks also due to Mr. Masayuki Nakano and Mr. Masatomo Kojoma who contributed to establish the questionnaire design.

References

1. Smith, R.L.: The Coming Revolution in Housing. Go Courseware (1987)
2. Koretsune, A., Kubo, S., Nakano, M., Tsuji, H., Jinno, Y., Mimura, E.: Causal Analysis for Thermal Comfort Votes. In: IEEE International Conference on Systems, Man & Cybernetics, pp. 1382–1387 (2006)
3. Aoki, S., Mukai, E., Tsuji, H., Inoue, S., Mimura, E.: Bayesian Networks for Thermal Comfort Analysis. In: IEEE International Conference on Systems, Man & Cybernetics, pp. 1919–1923 (2007)
4. Neapolitan, R.E.: Learning Bayesian Networks. Artificial Intelligence. Prentice-Hall, Englewood Cliffs (2004)
5. Russell, S., Norvig, P.: Artificial Intelligence A Modern Approach. Prentice Hall Series in Artificial Intelligence (1995)
6. Bayesian Network Construction System "BayoNet",
   `http://www.msi.co.jp/BAYONET/`
7. Motomura, Y., Iwasaki, H.: Bayesian Network Technology. Tokyo Denki University Press (in Japanese) (2006)

Learning Dance Movements by Imitation: A Multiple Model Approach

Axel Tidemann and Pinar Öztürk

IDI, Norwegian University of Science and Technology
{tidemann,pinar}@idi.ntnu.no

Abstract. Imitation learning is an intuitive and easy way of programming robots. Instead of specifying motor commands, you simply show the robot what to do. This paper presents a modular connectionist architecture that enables imitation learning in a simulated robot. The robot imitates human dance movements, and the architecture self-organizes the decomposition of movements into submovements, which are controlled by different modules. Modules both dominate and collaborate during control of the robot. Low-level examination of the inverse models (i.e. motor controllers) reveals a recurring pattern of neural activity during repetition of movements, indicating that the modules successfully capture specific parts of the trajectory to be imitated.

Keywords: Cognitive Modeling, Imitation Learning, Neural Networks.

1 Introduction

Learning by imitation is regarded as a cornerstone of human cognition. Humans can easily transfer motor knowledge by demonstration, making it a natural way for humans to program robots. The research presented in this paper focuses on making a simulated robot learn human dance movements, using visual input to guide the motor system of the robot. A modular connectionist architecture designed for motor learning and control by imitating movements was implemented. The modules work as experts, learning specific parts of the movement to be imitated through self-organization. The research agenda is to understand how and why such decompositions occur. These decompositions can be considered as motor primitives. How these motor primitives are represented and combined to form complex movements is examined. The architecture has a core of multiple paired inverse/forward models, implemented by Echo State Networks.

2 Imitation in Psychology, Neuroscience and AI

Developmental psychologists have extensively studied how infants learn by imitating adults. Piaget sees imitation as a continuing adaption of motor and perception stimuli to the external world [1]. Meltzoff and Moore propose an architecture that combines production and perception of actions (including a self-correcting

A. Dengel et al. (Eds.): KI 2008, LNAI 5243, pp. 380–388, 2008.

ability) called *active intermodal mapping* (AIM) [2], believing imitation learning is an innate mechanism. Some neuroscientists believe that mirror neurons form a neurological basis for imitation [3]. Mirror neurons are characterized by firing both when observing and producing the same action, and are considered to form the neural basis of imitation [4], language [5] and mind reading [6]. Research on imitation learning in AI is coarsely divided in two groups: solving the *correspondence problem* (i.e. the transformation from an extrinsic to an intrinsic coordinate system) or focusing on the perception-action system (perceptual stimuli has already been transformed to an internal representation) [4]. Schaal regards *model-based learning* (the focus of this brief background section) to be the most interesting approach to implement imitation learning. Model-based learning is achieved by pairing an inverse model (i.e. controller) with a forward model (i.e. predictor). Demiris [7] and Wolpert [8] use a model-based approach in their architectures to implement imitation learning. Wolpert argues that multiple paired inverse/forward models are located in the cerebellum [9]; thus this approach is suitable for architectures inspired from how the brain works.

3   The Multiple Paired Models Architecture

The architecture (abbreviated MPMA) presented in this paper is designed to equip an agent with imitative capabilities. It consists of multiple paired forward and inverse models, and is inspired by Demiris' HAMMER [7] and Wolpert's MOSAIC [8] architectures. The MPMA seeks to combine the best of both; the consistent inverse/forward model ordering of HAMMER (MOSAIC has a different ordering depending on whether it is recognizing or executing an action), and the focus on self-organization and the use of a responsibility predictor (explained shortly) from MOSAIC. The modular structure of the brain serves as inspiration to create an architecture with multiple models. Wolpert argues that the central nervous system computes internal simulations that can be modeled using inverse/forward models [8]. By spreading motor knowledge across several models the architecture can code redundancy, an important part of robust intelligent systems. Furthermore, it is an approach for motor control that is well understood in the literature [10]. In the following text, the term *module* is used to group three models together: the inverse model, forward model and the responsibility predictor. The MPMA is shown in figure 1. The dashed arrow shows the error signal for all models. The forward model is a *predictor*; it learns to predict the next state $\hat{x}^i_{t+1}$ given the current state x_t and the motor command u^i_t applied to the robot. The error signal is the difference between the predicted and the actual next state.

The inverse model is a *motor controller*; it learns what motor commands u^i_t (issued as *joint angle velocities*) will achieve a desired state x'_{t+1} given the current state x_t. The error signal is based on the difference between x'_{t+1} and x_{t+1}, called the *feedback motor error command*, u_{feedback}. The u_{feedback} is also added to the final motor output. Using the feedback controller is a simple way to pull the system in the right direction when bad motor commands are issued [11],

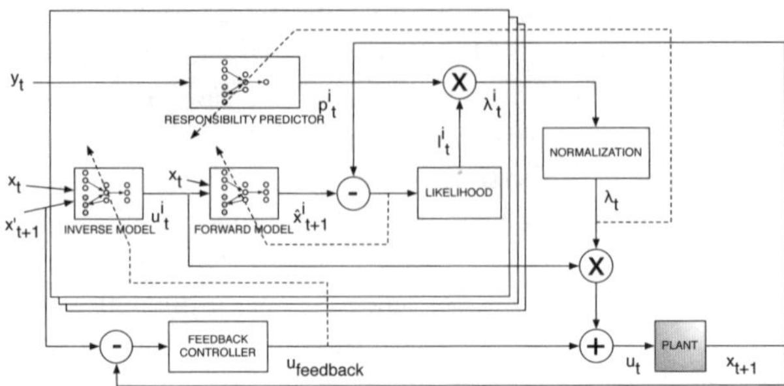

Fig. 1. The MPMA, inspired from [8] and [7]. Details described in the text.

guaranteeing a solution when training an inverse model in a redundant system. Additionally, it increases the robustness of the system by acting as an online correction controller. The responsibility predictor (RP) predicts p_t^i indicating how well the module is suited to control the robot *prior* to movement, based on the context signal y_t. The λ_t^i signal is based upon the p_t^i and the likelihood l_t^i, which is a measure of how well the forward model predicts the next state (see [9] for details). The final lambda vector is normalized, which also serves as the error signal for the RP. The λ_t vector is responsible for switching control of the robot between modules. The motor commands from each module is multiplied with the corresponding λ value before summation and application to the robot. Modules that make good predictions will have more influence over the final motor command. The λ value also gates the learning of each model by multiplying the error signal with the corresponding λ value, promoting modules with good predictions. The desired state was the normalized 3D coordinates of the elbow and wrist position of both arms of the demonstrator. The elbow coordinates were in the range $[-1, 1]$, with the shoulder as the origo. The wrist position was in the range $[-1, 1]$, with the elbow as origo. The state of the robot was defined in the same way to overcome the correspondence problem [12]. Findings in neuroscience anticipate a geometric stage where sensory input is transformed to postural trajectories that are meaningful to the motor system of the observer [13]. A simpler approach has been taken in previous work [14,15], using *joint angles* as input instead of 3D coordinates. The modules use the context signal to determine which module is more suitable to control the robot prior to movement (i.e. when lifting a cup, the context full/empty determines the appropriate inverse model). Since a dancer must continuously listen to the music while dancing, it is chosen as the context signal in the current experiment.

4 Experimental Setup

The experiment consisted of imitating the dance to the song YMCA by The Village People (see figure 2). The dance was chosen since it is rather well-known and easy to explain verbally, and sufficiently complex to make it an interesting

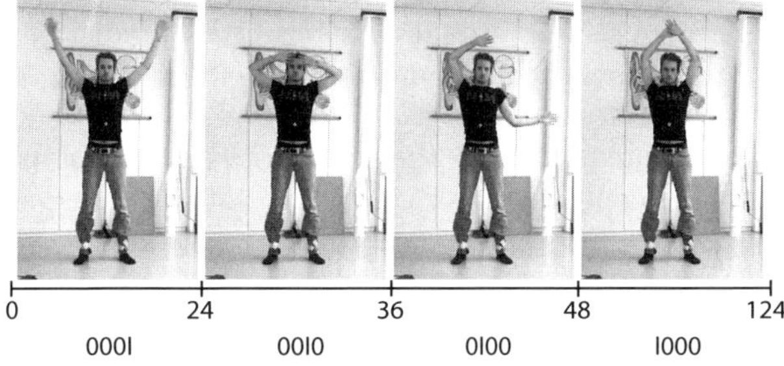

Fig. 2. Dancing the YMCA, by forming the letters $Y\ M\ C\ A$ using arm movements. The numbers show at which timestep the next letter is formed in the sequence. The four-digit vector is the context signal. This movement was repeated three times.

imitation task. Movement data was gathered with a Pro Reflex 3D motion tracking system, where markers were put on the shoulder, elbow and wrist of the dancer. The noisy data sampled at 20Hz was used as the desired state to the MPMA, forcing the models to predict 0.05 seconds into the future. The recorded YMCA movement was repeated three times and added small amounts of noise (1%) during training. Since the experiment was to imitate arm movements, a four degree of freedom model of a human arm was implemented [16]. The simulated robot was thus described by 8 degrees of freedom. To implement the inverse/forward models and the RP, Echo State Networks (ESNs) were used [17], exploiting the massive memory capacity and fast training algorithm.

Each inverse model had 24 inputs: 12 signals for the current state corresponding to the 3D coordinates of the elbow and wrist of each arm, same for the desired state. The 8 outputs with range $[-1, 1]$ corresponded to the degrees of freedom of the robot. There were 20 inputs for each forward model: 8 signals from the inverse model output (i.e. the suggested motor command), and 12 for the current state. There were 12 output signals (range $[-1, 1]$) to predict the next state. The RPs had four inputs from the context signal, with a single output in the range $[0, 1]$. Performance of the system was evaluated with different number of nodes in the hidden layer. All networks had spectral radius $\alpha = 0.1$ (range $[0, 1]$, determining the length of the memory with increasing α) and noise level $v = 0.2$ (effectively adding 10% noise to the internal state of the network).

Good error signals are crucial to ensure convergence in such a high dimensional system. An advantage of using the arm model in [16] is that joint angle rotations can be found analytically from positions of the elbow and wrist. The difference in desired and actual state can thus be expressed as differences in rotational angles, which the feedback controller adds to the final motor command to pull the system in the right direction. In the early stages of training, the u_{feedback} gain K was stronger than the output gain L of the inverse models (i.e. $K = 1$, $L = 0.01$) to force the system towards the desired trajectory. L was increased and K decreased linearly with increasing performance of the system, until $L = 1$ and $K < 0.15$. There were two stopping criteria: the prediction error p_i^t with respect to λ_i^t had to be less than 3%, and the actual trajectory could not differ more than 3% from the desired trajectory. The architecture had four modules. It was the designer's intention that the modules would decompose the movement according to the context signal, coinciding with the melody playing (see figure 2). The system was implemented in MatLab.

5   Results

Four different sizes of hidden layer were examined (50, 100, 200 and 400 nodes). Each of the network configurations were run 20 times. Figure 4 shows the close match between the desired and actual trajectory typical for the experiments. Table 1 shows the performance of the different runs. The $\Sigma u_{\text{feedback}}/\Sigma u_t$ ratio indicates to what extent the feedback controller influenced the total motor command at the last epoch. Being slightly less than $1/4$ on average, it shows that the modules control most of the total motor output; however online corrections are needed to ensure robustness. Table 1 also shows how many modules were active on average during each of the letters. Being active was defined as $\lambda_t^i > 0.1$ for at least 25% of the context signal. This eliminates small bursts of activity from the modules but includes persisting small contributions. The average active modules tell to what extent the modules would dominate or collaborate during control of the movement. Figure 3a gives an example of how the modules collaborate and compete when controlling the robot, and how the architecture is capable of decomposing the movement into smaller submovements. Figure 3b shows the motor output of the inverse models of each module. This figure allows for a visual inspection of *what* each module does, complementing the average number of active modules. Table 1 indicates that 200 nodes might be the best configuration with respect to the number of training epochs and the $\Sigma u_{\text{feedback}}/\Sigma u_t$ ratio, however a performance-wise a clear preference cannot be made.

Table 1. The p_e tells how much the actual state deviated from the desired state (in percent) at the final epoch. The $\Sigma u_{\text{feedback}}/\Sigma u_t$ ratio is an indication of how strong the feedback error motor signal was relative to the total motor command at the last epoch. How many modules were active during each letter show how often modules would collaborate or dominate when controlling the robot. The μ and σ are parameters of the normal distribution.

Nodes in hidden layer	Epochs μ/σ	p_e μ/σ	$\Sigma u_{\text{feedback}}/\Sigma u_t$ μ/σ	Y	M	C	A
50	18/3.78	2.76%/0.23%	0.23/0.02	1	1.6	2.7	1.05
100	17/5.07	2.73%/0.27%	0.24/0.09	1	1.95	2.25	1.05
200	14/3.89	2.76%/0.17%	0.20/0.02	1.05	1.8	2.25	1
400	15/4.71	2.79%/0.16%	0.23/0.06	1	2.15	2.65	1.05

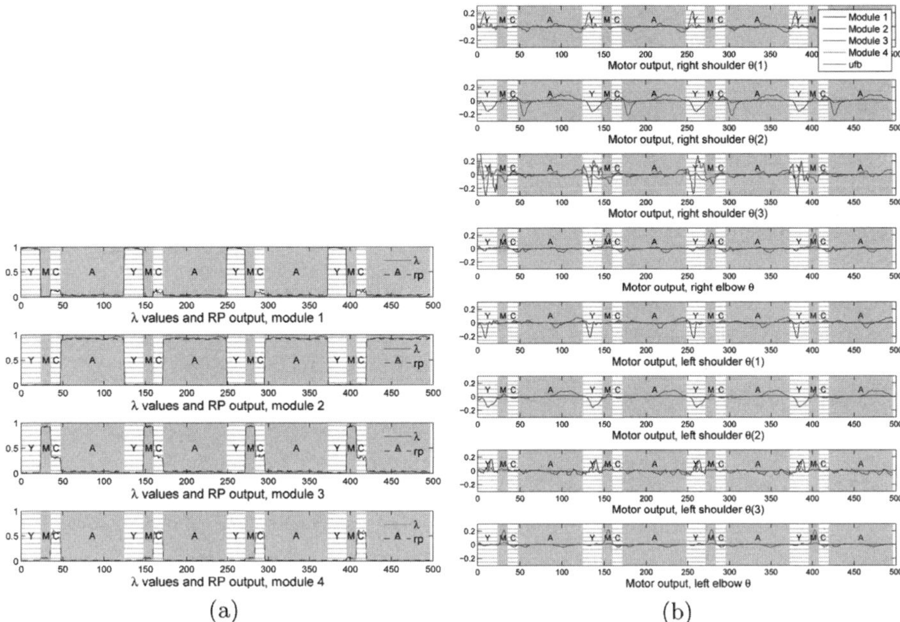

Fig. 3. (a): Collaboration and competition between modules controlling the robot (400 nodes), showing that the MPMA is capable of self-organizing the decomposition of a movement into smaller submovements. Overlapping RP output and λ indicate stability, since the RP correctly predicted how much it would influence the control of the robot. The gray background (stripes/fill) along with the corresponding letters shows the boundaries of the context signal. (b): The recurring patterns of motor output activity sent to the robot indicate that the modules have successfully captured parts of the movement to be imitated. The specific details of which modules captures what is subordinate to the fact that the modules have repeating activation patterns.

6   Discussion

The results reveal how each module becomes specialized on certain parts of the movement. When the movement to be imitated is repeated, the modules dominate and collaborate in controlling certain parts of the movement in accordance with the repetition of the movement (figure 3a). This is also the case on the neural activity layer, as low-level examination of the output of the inverse models reveal the same repeating pattern of activation (figure 3b). Note that the focus is not on what was learned where; the emphasis is examining and understanding the self-organizing capabilities of the architecture.

For all experiments, most collaboration occurs when there is a break in symmetry of the movement, i.e. at the letter C (see table 1). Y, M and A are symmetrical. It should be noted that there is more collaboration during M than for Y and A; this could be due to introduction of control of the elbow joints, whereas Y and A are controlled mostly by the shoulder joints (see figure 4).

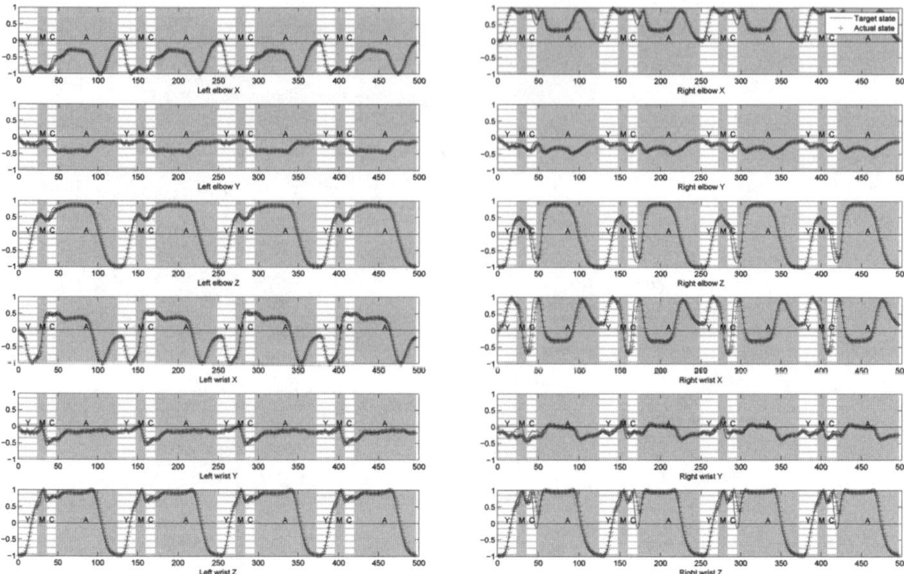

Fig. 4. Desired trajectory versus actual trajectory, same experiment as figure 3. The actual trajectory differs less than 3% from the desired trajectory.

The lack of symmetry could be a reason why the letter C is harder to learn, and therefore more modules collaborate to control the robot. This is backed up by neuroscientific findings, where nonsymmetric action in bimanual movements interfere and takes longer time to execute than symmetrical movements [18]. [19] argues for the presence of both shared and separate motor codes within the brain, explaining the varying degrees of coupling observed in neural activity when performing symmetric (strong correlation) and asymmetric (weak correlation) movements. This model accounts for the observation of active modules in the MPMA with respect to symmetry and complexity as discussed above.

There tends to be a switch between modules that is fairly coincident with the context signal, regardless of domination or collaboration between modules. The design of the context signal indicates *where* the division should be, but it is the modules that determine *how* to represent these motor primitives. Regarding the representation of the motor primitives, our results exhibit an encoding fashion resembling d'Avellas muscle synergy notion [20], where the representation is grounded on muscle or joint synergies, not in a single module. Observing these phenomena in the MPMA which is inspired from how the brain works is an indication that the architecture inhibits certain desirable properties (keeping in mind its limited scope and complexity compared to the brain). We believe it is a good starting point for modeling the important cognitive function that is imitation learning.

7   Future Work

Future work will investigate how the MPMA scales with increasing lengths and complexities of movements to be imitated, to see if there are saturation points in the architecture where more neural resources must be added. Another focus is investigating how the MPMA captures attractors through self-organization [21], along with methodology for evaluating which modules capture what in the motor control space.

References

1. Piaget, J.: Play, dreams and imitation in childhood. W. W. Norton, New York (1962)
2. Meltzoff, A.N., Moore, M.K.: Explaining facial imitation: A theoretical model. Early Development and Parenting 6, 179–192 (1997)
3. Rizzolatti, G., Fadiga, L., Gallese, V., Fogassi, L.: Premotor cortex and the recognition of motor actions. Cognitive Brain Research 3, 131–141 (1996)
4. Schaal, S.: Is imitation learning the route to humanoid robots? Trends in Cognitive Sciences 3(6), 233–242 (1999)
5. Arbib, M.: The Mirror System, Imitation, and the Evolution of Language. In: Imitation in animals and artifacts, pp. 229–280. MIT Press, Cambridge (2002)
6. Gallese, V., Goldman, A.: Mirror neurons and the simulation theory of mind-reading. Trends in Cognitive Sciences 2(12) (1998)
7. Demiris, Y., Khadhouri, B.: Hierarchical attentive multiple models for execution and recognition of actions. Robotics and Autonomous Systems 54, 361–369 (2006)
8. Wolpert, D.M., Doya, K., Kawato, M.: A unifying computational framework for motor control and social interaction. Philosophical Transactions: Biological Sciences 358(1431), 593–602 (2003)
9. Wolpert, D.M., Miall, R.C., Kawato, M.: Internal models in the cerebellum. Trends in Cognitive Sciences 2(9) (1998)
10. Jordan, M.I., Rumelhart, D.E.: Forward models: Supervised learning with a distal teacher. Cognitive Science 16, 307–354 (1992)
11. Kawato, M.: Feedback-error-learning neural network for supervised motor learning. In: Eckmiller, R. (ed.) Advanced neural computers, pp. 365–372 (1990)
12. Nehaniv, C.L., Dautenhahn, K.: The Correspondence Problem. In: Imitation in Animals and Artifacts, pp. 41–63. MIT Press, Cambridge (2002)
13. Torres, E.B., Zipser, D.: Simultaneous control of hand displacements and rotations in orientation-matching experiments. J. Appl. Physiol. 96(5), 1978–1987 (2004)
14. Demiris, Y., Hayes, G.: Imitation as a dual-route process featuring predictive and learning components: a biologically-plausible computational model. In: Imitation in animals and artifacts, pp. 327–361. MIT Press, Cambridge (2002)
15. Tidemann, A., Öztürk, P.: Self-organizing multiple models for imitation: Teaching a robot to dance the YMCA. In: Okuno, H.G., Ali, M. (eds.) IEA/AIE 2007. LNCS (LNAI), vol. 4570, pp. 291–302. Springer, Heidelberg (2007)
16. Tolani, D., Badler, N.I.: Real-time inverse kinematics of the human arm. Presence 5(4), 393–401 (1996)
17. Jaeger, H., Haas, H.: Harnessing Nonlinearity: Predicting Chaotic Systems and Saving Energy in Wireless Communication. Science 304(5667), 78–80 (2004)

18. Diedrichsen, J., Hazeltine, E., Kennerley, S., Ivry, R.B.: Moving to directly cued locations abolishes spatial interference during bimanual actions. Psychological Science 12(6), 493–498 (2001)
19. Cardoso de Oliveira, S.: The neuronal basis of bimanual coordination: Recent neurophysiological evidence and functional models. Acta Psychologica 110, 139–159 (2002)
20. d'Avella, A., Bizzi, E.: Shared and specific muscle synergies in natural motor behaviors. PNAS 102(8), 3076–3081 (2005)
21. Kuniyoshi, Y., Yorozu, Y., Ohmura, Y., Terada, K., Otani, T., Nagakubo, A., Yamamoto, T.: From humanoid embodiment to theory of mind. In: Pierre, S., Barbeau, M., Kranakis, E. (eds.) ADHOC-NOW 2003. LNCS, vol. 2865, pp. 202–218. Springer, Heidelberg (2003)

Ontology-Based Information Extraction and Reasoning for Business Intelligence Applications

Thierry Declerck, Christian Federmann,
Bernd Kiefer, and Hans-Ulrich Krieger

DFKI GmbH, Language Technology Lab, Stuhlsatzenhausweg, 3
66123 Saarbrücken, Germany
{declerck,christian.federmann,kiefer,krieger}@dfki.de

Abstract. In this demo we present the actual state of development of ontology-based information extraction in real world applications, as they are defined in the context of the MUSING European R&D project dealing with Business Intelligence applications. We present in some details the actual state of ontology development, including a time and domain ontologies, for guiding information extraction onto an ontology population task. We then show how the information is stored in the MUSING knowledge repository and how reasoning can act on this repository for generating new knowledge and also applications specific statistical models for supporting decision procedures.

Keywords: Linguistics, domain knowledge, information extraction, ontologies and ontology population, (temporal) reasoning. Business intelligence.

1 Introduction

MUSING is an R&D European project[1] dedicated to the development of Business Intelligence (BI) tools and modules founded on semantic-based knowledge and content systems. MUSING integrates Semantic Web and Human Language technologies for enhancing the technological foundations of knowledge acquisition and reasoning in BI applications. The impact of MUSING on semantic-based BI is being measured in three strategic domains:

- **Financial Risk Management (FRM),** providing services for the supply of information to build a creditworthiness profile of a subject -- from the collection and extraction of data from public and private sources up to the enrichment of these data with indices, scores and ratings;
- **Internationalization (INT),** providing an innovative platform, which an enterprise may use to support foreign market access and to benefit from resources originating in other markets;
- **IT Operational Risk & Business Continuity (ITOpR),** providing services to assess IT operational risks that are central for Financial Institutions -- as a

---

[1] See www.musing.eu for more details.

A. Dengel et al. (Eds.): KI 2008, LNAI 5243, pp. 389–390, 2008.

consequence of the Basel-II Accord – and to asses risks arising specifically from enterprise's IT systems -- such as software, hardware, telecommunications, or utility outage/disruption.

2 The Demo

In the demo we will show how Information Extraction (IE) is now grounded in (domain specific) ontologies, which are now giving a much larger and valid base as the "templates" used in the past in IE applications for guiding the identification and extraction of relevant parts from textual documents. The use of ontologies as the guide for information extraction leads to what we call Ontology-Based Information Extraction (OBIE) is certainly providing for en enrichment of IE, but also introduces a much higher level of complexity in defining the applications in which IE is playing a role. A modular and multi-layered approach to ontology design is here central for keeping this complexity within in a reasonable range.

In the demo will concentrate on an application dealing with Financial Risk Management and the monitoring of relevant entities in the economical field. We will show how operators in FRM can select relevant documents, submit those to OBIE, decide on which relevant extracted information should be used for updating the underlying knowledge repository, and the mechanisms implemented for this update. We will then also describe the integrated reasoning architecture, also considering temporal aspects, which are in MUSING a central aspect of interest, and how then the information generated so far can be used for generating models that can support decision procedures in the financial domain, so for example the rating of companies, or the possibility to have self-assessment of SMEs.

Acknowledgments. The research presented in this demo has been partially financed by the European Integrated Project MUSING, with contract number. FP6-027097.

References

1. Baader, F., Calvanese, D., McGuinness, D.L., Nardi, D., Patel-Schneider, P.: The Description Logic Handbook. Cambridge University Press, Cambridge (2003)
2. Declerck, T., Krieger, H.-U.: Translating XBRL Into Description Logic. In: An Approach Using Protégé, Sesame & OWL, BIS 2006, pp. 455–467 (2006)
3. Krieger, H.-U., Kiefer, B., Declerck, T.: A Framework for Temporal Representation and Reasoning in Business Intelligence Applications. In: AAAI 2008 Spring Symposium on AI Meets Business Rules and Process Management (2008)
4. McGuinness, D.L., van Harmelen, F.: OWL Web Ontology Language Overview. W3C Recommendation February 10 (2004), http://www.w3.org/TR/owl-features/

A Scalable Architecture for
Cross-Modal Semantic Annotation and Retrieval

Manuel Möller and Michael Sintek

German Research Center for Artificial Intelligence (DFKI) GmbH
Kaiserslautern, Germany
`manuel.moeller@dfki.de`, `michael.sintek@dfki.de`

1   Introduction

Even within constrained domains like medicine there are no truly generic methods for automatic image parsing and annotation. Despite the fact that the precision and sophistication of image understanding methods have improved to cope with the increasing amount and complexity of the data, the improvements have not resulted in more flexible or generic image understanding techniques. Instead, the analysis methods are object specific and modality dependent. Consequently, current image search techniques are still dependent on the manual and subjective association of keywords to images for retrieval. Manually annotating the vast numbers of images which are generated and archived in the medical practice is not an option.[1]

In automatic image understanding there is a semantic gap between low-level image features and techniques for automatic reasoning about the extracted features. Existing work aims to bridge this gap by ad-hoc and application specific knowledge. During diagnosis medical experts not only look on the visual examination results in form of an image but also take into account context like demographic data about a patient, knowledge about the medical domain in general (such as anatomy and pathology) and the special features of the examination method which resulted in the current image.

2   Approach

Our approach aims to integrate existing pattern recognition algorithms with formal ontologies. The benefit of this integration is twofold: Firstly, ontologies are used to improve the image analysis through checking the output from the pattern recognition algorithms for their medical consistency. Secondly, the mapping of images to concepts in an ontology allows semantic and cross-modal information retrieval. We propose to use a scientific workflow engine to make the complexity

[1] This research has been supported in part by the THESEUS Program in the MEDICO Project, which is funded by the German Federal Ministry of Economics and Technology under the grant number 01MQ07016. The responsibility for this publication lies with the authors.

A. Dengel et al. (Eds.): KI 2008, LNAI 5243, pp. 391–392, 2008.

of our approach tractable and aid the collaboration of medical experts and computer scientists. For the overall system architecture we refer the reader to [1]. Here we will limit ourselves to the feature extraction system.

3   Demonstration

The interaction between ontological background knowledge and visual object recognition can be illustrated using the workflow in Fig. 1. It consists of the following steps:

The *Metadata Extraction Module* retrieves the Body Region from the DICOM header. We are aware of studies like [2] showing that these entries are not always correct. For now, we use this information as it is and plan to put a weight on it later on signifying its limited reliability instead of disregarding it completely.

Retrieval of Anatomical Entities in Body Region Based on the output of the previous step the Foundational Model of Anatomy Ontology [3] is used to retrieve a list of anatomical entities that could possibly be detected in this body region.

Retrieval of Visual Classifiers Using a separate annotation ontology we retrieve a list of trained classifiers which for the modality of the given image which match entries from the previous list.

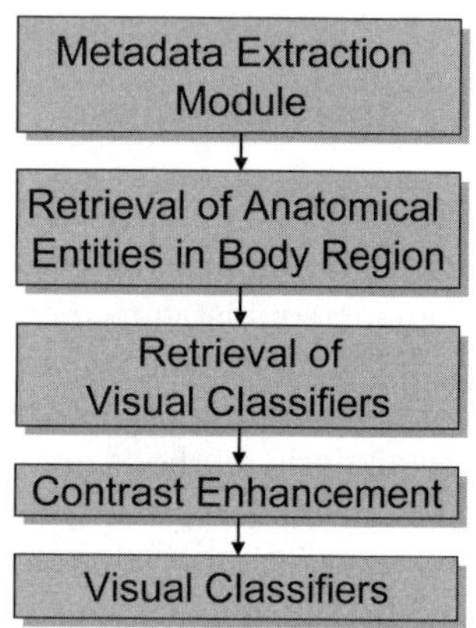

Fig. 1. Workflow

Contrast Enhancement By using additional annotations we can check whether these classifiers detect bone or soft tissue and adjust the contrast level and window accordingly.

Visual Classifiers In the final step the visual pattern matching algorithms are applied to the preprocessed image and mark detected anatomical entities with the FMA concept ID and a confidence value reflecting the uncertain nature of this detection technique.

References

1. Möller, M., Sintek, M.: A generic framework for semantic medical image retrieval. In: Proceedings of the Knowledge Acquisition from Multimedia Content (KAMC) Workshop, held at 2nd international conference on Semantics And digital Media Technologies (SAMT) (November 2007)
2. Güld, M., Kohnen, M., Keysers, D., Schubert, H., Wein, B.B., Bredno, J., Lehmann, T.M.: Quality of DICOM header information for image categorization. In: Proceedings SPIE, vol. 4685(39), pp. 280–287 (2002)
3. Rosse, C., Mejino, R.L.V.: A reference ontology for bioinformatics: The foundational model of anatomy. Journal of Biomedical Informatics 36, 478–500 (2003)

Prototype Prolog API for Mindstorms NXT[*]

Grzegorz J. Nalepa

Institute of Automatics,
AGH University of Science and Technology,
Al. Mickiewicza 30, 30-059 Kraków, Poland
`gjn@agh.edu.pl`

Abstract. The paper presents a Prolog API for controlling the LEGO Mindstorms NXT robot platform. It uses a multilayer architecture, composed of a behavioral, sensomotoric, and connection layer. The platform can be used as a generic solution for programming the NXT in Prolog.

1 Introduction to Mindstorms NXT Programming

Building intelligent robots [1] has always been one of the most important areas of research in Artificial Intelligence [2]. These days the field became more accessible to non-experts, thanks to number of ready robotics solutions. LEGO Mindstorms NXT is a universal robotics platform, that offers advanced robot construction possibilities, as well as programming solutions [3].

When it comes to Mindstorms NXT programming, LEGO offers a visual programming environment, that allows for algorithm synthesis using simple flowchart-like visual language. Thanks to the openness of the platform number of open programming solutions emerged, for languages such as C and C++, Java, etc. The examples include LeJOS/iCommand (`lejos.sourceforge.net`), and NQC (`bricxcc.sourceforge.net/nqc`). Some early attempts for providing a Prolog-based solution are within the LegoLog Project (`www.cs.toronto.edu/cogrobo/Legolog`). Unfortunately the project did not offer a general API, and supported only the older Mindstorms RCX version.

While numerous programming solutions exist, they fail to provide a clean high-level *declarative* programming solution for NXT. Programming robots, especially mobile ones, is a complex task, involving some typical AI problems, such as knowledge representation and processing, planning, etc.

The paper presents research developed within the *HeKatE* project (`hekate.ia.agh.edu.pl`) aimed at providing a high-level rule-based programming solution for Mindstorms NXT, based on Prolog API for the NXT platform.

2 Prolog NXT API Prototype

Basing on the review of existing solutions, the requirements of a new Prolog API for NXT has been formulated: support for all functions of the standard

---

[*] The paper is supported by the Hekate Project funded from 2007–2009 resources for science as a research project.

A. Dengel et al. (Eds.): KI 2008, LNAI 5243, pp. 393–394, 2008.

NXT components, crossplatform solution, for both Windows and GNU/Linux environments, reuse some of the available solutions, and provide compatibility where possible, ultimately integrate with the logical XTT layer [4]. The prototype Prolog library provides the following features: it is executed on a PC controlling an NXT robot, the control is performed with the use of the Bluetooth/USB, the low-level communication is provided by some well-tested existing communications modules, at the functional level the API is coherent with other available solutions. The following three layer API architecture has been designed.

The *behavioral layer* exposes to the programmer some high level functions and services. It provides abstract robot control functions, such as *go*, or *turn*. Ultimately a full navigation support for different robots is to be provided.

The *sensomotoric layer* controls the components of the Mindstorms NXT set motors, all the sensors, as well as Brick functions. This layer can be used to directly read the sensors, as well as program the motors. This is a layer, that can be used by a programmer to enhance high-level behavioral functions.

The goal of the *communication layer* is to execute the actions of the sensomotoric layer and communicate with the NXT Brick. Currently in this layer several modules are present, providing different means of communication: a pure Prolog module, using a serial port communication, and the NXT protocol commands, a hybrid solution based on the Java-based iCommand library, a hybrid socket-based solution, using the NXT++ library, that communicates with the robot. All of these actually wrap the *Mindstorms NXT Communication Protocol* [5].

Currently a prototype implementation of the API is available. The implementation for the SWI-Prolog environment (`www.swi-prolog.org`) has been provided by Masters students: Piotr Hołownia, with help of Paweł Gutowski, and Marcin Ziołkowski. For the full information see `https://ai.ia.agh.edu.pl/wiki/mindstorms:nxt_prolog_api`. The API has been successfully tested on number of simple control algorithms.

3   Future Work

An ultimate goal is to to provide an integration layer with the visual design with the XTT knowledge representation. This would provide a complete open robot design environment based on the rule-based programming, based on the XTT interpreter built with Prolog.

References

1. Holland, J.M.: Designing Mobile Autonomous Robots. Elsevier, Amsterdam (2004)
2. Russell, S., Norvig, P.: Artificial Intelligence: A Modern Approach, 2nd edn. Prentice-Hall, Englewood Cliffs (2003)
3. Ferrari, M., Guilio Ferrari, D.A.: Building Robots with LEGO Mindstorms NXT. Syngress (2007)
4. Nalepa, G.J., Ligęza, A.: A graphical tabular model for rule-based logic programming and verification. Systems Science 31(2), 89–95 (2005)
5. The LEGO Group: LEGO MINDSTORMS NXT Communication Protocol (2006)

VARDA Rule Design and Visualization Tool-Chain*

Grzegorz J. Nalepa and Igor Wojnicki

Institute of Automatics,
AGH University of Science and Technology,
Al. Mickiewicza 30, 30-059 Kraków, Poland
`gjn@agh.edu.pl`, `wojnicki@agh.edu.pl`

Abstract. A prototype design tool-chain (VARDA) for the ARD hierarchical rule design method is presented in the paper. It is implemented in the Unix environment using Prolog and Graphviz for design visualization.

1 Rule Design with ARD+

The HeKatE project (`hekate.ia.agh.edu.pl`) aims at providing design methods and tools that support a rule-based systems design process. Currently HeKatE supports a *conceptual design* with the ARD+ method (Attribute Relationships Diagrams). The main, so-called *logical design*, is carried out with the use of the XTT method (eXtended Tabular Trees) [1]. This paper is dedicated to the presentation of VARDA (*Visual ARD Rapid Development Alloy*), a rapid prototyping environment for the ARD+ method.

The main objective of ARD+ is to capture relationships between *attributes* in terms of *Attributive Logic* [2]. *Attributes* denote certain system *property*. ARD+ captures *functional dependencies* among these *properties*. Such dependencies form a directed graph with nodes being properties. There are two kinds of attributes: a *conceptual attribute* is an attribute describing some general, abstract aspect of the system to be specified and refined; a *physical attribute* is an attribute describing a well-defined, atomic aspect of the system. There are two transformations allowed during the ARD+ design process. *Finalization* transforms a simple property described by a conceptual attribute into a property described by one or more conceptual or physical attributes. It introduces a more specific knowledge about the given property. *Split* transforms a complex property (source property) into a number of properties (resulting properties) and defines functional dependencies among them. Attributes are unique, the same attribute cannot describe more than a single property. The evolution of the hierarchical ARD+ design is captured within the *Transformation Process History*. The TPH forms a tree that allows for recreation of any stage of the design.

---

* The paper is supported by the Hekate Project funded from 2007–2009 resources for science as a research project.

A. Dengel et al. (Eds.): KI 2008, LNAI 5243, pp. 395–396, 2008.

2   VARDA Tool

VARDA has a multi-layer architecture that consists of: knowledge base to represent the design, low-level primitives (adding and removing attributes, properties and dependencies), transformations (finalization and split including defining dependencies and automatic TPH creation), low-level visualization primitives (generating data for the visualization tool-chain, so-called DOT data), high-level visualization primitives (drawing actual dependency graph among properties and the TPH), and interoperability primitives (using XML). VARDA is implemented in Prolog, it has currently over 900 lines of Prolog code. Attributes, properties, dependencies and TPH are represented as Prolog facts.

A proper visualization of the current design state is the key element. It allows to browse the design more swiftly and identify gaps, misconceptions or mistakes more easily. A tool-chain of well proved tools is assembled to provide actual visualization of the ARD+ and TPH graphs. The tool-chain is based on three components: SWI-Prolog (`www.swi-prolog.org`), GraphViz (`www.graphviz.org`), and ImageMagick (`www.imagemagick.org`). The detailed interaction between the tool-chain components is given in Fig. 1. There are two scenarios the tool-chain is used, generating diagrams for an already designed system described in Prolog, and on-line during the design process.

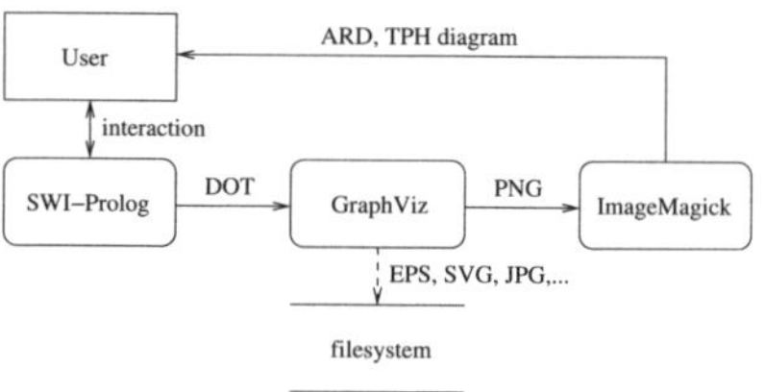

Fig. 1. Visualization tool-chain

3   Future Work

VARDA is currently being extended with a native, interactive, full-fledged shell, supporting the ARD+ design process based on a text-oriented environment with heavy hinting. The next step is developing a visual design environment for ARD+ based on VARDA using GUI toolkits in Java and Prolog XPCE.

References

1. Nalepa, G.J., Ligęza, A.: A graphical tabular model for rule-based logic programming and verification. Systems Science 31(2), 89–95 (2005)
2. Ligęza, A.: Logical Foundations for Rule-Based Systems. Springer, Berlin (2006)

COSAIR: A Platform for AI Education and Research in Computer Strategy Games

Darko Obradovič

German Research Center for Artificial Intelligence (DFKI)&
Technical University of Kaiserslautern
`obradovic@dfki.uni-kl.de`

Abstract. Computer strategy games are very challenging for AI agents
and highly interesting for students. We present a game platform specifically developed to support practical AI education of students. Featuring
all the complexity criteria in a simplified form, it easily supports the evaluation of bots, the reuse of existing AI modules, the inclusion of learning
techniques and the sharing of results and code. The platform has been
used by a number of students and proved to be in a fairly mature state.

1 Motivation

Current AI education often includes programming for board strategy games
like four-in-a-row, mastermind or chess, which all have a strong game-theoretic
component and whose good or best solutions are known and can easily be found
on the Internet. Furthermore, it is difficult for students to pursue continuative
ideas. In contrast, computer strategy games are challenging for AI agents [1]
and an attractive topic for students. Hence, the idea to use such a setting for AI
education and initial research activities is evident, but hardly practiced, due to
the overwhelming extent of the game rules and the inaccessibility of the systems,
which are proprietary commercial products in most cases.

We present the COSAIR platform (**C**omplex **S**trategy **AI** **R**easearch) that
has been designed to provide a challenging computer strategy game scenario and
to be highly accessible for AI programming at the student-level, while avoiding
the known problems of the existing solutions. Following the ideas in [2], we put
a strong emphasis on the publication and comparison of results and the sharing
of resources and ideas by setting up a central website[1].

2 Setting

The game is a simplified version of an existing web-based multiplayer strategy
game and the theme should be familiar to most players with some experience
in the field. It is located in space, players control a nation with several planets,
fleets and agents, each with their individual characteristics and possibilities. A

---

[1] http://www.cosair.org/

A. Dengel et al. (Eds.): KI 2008, LNAI 5243, pp. 397–398, 2008.

complete AI agent for the game has to control research and economy, use agents for espionage and counterespionage, allocate fleets for defense and attacks and manage planets. A detailed description of the game rules can be found on the project website. Altogether, these rules compose a highly complex game, while still being very compact for its genre.

3   Technical Realisation

The platform is programmed in Perl and is running as a web server process. The game engine can be accessed via a fully documented API. There also exists a special manual that explains the basics about bot programming. Bots are programmed in form of a class and thus can inherit from or be combined with each other. The source code is uploaded to the web server and is executed each turn in the course of a game, allowing to update or correct it at any time. Bot authors can include any of the numerously available AI modules from CPAN[2] in order to have immediate support for neural networks and the like.

In order to support AI development and evaluation as good as possible, the platform offers a number of additional features beyond those found in normal games. A simulation mechanism allows to schedule and execute the evaluation of a bot in a scenario and stores the results. There exists a well-defined standard test scenario and a ranking system that gives consideration to the difficult performance measurement in multiplayer games. Games can be completely recorded and the resulting game traces can be used for later analyses and learning. There already exists a database of 137 game traces of human players in the standard test scenario.

4   User Feedback

The platform has been used by several students in different types of projects with good results. Prior theoretical instruction, a phase of conceptual work and corresponding literature [3,4] are recommended. The platform proved to be fairly mature and to run stable, while there are still some suggested additional features waiting to be implemented in the future.

References

1. Laird, J.E., van Lent, M.: Human-level ai's killer application: Interactive computer games. AI Magazine 22, 15–25 (2001)
2. Cowling, P.: Writing ai as sport. In: Rabin, S. (ed.) AI Game Programming Wisdom, Charles River Media, vol. 3, pp. 89–96 (2006)
3. Rabin, S.: Common game ai techniques & promising game ai techniques. In: Rabin, S. (ed.) AI Game Programming Wisdom, vol. 2, pp. 3–28. Charles River Media (2003)
4. Davis, I.L.: Strategies for strategy game ai. In: AAAI Spring Symposium on Artificial Intelligence and Computer Games, pp. 24–27. AAAI Press, Menlo Park (1999)

---

[2] http://www.cpan.org/

Research Center Ambient Intelligence:
Assisted Bicycle Team Training

Bernd Schürmann and Roland Volk

Research Center "Ambient Intelligence", University of Kaiserslautern,
67653 Kaiserslautern, Germany
`http://www.eit.uni-kl.de/ami`

Abstract. Today's bicycle computers and power meters allow cyclists to control
the training at any time. But what should be done if several cyclists ride in a
group? Here, the cyclists will be stressed differently due to the exploitation of
lee. To take maximum advantage of the group, the position of the cyclists has to
be changed at the right time. The speed must be in accordance with the overall
training or racing goal as well as with the current fitness of all group members.
A bicycle demonstrator of the research center "Ambient Intelligence" (AmI) at
the University of Kaiserslautern shall optimize this problem.

1 Scenario

Cyclists with different conditions and strengths can ride together within a group because weaker cyclists can exploit lee. Headwind mainly slows down the leader of a
group while all other group members need less pedal power and energy within lee.
The power and energy saving depends on the group formation, the direction of wind
and the current strenth of cyclists as well as their topography preferences, too.

Our AmI system has to plan and control the cycling tour foresighted and it has to react to situational changes, adequately. This requires an intelligent support of human
beings by more than an ordinary control strategy. Today, the system's task is done minor optimally by a trainer who knows the overall ability of the cyclists but not the actual situation in detail. There is much space for improvement by a supporting system.

In our scenario, each cyclist has an individual profile and a (training) a schedule of
objectives for the current tour. Our demonstrator which is based on various sensors has
to control the formation and speed of the group so that all cyclists will meet their actual
goal as good as possible. Each cyclist will be directed to an exact position in a selected
formation. As there is typically no sensor detecting the actual positions of the cyclists,
detecting the actual formation of the group is another challenge for the system.

The demonstrator provides each cyclist with situational personalized data which
also depends on the available hardware. Communication within the sensor network of
a single bicycle as well as between bicycles is wireless with limited communication
range and changing network topology.

2 Research Center "Ambient Intelligence"

In case of such intelligent control environments as described above, we talk of "Ambient Intelligence" (AmI). The primary AmI goal is to improve the quality of life,

A. Dengel et al. (Eds.): KI 2008, LNAI 5243, pp. 399–401, 2008.

covering different areas like recreation and sport, health and security, working, care for elderly people and entertainment, to name just a few of them.

All AmI applications are built-up on the same basic technical features. AmI systems have a variety of interfaces to people and their environment. This ranges from micro sensors, over image and speech input and output, up to the detection of human gestures. The diversity of the collected information must be transmitted and evaluated adequately. This is done by distributed digital signal processing of voice, image, movement, environmental conditions, etc. Both, the various sensors of different kind as well as the associated analysis and communication electronics need to be very small and integrated into the environment, clothes, etc.. A challenge is the extremely low energy consumption.

This realization of the "ambient part" already implies many challenges, but even more challenges are imposed to the "intelligence part". Thus, the development of AmI systems requires a real interdisciplinary cooperation between various scientific communities. Our AmI research center has been established as such an interdisciplinary team. Research groups from four departments of the university work together with groups of the Fraunhofer Institute for Experimental Software Engineering (IESE), the German Research Institute for Artificial Intelligence (DFKI), and from Budapest in Hungaria.

3   Bicycle Demonstrator

The demonstrator that has been developed in Kaiserslautern can be used outdoor as well as indoor. For the indoor demonstrator four bicycles are mounted to wheel stands and are attached to a simulation computer (see fig. 1). This simulator, in combination with a screen and the braking force of the wheel stands, replaces the environment of the outdoor demonstrator. By adapting the braking force individually for each bike, cyclists with totally different physical conditions and strengths can use the demonstrator together. A smart simulator control can inject test patterns for a large number of experiments.

Fig. 1. Indoor demonstrator

Each bicycle has several sensors which are connected to a central bike computer. At least, all sensors attached to the cyclist have wireless interfaces. The overall demonstrator has a hierarchical wireless network.

Currently, our bicycle demonstrator will be evaluated by young professional cyclists from the Heinrich-Heine-Gymnasium, an elite school for sports. The evaluation is coordinated by the federal coach for young professional cyclists together with the research group for sport science at the University of Kaiserslautern.

Author Index

Printing: Mercedes-Druck, Berlin
Binding: Stein+Lehmann, Berlin

Lecture Notes in Artificial Intelligence (LNAI)

Vol. 5012: T. Washio, E. Suzuki, K.M. Ting, A. Inokuchi (Eds.), Advances in Knowledge Discovery and Data Mining. XXIV, 1102 pages. 2008.

Vol. 5009: G. Wang, T. Li, J.W. Grzymala-Busse, D. Miao, A. Skowron, Y. Yao (Eds.), Rough Sets and Knowledge Technology. XVIII, 765 pages. 2008.

Vol. 5003: L. Antunes, M. Paolucci, E. Norling (Eds.), Multi-Agent-Based Simulation VIII. IX, 141 pages. 2008.

Vol. 5001: U. Visser, F. Ribeiro, T. Ohashi, F. Dellaert (Eds.), RoboCup 2007: Robot Soccer World Cup XI. XIV, 566 pages. 2008.

Vol. 4999: L. Maicher, L.M. Garshol (Eds.), Scaling Topic Maps. XI, 253 pages. 2008.

Vol. 4994: A. An, S. Matwin, Z.W. Raś, D. Ślęzak (Eds.), Foundations of Intelligent Systems. XVII, 653 pages. 2008.

Vol. 4953: N.T. Nguyen, G.S. Jo, R.J. Howlett, L.C. Jain (Eds.), Agent and Multi-Agent Systems: Technologies and Applications. XX, 909 pages. 2008.

Vol. 4946: I. Rahwan, S. Parsons, C. Reed (Eds.), Argumentation in Multi-Agent Systems. X, 235 pages. 2008.

Vol. 4944: Z.W. Raś, S. Tsumoto, D.A. Zighed (Eds.), Mining Complex Data. X, 265 pages. 2008.

Vol. 4938: T. Tokunaga, A. Ortega (Eds.), Large-Scale Knowledge Resources. IX, 367 pages. 2008.

Vol. 4933: R. Medina, S. Obiedkov (Eds.), Formal Concept Analysis. XII, 325 pages. 2008.

Vol. 4930: I. Wachsmuth, G. Knoblich (Eds.), Modeling Communication with Robots and Virtual Humans. X, 337 pages. 2008.

Vol. 4929: M. Helmert, Understanding Planning Tasks. XIV, 270 pages. 2008.

Vol. 4924: D. Riaño (Ed.), Knowledge Management for Health Care Procedures. X, 161 pages. 2008.

Vol. 4923: S.B. Yahia, E.M. Nguifo, R. Belohlavek (Eds.), Concept Lattices and Their Applications. XII, 283 pages. 2008.

Vol. 4914: K. Satoh, A. Inokuchi, K. Nagao, T. Kawamura (Eds.), New Frontiers in Artificial Intelligence. X, 404 pages. 2008.

Vol. 4911: L. De Raedt, P. Frasconi, K. Kersting, S. Muggleton (Eds.), Probabilistic Inductive Logic Programming. VIII, 341 pages. 2008.

Vol. 4908: M. Dastani, A. El Fallah Seghrouchni, A. Ricci, M. Winikoff (Eds.), Programming Multi-Agent Systems. XII, 267 pages. 2008.

Vol. 4898: M. Kolp, B. Henderson-Sellers, H. Mouratidis, A. Garcia, A.K. Ghose, P. Bresciani (Eds.), Agent-Oriented Information Systems IV. X, 292 pages. 2008.

Vol. 4897: M. Baldoni, T.C. Son, M.B. van Riemsdijk, M. Winikoff (Eds.), Declarative Agent Languages and Technologies V. X, 245 pages. 2008.

Vol. 4894: H. Blockeel, J. Ramon, J. Shavlik, P. Tadepalli (Eds.), Inductive Logic Programming. XI, 307 pages. 2008.

Vol. 4885: M. Chetouani, A. Hussain, B. Gas, M. Milgram, J.-L. Zarader (Eds.), Advances in Nonlinear Speech Processing. XI, 284 pages. 2007.

Vol. 4884: P. Casanovas, G. Sartor, N. Casellas, R. Rubino (Eds.), Computable Models of the Law. XI, 341 pages. 2008.

Vol. 4874: J. Neves, M.F. Santos, J.M. Machado (Eds.), Progress in Artificial Intelligence. XVIII, 704 pages. 2007.

Vol. 4870: J.S. Sichman, J. Padget, S. Ossowski, P. Noriega (Eds.), Coordination, Organizations, Institutions, and Norms in Agent Systems III. XII, 331 pages. 2008.

Vol. 4869: F. Botana, T. Recio (Eds.), Automated Deduction in Geometry. X, 213 pages. 2007.

Vol. 4865: K. Tuyls, A. Nowe, Z. Guessoum, D. Kudenko (Eds.), Adaptive Agents and Multi-Agent Systems III. VIII, 255 pages. 2008.

Vol. 4850: M. Lungarella, F. Iida, J.C. Bongard, R. Pfeifer (Eds.), 50 Years of Artificial Intelligence. X, 399 pages. 2007.

Vol. 4845: N. Zhong, J. Liu, Y. Yao, J. Wu, S. Lu, K. Li (Eds.), Web Intelligence Meets Brain Informatics. XI, 516 pages. 2007.

Vol. 4840: L. Paletta, E. Rome (Eds.), Attention in Cognitive Systems. XI, 497 pages. 2007.

Vol. 4830: M.A. Orgun, J. Thornton (Eds.), AI 2007: Advances in Artificial Intelligence. XIX, 841 pages. 2007.

Vol. 4828: M. Randall, H.A. Abbass, J. Wiles (Eds.), Progress in Artificial Life. XII, 402 pages. 2007.

Vol. 4827: A. Gelbukh, Á.F. Kuri Morales (Eds.), MICAI 2007: Advances in Artificial Intelligence. XXIV, 1234 pages. 2007.

Vol. 4826: P. Perner, O. Salvetti (Eds.), Advances in Mass Data Analysis of Signals and Images in Medicine, Biotechnology and Chemistry. X, 183 pages. 2007.

Vol. 4819: T. Washio, Z.-H. Zhou, J.Z. Huang, X. Hu, J. Li, C. Xie, J. He, D. Zou, K.-C. Li, M.M. Freire (Eds.), Emerging Technologies in Knowledge Discovery and Data Mining. XIV, 675 pages. 2007.

Vol. 4811: O. Nasraoui, M. Spiliopoulou, J. Srivastava, B. Mobasher, B. Masand (Eds.), Advances in Web Mining and Web Usage Analysis. XII, 247 pages. 2007.

Vol. 4798: Z. Zhang, J.H. Siekmann (Eds.), Knowledge Science, Engineering and Management. XVI, 669 pages. 2007.

Vol. 4795: F. Schilder, G. Katz, J. Pustejovsky (Eds.), Annotating, Extracting and Reasoning about Time and Events. VII, 141 pages. 2007.

Vol. 4790: N. Dershowitz, A. Voronkov (Eds.), Logic for Programming, Artificial Intelligence, and Reasoning. XIII, 562 pages. 2007.

Vol. 4788: D. Borrajo, L. Castillo, J.M. Corchado (Eds.), Current Topics in Artificial Intelligence. XI, 280 pages. 2007.

Vol. 4775: A. Esposito, M. Faundez-Zanuy, E. Keller, M. Marinaro (Eds.), Verbal and Nonverbal Communication Behaviours. XII, 325 pages. 2007.

Vol. 4772: H. Prade, V.S. Subrahmanian (Eds.), Scalable Uncertainty Management. X, 277 pages. 2007.